Series on Concrete and Applicable Mathematics Vol. 19

Stochastic Models with Applications to Genetics, Cancers, AIDS and Other Biomedical Systems

Second Edition

Series on Concrete and Applicable Mathematics Vol. 19

Stochastic Models with Applications to Genetics, Cancers, AIDS and Other Biomedical Systems

Second Edition

Tan Wai-Yuan

University of Memphis, USA

World Scientific

NEW JERSEY · LONDON · SINGAPORE · BEIJING · SHANGHAI · HONG KONG · TAIPEI · CHENNAI · TOKYO

Published by

World Scientific Publishing Co. Pte. Ltd.
5 Toh Tuck Link, Singapore 596224
USA office: 27 Warren Street, Suite 401-402, Hackensack, NJ 07601
UK office: 57 Shelton Street, Covent Garden, London WC2H 9HE

British Library Cataloguing-in-Publication Data
A catalogue record for this book is available from the British Library.

Series on Concrete and Applicable Mathematics — Vol. 19
STOCHASTIC MODELS WITH APPLICATIONS TO GENETICS, CANCERS,
AIDS AND OTHER BIOMEDICAL SYSTEMS
2nd Edition

ISBN 978-981-4390-94-1

Printed in Singapore

This book is dedicated to
my wife Shiow-Jen,
my daughter Emy and my son Eden

Preface

The purpose of this book is to present a systematic treatment of stochastic models which arise from genetics, carcinogenesis, AIDS epidemiology and HIV pathogenesis. It is meant to provide basic methodological tools to analyze these processes and to study the stochastic behavior of these processes. This book is useful because cancer and AIDS are the most dangerous diseases threatening the survival of human beings and because the genetic principle has been used in developing computer algorithms by computer scientists. Also, the genome project has made the genetic theories one of the most important disciplines in scientific research.

This book is organized into 9 chapters. To illustrate the basic stochastic processes which arise from genetics, cancer and AIDS, in Chapter 1, numerous examples from these areas are presented. These processes include univariate and multivariate Markov chains with discrete time and with continuous time, diffusion processes, state space models and hidden Markov models. Having introduced these processes, the rest of the book is then devoted to develop basic theories of these processes and applications of these processes to genetic, cancer and AIDS. Thus, in Chapter 2, we present the basic theories of Markov chains with discrete time and describe the stochastic dynamic behavior of these processes. In Chapter 3, we present some basic theories of limiting results and stationary distributions in Markov chains with discrete time; as applications of stationary distributions, in Chapter 3, we also present some MCMC (Markov Chain Monte Carlo) methods to develop optimal computer algorithms to estimate unknown parameters in the models and illustrate its applications.

Chapters 4 and 5 are devoted to develop basic theories of Markov chains with continuous time and describe the stochastic dynamic behavior of these processes. In Chapters 6 and 7, basic theories and methodologies of diffusion processes are presented and illustrated by examples from genetics and biomedical problems. Finally in Chapters 8 and 9, we present some basic theories of state space models and describe how to construct state space models in cancer and AIDS and illustrate applications in these areas.

This book is unique and differs from other books on stochastic processes and stochastic models in several ways: First, it has presented and developed approaches which are not discussed in other books of stochastic processes. This includes MCMC methods and stochastic difference and differential equation approaches to Markov chains. Second, the book describes how to apply the theories to solve problems in genetics, cancer and AIDS. Third, it has presented and discussed state space models and illustrate its applications to cancer and AIDS problems which are not discussed in other books of stochastic processes.

I originally compiled this book for students in the Department of Mathematical Sciences at the University of Memphis, Memphis, Tennessee, when I was offering a graduate course in applied stochastic models. These lecture notes have then been up-dated and expanded to include stochastic and state space models of carcinogenesis, AIDS epidemiology and HIV pathogenesis in HIV-infected individuals. Thus, the book may be used as a text for applied stochastic processes or applied stochastic models. It may also be used as a reference book for courses in mathematical modeling and in stochastic models of biomedical systems and as a reference for research tools by medical doctors and researchers.

I would like to express my sincere appreciation to Professor George Anastassiou of University of Memphis for inviting me to submit my book to World Scientific for the series edited by him. I want also to express my thanks to my students Mr. Xiangke Huang, Ms. Ping Zhang and Mr. J. H. Zhu for drawing many of the figures in Chapters 2–3 and 8–9.

Finally I wish to thank Ms. Diane Mittelmeier, Mr. G. Luo and Mr. Weiming Ke for typing some of the chapters and to Dr. Sen Hu and Mr. Ye Qiang of the World Scientific Publication Company for assistance in the publication of my book.

Wai-Yuan Tan, 2001

Contents

7   Asymptotic Distributions, Stationary Distributions and Absorption Probabilities in Diffusion Models   303

8   State Space Models and Some Examples from Cancer and AIDS   337

Chapter 1

Introduction

In the studies of natural systems, the main problem is usually to derive mathematical models for the response or responses in terms of input and/or risk variables as well as time. Almost all of these models are stochastic models because most of the risk variables are subject to random variations and most of the measurements of the responses and input variables are subject to random measurement errors. This is true for medical systems such as AIDS, cancer and genetics as well as for biological sciences, engineering sciences and social and economic systems. To study these systems, therefore, it is important to examine the probability laws governing the behavior of these systems. The purpose of this book is to present a systematic treatment on stochastic models which have been used in genetics, cancer, AIDS and some other medical systems.

To set up the basic background, in this chapter we will define some basic terminologies and give examples from genetic, cancer and AIDS to illustrate some basic concepts of stochastic processes.

1.1. Some Basic Concepts of Stochastic Processes and Examples

Definition 1.1. A *stochastic process* $\{X(t), t \in T\}$ is a family of random variables indexed by the parameter t in T.

In biomedical sciences as well as in many other areas, the parameter t is usually related to time and the set T is referred to as the parameter space. The sample space S_t of $X(t)$ is referred to as the state space and the elements of S_t the states. The space S_t may be discrete in which case the number of elements of S_t is either finite or countable infinite, or continuous in which case the number of elements of S_t is uncountable infinitely many. Similarly, T may either be discrete or continuous. It follows that there are four types of stochastic processes:

(a) A stochastic process with discrete state space and discrete time t.
(b) A stochastic process with discrete state space and continuous time t.
(c) A stochastic process with continuous state space and discrete time t.
(d) A stochastic process with continuous state space and continuous time t.

For given n and for given $t_0 < t_1 < \cdots < t_n$, the observed values $\{X(t_0), X(t_1), \ldots, X(t_n)\}$ of $X(t)$ at $\{t_0, t_1, \ldots, t_n\}$ is referred to as a sample path of the process. We will refer the stochastic process as a finite stochastic process if the state space S contains only a finite number of states. To simplify notations, in what follows, we will let $T = \{t \geq 0\}$ if T is continuous and let $T = \{0, 1, \ldots, \}$ if T is discrete, unless otherwise stated (This can be achieved by defining the starting time of the process as 0). Similarly, in what follows, we will let $\{S = [a, b], -\infty \leq a < b \leq \infty\}$ if S is continuous and let $S = \{0, 1, \ldots, \}$ if S is discrete, unless otherwise stated (This can be achieved by defining the ith element of S as $i - 1$ $(i = 1, \ldots, \infty))$.

The above definition of stochastic process can also be extended to k-dimensional stochastic processes with $k \geq 1$ being a positive integer.

Definition 1.2. A *k-dimensional stochastic process* $\{\underset{\sim}{X}(t), t \in T\}$ is a family of k-dimensional random vectors indexed by the parameter t in T.

In k-dimensional stochastic processes, the state space S is then a subset of the k-dimensional Euclidean space $E^{(k)}$. Also, some of the variables of $\underset{\sim}{X}(t)$ may be discrete while other random variables of $\underset{\sim}{X}(t)$ may be continuous. These are mixed-type random vectors. In this book we will not consider cases with mixed types of random variables in $\underset{\sim}{X}(t)$, unless otherwise stated; thus we will only consider cases in which either all random variables in $\underset{\sim}{X}(t)$ are discrete or all random variables in $\underset{\sim}{X}(t)$ are continuous.

Example 1.1.   The frequency of mating types under full-sib mating in natural populations.   In animal breeding, the breeders are usually confronted with the problem of sib mating (brother-sister mating) leading to inbreed lines. Sib mating also are very common in wild animal populations. Hence, it is often of interest to compute the frequency of different mating types in sib mating in domestic as well as in wild populations. In a large population of diploid individuals, if we focus on a single locus with two alleles A and a, then there are three genotypes $\{AA, Aa, aa\}$ and there are 6 different mating types (see Remark 1.1): $\{AA \times AA, aa \times aa, AA \times aa, AA \times Aa, aa \times Aa, Aa \times Aa\}$ which we denote by $\{1, \ldots, 6\}$ respectively. (As a convention, the genotype on the left denotes the genotype of the mother whereas the genotype on the right denotes the genotype of the father). Let t denote generation and let $X(t)$ denote the frequency of mating types at time t under sib-mating. Then $\{X(t), t \in T = (0, 1, \ldots,)\}$ is a stochastic process with discrete time and with state space $S = \{1, \ldots, 6\}$. This is an example of stochastic process with discrete time and discrete state space.

Remark 1.1.   In most of the plants, animals and human beings, the chromosomes are grouped into a fixed number of pairs of homologous chromosomes, one from the mother and the other from the father. This type of individuals has been referred to as diploid. For example, in human being, there are 23 pairs of chromosomes and hence human being are diploid individuals. Biologists have also shown that all characters are controlled by genes which are segments of DNA in the chromosomes. This segment has been referred to as locus and different genes in the same locus are referred to as alleles.

Example 1.2.   Survival of mutant genes in natural population-branching processes.   In human beings, many of the inherited disease are caused by mutation of certain genes [1, 2]. Suppose that at a certain time, a mutant gene is introduced into the population. Suppose further that each mutant gene produces j mutant genes with probability p_j $(j = 0, 1, \ldots, \infty)$ in the next generation independently of other genes. Let $X(t)$ be the number of mutant genes at generation t. Then $X(t)$ is a stochastic process with discrete time and with state space $S = \{0, 1, \ldots, \infty\}$. As we shall see, this type of processes belongs to a class of stochastic processes referred to as Galton–Watson branching processes [3]. This is an example of stochastic process with discrete time and discrete state space.

Example 1.3. The change of frequency of genes in natural populations. In studying the theory of evolution, it is of interest to find the probability law in natural populations governing the changes of the frequencies of types or genes. Thus, in a large population of diploid individuals, for a single locus with two alleles A and a, one would need to find the probability law for the number of A allele over time. Let N be the population size and let t denote generation. Then the number $X(t)$ of A allele at time t is a stochastic process with discrete time and with state space $S = \{0, 1, \ldots, 2N\}$. (Note that in a diploid population, each individual has two alleles for each locus; hence the total number of alleles for each locus in the population is 2N.) This is an example of stochastic process with discrete time and discrete state space.

Let $Y(t) = \frac{X(t)}{2N}$. Since the population size N is usually very large and since the evolution process is an extremely slow process taking place over millions and millions of years, as we shall see in Chap. 6, $Y(t)$ can be closely approximated by a stochastic process with continuous time and continuous state space $S = [0, 1]$.

Example 1.4. The number of drug-resistant cancer tumor cells. In treating cancer by chemotherapy, a major difficulty is the development of drug-resistant cancer tumor cells. Thus, questions of the possible efficiency and optimal timing of cancer chemotherapy can be studied by mathematical models for the development of drug-resistant cancer tumor cells. Let $X_1(t)$ be the number of sensitive cancer tumor cells at time t and $X_2(t)$ the number of resistant cancer tumor cells at time t. Let 0 be the time starting treatment. Then $\{[X_1(t), X_2(t)], t > 0\}$ is a two-dimensional stochastic process with parameter space $T = \{t > 0\}$ and state space $S = \{(i, j), i, j = 0, 1, \ldots, \}$. Stochastic process of this type has been studied in [4]. This is an example of two-dimensional stochastic process with discrete state space and continuous parameter space.

Example 1.5. The multi-stage model of carcinogenesis. Cancer tumors develop from normal stem cells by going through a finite number of genetic changes and/or epigenetic changes with intermediate cells subjecting to stochastic proliferation (birth) and differentiation (death). That is, cancer tumors develop from normal stem cells by a multistage model with intermediate cells subjecting to stochastic birth and death. Assume that there are k $(k \geq 2)$ intermediate stages. For $t \geq 0$ with 0 being the time of birth of the individual, let $X_0(t)$ denote the number of normal stem cells at time t, $X_i(t)$ ($i = 1, \ldots, k - 1$) the number of the ith stage intermediate cells at time t and $T(t)$ the number of malignant cancer tumors at time t. Then $\{[X_i(t), i = 0, 1, \ldots, k - 1, T(t)], t \geq 0\}$ is a $(k + 1)$-dimensional stochastic process with parameter space $T = \{t > 0\}$ and with state space $S = \{(i_0, i_1, \ldots, i_k), i_r = 0, 1, \ldots, ; r = 0, 1, \ldots, k\}$; see **Remark 1**. In general, $\{X_i(t), i = 0, 1, \ldots, k - 1, T(t)\}$ involves both stochastic birth-death processes for cell proliferation and differentiation of normal stem cells, intermediate cells and cancer tumors and Poisson processes for generating intermediate cells through genetic changes or mutations. This is an example of multi-dimensional stochastic processes with discrete state space and continuous time. These processes have been discussed in detail in [5, 6].

Remark 1: As illustrated in [5, 6], because tumors develop from primary I_k cells, $I_k(t)$ is not identifiable.

Example 1.6. **The AIDS epidemiology in homosexual populations.** Consider a large population of homosexual men such as the San Francisco homosexual population which is at risk for AIDS. Then there are three types of people regarding HIV epidemic in the population: The S (susceptible) people, the I (infective) people and the A (clinical AIDS cases) people. A S person does not carry the AIDS virus but can contract it through sexual contact with I people or AIDS cases or by sharing needles in IV drug use or through blood transfusion of contaminated blood. An I person carries the AIDS virus and can transmit the virus to S people through sexual contact or sharing contaminated needles with I people; there is a chance that he/she will develop AIDS symptoms to become an AIDS case. An AIDS case (An A person) is a person who has developed AIDS symptoms or who has CD4$^{(+)}$ T cell counts in the blood falling below $200/\text{mm}^3$ [7].

Let $S(t), I(t)$ and $A(t)$ denote the numbers of susceptible people (S people), infected people (I people) and AIDS cases at time t respectively and write $\underset{\sim}{X}(t) = \{S(t), I(t), A(t)\}'$, where $\prime$ denotes transpose. Let $t_0 = 0$ be the time at which a few HIV were introduced into the population to start the AIDS epidemic. Then $\{\underset{\sim}{X}(t), t \geq 0\}$ is a three-dimensional stochastic process with parameter space $T = \{t \geq 0\}$ and with state space $\Omega = \{(i, j, k), i, j, k$ being non-negative integers$\}$. This is an example of multi-dimensional stochastic process with discrete state space and continuous parameter space [8, Chap. 3].

Example 1.7.   The HIV pathogenesis in HIV-infected individuals.
In a HIV-infected individual, let time 0 denote the time of HIV infection.
Then, there are three types of $CD4^{(+)}$ T cells, the uninfected $CD4^{(+)}$ T cells
(denoted by T_1), the latently infected $CD4^{(+)}$ T cells (denoted by T_2) and
the productively HIV-infected $CD4^{(+)}$ T cells (denoted by T_3, also referred
to as actively HIV-infected T cells).   Let $T_i(t)$ $(i = 1, 2, 3)$ denote the
number of T_i $(i = 1, 2, 3)$ cells at time t per mm^3 of blood and let $V(t)$
denote the number of free HIV at time t per mm^3 of blood.   Denote by
$\underset{\sim}{X}(t) = \{T_i(t), i = 1, 2, 3, V(t)\}'$. Then $\{\underset{\sim}{X}(t), t \geq 0\}$ is a four-dimensional
stochastic process with parameter space $T = \{t \geq 0\}$ and with discrete state
space $S = \{(i, j, k, l), i, j, k, l$ being non-negative integers$\}$; for more detail, see
[8, Chaps. 7–8] and [9].

1.2.   Markovian and Non-Markovian Processes, Markov Chains and Examples

In genetics, carcinogenesis, AIDS as well as in many other stochastic systems,
many processes can be characterized by a *dependence condition* referred to as
the *Markov condition*. These processes are classified as Markov processes.

Definition 1.3.   Let $\{X(t), t \in T\}$ be a stochastic process with parameter
space T and with state space S. Then $X(t)$ is called a *Markov process* iff (if
and only if) for every n and for every $t_1 < \cdots < t_n \leq t$ in T,

$$\Pr\{X(t) \in A | X(t_1) = x_1, \ldots, X(t_n) = x_n\} = P\{X(t) \in A | X(t_n) = x_n\},$$
$$\text{for any event } A \subset S. \tag{1.1}$$

where $\Pr\{X(t) \in A | X(t_1) = x_1, \ldots, X(t_n) = x_n\}$ is the conditional probability
of $X(t) \in A$ given $\{X(t_1) = x_1, \ldots, X(t_n) = x_n\}$ and $P\{X(t) \in A | X(t_n) = x_n\}$ the conditional probability of $X(t) \in A$ given $X(t_n) = x_n$.

The above definition is equivalent to stating that the probability distribution of $X(t)$ depends only on results in the most recent time and is independent
of past history. From this definition, it is then seen that most of the processes
in genetics and in evolution theory are Markov processes. Similarly, many process in carcinogenesis [5] and in AIDS epidemiology [8] are Markov processes.
Thus, Examples 1.1–1.4 are Markov processes. However, there are also many

processes in nature which are not Markov. An example from AIDS epidemiology is given in Example 1.10 whereas an example from cancer is given in Example 1.12 below. A sufficient condition for which $X(t)$ is Markov is that for every $t_1 < \cdots < t_n$, $X(t_2) - X(t_1), X(t_3) - X(t_2), \ldots, X(t_n) - X(t_{n-1})$ are independently distributed of one another. This latter condition has been referred to as "independent increment"; see Exercise 1.1.

Definition 1.4. A Markov process $\{X(t), t \in T\}$ with state space S is called a *Markov chain* iff S is discrete. (With no loss of generosity, one may assume $S = \{0, 1, \ldots, \infty\}$.) A *Markov chain* $\{X(t), t \in T\}$ is called a finite Markov chain iff the state space S contains only a finite number of states.

By this definition, Examples 1.1 and 1.3 are finite Markov chains whereas Examples 1.2 and 1.4 are Markov chains with infinite state space. Examples 1.1–1.3 are Markov chains with discrete time whereas Example 1.4 is a Markov chain with continuous time. General theories and its applications of Markov chains with discrete times will be discussed in detail in Chaps. 2 and 3 whereas general theories and its applications of Markov chains with continuous times will be discussed in detail in Chaps. 4 and 5. Notice that these general theories are characterized by the transition probabilities

$$p_{ij}(s, t) = \Pr\{X(t) = j | X(s) = i\}, \quad i \in S, j \in S.$$

For Markov chains with discrete times, the $p_{ij}(s, t)$'s are further characterized and derived by the one step transition probabilities

$$p_{ij}(t) = p_{ij}(t, t+1) = \Pr\{X(t+1) = j | X(t) = i\}, \quad i \in S, j \in S.$$

The analog of the one-step transition probabilities in Markov chains with continuous times are

$$p_{ij}(t, t+\Delta t) = \Pr\{X(t+\Delta t) = j | X(t) = i\} = \alpha_{ij}(t)\Delta t + o(\Delta t), \quad i \in S, j \in S,$$

where $o(\Delta t)$ is defined by $\lim_{\Delta t \to 0} \frac{o(\Delta t)}{\Delta t} = 0$. In the literature, the $\alpha_{ij}(t)$ have been referred to as the transition rates or infinitesimal parameters. Thus, for Markov chains with continuous time, the processes are characterized by the infinitesimal parameters or transition rates.

Definition 1.5. A Markov chain $\{X(t), t \in T\}$ with state space $S = \{0, 1, \ldots, \infty\}$ is a *homogeneous Markov chain* iff $p_{ij}(s, t) = p_{ij}(t - s)$ for all $i \in S, j \in S$.

From Definition 1.5, notice that homogeneous Markov chains depend on the time parameters only through the difference of times. It follows that if the chain is homogeneous, then $p_{ij}(t) = p_{ij}(s, s + t) = P\{X(s + t) = j | X(s) = i\}$ $= P\{X(t) = j | X(0) = i\}$ for all $s \geq 0$. Hence, for Markov chains with discrete time, the 1-step transition probabilities are given by $p_{ij}(1) = P\{X(n + 1) = j | X(n) = i\} = p_{ij}$ which are independent of n.

In natural systems, homogeneous Markov chains are very common although there are also nonhomogeneous Markov chains. For instance, Examples 1.1–1.4 given above are homogeneous Markov chains. In Example 1.2, however, if the progeny distribution of $X(n)$ depends on n, then the chain is not homogeneous although it is Markov. For ease of illustration, in what follows, we will assume that the chain $X(t)$ is homogeneous although many of the results hold also for some nonhomogeneous Markov chains, unless otherwise stated.

Remark 1.2. Homogeneous Markov chains are not stationary chains. In fact, as shown in Example 5.4, stationary distributions may not exist in some homogeneous Markov chains. On the other hand, the *homogeneity* condition is a pre-condition for defining stationary distributions.

Example 1.8. The full-sib mating model for one locus with two alleles in natural populations. In Example 1.1, we have considered a large diploid population under full-sib mating. In this example, we have focused on one locus with two alleles $A : a$ and let $X(t)$ denote the mating types at time t. Then, the state space consists of the six mating types $AA \times AA, aa \times aa, AA \times aa, AA \times Aa, aa \times Aa, Aa \times Aa$ which are denoted by $(1, \ldots, 6)$ respectively. Thus, $\{X(t), t \in T = (0, 1, 2, \ldots)\}$ is a finite homogeneous Markov chain with state space $S = \{1, \ldots, 6\}$. For this Markov chain, the matrix of the one-step transition probabilities is given by:

	$AA \times AA$	$aa \times aa$	$AA \times aa$	$AA \times Aa$	$aa \times Aa$	$Aa \times Aa$
$AA \times AA$	1	0	0	0	0	0
$aa \times aa$	0	1	0	0	0	0
$AA \times aa$	0	0	0	0	0	1
$AA \times Aa$	$\dfrac{1}{4}$	0	0	$\dfrac{1}{2}$	0	$\dfrac{1}{4}$
$aa \times Aa$	0	$\dfrac{1}{4}$	0	0	$\dfrac{1}{2}$	$\dfrac{1}{4}$
$Aa \times Aa$	$\dfrac{1}{16}$	$\dfrac{1}{16}$	$\dfrac{1}{8}$	$\dfrac{1}{4}$	$\dfrac{1}{4}$	$\dfrac{1}{4}$

The above matrix of one-step transition probabilities are derived by noting that matings occur only between brother and sister within the family. For example, the mating type $AA \times Aa$ gives only progenies AA and Aa with relative frequencies $\{\frac{1}{2} AA, \frac{1}{2} Aa\}$; hence the frequencies of mating types in the next generation is $\{\frac{1}{4} AA \times AA, \frac{1}{4} Aa \times Aa, \frac{1}{2} AA \times Aa\}$. As another example, notice that the mating type $Aa \times Aa$ gives progenies $\{AA, aa, Aa\}$ with relative frequencies $\{\frac{1}{4} AA, \frac{1}{2} Aa, \frac{1}{4} aa\}$; hence the frequencies of mating types in the next generation is $\{\frac{1}{16} AA \times AA, \frac{1}{16} aa \times aa, \frac{1}{8} AA \times aa, \frac{1}{4} AA \times Aa, \frac{1}{4} aa \times Aa, \frac{1}{4} Aa \times Aa\}$.

Example 1.9.   The simple Galton–Watson branching processes. In genetics, in biological problems as well as in many other stochastic systems, an important class of Markov processes is the branching process; see [3]. This includes the simple Galton–Watson process which is a homogeneous Markov chain with discrete time. This latter process has been used to examine the stochastic behavior of mutant genes in populations; in particular, the survival of mutant genes as time progresses.

Definition 1.6.   A Markov chain $\{X(t), t \in T = (0, 1, 2 \ldots)\}$, with state space $S = \{0, 1, 2 \ldots\}$ is called a *simple branching process* (or *Galton–Watson process*) with progeny distribution $\{p_k, k = 0, 1, 2, \ldots (p_k \geq 0, \sum_{k=0}^{\infty} p_k = 1)\}$ iff $P\{X(0) = 1\} = 1$ and the one-step transition probabilities $p_{ij} = \Pr\{X(n+1) = j | X(n) = i\}$ are given by:

(i) If $i = 0, j \geq 0$, then $p_{ij} = \delta_{ij}$, where, $\delta_{ij} = 1$ if $i = j$ and $\delta_{ij} = 0$ if $i \neq j$.
(ii) If $i > 0$, then

$$p_{ij} = \Pr\{Z_1 + Z_2 + \cdots + Z_i = j\},$$

where $Z_1, Z_2, \ldots$ are independently and identically distributed with probability density function (pdf) given by $p_j, j = 0, 1, 2 \ldots$.

From the above definition, we see that the Galton–Watson process is a homogeneous Markov chain. (Because $p_0 = 1$ indicates that the mutant is certainly to be lost in the next generation while $p_0 = 0$ is the situation that the mutant will never get lost, to avoid trivial cases we will assume $0 < p_0, p_1 < 1$ in what follows).

To obtain p_{ij} for $i > 0$, let $f(s)$ denote the *probability generating function* (pgf) of the progeny distribution $\{p_j, j = 0, \ldots, \infty\}$, $f_n(s)$ the pgf of $X(n)$ given $X(0) = 1$ and $g_i(s)$ the pgf of $\{p_{ij}, i > 0, j = 0, \ldots, \infty\}$. Then, by

(ii) above, $g_i(s) = [f(s)]^i$ and $p_{ij}(n) = [p_{1j}(n)]^i$. By definition of pgf, we have:

$$p_{ij} = \frac{1}{j!} \left\{ \frac{d^j}{ds^j} g_i(s) \right\}_{s=0} ;$$

$$p_{1j}(n) = \frac{1}{j!} \left\{ \frac{d^j}{ds^j} f_n(s) \right\}_{s=0} .$$

As an example, consider a diploid population with only one allele A at the A locus before the t_0th generation (With no loss of generality, one may assume that $t_0 = 0$). Suppose that at the 0th generation, an A allele has mutated to a so that at the 0th generation there is an individual with genotype Aa in the population. Let $X(t)$ be the number of mutant a at generation t. Assume that each mutant a reproduces itself independently of one another and that each mutant has probability p_j of giving j mutants in the next generation. Then, barring further mutations from A to a in the future, $\{X(t), t \in T = \{0, 1, 2 \ldots\}\}$ is a Galton–Watson process.

To specify p_j, let the fitness (i.e., average number of progenies per generation) of AA and Aa genotypes be given by μ and $\mu(1+v)$ ($\mu > 0$) respectively. Let N be the population size. Then in the 0th generation, the frequency of the a allele is

$$\frac{(1+v)\mu}{(2N - 1 + 1 + v)\mu} = \frac{1}{2N}(1+v) + o((2N)^{-1}) = p + o((2N)^{-1})$$

for finite v, where $p = \frac{1}{2N}(1+v)$. When N is sufficiently large, and if the mating is random, then to order of $o((2N)^{-1})$, the probability that there are j "a" mutants in the next generation is

$$p_j = \binom{2N}{j} p^j (1-p)^{2N-j} .$$

Since $\lambda = 2Np = (1+v) + (2N)o((2N)^{-1}) \to (1+v)$ as $N \to \infty$, when N is sufficiently large, $1+v$ is then the average number of progenies of the a allele and

$$p_j \sim e^{-(1+v)} \frac{(1+v)^j}{j!}, \quad j = 0, 1, 2, \ldots .$$

(Notice that the Poisson distribution is the limit of the binomial distribution if $N \to \infty$ and if $\lim_{N \to \infty}(2Np)$ is finite.)

Using the Poisson progeny distribution as above, we have that $f(s) = e^{\lambda(s-1)}$ and $g_i(s) = [f(s)]^i = e^{i\lambda(s-1)}$, $i = 1, 2, \ldots$. Hence,

$$p_{0j} = \delta_{0j}, \quad j = 0, 1, \ldots, \infty,$$

$$p_{ij} = e^{-i\lambda} \frac{(i\lambda)^j}{j!}, \quad i = 1, 2, \ldots; \quad j = 0, 1, 2, \ldots, \infty.$$

Example 1.10. Nonhomogeneous Galton–Watson processes. In the Galton–Watson processes, the progeny distribution may change as time progresses. This is true for new mutants which are usually selectively disadvantageous comparing with wild allele when they were first introduced into the population; however, environmental changes at latter times may make the mutants selectively more advantageous over the wild allele. In these cases, the branching processes become nonhomogeneous. To illustrate how to derive transition probabilities in these cases, assume that the progeny distributions of the mutant are given by $\{p_j^{(i)}, \ i = 1, 2\}$ for $n \leq t_1$ and $t_1 < n$ respectively, where

$$p_j^{(i)} = e^{-\lambda_i} \frac{\lambda_i{}^j}{j!}, \quad i = 1, 2, \quad j = 0, 1, \ldots, \infty,$$

and where $\lambda_i = 1 + s_i$, $i = 1, 2$.

$(1 + s_i$ is the relative fitness of the mutant comparing with the wild allele over time with 1 for time $n \leq t_1$ and 2 for time $n > t_1$.)

Let $p_{ij}(n, n+1) = \Pr\{X(n+1) = j | X(n) = i\}$. Then, for $j = 0, 1, \ldots, \infty$,

$$p_{0j}(n, n+1) = \delta_{0j} \quad \text{for all } n \in T = (0, 1, \ldots, \infty);$$

$$p_{ij}(n, n+1) = e^{-i\lambda_1} \frac{(i\lambda_1)^j}{j!} \quad \text{for } n \leq t_1 \text{ and for all } i = 1, 2;$$

$$p_{ij}(n, n+1) = e^{-i\lambda_2} \frac{(i\lambda_2)^j}{j!} \quad \text{for } n > t_1 \text{ and for all } i = 1, 2.$$

Example 1.11. The Wright model in population genetics. In Example 1.3, we have considered a large diploid population and have focused on one locus with two alleles, say A and a. Let the population size be N (In reality, N is the number of individuals who mature to produce progenies). Denote by $\{X_1(t), X_2(t)\}$ the numbers of the genotypes AA and Aa at generation t respectively (The number of the genotype aa at generation t is

$X_3(t) = N - X_1(t) - X_2(t)$ as the population size is N). Since the genotype AA contributes 2 A alleles while the genotype Aa contributes only one A allele, $X(t) = 2X_1(t) + X_2(t)$ is then the number of A allele at generation t. Let $p(i,t)$ denote the frequency of A allele at generation $t+1$ given $X(t)$. Since the total number of alleles in the population is $2N$ as each individual has two alleles, hence $p(i,t)$ is a function of $\frac{X(t)}{2N} = \frac{i}{2N}$ given $X(t) = i$. Under the assumption that the mating is random among individuals, the conditional probability that $\{X_1(t+1) = m, X_2(t+1) = n\}$ given $X(t) = i$ is then given by:

$$\Pr\{X_1(t+1) = m, X_2(t+1) = n | X(t) = i\}$$

$$= \frac{N!}{m!n!(N-n-m)!}[p(i,t)^2]^m$$

$$\times [2p(i,t)q(i,t)]^n [q(i,t)^2]^{N-m-n} ,$$

where $q(i,t) = 1 - p(i,t)$.

The probability generating function (pgf) of $X(t+1) = 2X_1(t+1) + X_2(t+1)$ given $X(t) = i$ is

$$\theta(s) = \sum_{r=0}^{2N} s^r \Pr\{X(t+1) = r | X(t) = i\}$$

$$= \sum_{m=0}^{N} \sum_{n=0}^{N-m} s^{2m+n} \frac{N!}{m!n!(N-n-m)!}[p(i,t)^2]^m [2p(i,t)q(i,t)]^n$$

$$\times [q(i,t)^2]^{N-m-n}$$

$$= \{[sp(i,t)]^2 + 2sp(i,t)q(i,t) + [q(i,t)]^2\}^N$$

$$= \{sp(i,t) + q(i,t)\}^{2N}$$

$$= \sum_{j=0}^{2N} s^j \binom{2N}{j} [p(i,t)]^j [q(i,t)]^{2N-j} .$$

It follows that the process $\{X(t), t \in T\}$ is a Markov chain with discrete time $T = (0, 1, \ldots, \infty)$ and with state space $S = \{0, 1, \ldots, 2N\}$. The one step transition probability is:

$$\Pr\{X(t+1) = j | X(t) = i\} = \binom{2N}{j} [p(i,t)]^j [q(i,t)]^{2N-j},$$

where $q(i,t) = 1 - p(i,t)$.

The above model has been referred to as the Wright model in population genetics [10]. Whether or not this chain is homogeneous depending on $p(i,t)$. The following cases have been widely considered in the literature of population genetics.

(i) If there are no mutation, no selection among the individuals and no immigration and migration, then $p(i,t) = \frac{i}{2N}$ if $X(t) = i$. This case has been referred to as the *Random Genetic Drift* in population genetics; see Chaps. 6 and 7. In this case, the chain is homogeneous.

(ii) Suppose that there are mutations from A to a and from a to A in each generation but there are no selection, no immigration and no migration. Let the mutation rates per generation from A to a be u and from a to A be v. Then, given $X(t) = i$, $p(i,t) = \frac{i}{2N}(1-u) + (1 - \frac{i}{2N})v$. If both u and v are independent of time t, then the chain is homogeneous. However, because of the changing environment, it is expected that both u and v are functions of time. In this latter case, the chain is not homogeneous.

(iii) Suppose that there are no mutations, no immigration and no migration but there are selections among the individuals. Let the fitness (i.e., the expected number of progenies) of the three genotypes $\{AA, Aa, aa\}$ be given by $c(1 + s_1), c(1 + s_2)$ and c respectively ($c > 0$); see Remark 1.3. Then, given $X(t) = i$, with $x = \frac{i}{2N}$, $p_{t+1} = p(i,t)$ is given by:

$$p(i,t) = \frac{2x^2 c(1 + s_1) + 2x(1 - x)c(1 + s_2)}{2x^2 c(1 + s_1) + 2 \times 2x(1 - x)c(1 + s_2) + 2(1 - x)^2 c}$$

$$= \frac{x[1 + xs_1 + (1 - x)s_2]}{1 + x[xs_1 + 2(1 - x)s_2]}.$$

Hence, if both s_1 and s_2 are independent of time t, then the chain is homogeneous; if any of s_1 and s_2 depend on time t, then the chain is not homogeneous.

(vi) Suppose that there are no immigration and no migration but there are selections among the individuals and there are mutations from A to a and from a to A. Let the mutation rates per generation be given in (ii) and let the fitness of the three genotypes be given in (iii). Then, given $X(t) = i$, with

$x = \frac{i}{2N}$, $p_{t+1} = p(i,t)$ is given by:

$$p(i,t) = (1-u)\frac{x^2(1+s_1) + x(1-x)(1+s_2)}{x^2(1+s_1) + 2x(1-x)(1+s_2) + (1-x)^2}$$

$$+ v\frac{x(1-x)(1+s_2) + (1-x)^2}{x^2(1+s_1) + 2x(1-x)(1+s_2) + (1-x)^2}$$

$$= (1-u)\frac{x[1 + xs_1 + (1-x)s_2]}{1 + x[xs_1 + 2(1-x)s_2]}$$

$$+ v\frac{(1-x)[1 + xs_2]}{1 + x[xs_1 + 2(1-x)s_2]}.$$

(v) Suppose that there are immigration and migration but there are no mutations and no selections among the individuals. To model this, we follow Wright [11] to assume that the population exchanges the A alleles with those from outside populations at the rate of m per generation. Let x_I denote the frequency of A allele among the immigrants. Then, given $X(t) = i$, with $x = \frac{i}{2N}$, $p_{t+1} = p(i,t)$ is given by:

$$p(i,t) = x + m(x_I - x).$$

The chain is homogeneous if m and/or x_I are independent of time t; otherwise, the chain is not homogeneous.

Remark 1.3. Because the frequency of the alleles are the major focus in population genetics, one may with no loss of generality assume relative fitness for the genotypes. This is equivalent to delete the constant c from the fitness of the genotypes.

Example 1.12. The staged model of the AIDS epidemiology. In the AIDS epidemic, for clinical management and for taking into account the effects of infection duration, the infective stage (I stage) is usually divided into substage $I_1, \ldots, I_k$ with stochastic transitions between these substage [12–17]. For example, based on the total number of $CD4^{(+)}$ T cell counts per mm^3 of blood, Satten and Longini [16–17] have classified the I stage into 6 substage given by: I_1, CD4 counts $\geq 900/mm^3$; I_2, $900/mm^3 >$ CD4 counts $\geq 700/mm^3$; I_3, $700/mm^3 >$ CD4 counts $\geq 500/mm^3$; I_4, $500/mm^3 >$ CD4 counts $\geq 350/mm^3$; I_5, $350/mm^3 >$ CD4 counts $\geq 200/mm^3$; I_6, $200/mm^3 >$ CD4 counts. (Because of the 1993 AIDS definition by CDC [7], we will merge

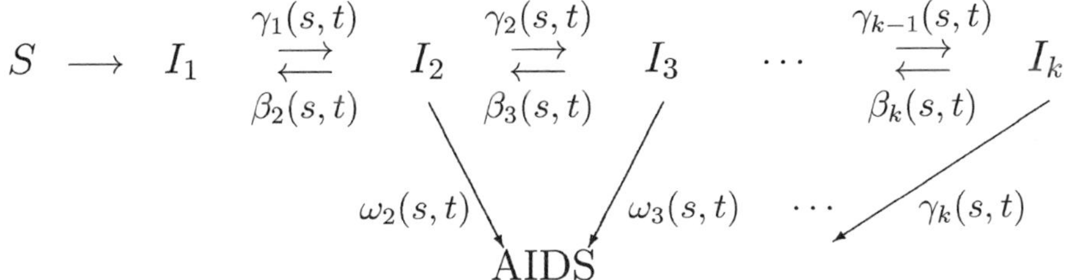

Fig. 1.1. A general model of the HIV epidemic with reverse transition.

the I_6 stage with the AIDS stage (A stage).) Let S denote the susceptible stage. Then the model and the transition is expressed schematically in Fig. 1.1.

Let $S(t)$ and $I_i(t)$ $(i = 1, \ldots, k)$ denote the numbers of S people and I_i $(i = 1, \ldots, k)$ people at time t respectively and $A(t)$ the number of AIDS cases at time t. Let $t = 0$ be the time to start the epidemic. Then we have a $(k+2)$-dimensional stochastic process $\underset{\sim}{U}(t) = \{S(t), I_i(t), i = 1, \ldots, k, A(t)\}$ with parameter space $T = \{t \geq 0\}$ and with sample space Ω which is a subset of an $(k+2)$-dimensional Euclidean space with non-negative integers as components.

Let $p_S(t)dt$ be the probability of $S \to I_1$ during $[t, t+dt)$. With $I_0 = S$ and $I_{k+1} = A$, let the transition rates of $I_i \to I_{i+1}$, $I_i \to I_{i-1}$ and $I_i \to A$ at time t be given by $\gamma_i(s,t), \beta_i(s,t)$ and $\omega_i(s,t)$, respectively, for I_i $(i = 1, \ldots, k)$ people who have arisen from I_{i-1} at time s. (Note that $\beta_1(s,t) = 0, \gamma_k(s,t) = \omega_k(s,t)$ from Fig. 1.1.) If $\gamma_i(s,t) = \gamma_i(t), \beta_i(s,t) = \beta_i(t)$ and $\omega_i(s,t) = \omega_i(t)$ are independent of s, then the process $\underset{\sim}{U}(t)$ is Markov. This process is in fact a Markov chain since the number of states is countable infinite. These are the processes considered by Longini and his associates [12]. On the other hand, if $\gamma_i(s,t), \beta_i(s,t)$ and $\omega_i(s,t)$ are dependent on s, then the process $\underset{\sim}{U}(t)$ is not Markov [13, 14, 18]. The non-Markovian processes arise because of treatment of HIV-infected individuals by anti-viral drugs such as AZT. These are the processes considered by Longini *et al.* [13, 14]; see also [18].

Example 1.13. The MVK two-stage model of carcinogenesis. The two-stage model of carcinogenesis which fits the cancer biological mechanism was first proposed by Knudson [19], Moolgavkar and Venzon [20] and Moolgavkar and Knudson [21] and has been referred to as the MVK two-stage model. This model assumes that a cancer tumor develops from a single normal stem cell by clonal expansion and views carcinogenesis as the end result of two discrete, heritable and irreversible events in normal stem cells; each event occurs during cell division. The MVK two-stage model has assumed that the parameter values are constant independent of time and has made the assumption that with probability one each cancer tumor cell grows instantaneously into a malignant tumor. Furthermore, it is assumed that the proliferation and differentiation of normal stem cells follow deterministic growth.

Let $N(t)$ denote the number of normal stem cells at time t and $\{I(t), T(t)\}$ the numbers of initiated cells and cancer tumor cells. (Note that $I(t)$ is actually the number of the first initiated tumor cells and $T(t)$ the number of second initiated cells under the assumption that each second initiated cell grows instantaneously into a malignant tumor; see [5, Chap. 3].) Then $N(t)$ is a deterministic function of time t and $\{I(t), T(t)\}$ is a two-dimensional Markov chain with continuous time; see Tan [5, Chap. 3]. To find the transition rates (or incidence functions) of this process, let $M_I(t)$ be the number of mutations from normal stem cells to I cells during $[t, t + \Delta t)$ and denote by $\lambda(t) = N(t)\alpha_N(t)$, where $\alpha_N(t)$ is the mutation rate per cell division from N to I at time t. Then, to order of $0(N(0)^{-1})$, $M_I(t)$ follows a Poisson distribution with mean $\lambda(t)\Delta t$; see Exercise 1.13. Let $b_I(t), d_I(t)$ and $\alpha_I(t)$ be the birth rate, the death rate and the mutation rate of the I cells. Then, during $[t, t + \Delta t)$, the probabilities that an I cell will yield two I cells, 0 I cells and 1 I cell and 1 T cell are given respectively by $b_I(t)\Delta t + o(\Delta t), d_I(t)\Delta t + o(\Delta t)$ and $\alpha_I(t)\Delta t + o(\Delta t)$, respectively. It follows that as defined in Chap. 4, $\{I(t), t \leq 0\}$ is a stochastic birth-death process with birth rate $jb_I(t) + \lambda(t)$ and death rate $jd_I(t)$. This is a nonhomogeneous stochastic Feller–Arley birth-death process with immigration as defined in [22].

For the above process, notice that because the number of stem cells after birth is usually very large ($10^6 \sim 10^8$), it is a good approximation to assume that $N(t)$ is a deterministic function of t; in fact it has been shown by Tan and Brown [23] through continuous multiple branching process that to order of $O(N(0)^{-1})$, $N(t)$ is indeed a deterministic function. However, it has been recognized that the assumption that with probability one each cancer tumor cell grows instantaneous into a malignant tumor does not hold in many real world situations [6, 24, 25]; in fact it has been shown by Yang and Chen [24] and Tan and Chen [6] that malignant tumor cells develop by clonal expansion from primary second initiated cells. It follows that conditional on the number of $I(s)$

cells for all $s \leq t$, $T(t)$ follows a Poisson distribution with conditional mean given by $\lambda_T(t) = \int_0^t I(x)\alpha_I(s)P_T(s,t)dx$, where $P_T(s,t)$ is the probability that a second initiated cell arising at time s will develop into a malignant cancer tumor by time t; for proof of this, see Chap. 8. Since the distribution of $T(t)$ depends on the $I(s)$ for all $s \leq t$, $T(t)$ is not even a Markov process; for more detail, see Chap. 8.

1.3. Diffusion Processes and Examples

Let $\{X(t), t \geq 0\}$ be a stochastic process with continuous parameter space $T = \{t \geq 0\}$ and with continuous state space $S = [a, b]$. (a can be $-\infty$ and b can be ∞.) Suppose that the increment $dX(t) = X(t+dt) - X(t)$ changes continuously in probability when dt is very small so that the probability of any jump (say $\epsilon > 0$) would be nil. Also, in many practical problems, it is reasonable to assume that if $dt \cong 0$, one may practically ignore higher order moments (i.e., with order ≥ 3) of $dX(t)$. This leads to a class of stochastic processes which involve only the first and second moments of $dX(t)$. If these processes are Markov processes, then they are classified as *Diffusion Processes*.

Definition 1.7. Let $X(t)$ be a Markov stochastic process with parameter space $T = \{t \geq 0\}$ and with state space $S = [a, b]$. Then $X(t)$ is called a *diffusion process with coefficients* $\{m(x,t), v(x,t)\}$ if and only if the following conditions hold:

(i) For every $\epsilon > 0$ given, for every $x \in S$ and for every $t \geq 0$,

$$P\{|X(t+dt) - X(t)| \geq \epsilon | X(t) = x\} = o(dt),$$

where $o(dt)$ is defined by $\lim_{dt \to 0} o(dt)/dt = 0$.

(ii) There exists a continuous function $m(x,t)$ of $x \in S$ and $t \geq 0$ satisfying the condition:

$$E[X(t+dt) - X(t)|X(t) = x] = m(x,t)dt + o(dt).$$

(iii) There exists a positive continuous function $v(x,t)$ of $x \in S$ and $t \geq 0$ satisfying

$$E\{[X(t+dt) - X(t)]^2 | X(t) = x\} = v(x,t)dt + o(dt).$$

(iv) For every $x \in S$ and every $t \geq 0$,

$$E\{|X(t+dt) - X(t)|^r | X(t) = x\} = o(dt) \text{ if } r \geq 3.$$

Any stochastic process $\{X(t), t \in T\}$ which has continuous parameter space T and continuous state space S and which satisfies condition (i) is called a continuous stochastic process. Notice that condition (i) in Definition 1.7 implies that the probability of any significant change of state is very small in a small time interval (Convergence to 0 faster than the time interval dt). That is, with probability one the process $X(t)$ will not have jumps as time increases. Condition (iii) implies, however, that for any time interval dt, no matter how small, with positive probability changes do occur. Thus, condition (iii) guarantees that the process is a dynamic process unless it has been absorbed into some absorbing states.

Let $f(x, y; s, t)$ be the conditional probability density function (pdf) of $X(t)$ at y given $X(s) = x$. Then, condition (i) can be expressed alternatively as: For any $\epsilon > 0$,

$$\int_{|y-x| \geq \epsilon} f(x, y; t, t+dt) dy = o(dt).$$

To be more precise, we notice that conditions (i)–(iv) are also equivalent to conditions (i), (ii)$'$, (iii)$'$ and (iv), where conditions (ii)$'$ and (iii)$'$ are given by the following:

(ii)$'$. For any $\epsilon > 0$,

$$\int_{|y-x| \leq \epsilon} (y - x) f(x, y; t, t+dt) dy = m(x, t) dt + o(dt).$$

(iii)$'$. For any $\epsilon > 0$,

$$\int_{|y-x| \leq \epsilon} (y - x)^2 f(x, y; t, t+dt) dy = v(x, t) dt + o(dt).$$

Condition (iii)$'$ follows easily from the observation that if $|y - x| > \epsilon \geq 1$, then $|y - x|^3 \geq (y - x)^2$ so that

$$\int_{|y-x| \geq \epsilon} (y-x)^2 f(x, y; t, t+dt) dy \leq \int_{|y-x| \geq \epsilon} |y-x|^3 f(x, y; t, t+dt) dy = o(dt);$$

on the other hand, if $1 \geq |y - x| > \epsilon$, then $(y - x)^2 \leq 1$ so that

$$\int_{|y-x| \geq \epsilon} (y - x)^2 f(x, y; t, t + dt) dy \leq \int_{|y-x| \geq \epsilon} f(x, y; t, t + dt) dy = o(dt) \,.$$

Condition (ii)$'$ follows readily from the Schwarz inequality given by:

$$\int_{|y-x| \geq \epsilon} |y - x| f(x, y; t, t + dt) dy \leq \left\{ \int_{|y-x| \geq \epsilon} (y - x)^2 f(x, y; t, t + dt) dy \right\}^{1/2}$$

$$\times \left\{ \int_{|y-x| \geq \epsilon} f(x, y; t, t + dt) dy \right\}^{1/2} \,.$$

Using the conditional pdf $f(x, y; s, t)$, one may also define the diffusion process as homogeneous iff $f(x, y; s, t) = f(x, y; 0, t - s) = f(x, y; t - s)$: That is, $f(x, y; s, t)$ depends on the times s and t only through the difference $t - s$. Notice that, in order for the diffusion process to be homogeneous, a precondition is that $m(x, t) = m(x)$ and $v(x, t) = v(x)$ must be independent of time t. In Chaps. 6 and 7, we will provide some general theories of diffusion processes and illustrates its applications in detail.

Example 1.14. Diffusion approximation of population growth models. Let $\{X(t), t \geq 0\}$ denote the number of bacteria at time t with M being the maximum population size. Then, under some general conditions, it is shown in Chap. 6 that to the order of $O(M^{-2})$, $Y(t) = X(t)/M$ follows a diffusion process with state space $S = [0, 1]$. For the stochastic logistic growth process, this was proved in [26, 27] by using alternative methods.

Example 1.15. Diffusion approximation of the Galton–Watson branching processes. Let $\{X(t), t \in T = (0, 1, \ldots, \infty)\}$ be a Galton–Watson branching process with progeny distribution $\{p_j, j = 0, 1, \ldots, \infty\}$. Assume that the mean and the variance of the progeny distribution are given respectively by $1 + \frac{1}{N}\alpha + O(N^{-2})$ and σ^2, where N is very large. Then it is shown in Example 6.6 that to the order of $O(N^{-2})$, $Y(t) = X(t)/N$ is a diffusion process with state space $S = [0, \infty)$ and with coefficients $\{m(x, t) = x\alpha, v(x, t) = x\sigma^2\}$.

Example 1.16. Diffusion approximation of the Wright model in population genetics. In Example 1.11, we have considered the Wright model in population genetics. In this model, $\{X(t), t \in T\}$ is the number of A allele in a large diploid population of size N, where $T = \{0, 1, \ldots, \infty\}$. It is shown in

Example 1.11 that this is a Markov chain with state space $S = \{0, 1, \ldots, 2N\}$ and with one-step transition probabilities given by:

$$\Pr\{X_1(t+1) = j | X(t) = i\} = \binom{2N}{j} p_{t+1}^j q_{t+1}^{2N-j},$$

where p_{t+1} is the frequency of A allele at generation $t+1$ and $q_{t+1} = 1 - p_{t+1}$.

Denote by $x = i/2N$, $m(x,t) = (2N)(x - p_{t+1})$, and

$$v(x,t) = (2N)(x - p_{t+1})^2 + p_{t+1}(1 - p_{t+1}).$$

If $m(x,t)$ and $v(x,t)$ are bounded functions of x and t for all $i \geq 0$ and for all $t \geq 0$, then, it is shown in Theorem 6.6, that to order of $O(N^{-2})$, $\{Y(t) = X(t)/(2N), t \geq 0\}$ is a diffusion process with state space $S = [0, 1]$ and with diffusion coefficients $\{m(x,t), v(x,t)\}$.

Example 1.17. Diffusion approximation of the SIR model in infectious disease. Consider the SIR model in infectious diseases. Let $S(t)$ and $I(t)$ denote the number of S people and I people at time t. Let $c(t)\Delta t$ be the average number of partners of each S person during $[t, t + \Delta t)$. Let the transition rates of $I \to R$ be $\gamma(t)$ and $q(t)$ the per partner transmission probability of the disease given contacts between a S person and an I person during $[t, t + \Delta t)$. Let the death rate and the immigration and recruitment rate of I people be $\mu_I(t)$ and $\nu_I(t)$ respectively. Suppose that the following conditions hold:

(1) There is no contact between S people and R people.

(2) The population size changes very little over time so that $S(t) + I(t) = N(t) \sim N$ is approximately independent of time t.

(3) There is only one sexual activity level and the mixing pattern is random mixing.

Then it is shown in Example 6.8 that to the order of $O(N^{-2})$, $\{Y(t) = \frac{I(t)}{N}, t \geq 0\}$ is a diffusion process with state space $S = [0, 1]$ and with coefficients $\{m(x,t) = \alpha(t)x(1-x) + \mu_I(t) - x[\gamma(t) + \nu_I(t)], v(x,t) = \alpha(t)x(1-x) + \mu_I(t) + x[\gamma(t) + \nu_I(t)]\}$, where $\alpha(t) = c(t)q(t)$.

Example 1.18. Diffusion approximation of initiated cancer cells in carcinogenesis. Consider the MVK two-stage model of carcinogenesis as described in Example 1.13. Let N_0 denote the number of normal stem cells at time 0 and denote by $X(t) = \frac{1}{N_0}I(t)$. For large N_0, it is shown in Example 6.7

that to the order of $O(N_0^{-2})$, $\{X(t), t \geq 0\}$ is a diffusion process with state space $\Omega = [0, \infty)$ and with coefficients

$$\{m(x,t) = \alpha_N(t) + x\gamma(t), v(x,t) = \frac{1}{N_0}x\omega(t)\},$$

where $\gamma(t) = b_I(t) - d_I(t)$, $\omega(t) = b_I(t) + d_I(t)$ and $\alpha_N(t)$ is the mutation rate from normal stem cells to initiated cells.

1.4. State Space Models and Hidden Markov Models

To validate the stochastic models and to estimate unknown parameters in the model, one usually generates observed data from the system. Based on these data sets, statisticians have constructed statistical models to make inferences about the unknown parameters and to validate the model. To combine information from both the mechanism and the data, the state space model then combines the stochastic model and the statistical model into one model. Thus, the state space model has two sub-models:

(1) The stochastic system model which is the stochastic model of the system, and

(2) the observation model which is the statistical model based on some observed data from the system.

Definition 1.8.   Let $X(t)$ be a stochastic process with parameter space T and with state space S. Let $\{Y(t_j) = Y_j, j = 1, \ldots, n\}$ be the observed values on $X(t)$ at the time points $t_1 \leq t_2 \leq \cdots \leq t_{n-1} \leq t_n$. Suppose that $Y_j = f[X(t), t \leq t_j] + e_j$ for some function $f()$ of $X(t), t \leq t_j$, where e_j is the random measurement error for measuring Y_j. Then the combination $\{X(t), t \in T; Y_j, j = 1, \ldots, n\}$ is called *a state space model of the stochastic system* with stochastic system model given by the stochastic process $\{X(t), t \in T\}$ and with the observation model given by the statistical model $Y_j = f[X(t), t \leq t_j] + e_j$ for the system. In other word, a state space model of a stochastic system is the stochastic model of the system plus some statistical model based on some observed data from the system.

From this definition, it appears that if some data are available on the system, then one may always construct a state space model for the system. For this state space model, the stochastic process of the system is the stochastic system model whereas the statistical model of the system is the observation model. As such, one may look at the state space model as a device to combine information from both sources: The mechanism of the system via stochastic models and the information from the data on the system. It is advantageous over both the stochastic model and the statistical model used alone as it combines advantages and information from both models.

The state space model was originally proposed by Kalman in the 60's for engineering control and communication [28]. Since then it has been successfully used in satellite research and military missile research. It has also been used by economists in econometric research [29] and by mathematician and statisticians in time series research [30] for solving many difficult problems which appear to be extremely difficult from other approaches. In 1995, the state space model was first proposed by Wu and Tan in AIDS research [31, 32]. Since then it has been used by Cazelles and Chau [33] and by Tan and his associates for modeling AIDS epidemic [34, 35]; it has also been used by Tan and his associates for studying the HIV pathogenesis in HIV-infected individuals [36–39]. Recently, Tan and his associates [40–42] have developed state space models for carcinogenesis. In Chaps. 8 and 9, we will illustrate and discuss these models and demonstrate some of its applications to cancer and AIDS.

Definition 1.9. A state space model is called a *hidden Markov model* if the stochastic system model is a Markov process.

Hidden Markov models usually apply to observed data on a Markov process because the observed data are usually masked by random measurement errors in measuring the observations. As such, it is appropriate to define hidden Markov models as above because the Markov process is hidden in the observed equations. In this section we will illustrate this by an example from the AIDS epidemiology. This example has been used by Satten and Longini [17] to estimate the transition rates in the San Francisco homosexual population.

Example 1.19. The hidden Markov models of HIV epidemic as state space models. Consider a population involving only HIV-infected individuals and AIDS cases. Following Satten and Longini [17], we partition the HIV-infected individuals into 6 sub-stages by the number of CD4$^{(+)}$ T cells per mm^3 of blood as given in Example 1.12. Let i stand for the I_i stage with I_6 denoting the AIDS stage. Let $X(t)$ denote the stochastic process representing the infective stages with state space $\Omega = \{1, \ldots, 6\}$ and with parameter space $T = \{0, 1, \ldots, \infty\}$ with 0 denoting the starting time. Let γ_{ji} be the one-step transition probability from I_j to I_i $(j, i = 1, \ldots, 6)$ and with $\gamma_{6i} = \delta_{6i}, i = 1, \ldots, 6$. Then $X(t)$ is a homogeneous Markov chain with 6 states and with discrete time. In this Markov chain, the state I_6 is the absorbing state (persistent state) and all other states are transient states (For definition of persistent states and transient states, see, Definition 2.3.) Let $Y_i(t)$ be the observed number of the I_i people at time t. Then, because the CD4$^{(+)}$ T cell counts are subjected to measurement errors, in terms of the observed numbers, the process is a hidden Markov chain. In this section, we proceed to show that this hidden Markov chain can be expressed as a state space model which consists of the stochastic system model and the observation model (For more detail about state space models, see Chaps. 8 and 9.) To this end, let $W_{ij}(t)$ denote the number of I_j people at time $t+1$ given $I_i(t)$ I_i people at time t for $i = 1, \ldots, 5$; $j = 1, \ldots, 6$ and $Z_{ij}(r, t)$ the observed number of I_r people at time $t + 1$ counted among the $W_{ij}(t)$ people. Assume now that the death rate is very small for people other than AIDS and that there are no immigration and no migration in the population. Then, given $I_i(t)$ for $(i = 1, \ldots, 5)$, the probability distribution of $W_{ij}(t), j = 1, \ldots, 6$ follows a five-dimensional multinomial distribution with parameters $\{I_i(t); \gamma_{ij}, j = 1, \ldots, 5\}$. That is, with $W_{i6}(t) = I_i(t) - \sum_{j=1}^{5} W_{ij}(t)$, we have that, for $i = 1, \ldots, 5$,

$$\{W_{ij}(t), j = 1, \ldots, 5\} | I_i(t) \sim ML\{I_i(t); \gamma_{ij}, j = 1, \ldots, 5\}.$$

Note that $\sum_{j=1}^{6} \gamma_{ij} = 1$ for $i = 1, \ldots, 5$ and $I_6 \to I_6$ only.

Let $I_6(t)$ include people who died from AIDS during $[t, t + 1)$. Then, we have the following stochastic equations for $I_j(t), j = 1, \ldots, 6$:

$$I_j(t + 1) = \sum_{i=1}^{5} W_{ij}(t) + \delta_{j6} I_6(t)$$

$$= \sum_{i=1}^{5} I_i(t)\gamma_{ij} + \delta_{j6} I_6(t) + \epsilon_j(t + 1), \tag{1.2}$$

where

$$\epsilon_j(t + 1) = \sum_{i=1}^{5} [W_{ij}(t) - I_i(t)\gamma_{ij}].$$

Denote by $F' = (\gamma_{ij})$ the one-step transition matrix, $\underset{\sim}{I}(t) = \{I_1(t), \ldots, I_6(t)\}'$ and $\underset{\sim}{\epsilon}(t+1) = \{\epsilon_1(t+1), \ldots, \epsilon_6(t+1)\}'$. Then in matrix notation, the above system of equations become:

$$\underset{\sim}{I}(t+1) = F\,\underset{\sim}{I}(t) + \underset{\sim}{\epsilon}(t+1)\,. \tag{1.3}$$

This is the stochastic system model for the state space model associated with the above hidden Markov chain.

To account for the random measurement error, we assume that the measurement errors follow Gaussian distributions and that measurement errors for AIDS cases is very small to be ignored. Let ν_i $(i = 1, \ldots, 5)$ denote the mean number of CD4$^{(+)}$ T cells per mm^3 of blood for the I_i stage. (One may take $\nu_1 = 1000/\text{mm}^3$, $\nu_2 = 800/\text{mm}^3$, $\nu_3 = 600/\text{mm}^3$, $\nu_4 = 425/\text{mm}^3$, $\nu_5 = 275/\text{mm}^3$.) Let $X_i = \frac{Z - \nu_i}{100}$ $i = 1, \ldots, 5$, where Z is the observed number of CD4$^{(+)}$. Then, given the I_i stage $(i = 1, \ldots, 5)$, the conditional distribution of X_i is a truncated Gaussian with mean 0 and variance σ^2 and with state space $[-\frac{\nu_i}{100}, \frac{1100 - \nu_i}{100}]$ independently for $i = 1, \ldots, 5$. For $(i = 1, \ldots, 5)$, let

$$a_{i,0} = \frac{1100 - \nu_i}{100}, \quad a_{i,1} = \frac{900 - \nu_i}{100}, \quad a_{i,2} = \frac{700 - \nu_i}{100}, \quad a_{i,3} = \frac{500 - \nu_i}{100},$$

$$a_{i,4} = \frac{350 - \nu_i}{100}, \quad a_{i,5} = \frac{200 - \nu_i}{100}, \quad a_{i,6} = -\frac{\nu_i}{100}\,.$$

Denote, for $(i = 1, \ldots, 5; j = 1, \ldots, 6)$,

$$p_{ij} = C_i \int_{a_{i,j}}^{a_{i,j-1}} f(x)dx\,.$$

where $f(x)$ is the pdf of the Gaussian distribution with mean 0 and variance σ^2 and $C_i^{-1} = \int_{a_{i,6}}^{a_{i,0}} f(x)dx$.

Then for $(i = 1, \ldots, 5; j = 1, \ldots, 5)$, the conditional probability distribution of $\{Z_{ij}(r, t), r = 1, \ldots, 5\}$ given $W_{ij}(t)$ is:

$$\{Z_{ij}(r, t), r = 1, \ldots, 5\} | W_{ij}(t) \sim ML\{W_{ij}(t); \quad p_{jr}, r = 1, \ldots, 5\}\,.$$

Note that $\sum_{r=1}^{6} p_{jr} = 1$ for $j = 1, \ldots, 5$.

It follows that we have, with $p_{6i} = \gamma_{6i} = \delta_{6i}$:

$$Y_i(t+1) = \sum_{u=1}^{5}\left\{\sum_{v=1}^{5} Z_{uv}(i,t) + W_{u6}p_{6i}\right\} + \delta_{i6}I_6(t)$$

$$= \sum_{u=1}^{5}\sum_{v=1}^{6} W_{uv}(t)p_{vi} + \delta_{i6}I_6(t) + e_{i1}(t+1)$$

$$= \sum_{u=1}^{5} I_u(t)\sum_{v=1}^{6} \gamma_{uv}p_{vi} + \delta_{i6}I_6(t) + e_{i1}(t+1) + e_{i2}(t+1)$$

$$= \sum_{u=1}^{6} I_u(t)\sum_{v=1}^{6} \gamma_{uv}p_{vi} + e_i(t+1)\,,$$

where

$$e_{i1}(t+1) = \sum_{u=1}^{5}\left\{\sum_{v=1}^{5}[Z_{uv}(i,t) - W_{uv}(t)p_{vi}]\right\}\,,$$

and

$$e_{i2}(t+1) = \sum_{u=1}^{5}\left\{\sum_{v=1}^{6} p_{vi}[W_{uv}(t) - I_u(t)\gamma_{uv}]\right\}\,,$$

and $e_i(t+1) = e_{i1}(t+1) + e_{i2}(t+1)$.

Put $P = (p_{ij})$ and $H = P'F$. Denote by $\underset{\sim}{Y}(t) = \{Y_1(t),\ldots,Y_6(t)\}'$ and $\underset{\sim}{e}(t) = \{e_1(t),\ldots,e_6(t)\}'$. Then, in matrix notation, we have:

$$\underset{\sim}{Y}(t+1) = H\underset{\sim}{I}(t) + \underset{\sim}{e}(t+1)\,. \tag{1.4}$$

Equation (1.4) is the observation model for the state space model associated with the above hidden Markov chain.

1.5.   The Scope of the Book

The stochastic models described in Secs. 1.1–1.4 are the major models which arise from genetics, cancer and AIDS. In this book we will thus present a systematic treatment of these models and illustrate its applications to genetics, cancer and AIDS.

In Chap. 2, general theories of Markov chains with discrete time will be presented and discussed in detail. As a continuation of Chap. 2, in Chap. 3, we will present some general theories on stationary distributions of Markov chains with discrete time; as an application of stationary distributions, we also present some MCMC (Markov Chain Monte Carlo) methods to develop computer algorithms for estimating unknown parameters and state variables. In Chaps. 4 and 5, general theories of Markov chains with continuous time will be presented and discussed in detail. Applications of these theories to genetics, cancer and AIDS to solve problems in these areas will be discussed and demonstrated. In Chaps. 6 and 7, we will present and discuss in detail general theories of diffusion processes. We will show that most processes in genetics, cancer and AIDS can be approximated by diffusion processes. Hence, one may use theories of diffusion process to solve many problems in these areas. Finally in Chaps. 8 and 9, we present and discuss some general theories of state space models and illustrate its applications to cancer and AIDS.

This book is unique and differs from other books on stochastic processes and stochastic models in that it has presented many important topics and approaches which would not be discussed normally in other books of stochastic processes. This includes MCMC methods and applications, stochastic difference and differential equation approaches to Markov chains as well as state space models and applications. It follows that there are minimal overlaps with other books on stochastic processes and stochastic models. Also, the applications to cancer, AIDS and genetics as described in this book are unique and would normally not be available in other books of stochastic processes.

1.6. Complements and Exercises

Exercise 1.1. Let $\{X(t), t \geq 0\}$ be a stochastic process with state space $S = (0, 1, \ldots, \infty)$. Suppose that the following two conditions hold:

(a) $P\{X(0) = 0\} = 1$.

(b) $\{X(t), t \geq 0\}$ has independent increment. That is, for every n and for every $0 \leq t_1 < \cdots < t_n$, $\{Y_j = X(t_j) - X(t_{j-1}), j = 1, \ldots, n\}$ are independently distributed of one another.

Show that $X(t)$ is a Markov process.

Exercise 1.2. Let $\{X(j), j = 1, 2, \ldots, \infty\}$ be a sequence of independently distributed random variables. That is, for every n and for every set of integers $0 \leq j_1 < \cdots < j_n$, $\{X(j_i), i = 1, \ldots, n\}$ are independently distributed of one another. Define $Y(j) = \frac{1}{3}[X(j-1) + X(j) + X(j+1)], j = 1, \ldots,$ with $P\{X(0) = 0\} = 1$. Show that $\{Y(j), j = 1, 2, \ldots\}$ is not a Markov process.

Exercise 1.3. Consider the two-stage model described in Example 1.13. Let $\{B_N(t), D_N(t)\}$ denote the numbers of birth and death of normal stem cells during $[t, t + \Delta t)$ and $M_I(t)$ the number of mutations from $N \to I$ during $[t, t + \Delta t)$. Denote the birth rate, the death rate and the mutation rate of normal stem cells by $b_N(t)$, $d_N(t)$ and $\alpha_N(t)$ respectively. Then given $N(t)$, $\{B_N(t), D_N(t), M_I(t)\}$ follows a multinomial distribution with parameters $\{N(t), b_N(t)\Delta t, d_N(t)\Delta t, \alpha_N(t)\Delta t\}$. Show that if $N(t)$ is very large and if $\lambda(t) = N(t)\alpha_N(t)$ is finite for all $t > 0$, then to order of $O([N(t)]^{-1})$, $M_I(t)$ is a Poisson random variable with parameter $\lambda(t)\Delta t$ independent of $\{B_N(t), D_N(t)\}$.

Exercise 1.4. Let $\{X(n), n = 0, 1, \ldots, \infty\}$ be a simple branching process with progeny distribution $\{p_j = \binom{\alpha+j-1}{j}\theta^\alpha(1-\theta)^j, j = 0, 1, \ldots, \infty\}$, where $\alpha > 0$ and $0 < \theta < 1$. Derive the one-step transition probabilities. Show that this is a homogeneous Markov chain with discrete time.

Exercise 1.5. Let $\{X(n), n = 0, 1, \ldots, \infty\}$ be a simple branching process with progeny distribution given by:

$$p_j(n) = \binom{\alpha_1 + j - 1}{j} \theta_1^{\alpha_1}(1 - \theta_1)^j, \quad j = 0, 1, \ldots, \infty, \text{ if } n \leq n_1,$$

$$p_j(n) = \binom{\alpha_2 + j - 1}{j} \theta_2^{\alpha_2}(1 - \theta_2)^j, \quad j = 0, 1, \ldots, \infty, \text{ if } n > n_1,$$

where $\alpha_i > 0$ $(i = 1, 2)$ and $0 < \theta_i < 1$ $(i = 1, 2)$. Derive the one-step transition probabilities.

Exercise 1.6. Let $\{X(t), t \geq 0\}$ be a continuous stochastic process with state space $S = [a, b]$. Denote by $\Delta X(t) = X(t + \Delta t) - X(t)$ and suppose that the following conditions hold:

(a) $E\{\Delta X(t) | X(t) = x\} = m(x, t)\Delta t + o(\Delta t)$,
(b) $E\{[\Delta X(t)]^2 | X(t) = x\} = v(x, t)\Delta t + o(\Delta t)$,
(c) $E\{[\Delta X(t)]^k | X(t) = x\} = o(\Delta t)$ for $k = 3, 4, \ldots, \infty$,

where $\{m(x, t), v(x, t)\}$ are continuous functions of (x, t) with $v(x, t) > 0$.

Show that $\{X(t), t \geq 0\}$ is a diffusion process with state space $S = [a, b]$ and with diffusion coefficients $\{m(x, t), v(x, t)\}$.

Exercise 1.7. In the hidden Markov model given by Example 1.19, prove or derive the following results:

(a) The elements of $\underset{\sim}{\epsilon}(t)$ and of $\underset{\sim}{e}(t)$ have expected value 0.

(b) Using the basic result $\mathrm{Cov}(X, Y) = \underset{Z}{E}\,\mathrm{Cov}[E(X|Z), E(Y|Z)] + \underset{Z}{E}\,\mathrm{Cov}(X, Y|Z)$, show that

$$\mathrm{Cov}\{\underset{\sim}{I}(t), \underset{\sim}{\epsilon}(\tau)\} = \mathbf{0}, \quad \mathrm{Cov}\{\underset{\sim}{I}(t), \underset{\sim}{e}(\tau)\} = \mathbf{0}$$

for all $\{t \geq 0, \tau \geq 0\}$.

(c) Derive the variances and covariances of the random noises

$$\{\mathrm{Var}[\epsilon_i(t)], \mathrm{Cov}[\epsilon_i(t), \epsilon_j(t)], i \neq j\}\,.$$

References

[1] L. B. Jorde, J. C. Carey, M. J. Bamshad and R. J. White, *Medical genetics*, Second Edition, Moshy Inc., New York (1999).

[2] C. R. Scriver, A. I. Beaudet, W. S. Sly and D. Valle, eds., *The Metabolic and Molecular Basis of Inherited Diseases*, Vol. 3, McGraw-Hill, New York (1995).

[3] T. Harris, *The Theory of Branching Processes*, Springer-Verlag, Berlin (1963).

[4] W. Y. Tan and C. C. Brown, *A stochastic model for drug resistance and immunization*, I. *One drug case*, Math. Biosciences **97** (1989) 145–160.

[5] W. Y. Tan, *Stochastic Models of Carcinogenesis*, Marcel Dekker, New York (1991).

[6] W. Y. Tan and C. W. Chen, *Stochastic models of carcinogenesis, Some new insight*, Math Comput. Modeling **28** (1998) 49–71.

[7] CDC, 1993 *Revised Classification System for HIV Infection and Expanded Surveillance Case Definition for AIDS Among Adolescents and Adults*, MMWR **41** (1992), No. RR17.

[8] W. Y. Tan, *Stochastic Modeling of AIDS Epidemiology and HIV Pathogenesis*, World Scientific, Singapore (2000).

[9] W. Y. Tan and H. Wu, *Stochastic modeling of the dynamics of $CD4^{(+)}$ T cell infection by HIV and some Monte Carlo studies*, Math. Biosciences **147** (1998) 173–205.

[10] J. F. Crow and M. Kimura, *An Introduction to Population Genetics Theory*, Harper and Row, New York (1970).

[11] S. Wright, *Evolution and the Genetics of Population. Vol. 2, The Theory of Gene Frequencies*, University of Chicago Press Chicago (1969).

[12] I. M. Longini, W. S. Clark, L. I. Gardner and J. F. Brundage, *The dynamics of CD4+ T-lymphocyte decline in HIV-infected individuals: A Markov modeling approach*, J. AIDS **4** (1991) 1141–1147.

[13] I. M. Longini, R. H. Byers, N. A. Hessol and W. Y. Tan, *Estimation of the stage-specific numbers of HIV infections via a Markov model and backcalculation*, Statistics in Medicine **11** (1992) 831–843.

[14] I. M. Longini, W. S. Clark and J. Karon, *Effects of routine use of therapy in slowing the clinical course of human immunodeficiency virus (HIV) infection in a population based cohort*, Amer. J. Epidemiology **137** (1993) 1229–1240.

[15] I. M. Longini, W. S. Clark, G. A. Satten, R. H. Byers and J. M. Karon, *Staged Markov models based on $CD4^{(+)}$ T-lymphocytes for the natural history of HIV infection*, in *Models for Infectious Human Diseases: Their Structure and Relation to Data*, eds. V. Isham and G. Medley, pp. 439–459, Cambridge University Press (1996).

[16] G. Satten and I. M. Longini, *Estimation of incidence of HIV infection using cross-sectional marker survey*, Biometrics **50** (1994) 675–688.

[17] G. Satten and I. M. Longini, *Markov Chain With Measurement Error: Estimating the 'True' Course of Marker of the Progression of Human Immunodeficiency Virus Disease*, Appl. Statist. **45** (1996) 275–309.

[18] W. Y. Tan, *On the HIV Incubation Distribution under non-Markovian Models*, Statist. & Prob Lett. **21** (1994) 49–57.

[19] A. G. Knudson, *Mutation and cancer: Statistical study of retinoblastoma*, Proc. Natl. Acad. Sci. USA **68** (1971) 820–823.

[20] S. H. Moolgavkar and D. J. Venzon, *Two-Event Models for Carcinogenesis: Incidence Curve for Childhood and Adult Tumors*, Math. Biosciences **47** (1979) 55–77.

[21] S. H. Moolgavkar and A. G. Knudson, *Mutation and Cancer: A Model for Human Carcinogenesis*, Journal of the National Cancer Institute **66** (1981) 1037–1052.

[22] G. M. Grimmett and D. R. Stirzaker, *Probability and Random Processes*, 2nd Edition, Clarendon Press, Oxford (1992).

[23] W. Y. Tan and C. C. Brown, *A nonhomogeneous two stages model of carcinogenesis*, Math. Modeling **9** (1985) 631–642.

[24] G. L. Yang and C. W. Chen, *A stochastic two-stage carcinogenesis model: A new approach to computing the probability of observing tumor in animal bioassays*, Math. Biosci. **104** (1991) 247–258.

[25] A. Y. Yakovlev and A. D. Tsodikov, *Stochastic Models of Tumor Latency and Their Biostatistical Applications*, World Scientific, Singapore (1996).

[26] W. Y. Tan, *Stochastic logistic growth and applications*, in *Logistic Distributions*, ed. B. N. Balakrishnan, pp. 397–426, Marcel Dekker, Inc., New York (1991).

[27] W. Y. Tan and S. Piatadosi, *On stochastic growth process with application to stochastic logistic growth*, Statistica Sinica **1** (1991) 527–540.

[28] R. E. Kalman, *A new approach to linear filter and prediction problems*, J. Basic Eng. **82** (1960) 35–45.

[29] A. C. Harvey, *Forecasting, Structural Time Series Models and the Kalman Filter*, Cambridge University Press, Cambridge (1994).

[30] M. Aoki, *State Space Modeling of Time Series*, 2nd edition, Spring-Verlag, Berlin (1990).

[31] H. Wu and W. Y. Tan, *Modeling the HIV epidemic: A state space approach*, in *ASA 1995 Proceeding of the Epidemiology Section*, ASA, Alexdria, VA (1995) 66–71.

[32] H. Wu and W. Y. Tan, *Modeling the HIV epidemic: A state space approach*, Math. Computer Modelling **32** (2000) 197–215.

[33] B. Cazelles and N. P. Chau, *Using the Kalman filter and dynamic models to assess the changing HIV/AIDS epidemic*, Math. Biosciences **140** (1997) 131–154.

[34] W. Y. Tan and Z. H. Xiang, *State space models of the HIV epidemic in homosexual populations and some applications*, Math. Biosciences **152** (1998) 29–61.

[35] W. Y. Tan and Z. H. Xiang, *Modeling the HIV epidemic with variable infection in homosexual populations by state space models*, J. Statist. Inference and Planning **78** (1999) 71–87.

[36] W. Y. Tan and Z. H. Xiang, *Estimating and predicting the numbers of T cells and free HIV by non-linear Kalman filter*, in *Artificial Immune Systems and Their Applications*, ed. DasGupta, pp. 115–138, Springer-Verlag, Berlin (1998).

[37] W. Y. Tan and Z. H. Xiang, *State Space Models for the HIV pathogenesis*, in *Mathematical Models in Medicine and Health Sciences*, eds. M. A. Horn, G. Simonett and G. Webb, Vanderbilt University Press, Nashville (1998) 351–368.

[38] W. Y. Tan and Z. H. Xiang, *A state space model of HIV pathogenesis under treatment by anti-viral drugs in HIV-infected individuals*, Math. Biosciences **156** (1999) 69–94.

[39] W. Y. Tan and Z. H. Xiang, *Stochastic modeling of early response of HIV pathogenesis with lymph nodes under treatment by protease inhibitors*, in *Memory Volume of Sid. Yakowitz*, Kluwer Academic Publisher, Boston (2002).

[40] W. Y. Tan, C. W. Chen and W. Wang, *Some state space models of carcinogenesis*, in *Simulation in Medical Sciences*, eds. J. G. Anderson and M. Katzper, pp. 183–189, The Society of Computer Simulation International (1999).

[41] W. Y. Tan, C. W. Chen and W. Wang, *Some multiple pathways state space models of carcinogenesis*, in *Simulation in Medical Sciences*, eds. J. G. Anderson and M. Katzper, pp. 162–169, The Society of Computer Simulation International (2000).

[42] W. Y. Tan, C. W. Chen and W. Wang, *Stochastic modeling of carcinogenesis by state space models: A new approach*, Math and Computer Modeling **33** (2001) 1323–1345.

Chapter 2

Discrete Time Markov Chain Models in Genetics and Biomedical Systems

In many stochastic systems in natural sciences including genetics as well as in biomedical problems, one may treat both the state space and the parameter space as discrete spaces. In these cases, when the process is Markov, one is then entertaining Markov chains with discrete time. This is especially true in mating types and gene frequencies in natural populations with generation as time unit, see Examples 1.1, 1.2 and 1.11; it is also true in many branching processes involving new mutants and in AIDS epidemics; see Examples 1.9, 1.10 and 1.12. In the past 15 to 20 years, some general theories of Markov chains with discrete time have also been invoked to develop computer algorithms to solve many complicated computational problems in natural sciences. This has been referred to as the MCMC (Markov Chain Monte Carlo) method and has become a very popular method. By using examples from many genetic systems, in this chapter we will develop some general results of discrete time Markov chains and illustrate its applications. In the next chapter, we will develop theories for stationary distributions and illustrate the applications of some MCMC methods.

2.1. Examples from Genetics and AIDS

Example 2.1. The self-fertilization Markov chain. In many plants such as rice, wheat and green beans, the flowers have both the male and female organs. These plants are diploid and reproduce itself through self-fertilization.

To study the stochastic behavior in these populations, we consider a single locus with two alleles $A : a$ in the population and let $X(t)$ represent the three genotypes AA, Aa and aa at generation t. Let $\{1, 2, 3\}$ stand for the genotypes $\{AA, aa, Aa\}$ respectively. Then $X(t)$ is a homogeneous Markov chain with state space $S = \{1, 2, 3\}$. Under self-fertilization, the one-step transition probabilities $p_{ij} = \Pr\{X(t+1) = j | X(t) = i\}$ are:

$$
\begin{array}{c}
\quad\;\; AA \quad aa \quad Aa \\
\begin{array}{c} AA \\ aa \\ Aa \end{array}
\left(
\begin{array}{ccc}
1 & 0 & 0 \\
0 & 1 & 0 \\
\dfrac{1}{4} & \dfrac{1}{4} & \dfrac{1}{2}
\end{array}
\right).
\end{array}
$$

That is, the one-step transition matrix is

$$
P =
\left(
\begin{array}{ccc}
1 & 0 & 0 \\
0 & 1 & 0 \\
\dfrac{1}{4} & \dfrac{1}{4} & \dfrac{1}{2}
\end{array}
\right).
$$

Example 2.2. The frequency of genes in natural populations under steady state conditions. In population genetics, an important topic is the random changes of frequencies of different genes or genotypes as time progresses. This is the major thesis of evolution theory [1]. In these studies, one usually images a sufficiently large population of diploid. If the mating is random between individuals (i.e. random union of gametes to yield progenies) and if there are no selection, no mutation, no immigration and no migration, then as time progresses, the population will reach a steady-state condition under which the frequencies of genes and genotypes will not change from generation to generation. This steady state condition has been referred to as the Hardy-Weinberg law in population genetics. This is illustrated in some detail in Subsec. 2.10.1. In the case of one locus with two alleles (say A and a), the Hardy-Weinberg law states that the frequency p of the A allele is independent of time and at any generation, the frequencies of the three genotypes $\{AA, Aa, aa\}$ are given respectively by $\{p^2, 2pq, q^2\}$, where $q = 1 - p$.

To describe the stochastic changes of frequencies of the genotypes under steady state conditions, let $\{X(t), t = 1, \ldots,\}$ denote the three genotypes $\{AA, Aa, aa\}$ at generation t for the females. Then $\{X(t), t = 1, 2, \ldots,\}$ is a Markov chain with state space $\Omega = \{AA, Aa, aa\}$. Under steady-state conditions for which the Hardy-Weinberg law holds, the one-step transition matrix of this Markov chain is given by:

$$P = \begin{matrix} & \begin{matrix} AA & Aa & aa \end{matrix} \\ \begin{matrix} AA \\ Aa \\ aa \end{matrix} & \begin{pmatrix} p & q & 0 \\ \frac{1}{2}p & \frac{1}{2} & \frac{1}{2}q \\ 0 & p & q \end{pmatrix} \end{matrix}.$$

The above transition probability matrix can easily be derived by using argument of conditional probabilities as given below:

(i) Since daughter and mother must have one gene in common, so, given AA mother, the genotypes of the daughters must either be AA or Aa; given AA mother, the daughters have one gene fixed as A, the probabilities of the other gene being A or a are given respectively by p and q under Hardy-Weinberg law since these are the frequencies in the population and since the mating is random. This gives the first row of the transition matrix. Similarly, we obtain the third row of the above transition matrix.

(ii) Given Aa mother, then with $\frac{1}{2}$ probability each daughter has one gene fixed by A, and with $\frac{1}{2}$ probability each daughter has one gene fixed by "a". In the first case, each daughter will have probability p being AA and probability q being Aa; in the second case, the probability is p that each daughter has genotype Aa, and probability q that each daughter has genotype aa. Hence, given Aa mother, the probability is $\frac{1}{2} \times p = \frac{1}{2}p$ that the daughter genotype is AA whereas the probability is $\frac{1}{2}q + \frac{1}{2}p = \frac{1}{2}$ that the daughter genotype is Aa; similarly, given Aa mother, the probability is $\frac{1}{2}q$ that the daughter genotype is aa.

Example 2.3.   The inbreeding systems in natural populations. In wild natural populations, matings between individuals can hardly be expected to be random; see Subsec. 2.11.2. Hence, in studying evolution theories, it is of considerable interests to study many other mating systems than random mating in the population. Fisher [2] and Karlin [3, 4] have shown that many of these mating systems can be described by homogeneous Markov chains. In Example 1.8, we have described the full sib-mating systems (brother-sister matings); other mating types which have been considered by Fisher and Karlin include parent-offspring mating, mixtures of random mating and sib-mating as well as other mating types, assortative matings, first cousin mating, second cousin mating, etc.

Example 2.4.   The two-loci linkage Markov chain in self-fertilized populations. By using theories of finite Markov chain with discrete time, Tan [5] has developed a two-loci model with linkage to assess effects of selection in self-fertilized diploid populations such as rice and wheat. This model considers two linked loci (A and B) each with two alleles (A, a) and (B, b), respectively. Then, denoting by XY/ZW the genotype with XY on one chromosome and ZW on the other, there are altogether ten genotypes AB/AB, Ab/Ab, aB/aB, ab/ab, AB/Ab, AB/aB, Ab/ab, aB/ab, AB/ab, and aB/Ab. For the consideration of the effect of selection it is assumed that there is no difference in effect of selection regarding sex, and that the two loci act independently of each other with respect to selection. Then, the fitness of the genotypes are given by:

$$
\begin{array}{cccc}
 & BB & Bb & bb \\
AA & x_1 + x_2 & x_1 + 1 & x_1 + y_2 \\
Aa & 1 + x_2 & 1 + 1 & 1 + y_2 \\
aa & y_1 + x_2 & y_1 + 1 & y_1 + y_2
\end{array}
$$

where it is assumed that $x_i \geq 0$, $y_i \geq 0$, $i = 1, 2$.

Let $\{X(t), t = 1, 2, \dots, \}$ denote the above ten genotypes at generation t. Then $X(t)$ is a finite Markov chain with discrete time and with state space given by $S = \{AB/AB, Ab/Ab, aB/aB, ab/ab, AB/Ab, AB/aB, Ab/ab, aB/ab, AB/ab, aB/Ab\}$. Letting p be the recombination value between the two loci ($0 \leq p \leq \frac{1}{2}$), then under self-fertilization with selection, the one step transition matrix is given by:

$$
P = \left(\begin{array}{c|c}
I_4 & 0 \\
\hline
R & Q
\end{array} \right),
$$

where

$$R = (\underset{\sim}{R}_1,\ \underset{\sim}{R}_2,\ \underset{\sim}{R}_3,\ \underset{\sim}{R}_4)$$

$$= \begin{bmatrix}
\dfrac{x_1 + x_2}{c_1} & \dfrac{x_1 + y_2}{c_1} & 0 & 0 \\[2ex]
\dfrac{x_1 + x_2}{c_2} & 0 & \dfrac{y_1 + x_2}{c_2} & 0 \\[2ex]
0 & \dfrac{x_1 + y_2}{c_3} & 0 & \dfrac{y_1 + y_2}{c_3} \\[2ex]
0 & 0 & \dfrac{y_1 + x_2}{c_4} & \dfrac{y_1 + y_2}{c_4} \\[2ex]
\dfrac{q^2}{c_5}(x_1 + x_2) & \dfrac{p^2}{c_5}(x_1 + y_2) & \dfrac{p^2}{c_5}(y_1 + x_2) & \dfrac{q^2}{c_5}(y_1 + y_2) \\[2ex]
\dfrac{p^2}{c_5}(x_1 + x_2) & \dfrac{q^2}{c_5}(x_1 + y_2) & \dfrac{q^2}{c_5}(y_1 + x_2) & \dfrac{p^2}{c_5}(y_1 + y_2)
\end{bmatrix} ,$$

$$Q = \begin{bmatrix}
\dfrac{2(1 + x_1)}{c_1} & 0 & 0 & 0 & 0 & 0 \\[2ex]
0 & \dfrac{2(1 + x_2)}{c_2} & 0 & 0 & 0 & 0 \\[2ex]
0 & 0 & \dfrac{2(1 + y_2)}{c_3} & 0 & 0 & 0 \\[2ex]
0 & 0 & 0 & \dfrac{2(1 + y_1)}{c_4} & 0 & 0 \\[2ex]
2pq\dfrac{1 + x_1}{c_5} & 2pq\dfrac{1 + x_2}{c_5} & 2pq\dfrac{1 + y_2}{c_5} & 2pq\dfrac{1 + y_1}{c_5} & \dfrac{4q^2}{c_5} & \dfrac{4p^2}{c_5} \\[2ex]
2pq\dfrac{1 + x_1}{c_5} & 2pq\dfrac{1 + x_2}{c_5} & 2pq\dfrac{1 + y_2}{c_5} & 2pq\dfrac{1 + y_1}{c_5} & \dfrac{4p^2}{c_5} & \dfrac{4q^2}{c_5}
\end{bmatrix} ,$$

and where

$$c_1 = 4x_1 + x_2 + y_2 + 2 , \quad c_2 = 4x_2 + x_1 + y_1 + 2 ,$$
$$c_3 = 4y_2 + x_1 + y_1 + 2 , \quad c_4 = 4y_1 + x_2 + y_2 + 2 ,$$
$$c_5 = x_1 + x_2 + y_1 + y_2 + 4 .$$

The above transition matrix is derived by first considering the Mendelian segregation under self-fertilization and then imposing effects of selection to yield frequencies of different genotypes. As an illustration, consider the genotype $\frac{Ab}{aB}$ at generation t. Under Mendelian segregation with linkage, this genotype produces four types of gametes $\{AB, Ab, aB, ab\}$ with frequencies $\{\frac{1}{2}p, \frac{1}{2}q, \frac{1}{2}q, \frac{1}{2}p\}$, $q = 1 - p$, respectively. Under self-fertilization with no selection, this gives the frequencies of the above ten genotypes at generation $t + 1$ as:

$$
\begin{array}{cccccccccc}
\frac{AB}{AB} & \frac{Ab}{Ab} & \frac{aB}{aB} & \frac{ab}{ab} & \frac{AB}{Ab} & \frac{AB}{aB} & \frac{Ab}{ab} & \frac{aB}{ab} & \frac{AB}{ab} & \frac{aB}{Ab} \\[2mm]
\frac{1}{4}p^2 & \frac{1}{4}q^2 & \frac{1}{4}q^2 & \frac{1}{4}p^2 & \frac{1}{2}pq & \frac{1}{2}pq & \frac{1}{2}pq & \frac{1}{2}pq & \frac{1}{2}p^2 & \frac{1}{2}q^2 .
\end{array}
$$

The average fitness is

$$
\frac{1}{4}p^2(x_1 + x_2) + \frac{1}{4}q^2(x_1 + y_2) + \frac{1}{4}q^2(y_1 + x_2) + \frac{1}{4}p^2(y_1 + y_2)
$$

$$
+ \frac{1}{2}pq(x_1 + 1) + \frac{1}{2}pq(1 + x_2) + \frac{1}{2}pq(1 + y_2) + \frac{1}{2}pq(1 + y_1)
$$

$$
+ 2\frac{1}{2}p^2 + 2\frac{1}{2}q^2 = \frac{c_5}{4} .
$$

Hence the frequencies of the ten genotypes at generation $t + 1$ under selection are given by:

$$
\begin{array}{ccccc}
\frac{AB}{AB} & \frac{Ab}{Ab} & \frac{aB}{aB} & \frac{ab}{ab} & \frac{AB}{Ab} \\[3mm]
\frac{1}{c_5}p^2(x_1 + x_2) & \frac{1}{c_5}q^2(x_1 + y_2) & \frac{1}{c_5}q^2(y_1 + x_2) & \frac{1}{c_5}p^2(y_1 + y_2) & \frac{1}{c_5}2pq(x_1 + 1) \\[4mm]
\frac{AB}{aB} & \frac{Ab}{ab} & \frac{aB}{ab} & \frac{AB}{ab} & \frac{aB}{Ab} \\[3mm]
\frac{1}{c_5}2pq(1 + x_2) & \frac{1}{c_5}2pq(1 + y_2) & \frac{1}{c_5}2pq(1 + y_1) & \frac{4}{c_5}p^2 & \frac{4}{c_5}q^2 .
\end{array}
$$

This gives the elements of the last row of P above. Similarly, one may derive elements of other rows of P; see Exercise 2.1.

Example 2.5.   The base substitution model in DNA sequence. DNA strands are made up by four nitrogen bases A (adenine), G (guanine), C (cytosine) and T (thymine), a phosphate molecule (P) and a sugar (Deoxyribose to be denoted by dR). These strands are directional starting with a 5′ end (A phosphate) and walk down to the 3′ end (a hydroxyl, i.e. OH). A and G belong to the purine group whereas C and T are pirimidines. A is always linked to T by two hydrogen bonds (weak bone) and G always linked to C by three hydrogen bonds (strong bond). Hence, given the base sequence in one strand of DNA, the sequence in the other strand of DNA is uniquely determined. The DNA molecules have two strands coiled up in a double helix form. (RNA molecules differ from the DNA in three aspects: (1) The RNA's are single strand molecules. (2) The T base is replaced by U (uracil) in RNA. (3) The sugar in RNA is ribose.) Given in Fig. 2.1 is a schematic representation of the DNA molecule.

To analyze the DNA sequence in human genome, it has been discovered that the arrangement of nitrogen bases in the sequence are not independent; however, in many cases, the base substitution in the sequence as time progresses can be described by Markov chain models; see [6]. The discovery of restriction enzymes in bacteria has allowed biologists to estimate the substitution rates in successive bases in the DNA sequence in many cases. (Restriction enzymes are enzymes which recognize specific sequences of basis along the DNA strand and cut the DNA at these restriction sites.) Given below is a Markov chain for base substitution in the DNA dimers in humans considered by Bishop *et al.* [6]. The state space of this chain consists of the states

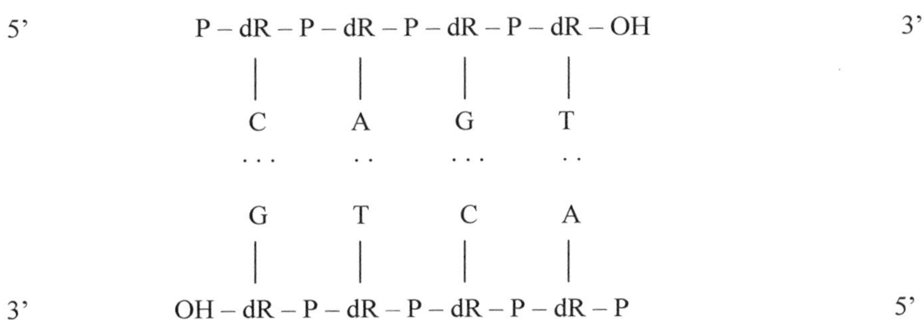

Fig. 2.1.   Schematic representation of double stranded DNA. P, phosphate; dR, deoxyribose; OH, hydroxyl; - , covalend bond; ..., weak bond; ..., strong bond; A, adenine; C, cytosine; T, thymine; G, guanine.

$\{A, C, G, T, AG, AGC, AGCT\}$. The matrix of the one step transition probabilities has been determined as:

$$P = \begin{array}{c c} & \begin{array}{c c c c c c c} A & C & G & T & AG & AGC & AGCT \end{array} \\ \begin{array}{c} A \\ C \\ G \\ T \\ AG \\ AGC \\ AGCT \end{array} & \left(\begin{array}{c c c c c c c} 0.32 & 0.18 & 0 & 0.27 & 0.23 & 0 & 0 \\ 0.37 & 0.23 & 0.05 & 0.35 & 0 & 0 & 0 \\ 0.30 & 0.21 & 0.25 & 0.24 & 0 & 0 & 0 \\ 0.23 & 0.19 & 0.25 & 0.33 & 0 & 0 & 0 \\ 0.30 & 0 & 0.25 & 0.24 & 0 & 0.21 & 0 \\ 0.37 & 0.23 & 0.05 & 0 & 0 & 0 & 0.35 \\ 0.23 & 0.19 & 0.25 & 0.33 & 0 & 0 & 0 \end{array} \right) \end{array}.$$

Example 2.6. The AIDS epidemic in homosexual or IV drug user populations. Consider a large population of homosexual men or IV drug users who are at risk for AIDS. Then, as illustrated in Example 1.6, there are three types of people: The S people, the I people and the AIDS cases. Denote by $I(u)$ the infective people with infection duration u (i.e. u is the time elapsed since the infective person contracted HIV). Let $\{S(t), I(u,t)\}$ denote the numbers of S people and of $I(u)$ people at the tth month respectively and $A(t)$ the total number of AIDS cases including those died from AIDS by the tth month. Put $\underset{\sim}{X}(t) = \{S(t), I(u,t), u = 0, 1, \ldots, t, A(t)\}'$. Then under some general conditions, $\{\underset{\sim}{X}(t), t \in T = (0, 1, \ldots, \infty)\}$ is a multi-dimensional Markov chain with discrete time. This type of Markov chains in HIV epidemic has been developed by Tan and his associates in terms of chain binomial and chain multinomial distributions; see [7–12].

2.2. The Transition Probabilities and Computation

Let $\{X(t), t \in T\}$ be a Markov chain with state space $S = \{0, 1, 2, \ldots\}$ and parameter space $T = \{0, 1, 2, \ldots\}$. Then the random behavior and properties of this Markov chain is characterized by the transition probabilities which is defined by $p_{ij}(s,t) = P\{X(t) = j | X(s) = i\}$ for $t \geq s$. From these transition probabilities, obviously, we have: $p_{ij}(s,s) = \delta_{ij}$, where δ_{ij} is the Kronecker's δ defined by $\delta_{ii} = 1$ and $\delta_{ij} = 0$ if $i \neq j$. To compute these probabilities and to prove some general theories, we first give the following theorem which has been referred to as the Chapman-Kolmogorov equation.

Theorem 2.1.   The Chapman-Kolmogorov equation. *Let $\{X(t), t = 0, 1 \ldots, \}$ be a Markov chain with state space $S = \{0, 1, 2 \ldots\}$. Let the transition probabilities be given by $p_{ij}(m, n) = \Pr\{X(n) = j | X(m) = i\}$. Then, for any $0 < m < r < n$, we have that*

$$p_{ij}(m, n) = \sum_{k=0}^{\infty} p_{ik}(m, r) p_{kj}(r, n) \,. \tag{2.1}$$

Proof.   By applying the Markov condition, we have

$$p_{ij}(m, n) = \Pr\{X(n) = j | X(m) = i\}$$

$$= \sum_{k=0}^{\infty} \Pr\{X(n) = j, X(r) = k | X(m) = i\}$$

$$= \sum_{k=0}^{\infty} \Pr\{X(n) = j | X(r) = k, X(m) = i\} \Pr\{X(r) = k | X(m) = i\}$$

$$= \sum_{k=0}^{\infty} \Pr\{X(n) = j | X(r) = k\} \Pr\{X(r) = k | X(m) = i\}$$

$$= \sum_{k=0}^{\infty} \Pr\{X(n) = j | X(r) = k\} \Pr\{X(r) = k | X(m) = i\}$$

$$= \sum_{k=0}^{\infty} p_{ik}(m, r) p_{kj}(r, n) \,. \qquad \square$$

Equation (2.1) is called the *Chapman-Kolmogorov equation*. As we shall see, this equation is the basic approach for proving many of the results in Markov chains.

Using the above Chapman-Kolmogorov equation, if the chain is finite (i.e. the state space S has only a finite number of elements), then, $P(m, n) = (p_{ij}(m, n)) = P(m, r)P(r, n) = \prod_{r=1}^{n-m} P(m + r - 1, m + r)$. If the chain is finite and homogeneous, then $P(n, n + 1) = P$ is independent of n so that $P(m, n) = P^{n-m}$.

The above results also extend to some non-homogeneous finite Markov chains. To illustrate, suppose that $p_{ij}(n, n+1) = p_{ij}(s)$ for $t_{s-1} \leq n < t_s$ $s = 1, \ldots, N$ with $t_0 = 0$ and $t_N = \infty$. Put $P_s = (p_{ij}(s))$ for $s = 1, \ldots, N$. Then, with $t_0 = 0$,

$$P(0, n) = P_1^n, \quad \text{if } n \leq t_1\,;$$

$$= \left\{ \prod_{i=0}^{s-1} P_{i+1}^{t_{i+1}-t_i} \right\} P_{s+1}^{n-t_s}, \quad \text{if } t_1 < t_2 < \cdots t_s \leq n < t_{s+1},$$

$$s = 1, \ldots, N-1\,.$$

From above, it follows that to compute $P(0, n)$ for finite Markov chains with discrete time, in many cases one would need to find the power of the one-step transition matrices. This problem is made easy if the one-step transition matrices have real distinct eigenvalues so that these matrices can be transformed into diagonal matrices. For example, if the matrix P has real distinct eigenvalues $\lambda_j, j = 1, \ldots, r$, then, as shown in Subsec. 2.11.3,

$$P = \sum_{j=1}^{r} \lambda_j E_j\,,$$

where $E_j = \prod_{i \neq j} \frac{1}{\lambda_j - \lambda_i}(P - \lambda_i I), j = 1, \ldots, r$.

It is shown in Subsec. 2.11.3 that $\sum_{j=1}^{r} E_j = I, E_j^2 = E_j$ and $E_i E_j = 0$ if $i \neq j$; hence

$$P(n) = P^n = \sum_{j=1}^{r} \lambda_j^n E_j\,.$$

The expansion $P = \sum_{j=1}^{r} \lambda_j E_j$ will be referred to as a spectral expansion of P and the E_i's the spectrum matrices of P.

Example 2.7. The self-fertilization Markov chain. In Example 2.1, we have considered a self-fertilized Markov chain for one locus with two alleles A and a. The state space of this chain consists of the three genotypes $\{AA, aa, Aa\}$. This is a homogeneous Markov chain with the one-step transition probability matrix given by:

$$P = \begin{pmatrix} 1 & 0 & 0 \\ 0 & 1 & 0 \\ \dfrac{1}{4} & \dfrac{1}{4} & \dfrac{1}{2} \end{pmatrix}.$$

The characteristic function of P is $\phi(x) = |P - xI_3| = (1-x)^2(\frac{1}{2} - x) = 0$. Thus the two eigenvalues of P are $\lambda_1 = 1$ and $\lambda_2 = \frac{1}{2}$ with multiplicities 2 and 1 respectively. Hence, the spectrum matrices are:

$$E_1 = \frac{1}{(\lambda_1 - \lambda_2)}(P - \lambda_2 I_3) = 2 \begin{pmatrix} 1 - \dfrac{1}{2} & 0 & 0 \\ 0 & 1 - \dfrac{1}{2} & 0 \\ \dfrac{1}{4} & \dfrac{1}{4} & \dfrac{1}{2} - \dfrac{1}{2} \end{pmatrix} = \begin{pmatrix} 1 & 0 & 0 \\ 0 & 1 & 0 \\ \dfrac{1}{2} & \dfrac{1}{2} & 0 \end{pmatrix},$$

and

$$E_2 = \frac{1}{(\lambda_2 - \lambda_1)}(P - \lambda_1 I_3) = (-2) \begin{pmatrix} 1 - 1 & 0 & 0 \\ 0 & 1 - 1 & 0 \\ \dfrac{1}{4} & \dfrac{1}{4} & \dfrac{1}{2} - 1 \end{pmatrix}$$

$$= \begin{pmatrix} 0 & 0 & 0 \\ 0 & 0 & 0 \\ -\dfrac{1}{2} & -\dfrac{1}{2} & 1 \end{pmatrix}.$$

It follows that we have

$$P = \lambda_1 E_1 + \lambda_2 E_2 = \begin{pmatrix} 1 & 0 & 0 \\ 0 & 1 & 0 \\ \dfrac{1}{2} & \dfrac{1}{2} & 0 \end{pmatrix} + \frac{1}{2} \begin{pmatrix} 0 & 0 & 0 \\ 0 & 0 & 0 \\ -\dfrac{1}{2} & -\dfrac{1}{2} & 1 \end{pmatrix},$$

$$P(t) = P^t = \lambda_1^t E_1 + \lambda_2^t E_2 = \begin{pmatrix} 1 & 0 & 0 \\ 0 & 1 & 0 \\ \frac{1}{2} & \frac{1}{2} & 0 \end{pmatrix} + \left(\frac{1}{2}\right)^t \begin{pmatrix} 0 & 0 & 0 \\ 0 & 0 & 0 \\ -\frac{1}{2} & -\frac{1}{2} & 1 \end{pmatrix}.$$

Denote by $\underset{\sim}{e}_3{}' = (0,0,1)$. Then we have

$$(P^t)' \underset{\sim}{e}_3 = \begin{pmatrix} \frac{1}{2} \\ \frac{1}{2} \\ 0 \end{pmatrix} + \left(\frac{1}{2}\right)^t \begin{pmatrix} -\frac{1}{2} \\ -\frac{1}{2} \\ 1 \end{pmatrix} = \begin{pmatrix} \frac{1}{2}\left(1 - \frac{1}{2^t}\right) \\ \frac{1}{2}\left(1 - \frac{1}{2^t}\right) \\ \frac{1}{2^t} \end{pmatrix}.$$

It follows that given at time $t = 0$ an individual with genotype Aa, at the t generation, the probabilities of the three types AA, aa and Aa are given by $\frac{1}{2}(1 - \frac{1}{2^t}), \frac{1}{2}(1 - \frac{1}{2^t})$ and $\frac{1}{2^t}$, respectively.

Example 2.8. **The sib mating Markov chain for one locus with two alleles in natural populations.** In Example 1.8, we have considered a Markov chain of mating types under full-sib mating in large diploid populations. The state space of this Markov chain consists of the six mating types $AA \times AA, aa \times aa, AA \times aa, AA \times Aa, aa \times Aa, Aa \times Aa$. The one-step transition probability matrix of this chain is given by $P = \begin{pmatrix} I_2 & 0 \\ R & Q \end{pmatrix}$, where

$$R = \begin{pmatrix} 0 & 0 \\ \frac{1}{4} & 0 \\ 0 & \frac{1}{4} \\ \frac{1}{16} & \frac{1}{16} \end{pmatrix}, \quad \text{and,} \quad Q = \begin{pmatrix} 0 & 0 & 0 & 1 \\ 0 & \frac{1}{2} & 0 & \frac{1}{4} \\ 0 & 0 & \frac{1}{2} & \frac{1}{4} \\ \frac{1}{8} & \frac{1}{4} & \frac{1}{4} & \frac{1}{4} \end{pmatrix}.$$

We have

$$P^n = \begin{pmatrix} I_2 & 0 \\ R_n & Q^n \end{pmatrix}, \quad \text{where} \quad R_n = \sum_{i=0}^{n-1} Q^i R = (I_4 - Q)^{-1}(I_4 - Q^n)R.$$

The characteristic function $\phi(x)$ of Q is

$$\phi(x) = |Q - xI_4| = (-x)\left\{ \left(\frac{1}{2} - x\right)^2 \left(\frac{1}{4} - x\right) - \frac{1}{8}\left(\frac{1}{2} - x\right)\right\} - \frac{1}{8}\left(\frac{1}{2} - x\right)^2$$

$$= -\left(\frac{1}{2} - x\right)\left\{ x\left(\frac{1}{2} - x\right)\left(\frac{1}{4} - x\right) - \frac{x}{8} + \frac{1}{8}\left(\frac{1}{2} - x\right)\right\}$$

$$= \frac{1}{4}\left(\frac{1}{2} - x\right)\left(\frac{1}{4} - x\right)\left(4x^2 - 2x - 1\right) = 0\,.$$

Hence the eigenvalues of Q are $\{\lambda_1 = \frac{1}{2}, \lambda_2 = \frac{1}{4}, \lambda_3 = \frac{1}{4}(1 + \sqrt{5}) = \frac{1}{4}\epsilon_1, \lambda_4 = \frac{1}{4}(1 - \sqrt{5}) = \frac{1}{4}\epsilon_2\}$. Since all eigenvalues are real and distinct, Q can be expressed in terms of spectral expansion. The spectrum matrices of Q are:

$$E_1 = \frac{1}{(\lambda_1 - \lambda_2)(\lambda_1 - \lambda_3)(\lambda_1 - \lambda_4)}(Q - \lambda_2 I_4)(Q - \lambda_3 I_4)(Q - \lambda_4 I_4)$$

$$= \begin{pmatrix} 0 & 0 & 0 & 0 \\ 0 & \dfrac{1}{2} & -\dfrac{1}{2} & 0 \\ 0 & -\dfrac{1}{2} & \dfrac{1}{2} & 0 \\ 0 & 0 & 0 & 0 \end{pmatrix},$$

$$E_2 = \frac{1}{(\lambda_2 - \lambda_1)(\lambda_2 - \lambda_3)(\lambda_2 - \lambda_4)}(Q - \lambda_1 I_4)(Q - \lambda_3 I_4)(Q - \lambda_4 I_4)$$

$$= \frac{4}{5}\begin{pmatrix} \dfrac{1}{2} & -1 & -1 & 1 \\ -\dfrac{1}{8} & \dfrac{1}{4} & \dfrac{1}{4} & -\dfrac{1}{4} \\ -\dfrac{1}{8} & \dfrac{1}{4} & \dfrac{1}{4} & -\dfrac{1}{4} \\ -\dfrac{1}{8} & -\dfrac{1}{4} & -\dfrac{1}{4} & \dfrac{1}{4} \end{pmatrix},$$

$$E_3 = \frac{1}{(\lambda_3 - \lambda_1)(\lambda_3 - \lambda_2)(\lambda_3 - \lambda_4)}(Q - \lambda_1 I_4)(Q - \lambda_2 I_4)(Q - \lambda_4 I_4)$$

$$= \frac{2}{5}\begin{pmatrix} \left(1 - \frac{\epsilon_1}{4}\right) & 1 & 1 & -\epsilon_2 \\ \frac{1}{8} & \frac{1}{4}\left(1 + \frac{\epsilon_1}{2}\right) & \frac{1}{4}\left(1 + \frac{\epsilon_1}{2}\right) & \frac{\epsilon_1}{4} \\ \frac{1}{8} & \frac{1}{4}\left(1 + \frac{\epsilon_1}{2}\right) & \frac{1}{4}\left(1 + \frac{\epsilon_1}{2}\right) & \frac{\epsilon_1}{4} \\ \frac{\epsilon_2}{8} & \frac{\epsilon_1}{4} & \frac{\epsilon_1}{4} & 1 \end{pmatrix},$$

and

$$E_4 = \frac{1}{(\lambda_4 - \lambda_1)(\lambda_4 - \lambda_2)(\lambda_4 - \lambda_3)}(Q - \lambda_1 I_4)(Q - \lambda_2 I_4)(Q - \lambda_3 I_4)$$

$$= \frac{2}{5}\begin{pmatrix} \left(1 - \frac{\epsilon_2}{4}\right) & 1 & 1 & -\epsilon_1 \\ \frac{1}{8} & \frac{1}{4}\left(1 + \frac{\epsilon_2}{2}\right) & \frac{1}{4}\left(1 + \frac{\epsilon_2}{2}\right) & \frac{\epsilon_2}{4} \\ \frac{1}{8} & \frac{1}{4}\left(1 + \frac{\epsilon_2}{2}\right) & \frac{1}{4}\left(1 + \frac{\epsilon_2}{2}\right) & \frac{\epsilon_2}{4} \\ \frac{\epsilon_1}{8} & \frac{\epsilon_2}{4} & \frac{\epsilon_2}{4} & 1 \end{pmatrix}.$$

It follows that $Q^n = \sum_{i=1}^{4} \lambda_i^n E_i$ and $R_n = (I_4 - Q)^{-1}(I_4 - Q^n)R = \sum_{i=1}^{4} \frac{1 - \lambda_i^n}{1 - \lambda_i} E_i R$.

Hence, we have

$$P^n = \begin{pmatrix} I_2 & 0 \\ \sum_{i=1}^{4} \frac{1 - \lambda_i^n}{1 - \lambda_i} E_i R & \sum_{i=1}^{4} \lambda_i^n E_i \end{pmatrix}.$$

Example 2.9. The two-loci linkage Markov chain in self-fertilized populations. In Example 2.4, we have considered a two-loci model with linkage with in self-fertilized diploid populations. This model is a homogeneous

Markov chain with one step transition matrix given by:

$$P = \left(\begin{array}{c|c} I_4 & 0 \\ \hline R & Q \end{array} \right),$$

where R is a 6×4 matrix and Q a 6×6 matrix.

For this transition matrix, we have

$$P^n = \left(\begin{array}{cc} I_4 & 0 \\ R_n & Q^n \end{array} \right), \quad \text{where} \quad R_n = \sum_{i=0}^{n-1} Q^i R = (I_4 - Q)^{-1}(I_4 - Q^n)R.$$

The eigenvalues of Q are readily obtained as:

$$\lambda_1 = \frac{2}{c_1}(1 + x_1), \ \lambda_2 = \frac{2}{c_2}(1 + x_2), \ \lambda_3 = \frac{2}{c_3}(1 + y_2), \ \lambda_4 = \frac{2}{c_4}(1 + y_1),$$

$$\lambda_5 = \frac{4}{c_5}(q^2 + p^2), \quad \text{and} \quad \lambda_6 = \frac{4}{c_5}(q^2 - p^2).$$

The spectrum matrices E_i of Q are obtained as:

$$E_1 = \begin{bmatrix} 1 & 0 & 0 & 0 & 0 & 0 \\ 0 & 0 & 0 & 0 & 0 & 0 \\ 0 & 0 & 0 & 0 & 0 & 0 \\ 0 & 0 & 0 & 0 & 0 & 0 \\ z_1 & 0 & 0 & 0 & 0 & 0 \\ z_1 & 0 & 0 & 0 & 0 & 0 \end{bmatrix}, \quad E_2 = \begin{bmatrix} 0 & 0 & 0 & 0 & 0 & 0 \\ 0 & 1 & 0 & 0 & 0 & 0 \\ 0 & 0 & 0 & 0 & 0 & 0 \\ 0 & 0 & 0 & 0 & 0 & 0 \\ 0 & z_2 & 0 & 0 & 0 & 0 \\ 0 & z_2 & 0 & 0 & 0 & 0 \end{bmatrix},$$

$$E_3 = \begin{bmatrix} 0 & 0 & 0 & 0 & 0 & 0 \\ 0 & 0 & 0 & 0 & 0 & 0 \\ 0 & 0 & 1 & 0 & 0 & 0 \\ 0 & 0 & 0 & 0 & 0 & 0 \\ 0 & 0 & z_3 & 0 & 0 & 0 \\ 0 & 0 & z_3 & 0 & 0 & 0 \end{bmatrix}, \quad E_4 = \begin{bmatrix} 0 & 0 & 0 & 0 & 0 & 0 \\ 0 & 0 & 0 & 0 & 0 & 0 \\ 0 & 0 & 0 & 0 & 0 & 0 \\ 0 & 0 & 0 & 1 & 0 & 0 \\ 0 & 0 & 0 & z_4 & 0 & 0 \\ 0 & 0 & 0 & z_4 & 0 & 0 \end{bmatrix},$$

$$
E_5 = \begin{bmatrix}
0 & 0 & 0 & 0 & 0 & 0 \\
0 & 0 & 0 & 0 & 0 & 0 \\
0 & 0 & 0 & 0 & 0 & 0 \\
0 & 0 & 0 & 0 & 0 & 0 \\
-z_1 & -z_2 & -z_3 & -z_4 & \dfrac{1}{2} & \dfrac{1}{2} \\
-z_1 & -z_2 & -z_3 & -z_4 & \dfrac{1}{2} & \dfrac{1}{2}
\end{bmatrix}, \qquad
E_6 = \begin{bmatrix}
0 & 0 & 0 & 0 & 0 & 0 \\
0 & 0 & 0 & 0 & 0 & 0 \\
0 & 0 & 0 & 0 & 0 & 0 \\
0 & 0 & 0 & 0 & 0 & 0 \\
0 & 0 & 0 & 0 & \dfrac{1}{2} & -\dfrac{1}{2} \\
0 & 0 & 0 & 0 & -\dfrac{1}{2} & \dfrac{1}{2}
\end{bmatrix},
$$

where

$$
z_1 = p(1-p)(1+x_1) \left/ \left[\frac{c_5}{c_1}(1+x_1) - 2(p^2 + q^2) \right] \right. ,
$$

$$
z_2 = p(1-p)(1+x_2) \left/ \left[\frac{c_5}{c_2}(1+x_2) - 2(p^2 + q^2) \right] \right. ,
$$

$$
z_3 = p(1-p)(1+y_2) \left/ \left[\frac{c_5}{c_3}(1+y_2) - 2(p^2 + q^2) \right] \right. ,
$$

$$
z_4 = p(1-p)(1+y_1) \left/ \left[\frac{c_5}{c_4}(1+y_1) - 2(p^2 + q^2) \right] \right. .
$$

It follows that $Q^n = \sum_{i=1}^{6} \lambda_i^n E_i$ and $R_n = (I_6 - Q)^{-1}(I_6 - Q^n)R = \sum_{i=1}^{6} \frac{1-\lambda_i^n}{1-\lambda_i} E_i R$.

Example 2.10. The Galton-Watson branching processes. Let $\{X(t), t \in T = (0, 1, \ldots, \infty)\}$ be a simple Galton-Watson branching process with progeny distribution $\{p_j, j = 0, 1, \ldots, \infty\}$. Then the one-step transition probabilities are given by:

$$
p_{0j} = \delta_{0j}, \; j = 0, 1, \ldots, \infty,
$$

$$
p_{ij} = \frac{1}{j!} \left\{ \frac{d^j}{ds^j} g_i(s) \right\}_{s=0} \qquad \text{for } i > 0,
$$

where $g_i(s) = f(s)^i$ with $f(s)$ being the probability generating function (pgf) of the progeny distribution.

To find $p_{ij}(n)$ for $n > 1$, let $f_t(s)$ be the pgf of $X(t)$ given $X(0) = 1$ for $t = 1, 2, \ldots$. Then

$$p_{1j}(n) = \frac{1}{j!}\left\{\frac{d^j}{ds^j}f_n(s)\right\}_{s=0} \quad \text{and} \quad p_{ij}(n) = \frac{1}{j!}\left\{\frac{d^j}{ds^j}[f_n(s)]^i\right\}_{s=0} \quad \text{for } i > 1.$$

To find $f_t(s)$, notice that $f_1(s) = f(s)$ and, for $t > 1$,

$$f_t(s) = \sum_{n=0}^{\infty} s^n \Pr\{X(t) = n | X(0) = 1\}$$

$$= \sum_{n=0}^{\infty} s^n \left\{\sum_{k=0}^{\infty} \Pr\{X(t) = n | X(t-1) = k\}\Pr\{X(t-1) = k | X(0) = 1\}\right\}$$

$$= \sum_{n=0}^{\infty}\sum_{k=0}^{\infty} s^n \Pr\{X(1) = n | X(0) = k\}\Pr\{X(t-1) = k | X(0) = 1\}$$

$$= \sum_{k=0}^{\infty}\left\{\sum_{n=0}^{\infty} s^n \Pr(Z_1 + Z_2 + \cdots + Z_k = n)\right\}\Pr\{X(t-1) = k | X(0) = 1\}$$

$$= \sum_{k=0}^{\infty} [f(s)]^k \Pr\{X(t-1) = k | X(0) = 1\}$$

$$= f_{t-1}[f(s)].$$

On continuing, we have that

$$f_t(s) = f_{t-1}[f(s)] = f_{t-2}\{f[f(s)]\} = f_{t-3}\{f[f[f(s)]]\}$$

$$= \cdots = f\{f[f \cdots [f(s)] \cdots]\} = f[f_{t-1}(s)].$$

The above formula for the pgf of the Galton-Watson branching process is the basic tool for deriving results for Galton-Watson processes. For example, if we let μ and σ^2 denote the mean and the variance of the progeny distribution respectively, then by using the above generating functions, it can easily be shown that (Exercise 2.5)

$$E[X(n)|X(0) = 1] = \mu^n \quad \text{and} \quad \text{Var}[X(n)|X(0) = 1] = \sigma^2 \mu^{n-1}(1 + \mu + \cdots + \mu^{n-1}).$$

In Example 2.16, we will use the above pgf formulae to find the first absorption probability and mean absorption times. In Example 6.6 we will use the above pgf formulae to show that $Y(t) = X(t)/N$ can be approximated by a diffusion process when N is very large.

Example 2.11. Non-homogeneous Galton-Watson processes. In the Galton-Watson processes, the progeny distribution may change as time progresses. This is true for new mutants which are usually selectively disadvantageous comparing with wild allele when they were first introduced into the population; however, environmental changes at latter times may make the mutants selectively more advantageous over the wild allele. In these cases, the branching processes become non-homogeneous. To illustrate how to derive transition probabilities in these cases, assume that the progeny distributions of the mutant are given by $\{p_j^{(i)}\ j = 0, 1, \ldots, \infty, i = 1, 2\}$ for $n \leq t_1$ and $t_1 < n$ respectively, where

$$p_j^{(i)} = e^{-\lambda_i} \frac{\lambda_i^j}{j!}, \ i = 1, 2, j = 0, 1, \ldots, \infty,$$

where $\lambda_i = 1 + s_i$, $i = 1, 2$. ($1 + s_i$ is the relative fitness of the mutant comparing with the wild allele over time with 1 for time $n \leq t_1$ and 2 for time $n > t_1$.)

Let $p_{ij}(n) = \Pr\{X(n + 1) = j | X(n) = i\}$. Then, for $j = 0, 1, \ldots, \infty$,

$$p_{0j}(n) = \delta_{0j} \quad \text{for all } n \in T = (0, 1, \ldots, \infty);$$

$$p_{ij}(n) = e^{-i\lambda_1} \frac{(i\lambda_1)^j}{j!} \quad \text{for } n \leq t_1;$$

$$p_{ij}(n) = e^{-i\lambda_2} \frac{(i\lambda_2)^j}{j!} \quad \text{for } n > t_1.$$

Let $f^{(i)}(s)$ denote the pgf of the progeny distribution $\{p_j^{(i)}, j = 0, \ldots, \infty\}$, $f_n^{(i)}(s)$ the pgf of the simple homogeneous Galton-Watson branching process with progeny distribution $\{p_j^{(i)}, j = 0, 1, \ldots, \infty\}$ and $f_n(s)$ the pgf of the above non-homogeneous Galton-Watson branching process. Then, it is obvious that

$$f_n(s) = f_n^{(1)}(s) = f^{(1)}[f_{n-1}^{(1)}(s)], \quad \text{if } n \leq t_1;$$

$$f_n(s) = f_{n-t_1}^{(2)}[f_{t_1}^{(1)}(s)], \quad \text{if } n > t_1.$$

It follows that for $n \leq t_1$,

$$p_{1j}(0,n) = \frac{1}{j!} \left\{ \frac{d^j}{ds^j} f_n^{(1)}(s) \right\}_{s=0} \quad \text{and} \quad p_{ij}(0,n) = \frac{1}{j!} \left\{ \frac{d^j}{ds^j} [f_n^{(1)}(s)]^i \right\}_{s=0} \quad \text{for } i > 1 \, ;$$

but for $n > t_1$,

$$p_{1j}(0,n) = \frac{1}{j!} \left\{ \frac{d^j}{ds^j} f_{n-t_1}^{(2)} [f_{t_1}^{(1)}(s)] \right\}_{s=0}, \quad \text{and}$$

$$p_{ij}(0,n) = \frac{1}{j!} \left\{ \frac{d^j}{ds^j} \{ f_{n-t_1}^{(2)} [f_{t_1}^{(1)}(s)] \}^i \right\}_{s=0}, \quad \text{for } i > 1 \, .$$

2.3. The Structure and Decomposition of Markov Chains

Consider a Markov chain $\{X(t), t \in T\}$ with parameter space $T = (0, 1, \ldots, \infty)$ and with state space $S = \{0, 1, \ldots, \}$. In this section we will illustrate the basic structure and decomposition of these chains. For simplicity of illustration, we will restrict ourselves to homogeneous Markov chains, unless otherwise stated, although many results also hold in non-homogeneous chains.

Definition 2.1. Let $j \in S$ and $k \in S$ be two states of S. Then we say j *leads to* k, denoted by $j \to k$, iff there exists an $n > 0$ such that $P_{jk}(0, n) > 0$; we say $j \leftrightarrow k$, iff $j \to k$ and $k \to j$.

For example, in the sib-mating Markov chain given by Example 1.8, $1 \to 1$, $4 \to 1$, $4 \to 4$, $4 \to 6$, $6 \to j$, $j = 1, \ldots, 6$, etc.

Definition 2.2. Let C be a set of states of the homogeneous Markov chain $\{X(t), t \in T = (0, 1, 2, \ldots)\}$. Then C is called *a closed set* iff for every $k \notin C$ and $j \in C$, $j \not\to k$. By $j \not\to k$, it means that it is not possible to find a n such that $p_{jk}(0, n) > 0$. A closed set consists of a single state is called an absorbing class and the element the absorbing state.

By Definition 2.2, the state space itself is a closed set. If the state space contains a proper closed subset, then the chain is said to be reducible; otherwise not reducible or irreducible. Examples 1.8, 2.1 and 2.4 given above are reducible homogeneous Markov chains but Examples 2.2 and 2.5 are irreducible homogeneous Markov chains. For example, in the full-sib mating Markov chain given in Example 2.2, the state $AA \times AA$ forms an absorbing class, so is the state $aa \times aa$; that is, in the sib-mating example, there are two absorbing classes $\{AA \times AA\}$ and $\{aa \times aa\}$. Similarly in Example 2.3, there are four closed sets each with only one state; these absorbing classes are $\{AB/AB, Ab/Ab, aB/aB, ab/ab\}$. The following theorem gives a necessary and sufficient condition for a Markov chain to be irreducible.

Theorem 2.2. *A Markov chain is irreducible iff for any two states j and $k, j \leftrightarrow k$.*

Proof. The "if" part is trivial; hence we prove only the "only if" part.

To prove the "only if" part, assume that the chain is irreducible and consider an arbitrary state $j \in S$. Let C_j be the set of all states that j leads to and $C_j^{(*)}$ the set of all states that j does not lead to. Then $C_j \cap C_j^{(*)} = \emptyset$ and $C_j \cup C_j^{(*)} = S$, the state space. We now show that for any $\ell \in C_j$ and for any $k \in C_j^{(*)}, \ell \nrightarrow k$ so that C_j is closed. Now, $j \to \ell$ and $j \nrightarrow k$ by definitions of C_j and $C_j^{(*)}$; thus, if $\ell \to k$, then $j \to k$, a contradiction. Hence $\ell \nrightarrow k$ and C_j is closed. It follows that $C_j^{(*)} = \emptyset$. Since j is an arbitrary state, the theorem is established. $\square$

By Theorem 2.2, the Markov chain in Example 2.2 and the base substitution model in Example 2.5 are irreducible since each state can be reached by other states. For the Wright model given in Example 1.11, if there are no mutation and no immigration and migration, then the chain is reducible with two absorbing states 0 and $2N$; on the other hand, if there are forward and backward mutations, then the chain is irreducible since each state can be reached from other states.

The following simple result gives a necessary and sufficient condition for a set of states in a Markov chain to be closed.

Theorem 2.3. *Let C be a set of states in a Markov chain. Then C is closed iff, for any $n > 0, \sum_{j \in C} p_{ij}(0, n) = 1$ for any $i \in C$.*

Proof. For any $i \in C, \sum_j p_{ij}(0, n) = 1 = \sum_{j \in C} p_{ij}(0, n) + \sum_{j \notin C} p_{ij}(0, n) = \sum_{j \in C} p_{ij}(0, n)$ iff $\sum_{j \notin C} p_{ij}(0, n) = 0$ iff $p_{ij}(0, n) = 0$ for $j \notin C$. This holds for any $n > 0$ so that the theorem is proved. $\square$

By Theorem 2.3, all absorbing states in Examples 1.8 and 2.4 are closed sets consisting of a single state.

From Theorem 2.3, if C is a closed set, then, by deleting all states not in C, we obtain another Markov chain with state space $= C$. Using Theorem 2.3, we have also the following theorem for the structure of transition matrix in finite chains. A Markov chain is said to be a finite Markov chain iff its state space S contains only a finite number of elements.

Theorem 2.4. *A finite homogeneous Markov chain* $\{X(t), t \in T\}$ *is reducible iff its one-step transition matrix* P *can be expressed as*

$$P = \begin{pmatrix} M & O \\ R & Q \end{pmatrix}.$$

Proof. (1) If the chain is reducible, then its state space contains a proper closed subset C. With no loss of generality we may let the elements of C be $0, 1, 2, \ldots, c$. Then $p_{ij}(n) = 0$ for all $n = 1, 2, \ldots$ and for all $j = c+1, c+2, \ldots$ if $0 \le i \le c$. Hence

$$P = \begin{pmatrix} M & O \\ R & Q \end{pmatrix}.$$

(2) Let $P = \begin{pmatrix} M & O \\ R & Q \end{pmatrix}$. Then $P_{ij}(n) = 0$ for all $n = 1, 2, \ldots$ and all $j = c+1, c+2, \ldots$ if $0 \le i \le c$. Thus $C = \{0, 1, 2, \ldots, c\}$ is a proper closed set of the state space. The chain is therefore reducible. $\square$

As an Corollary of Theorem 2.4, we have that if the finite homogeneous Markov chain contains k $(k \ge 1)$ closed sets and if these k closed sets do not exhaust all states, then by rearranging and renaming the states, the one-step transition matrix P can be expressed as:

$$P = \begin{pmatrix} Q & R_1 & R_2 & \cdots & R_k \\ & P_1 & & & 0 \\ & & P_2 & & \\ & & & \ddots & \\ & 0 & & & P_k \end{pmatrix},$$

where P_j is the one-step transition matrix for the jth closed set.

The above form is called the canonical form of P. Notice that $(R_1, R_2, \ldots, R_k) \ne 0$ for otherwise, the chain contains $k+1$ closed sets, contradicting the assumption of k closed sets.

As an illustration, consider the sib-mating Markov chain given by Example 1.8. Rearranging the states so that $\{1, 2, 3, 4, 5, 6\}$ stand for the states

$\{AA \times aa, AA \times Aa, aa \times Aa, Aa \times Aa, AA \times AA, aa \times aa\}$ respectively, then the one-step transition matrix is given by:

$$P = \begin{pmatrix} Q & R_1 & R_2 \\ 0 & P_1 & 0 \\ 0 & 0 & P_2 \end{pmatrix},$$

where Q and $R = (R_1, R_2)$ are given in Example 2.2 and $P_j = 1$ is the one-step transition matrix for the jth closed set.

Similarly, in Example 2.4, by rearranging the order of the states in the state space so that the first six states are transient states, then the one step transition matrix can be expressed in the above form.

2.4. Classification of States and the Dynamic Behavior of Markov Chains

To study the behavior of Markov chains, one needs to classify the states. For this purpose, define the event $A_{jk}(n)$ by $A_{jk}(n) = \{X(n) = k, X(m) \neq k, m = 1, 2, \ldots, n-1 | X(0) = j\}$. Then $A_{jk}(n)$ is the event of first passage to k at time n from j at time 0. With this definition, it is obvious that $A_{jk}(n) \cap A_{jk}(n') = 0$ if $n \neq n'$ and $A_{jk} = \bigcup_{n=1}^{\infty} A_{jk}(n)$ is the event that the chain ever reaches the state k starting at the state j initially. Thus,

$$f_{jk} = \Pr\{A_{jk}\} = \Pr\left[\bigcup_{n=1}^{\infty} A_{jk}(n)\right] = \sum_{n=1}^{\infty} \Pr(A_{jk}(n))$$

$$= \sum_{n=1}^{\infty} \Pr\{X(n) = k | X(m) \neq k, m = 1, 2, \ldots, n-1, X(0) = j\}$$

$$= \sum_{n=1}^{\infty} f_{jk}(n),$$

where f_{jk} is the probability that the chain ever reaches the state k starting with the state j at time 0 and $f_{jk}(n)$ the probability that the chain first reaches the state k at time n starting with the state j initially.

Definition 2.3. The state j is called *persistent* (or *recurrent*) iff $f_{jj} = 1$, and *transient* (or *nonrecurrent*) iff $f_{jj} < 1$.

From this definition, the states $AA \times AA$ and $aa \times aa$ in the full-sib mating chain given by Example 1.8 are persistent states whereas all other states are transient states. Similarly, in Example 2.4, the states $\{AB/AB, Ab/Ab, aB/aB, ab/ab\}$ are persistent states whereas all other states are transient states. On the other hand, all states in Examples 2.2 and 2.5 are persistent states.

If $f_{jj} = 1$, then $f_{jj}(n), n = 1, 2, \ldots$ forms a probability density function for the first return time T_j of the state j. The expected value of the first return time T_j, denoted by $\mu_j = \sum_{n=1}^{\infty} n f_{jj}(n)$, is called the mean return time of the state j.

Definition 2.4. Let j be a persistent state and μ_j the mean return time of j. Then j is called *positive* iff $\mu_j < \infty$ and *null* iff $\mu_j = \infty$.

In the next chapter, we will show that if j is persistent, then

$$\lim_{n \to \infty} \frac{1}{n} \sum_{m=1}^{n} p_{jj}(m) = \frac{1}{\mu_j} \, .$$

Thus, the limit is positive iff j is positive; the limit is 0 iff j is null.

Given Definition 2.3, an immediate question is: Given a state, say j, is it persistent or transient? To answer this question, we first prove the following lemma.

Lemma 2.1.

$$f_{ij} = \lim_{N \to \infty} \frac{\sum_{n=1}^{N} p_{ij}(n)}{1 + \sum_{n=1}^{N} p_{jj}(n)} \, ,$$

where $p_{ij}(n) = \Pr\{X(n) = j | X(0) = i\}$, for any homogeneous Markov chain $\{X(t)\}, t \in T = \{0, 1, 2, \ldots\}$, with state space $S = \{0, 1, 2, \ldots\}$.

Proof. We have $p_{ij}(n) = \sum_{m=1}^{n} f_{ij}(m) p_{jj}(n - m)$ for any $n \geq 1$. Hence,

$$\sum_{n=1}^{N} p_{ij}(n) = \sum_{n=1}^{N} \sum_{m=1}^{n} f_{ij}(m) p_{jj}(n - m) = \sum_{m=1}^{N} \sum_{n=m}^{N} f_{ij}(m) p_{jj}(n - m)$$

$$= \sum_{m=1}^{N} f_{ij}(m) \left\{ \sum_{n=0}^{N-m} p_{jj}(n) \right\} \leq \sum_{m=1}^{N} f_{ij}(m) \sum_{n=0}^{N} p_{jj}(n) \, .$$

Thus, since $p_{jj}(0) = 1$, we have that

$$\sum_{m=1}^{N} f_{ij}(m) \geq \frac{\sum_{n=1}^{N} p_{ij}(n)}{1 + \sum_{n=1}^{N} p_{jj}(n)}$$

and so

$$f_{ij} \geq \frac{\sum_{n=1}^{\infty} p_{ij}(n)}{1 + \sum_{n=1}^{\infty} p_{jj}(n)} \, .$$

Let now $N_1 = \left[\frac{N}{2}\right]$ be the largest integer $\leq \frac{N}{2}$. Then, $\frac{N}{2} - 1 < N_1 \leq \frac{N}{2} < N$ and $N_2 = N - N_1 \geq N - \frac{N}{2} = \frac{N}{2}$. Thus, $N \to \infty \Rightarrow N_1 \to \infty$ and $N_2 \to \infty$. Furthermore, $N_1 \geq m \Rightarrow N - m \geq N - N_1 = N_2$. Hence

$$\sum_{n=1}^{N} P_{ij}(n) = \sum_{m=1}^{N} f_{ij}(m) \left\{ \sum_{n=0}^{N-m} p_{jj}(m) \right\} \geq \sum_{m=1}^{N_1} f_{ij}(m) \sum_{n=0}^{N_2} p_{jj}(n) \, .$$

It follows that

$$\frac{\sum_{n=1}^{N} p_{ij}(n)}{1 + \sum_{n=1}^{N_2} p_{jj}(n)} \geq \sum_{m=1}^{N_1} f_{ij}(m) \, .$$

Let $N \to \infty$ (and hence $N_1 \to \infty$ and $N_2 \to \infty$). Then

$$f_{ij} \leq \frac{\sum_{n=1}^{\infty} p_{ij}(n)}{1 + \sum_{n=1}^{\infty} p_{jj}(n)} \, . \qquad \square$$

Using the above lemma, we have immediately the following results for homogeneous Markov chains with discrete time:

Theorem 2.5. *For homogeneous Markov chains with discrete time, the following results hold:*

(1) *j is persistent iff $\sum_{n=1}^{\infty} p_{jj}(n) = \infty$;*

(2) *j is transient iff $\sum_{n=1}^{\infty} p_{jj}(n) < \infty$;*

(3) *If j is transient, then $\sum_{n=1}^{\infty} p_{ij}(n) < \infty$; if $i \to j$ and if $\sum_{n=1}^{\infty} p_{ij}(n) < \infty$, then j is transient;*

(4) *If j persistent and $i \to j$, then $\sum_{n=1}^{\infty} p_{ij}(n) = \infty$; if $\sum_{n=1}^{\infty} p_{ij}(n) = \infty$; for some state i, then j must be persistent.*

Proof. (1) and (2) follow immediately from the lemma by putting $i = j$. (3) and (4) follow from Lemma 2.1 and the results: (a) $0 \leq f_{ij} \leq 1$, (b) $i \to j \Rightarrow p_{ij}(n) > 0$ for some $n > 0$ so that $\sum_{n=1}^{\infty} p_{ij}(n) > 0$, and (c) $i \to j \Rightarrow f_{ij} > 0$. $\qquad\square$

Notice that

$$\sum_{n=1}^{\infty} p_{ij}(n) < \infty \Rightarrow \lim_{n \to \infty} p_{ij}(n) = 0 \,.$$

As a consequence of the above theorem, if j is transient, $\lim_{n \to \infty} p_{ij}(n) = 0$ for any state i. If the number of transient states is finite and if $Q(n)$ is the n-step transition matrix for transient states into transient states, then we must have $\lim_{n \to \infty} Q(n) = 0$. This result implies that finite Markov chains must contain persistent states; for if not, then, because $\sum_{j=0}^{N} p_{ij}(n) = 1$ for all $n > 0$ and for all $i \in S$ if the state space of the chain is $S = \{0, 1, \ldots, N\}$, we obtain the result $0 = 1$ by letting $n \to \infty$, which is not possible.

Example 2.12. **Random walk on a fixed path.** Consider a random walk on a fixed path starting at time $t = 0$. Let $X(t)$ be the position at time t and assume that $\Pr\{X(t + 1) = j + 1 | X(t) = j\} = p$ $(0 < p < 1)$ and $\Pr\{X(t+1) = j-1 | X(t) = j\} = 1-p = q$. Then, $\{X(t), t \in T = (0, 1, \ldots, \infty)\}$ is a homogeneous Markov chain with discrete time and with state space $\Omega = \{-\infty, \ldots, -1, 0, 1, \ldots, \infty\}$.

For this Markov chain, obviously $i \leftrightarrow j$ for all $i \in \Omega$ and $j \in \Omega$ so that the chain is irreducible. Thus, if any state is transient, then the chain is transient in which case the chain contains no persistent states; on the other hand, if any state is persistent, then the chain is persistent and contains no transient states.

To find conditions for which the chain is transient or persistent, consider the state 0 and assume that $X(0) = 0$. Then, it takes an even number of steps to return to 0 given $X(0) = 0$ and $\Pr\{X(2n) = 0 | X(0) = 0\} = \binom{2n}{n}(pq)^n$ for all $n = 1, \ldots, \infty$. (That is, the period of 0 is 2 so that $\Pr\{X(2n + 1) = 0 | X(0) = 0\} = 0$ for all $n = 0, 1, \ldots, \infty$; see Definition 3.2.) We will show that if $p = \frac{1}{2}$, then 0 is persistent in which case the chain is persistent and contains no transient states; on the other hand, if $0 < p < \frac{1}{2}$, then 0 is transient in which case the chain is transient and contains no persistent states.

To prove the above, notice the Stirling formulae given by [13]:

$$(n!) = n^{n+\frac{1}{2}}\sqrt{2\pi}\,e^{-n}\left\{1 + \frac{1}{12n} + 0(n^{-2})\right\}.$$

Applying the above Stirling formulae, we obtain:

$$\binom{2n}{n}(pq)^n = \sqrt{2\pi}(2n)^{2n+\frac{1}{2}}e^{-2n}(pq)^n\{n^{n+\frac{1}{2}}\sqrt{2\pi}\,e^{-n}\}^{-2}$$

$$\times\left\{1 + \frac{1}{24n} + 0(n^{-2})\right\}\left\{1 + \frac{1}{12n} + 0(n^{-2})\right\}^{-2}$$

$$= \frac{1}{\sqrt{n\pi}}(4pq)^n\left\{1 - \frac{1}{8n} + 0(n^{-2})\right\} \cong \frac{1}{\sqrt{n\pi}}(4pq)^n.$$

It follows that with $r = 4pq$,

$$\sum_{n=1}^{\infty} P_{00}(2n) \cong \sum_{n=1}^{\infty}\frac{1}{\sqrt{n\pi}}r^n.$$

Now $r = 1$ iff $p = \frac{1}{2}$ and $0 < r < 1$ iff $0 < p < \frac{1}{2}$ or $\frac{1}{2} < p < 1$. Hence $\sum_{n=1}^{\infty}\frac{1}{\sqrt{n\pi}}r^n = \infty$ iff $p = \frac{1}{2}$ and $\sum_{n=1}^{\infty}\frac{1}{\sqrt{n\pi}}r^n < \infty$ iff $0 < p < \frac{1}{2}$. Or, 0 is persistent iff $p = \frac{1}{2}$ and 0 is transient iff $0 < p < \frac{1}{2}$.

Assume $p = \frac{1}{2}$ so that the chain is persistent and irreducible. The mean return time of 0 is $\nu_0 = \sum_{n=1}^{\infty}(2n)P_{00}(2n) \cong \sum_{n=1}^{\infty}(2n)/(\sqrt{n\pi}) = \infty$ and hence the state 0 is null. On the other hand, all persistent states in Examples 2.1–2.2 and 2.4 are aperiodic and have mean return time 1.

To further characterize the dynamic behavior of Markov chains, an immediate question is: Given a state j initially, how many times the state j will return to j as time progresses? One may also wonder how the probability of this event is related to the classification of the states! To answer these questions, we prove the following theorem which is referred to as the 0–1 law in homogeneous Markov chains.

Theorem 2.6. (The 0–1 law). *Let j be an arbitrary state in S and let g_{jj} be the probability that the chain will visit j an infinitely many times given j at time 0. Then,*

(1) $g_{jj} = 1$ *iff j is persistent;*
(2) $g_{jj} = 0$ *iff j is transient.*

Proof. Let $g_{ij}(m)$ be the probability that starting with the state i, the chain visits the state j at least m times. Then $g_{ij}(1) = f_{ij}$ and

$$g_{ij}(m) = f_{ij}(1)g_{jj}(m-1) + f_{ij}(2)g_{jj}(m-1) + \cdots$$

$$= \sum_{n=1}^{\infty} f_{ij}(n)g_{jj}(m-1) = f_{ij}g_{jj}(m-1).$$

Hence $g_{ij}(m) = f_{ij}f_{jj}^{m-1}$ for $m \geq 1$. Putting $i = j$ and letting $m \to \infty$, 0–1 law follows immediately. $\qquad\square$

The above 0–1 law leads immediately to the following results for the behavior of homogeneous Markov chains.

(i) If j persistent, then $g_{jj} = f_{jj} = 1 = g_{jj}(1)$; from the proof of the 0–1 law, we have also $g_{ij} = f_{ij} = g_{ij}(1)$, for any i.

(ii) If j transient, then $g_{ij} = 0$ for any i. Thus, if the chain is finite, it must contain persistent states as it can stay in transient states only in a finite number of times. However, if the chain contains an infinite number of states, then Example 2.12 shows that the chain may not contain persistent states.

The following theorem also shows that persistent states go only to persistent states in homogeneous Markov chains. Hence, once a transient state visits a persistent state, then the chain will stay forever in persistent states.

Theorem 2.7. *For homogeneous Markov chains, if i is persistent and if $i \to j$, then $j \to i$ and j is persistent; moreover $f_{ji} = g_{ji} = f_{ij} = g_{ij} = 1$.*

Proof. $i \to j \Rightarrow$ there exists a $n_0 > 0$ such that $p_{ij}(n_0) > 0$; and i persistent $\Rightarrow 1 = f_{ii} = g_{ii}$. But then $\sum_k p_{ik}(n_0) = 1 = f_{ii} = g_{ii} = \sum_k p_{ik}(n_0)g_{ki}$. Hence, $0 = \sum_k p_{ik}(n_0)(1 - g_{ki}) \Rightarrow 0 = p_{ik}(n_0)(1 - g_{ki})$ for all $k = 0, 1, 2, \ldots$ as $0 \leq g_{ki} \leq 1$. Hence, $p_{ij}(n_0) > 0 \Rightarrow g_{ji} = 1 = f_{ji}$. This implies further that $j \to i$. Thus there exists an m_0 such that $p_{ji}(m_0) > 0$.

To prove that j is persistent, notice that, by using the Chapman-Kolmogorov equation, one has:

$$p_{jj}(m_0 + n + n_0) \geq p_{ji}(m_0)p_{ii}(n)p_{ij}(n_0).$$

Hence $\sum_n p_{jj}(n) \geq \sum_n p_{jj}(m_0 + n + n_0) \geq p_{ij}(n_0)p_{ji}(m_0)\sum_n p_{ii}(n)$. Thus, i persistent $\Rightarrow \sum_n p_{jj}(n) = \infty \Rightarrow j$ persistent. This also implies that $f_{ij} = g_{ij} = 1$, as proved above. $\qquad\square$

2.5.   The Absorption Probabilities of Transient States

Consider a homogeneous Markov chain $\{X(t), t \in T\}$ with discrete time $T = \{0, 1, \ldots, \infty\}$ and with state space $S = \{0, 1, \ldots, \infty\}$. Assume that the chain consists of both persistent states and transient states and that the persistent states are grouped into k closed sets of persistent states. (Results in Sec. 2.4 shows that if the chain contains a finite number of transient states, it must contain persistent states; on the other hand, if the chain contains infinitely many transient states, it may or may not contain persistent states; see Example 2.12). Let $p_{ij}(n)$ denote the n-step ($n \geq 0$) transition probability from the state i to the state j with $p_{ij}(1) = p_{ij}$ and $p_{ij}(0) = \delta_{ij}$. Let C_T denote the set of transient states. If C_T is finite (i.e. C_T contains only a finite number of elements), then, as shown by Theorem 2.7, starting with any transient state $i \in C_T$ initially, with probability one the state i will eventually be absorbed into some closed sets C_j as time progresses. On the other hand, if C_T is infinite, then as illustrated in Example 2.12, the chain may not contain persistent states; also, even if there are persistent states, Example 2.16 shows that, under some conditions, with positive probability the chain may stay in transient states forever as time progresses; see Subsec. 2.5.2 and Example 2.16.

To derive formula for absorption probabilities of transient states, denote by:

(1) $F_{i,j}(n) = $ The probability that starting with $i \in C_T$ at time $t = 0$, the chain will be absorbed into the jth closed set C_j at or before time n ($n > 0$).

(2) $g_{i,j}(n) = $ The probability that starting with $i \in C_T$ at time $t = 0$, the chain will be absorbed into the jth closed set C_j at time n ($n > 0$) for the first time.

(3) $\rho_i(j) = $ The ultimate absorption probability of $i \in C_T$ into the closed set C_j as time progresses.

Then, $g_{i,j}(n) = F_{i,j}(n) - F_{i,j}(n-1)$, and noting that $F_{i,j}(0) = 0$,

$$\rho_i(j) = \lim_{n \to \infty} F_{i,j}(n) = \sum_{n=1}^{\infty} g_{i,j}(n) \,.$$

Further, $F_i(n) = \sum_{j=1}^{k} F_{i,j}(n)$ is the probability of absorption into a persistent state at or before time n starting with $i \in C_T$ at time $t = 0$;

$g_i(n) = \sum_{j=1}^{k} g_{i,j}(n)$ is the probability of first time absorption of $i \in C_T$ into a closed set at time n;

$\rho_i = \sum_{j=1}^{k} \rho_i(j)$ is the ultimate absorption probability of $i \in C_T$ into a closed set as time progresses.

We will show that if C_T is finite, then $\rho_i = 1$, for all $i \in C_T$. In this case $g_i(n)$ is the pdf of the first absorption time T_i of $i \in C_T$ and $F_i(n)$ the cdf of T_i.

To find these probabilities, denote by

$$a_i(j) = \sum_{m \in C_j} p_{im}, \ i \in C_T, \ j = 1, \ldots, k, \ \text{and} \ a_i = \sum_{j=1}^{k} a_i(j) \, .$$

Since with probability one persistent states go only to persistent states by Theorem 2.7, we have that with $F_{i,j}(0) = 0$,

$$F_{i,j}(n) = \sum_{m \in C_j} p_{im} + \sum_{r=1}^{n-1} \sum_{m \in C_T} p_{im}(r) a_m(j)$$

$$= a_i(j) + \sum_{m \in C_T} p_{im} F_{m,j}(n-1) \quad \text{for } n = 1, \ldots, \infty, i \in C_T, j = 1, \ldots, k$$

$$(2.2)$$

and

$$F_i(n) = a_i + \sum_{r=1}^{n-1} \sum_{m \in C_T} p_{im}(r) a_m$$

$$= a_i + \sum_{m \in C_T} p_{im} F_m(n-1), \quad \text{for } n = 1, \ldots, \infty, \ \text{and} \ i \in C_T \, . \quad (2.3)$$

It follows that with $p_{im}(0) = \delta_{im}$,

$$g_{i,j}(n) = \sum_{m \in C_T} p_{im}(n-1) a_m(j) \quad \text{for } n = 1, \ldots, \infty, \ i \in C_T, j = 1, \ldots, k \, .$$

$$(2.4)$$

Thus,

$$g_i(n) = \sum_{j=1}^{k} g_{i,j}(n) = \sum_{m \in C_T} p_{im}(n-1) a_m, i \in C_T, n = 1, \ldots, \infty \, ; \qquad (2.5)$$

$$\rho_i(j) = \lim_{n \to \infty} F_{i,j}(n) = \sum_{m \in C_T} p_{im}\rho_m(j) + a_i(j) \text{ for } i \in C_T, j = 1, \ldots, k \qquad (2.6)$$

and

$$\rho_i = \sum_{j=1}^{k} \rho_i(j) = \sum_{m \in C_T} p_{im}\rho_m + a_i, \quad \text{for } i \in C_T. \qquad (2.7)$$

In the Galton-Watson branching process with progeny distribution $\{p_j, j = 0, 1, \ldots\}$, there is only one persistent state 0 and $C_T = \{1, \ldots, \infty\}$. In this case, $a_1 = p_0, x_0 = \rho_1, \rho_i = x_0^i$ $(i = 1, \ldots)$ and $p_{1j} = p_j$ so that Eq. (2.7) becomes:

$$x_0 = \sum_{j=1}^{\infty} p_j x_0^j + p_0 = f(x_0).$$

This is the formulae given in Theorem 2.11.

2.5.1. *The case when C_T is finite*

Suppose now that the set C_T is finite. Assume that C_T has r elements and C_j has n_j elements. With no loss of generality, we assume that the first r states are the transient states and the other states are persistent states. Let H_j be the $r \times n_j$ matrix of the one-step transition probabilities from transient states to states in C_j and Q the $r \times r$ matrix of the one-step transition probabilities from transient states to transient states. Denote by $Q(n)$ the n-step transition matrix of transient states with $\underset{\sim}{q}'_i(n)$ $(i = 1, \ldots, r)$ as the ith row of $Q(n)$. Then $Q(n) = Q^n$. Further, as shown in Subsec. 2.7.1, the absolute values of eigenvalues of Q are less than one so that $Q^n \to 0$ as $n \to \infty$. Denoting by $\underset{\sim}{a}(j) = \{a_1(j), \ldots, a_r(j)\}'$, $j = 1, \ldots, k$ and $\underset{\sim}{a} = \{a_1, \ldots, a_r\}'$. Since the elements in each row of a transition matrix sum to 1, we have that $\underset{\sim}{a}(j) = H_j \underset{\sim}{1}_{nj}$ and $\underset{\sim}{a} = (I_r - Q)\underset{\sim}{1}_r$.

To express the above absorption probabilities in matrix notation, denote by:

(1) $\underset{\sim}{F}_j(n) = \{F_{1,j}(n), \ldots, F_{r,j}(n)\}'$, $j = 1, \ldots, k$ and $\underset{\sim}{F}(n) = \{F_1(n), \ldots, F_r(n)\}'$;

(2) $\underset{\sim}{g}_j(n) = \{g_{1,j}(n), \ldots, g_{r,j}(n)\}'$, $j = 1, \ldots, k$ and $\underset{\sim}{g}(n) = \{g_1(n), \ldots, g_r(n)\}'$;

(3) $\underset{\sim}{\rho}(j) = \{\rho_1(j), \ldots, \rho_r(j)\}'$, $j = 1, \ldots, k$ and $\underset{\sim}{\rho} = \{\rho_1, \ldots, \rho_r\}'$.

Then, in matrix notation, we have:

(1) The vectors of absorption probabilities into C_j and into any persistent state at or before n are given respectively by:

$$\underset{\sim}{F}_j(n) = \sum_{r=0}^{n-1} Q^r \underset{\sim}{a}(j) = (I_r - Q)^{-1}(I_r - Q^n)\underset{\sim}{a}(j)$$

$$= (I_r - Q)^{-1}(I_r - Q^n)H_j\underset{\sim}{1}_{nj} \quad \text{for } j = 1, \ldots, k, \qquad (2.8)$$

$$\underset{\sim}{F}(n) = (I_r - Q)^{-1}(I_r - Q^n)\underset{\sim}{a}$$

$$= (I_r - Q^n)(I_r - Q)^{-1}(I_r - Q)\underset{\sim}{1}_r = (I_r - Q^n)\underset{\sim}{1}_r. \qquad (2.9)$$

(2) Since $g_{i,j}(n) = \underset{\sim}{q}'_i(n-1)\underset{\sim}{a}(j)$, $g_i(t) = \underset{\sim}{q}'_i(n-1)\underset{\sim}{a}$, the vectors of first time absorption probabilities into C_j and into any persistent state are given respectively by:

$$\underset{\sim}{g}_j(n) = Q^{n-1}\underset{\sim}{a}(j) = Q^{n-1}H_j\underset{\sim}{1}_{nj}, j = 1, \ldots, k, \quad \text{for } n > 0, \qquad (2.10)$$

$$\underset{\sim}{g}(n) = \sum_{j=1}^{k} \underset{\sim}{g}_j(n) = Q^{n-1}\underset{\sim}{a} = Q^{n-1}(I_r - Q)\underset{\sim}{1}_r \quad \text{for } n > 0. \qquad (2.11)$$

(3) Since $\sum_{n=0}^{\infty} Q^n = (I_r - Q)^{-1}$, the vectors of ultimate absorption probabilities into C_j and into any persistent state are given respectively by:

$$\underset{\sim}{\rho}(j) = \sum_{n=1}^{\infty} Q^{n-1}\underset{\sim}{a}(j) = (I_r - Q)^{-1}\underset{\sim}{a}(j) = (I_r - Q)^{-1}H_j\underset{\sim}{1}_{nj} \qquad (2.12)$$

and

$$\underset{\sim}{\rho} = \sum_{n=1}^{\infty} Q^{n-1}\underset{\sim}{a} = (I_r - Q)^{-1}\underset{\sim}{a} = (I_r - Q)^{-1}(I_r - Q)\underset{\sim}{1}_r = \underset{\sim}{1}_r. \qquad (2.13)$$

The result $\underset{\sim}{\rho} = \underset{\sim}{1}_r$ is equivalent to stating that the probability that $i \in C_T$ will eventually be absorbed into a persistent state is one. This also implies that the element $g_i(n)$ in $\underset{\sim}{g}(n)$ is the discrete probability density function (pdf) over the space $\{1, \ldots, \infty\}$ of the first passage time T_i of $i \in C_T$ and $F_i(n)$ the cumulative distribution function (cdf) of T_i.

If Q has real distinct eigenvalues $\lambda_1, \ldots, \lambda_u, u \le r$, then, as shown in Subsec. 2.11.3, $Q = \sum_{i=1}^{u} \lambda_i E_i$, where $E_i = \prod_{j \ne i} \frac{1}{\lambda_i - \lambda_j}(Q - \lambda_j I_r)$ $i = 1, \ldots, u$. The E_i's satisfy the conditions $E_i^2 = E_i, E_i E_j = 0$ if $j \ne i$ and $\sum_{i=1}^{u} E_i = I_r$. It follows that $Q(n) = Q^n = \sum_{i=1}^{u} \lambda_i^n E_i$ and $(I_r - Q)^{-1} = \sum_{i=1}^{u} \frac{1}{1 - \lambda_i} E_i$. (Note that $\lambda_i < 1$ so that $1 - \lambda_i > 0$.)

Hence, noting that $\underset{\sim}{a}(j) = H_j \underset{\sim}{1}_{nj}$ and $\underset{\sim}{a} = (I_r - Q) \underset{\sim}{1}_r = \sum_{i=1}^{u} (1 - \lambda_i) E_i \underset{\sim}{1}_r$,

$$\underset{\sim}{g}_j(n) = \sum_{i=1}^{u} \lambda_i^{n-1} E_i \underset{\sim}{a}(j), n = 1, \ldots, \infty;$$

$$\underset{\sim}{g}(n) = \sum_{i=1}^{u} \lambda_i^{n-1} E_i \underset{\sim}{a} = \sum_{i=1}^{u} \lambda_i^{n-1} (1 - \lambda_i) E_i \underset{\sim}{1}_r, n = 1, \ldots, \infty.$$

Thus, since $\sum_{t=0}^{\infty} Q^n = (I_r - Q)^{-1}$,

$$\underset{\sim}{\rho}(j) = \sum_{n=1}^{\infty} Q^{n-1} \underset{\sim}{a}(j) = (I_r - Q)^{-1} \underset{\sim}{a}(j)$$

$$= \sum_{i=1}^{u} (1 - \lambda_i)^{-1} E_i \underset{\sim}{a}(j) = \sum_{i=1}^{u} (1 - \lambda_i)^{-1} E_i H_j \underset{\sim}{1}_{nj}.$$

2.5.2.   *The case when C_T is infinite*

When C_T is infinite, the ultimate absorption probabilities of transient states into persistent states may or may not be one. In this section we derive the probability $\sigma_i (i \in C_T)$ that the chain will stay forever in transient states starting with transient state $i \in C_T$, initially. To derive these probabilities, denote by $\omega_i(n) = \Pr\{X(n) \in C_T | X(0) = i\}$, $\omega_i(0) = 1$, for $i \in C_T$. Then these probabilities are given by $\sigma_i = \lim_{n \to \infty} \omega_i(n), i \in C_T$. (As shown in Theorem 2.9, such limits always exist.) To find σ_i, since with probability one persistent states go only to persistent states, we have, for $n = 1, \ldots, \infty$:

$$\omega_i(n) = \sum_{m \in C_T} p_{im} \omega_m(n - 1).$$

If the limit $\lim_{n \to \infty} \omega_i(n) = \sigma_i$ exists, then by Lebesque dominated convergence theorem (see Lemma 3.3), we obtain by taking limit on both sides:

$$\sigma_i = \sum_{m \in C_T} p_{im} \sigma_m .$$

This leads to the following theorem for computing σ_i.

Theorem 2.8. *Let σ_i be the probability that the chain stays forever in the transient states given initially $X(0) = i \in C_T$. Let $\omega_i(n)$ be defined as above. Then,*

(i) *The limit $\lim_{n \to \infty} \omega_i(n)$ exists and $\lim_{n \to \infty} \omega_i(n) = \sigma_i$.*

(ii) *σ_i satisfies the equation*

$$\sigma_i = \sum_{v \in C_T} p_{iv} \sigma_v, i \in C_T . \tag{2.14}$$

(iii) *If $v_i, i \in C_T$, satisfies the above system of equations and if $|v_i| \leq 1, i \in C_T$, then $|v_i| \leq \sigma_i, i \in C_T$.*

(iv) *For $i \in C_T$, let x_i be the probability that starting with $X(0) = i \in C_T$, the chain will eventually be absorbed into a persistent state as time progresses. Then $x_i = 1$ for all $i \in C_T$ iff $\sigma_i = 0$ for all $i \in C_T$ is the only solution of the system of Eqs. (2.14).*

Proof. By definition, for every $i \in C_T, \omega_i(n) = \Pr\{X(n) \in C_T | X(0) = i\}$. Since the chain is in C_T at time n implies that the chain must be in C_T at times $1, 2 \ldots, n - 1$, so, $\sigma_i = \lim_{n \to \infty} \omega_i(n)$ if the limit exists. To show that the limit does exist, note first that $0 \leq \omega_i(n) \leq 1$ for all $i \in C_T$ and for all $n = 1, 2, \ldots$ as they are probabilities. Thus, for every $i \in C_T$, $0 \leq \omega_i(1) \leq 1$, $0 \leq \omega_i(2) = \sum_{v \in C_T} p_{iv} \omega_v(1) \leq \sum_{v \in C_T} p_{iv} = \omega_i(1) \leq 1$, $0 \leq \omega_i(3) = \sum_{v \in C_T} p_{iv} \omega_v(2) \leq \sum_{v \in C_T} p_{iv} \omega_v(1) = \omega_i(2) \leq \omega_i(1) \leq 1$; by induction, $0 \leq \omega_i(n + 1) \leq \omega_i(n) \leq \cdots \leq \omega_i(1) \leq 1$. This shows that for each $i \in C_T, \{\omega_i(n)\}$ is a bounded monotonic non-increasing sequence so that the $\lim_{n \to \infty} \omega_i(n) = \sigma_i$ exists for all $i \in C_T$. This proves (i).

Now, $0 \leq \sum_{v \in C_T} p_{iv} \leq 1, p_{iv} \geq 0$ and $0 \leq \omega_v(n) \leq 1$; by Lebesque dominated convergence theorem,

$$\sigma_i = \lim_{n \to \infty} \omega_i(n + 1) = \lim_{n \to \infty} \sum_{v \in C_T} p_{iv} \omega_v(n)$$

$$= \sum_{v \in C_T} p_{iv} [\lim_{n \to \infty} \omega_v(n)] = \sum_{v \in C_T} p_{iv} \sigma_v, i \in C_T .$$

This shows that the $\sigma_i, i \in C_T$ satisfy Eq. (2.14). This proves not only (ii) and also that the solution of $\sigma_i = \sum_{v \in C_T} p_{iv}\sigma_v$ exists as the limit exists.

To prove (iii), suppose that v_j satisfy $v_i = \sum_{k \in C_T} p_{ik}v_k, i \in C_T$ and $|v_i| \leq 1, i \in C_T$. Then, $|v_i| \leq \sum_{k \in C_T} p_{ik}|v_k| \leq \sum_{k \in C_T} p_{ik} = \omega_i(1)$, $|v_i| \leq \sum_{k \in T} p_{ik}|v_k| \leq \sum_{k \in C_T} p_{ik}\omega_k(1) = \omega_i(2)$, and by induction, $|v_i| \leq \omega_i(n)$ for all $n = 1, 2, \ldots, \infty$ and for all $i \in C_T$. Hence, $|v_i| \leq \sigma_i$ for all $i \in C_T$.

To prove (vi), denote by $\beta_i = \sum_{j=1}^{k} \sum_{m \in C_j} p_{im}$. Then, obviously, the $\{x_i, i \in C_T\}$ satisfies the system of equations:

$$x_i = \sum_{v \in C_T} p_{iv}x_v + \beta_i, i \in C_T.$$

The general solution of this system of equations is given by $x_i = x_i^{(p)} + x_i^{(q)}$, where $0 \leq x_i \leq 1; 0 \leq x_i^{(q)} \leq 1$ is the general solution of $x_i = \sum_{v \in C_T} p_{iv}x_v$ and $x_i^{(p)}$ is a particular solution of $x_i = \sum_{v \in C_T} p_{iv}x_v + \beta_i$. Since 0 is the solution of $x_i = \sum_{v \in C_T} p_{iv}x_v, i \in C_T$, so the above system of equations has unique solution iff $\{x_i^{(q)} = 0, i \in C_T\}$ is the only solution of $x_i = \sum_{v \in C_T} p_{iv}x_v, i \in C_T$ or iff the probability is 0 that starting with $X(0) = i \in C_T$ initially, the chain stays forever in transient states. $\qquad\qquad\qquad\square$

2.6.   The Moments of First Absorption Times

Assuming that starting with $i \in C_T$, with probability one the chain will eventually be absorbed into a persistent state. Then $g_i(t)$ is the probability density function of the first absorption time T_i of $i \in C_T$. In this section we proceed to find the moments of $T_i, i \in C_T$. In particular we will find the mean μ_i of T_i and the variance V_i of T_i.

Now by definition, for each $i \in C_T$, $\Pr(T_i = t) = g_i(t) = \omega_i(t-1) - \omega_i(t)$. Hence,

$$\mu_i = \sum_{n=1}^{\infty} n g_i(n) = \sum_{n=1}^{\infty} n[\omega_i(n-1) - \omega_i(n)]$$

$$= \sum_{n=1}^{\infty}(n-1)\omega_i(n-1) - \sum_{n=1}^{\infty} n\omega_i(n) + \sum_{n=1}^{\infty} \omega_i(n-1) = \sum_{n=0}^{\infty} \omega_i(n). \quad (2.15)$$

Let $\eta_i = \sum_{n=1}^{\infty} n^2 g_i(t)$. Then $V_i = \eta_i - \mu_i^2$. On substituting $g_i(n) = \omega_i(n-1) - \omega_i(n)$, we obtain:

$$\eta_i = \sum_{n=1}^{\infty} n^2 g_i(n) = \sum_{n=1}^{\infty} n^2 [\omega_i(n-1) - \omega_i(n)]$$

$$= \sum_{n=1}^{\infty} (n-1)^2 \omega_i(n-1) - \sum_{n=1}^{\infty} n^2 \omega_i(n) + 2 \sum_{n=1}^{\infty} (n-1)\omega_i(n-1)$$

$$+ \sum_{n=1}^{\infty} \omega_i(n-1) = 2 \sum_{n=0}^{\infty} n\omega_i(n) + \sum_{n=0}^{\infty} \omega_i(n) = 2 \sum_{n=0}^{\infty} n\omega_i(n) + \mu_i. \quad (2.16)$$

2.6.1. *The case when C_T is finite*

If C_T is finite with r elements, then we may express the above equations in matrix notations. For this purpose, denote by:

$$\underset{\sim}{U} = \{\mu_1, \ldots, \mu_r\}', \ \underset{\sim}{V} = \{V_1, \ldots, V_r\}',$$

$$\underset{\sim}{\eta} = \{\eta_1, \ldots, \eta_r\}', \ \text{and} \ \underset{\sim}{U}_{sq} = \{\mu_1^2, \ldots, \mu_r^2\}'.$$

Then, since $\underset{\sim}{\omega}(n) = \{\omega_1(n), \ldots, \omega_r(n)\}' = Q(n)\underset{\sim}{1}_r = Q^n \underset{\sim}{1}_r$, we have:

$$\underset{\sim}{U} = \sum_{n=0}^{\infty} \underset{\sim}{\omega}(n) = \sum_{n=0}^{\infty} Q^n \underset{\sim}{1}_r = (I_r - Q)^{-1} \underset{\sim}{1}_r. \quad (2.17)$$

Since $\sum_{i=0}^{\infty} iQ^{i-1} = (I_r - Q)^{-2}$, we obtain:

$$\underset{\sim}{\eta} = 2 \sum_{n=0}^{\infty} n\underset{\sim}{\omega}(n) + \underset{\sim}{U} = 2Q \sum_{n=0}^{\infty} nQ^{n-1} \underset{\sim}{1}_r + \underset{\sim}{U}$$

$$= 2Q(I_r - Q)^{-2} \underset{\sim}{1}_r + \underset{\sim}{U} = (I_r + Q)(I_r - Q)^{-1} \underset{\sim}{U}. \quad (2.18)$$

If Q has real distinct eigenvalues $\lambda_1, \ldots, \lambda_u, u \leq r$, then

$$\underset{\sim}{U} = (I_r - Q)^{-1} \underset{\sim}{1}_r = \sum_{i=1}^{u} (1 - \lambda_i)^{-1} E_i \underset{\sim}{1}_r; \quad (2.19)$$

and

$$\underset{\sim}{\eta} = (I_r + Q)(I_r - Q)^{-1} \underset{\sim}{U} = \sum_{i=1}^{u} (1 + \lambda_i)(1 - \lambda_i)^{-2} E_i \underset{\sim}{1}_r . \tag{2.20}$$

2.7. Some Illustrative Examples

In this section we illustrate the applications of the above results by some examples from genetics.

Example 2.13. The full-sib mating model for one locus with two alleles in natural populations. In Example 2.4, we have considered a full-sib mating chain for one locus with two alleles in a large diploid population. In this example, the one-step transition matrix is:

$$P = \begin{pmatrix} 1 & 0 & \underset{\sim}{0}' \\ 0 & 1 & \underset{\sim}{0}' \\ \underset{\sim}{R}_1 & \underset{\sim}{R}_2 & Q \end{pmatrix},$$

where

$$\underset{\sim}{R}'_1 = \left(0, \frac{1}{4}, 0, \frac{1}{16}\right), \quad \underset{\sim}{R}_2 = \left(0, 0, \frac{1}{4}, \frac{1}{16}\right), \quad Q = \begin{pmatrix} 0 & 0 & 0 & 1 \\ 0 & \frac{1}{2} & 0 & \frac{1}{4} \\ 0 & 0 & \frac{1}{2} & \frac{1}{4} \\ \frac{1}{8} & \frac{1}{4} & \frac{1}{4} & \frac{1}{4} \end{pmatrix}.$$

Thus,

$$N = (I_4 - Q)^{-1} = \begin{pmatrix} 1 & 0 & 0 & -1 \\ 0 & \frac{1}{2} & 0 & -\frac{1}{4} \\ 0 & 0 & \frac{1}{2} & -\frac{1}{4} \\ -\frac{1}{8} & -\frac{1}{4} & -\frac{1}{4} & \frac{3}{4} \end{pmatrix}^{-1} = \frac{1}{24} \begin{pmatrix} 32 & 32 & 32 & 64 \\ 4 & 64 & 16 & 32 \\ 4 & 16 & 64 & 32 \\ 8 & 32 & 32 & 64 \end{pmatrix}.$$

Hence, the vector of the probabilities of ultimate absorption into $AA \times AA$ given the transient states is

$$\underset{\sim}{\rho}(1) = (I_4 - Q)^{-1} \underset{\sim}{R}_1 = N \underset{\sim}{R}_1 = \left(\frac{1}{2}, \frac{3}{4}, \frac{1}{4}, \frac{1}{2} \right)',$$

while the vector of the probabilities of ultimate absorption into $aa \times aa$ given the transient states is

$$\underset{\sim}{\rho}(2) = N \underset{\sim}{R}_2 = \left(\frac{1}{2}, \frac{1}{4}, \frac{3}{4}, \frac{1}{2} \right)'.$$

The vectors of mean absorption times and variances of first absorption times of the mating types $AA \times aa$, $AA \times Aa$, $aa \times Aa$, $Aa \times Aa$ are given respectively by:

$$\underset{\sim}{U} = (I_r - Q)^{-1} \underset{\sim}{1}_r = N \underset{\sim}{1}_4$$

$$= \left(\frac{160}{24} = 6.67, \ \frac{116}{24} = 4.5, \ \frac{116}{24} = 4.5, \ \frac{136}{24} = 5.67 \right)',$$

and

$$\underset{\sim}{V} = \{2(I_r - Q)^{-1} - I_r\} \underset{\sim}{U} - \underset{\sim}{U}_{st} = (2N - I_4) \underset{\sim}{U} - \underset{\sim}{U}_{st}$$

$$= \frac{1}{576}(13056, \ 12304, \ 12304, \ 13056)' = (22.67, \ 21.36, \ 21.36, \ 22.67)'.$$

Using the eigenvalues and spectrum expansion matrices of Q from Example 2.8, we obtain the probability of first absorption into the type $AA \times AA$ at time n given the mating types $\{AA \times aa, AA \times Aa, aa \times Aa, Aa \times Aa\}$ is, with $\underset{\sim}{R}'_1 = (0, \frac{1}{4}, 0, \frac{1}{16})$,

$$\underset{\sim}{g}_1(n) = \frac{1}{2^{n-1}} E_1 \underset{\sim}{R}_1 + \frac{1}{4^{n-1}} E_2 \underset{\sim}{R}_1 + \left(\frac{\epsilon_1}{4} \right)^{n-1} E_3 \underset{\sim}{R}_1 + \left(\frac{\epsilon_2}{4} \right)^{n-1} E_4 \underset{\sim}{R}_1$$

$$= \frac{1}{2^{n+2}} \begin{pmatrix} 0 \\ 1 \\ -1 \\ 0 \end{pmatrix} + \frac{1}{4^{n+2}} \begin{pmatrix} \dfrac{4^4}{5} \\ -12 \\ 3 \\ 3 \end{pmatrix}$$

$$+ \left(\frac{\epsilon_1}{4}\right)^{n-1} \times \frac{1}{40} \begin{pmatrix} 3 - \sqrt{5} \\ 1 + \frac{3}{4}\epsilon_1 \\ 1 + \frac{3}{4}\epsilon_1 \\ 1 + \epsilon_1 \end{pmatrix}$$

$$+ \left(\frac{\epsilon_2}{4}\right)^{n-1} \times \frac{1}{40} \begin{pmatrix} 3 + \sqrt{5} \\ 1 + \frac{3}{4}\epsilon_2 \\ 1 + \frac{3}{4}\epsilon_2 \\ 1 + \epsilon_2 \end{pmatrix}.$$

Similarly, the probability of first absorption into the type $aa \times aa$ at time n given the mating types $\{AA \times aa, AA \times Aa, aa \times Aa, Aa \times Aa\}$ is, with $\underset{\sim}{R}_{2'} = (0, 0, \frac{1}{4}, \frac{1}{16})$,

$$\underset{\sim}{g}_2(n) = \frac{1}{2^{n-1}} E_1 \underset{\sim}{R}_2 + \frac{1}{4^{n-1}} E_2 \underset{\sim}{R}_2 + \left(\frac{\epsilon_1}{4}\right)^{n-1} E_3 \underset{\sim}{R}_2 + \left(\frac{\epsilon_2}{4}\right)^{n-1} E_4 \underset{\sim}{R}_2$$

$$= \frac{1}{2^{n+2}} \begin{pmatrix} 0 \\ 1 \\ -1 \\ 0 \end{pmatrix} + \frac{1}{4^{n+2}} \begin{pmatrix} \frac{4^4}{5} \\ -12 \\ 3 \\ 3 \end{pmatrix} + \left(\frac{\epsilon_1}{4}\right)^{n-1} \times \frac{1}{40} \begin{pmatrix} 3 - \sqrt{5} \\ 1 + \frac{3}{4}\epsilon_1 \\ 1 + \frac{3}{4}\epsilon_1 \\ 1 + \epsilon_1 \end{pmatrix}$$

$$+ \left(\frac{\epsilon_2}{4}\right)^{n-1} \times \frac{1}{40} \begin{pmatrix} 3 + \sqrt{5} \\ 1 + \frac{3}{4}\epsilon_1 \\ 1 + \frac{3}{4}\epsilon_1 \\ 1 + \epsilon_1 \end{pmatrix}.$$

Example 2.14. The linkage model in self-fertilized populations. In Example 2.4, we have considered a two loci linkage model in large self-fertilized diploid populations. To illustrate the applications of Sec. 2.6, for simplicity we assume that there are no selection (i.e. $x_i = y_i = 1, i = 1, 2$); for general results under selection, we refer the readers to Tan [4]. In this case, the 1-step transition matrix is

$$
P = \begin{pmatrix}
1 & 0 & 0 & 0 & \underset{\sim}{0}' \\
0 & 1 & 0 & 0 & \underset{\sim}{0}' \\
0 & 0 & 1 & 0 & \underset{\sim}{0}' \\
0 & 0 & 0 & 1 & \underset{\sim}{0}' \\
\underset{\sim}{R}_1 & \underset{\sim}{R}_2 & \underset{\sim}{R}_3 & \underset{\sim}{R}_4 & Q
\end{pmatrix},
$$

where

$$
\underset{\sim}{R}_{1'} = \left(\frac{1}{4}, \frac{1}{4}, 0, 0, \frac{q^2}{4}, \frac{p^2}{4} \right), \quad
\underset{\sim}{R}_{2'} = \left(\frac{1}{4}, 0, \frac{1}{4}, 0, \frac{p^2}{4}, \frac{q^2}{4} \right)
$$

$$
\underset{\sim}{R}_{3'} = \left(0, \frac{1}{4}, 0, \frac{1}{4}, \frac{p^2}{4}, \frac{q^2}{4} \right), \quad
\underset{\sim}{R}_{4'} = \left(0, 0, \frac{1}{4}, \frac{1}{4}, \frac{q^2}{4}, \frac{p^2}{4} \right) ;
$$

and

$$
Q = \begin{pmatrix}
\frac{1}{2} & 0 & 0 & 0 & 0 & 0 \\
0 & \frac{1}{2} & 0 & 0 & 0 & 0 \\
0 & 0 & \frac{1}{2} & 0 & 0 & 0 \\
0 & 0 & 0 & \frac{1}{2} & 0 & 0 \\
\frac{2pq}{4} & \frac{2pq}{4} & \frac{2pq}{4} & \frac{2pq}{4} & \frac{q^2}{2} & \frac{p^2}{2} \\
\frac{2pq}{4} & \frac{2pq}{4} & \frac{2pq}{4} & \frac{2pq}{4} & \frac{p^2}{2} & \frac{q^2}{2}
\end{pmatrix}.
$$

The eigenvalues of Q are $\lambda_1 = \frac{1}{2}$ with multiplicity 4, $\lambda_2 = \frac{1}{2}(p^2 + q^2)$ and $\lambda_3 = \frac{1}{2}(q - p)$ and the spectral matrices of Q have been obtained and given in Example 2.9 with $x_i = y_i = 1, i = 1, 2$.

(i) The vectors of the probabilities of ultimate absorption into the four absorbing types $AB/AB, Ab/Ab, aB/aB$ and ab/ab given the types

$(AB/Ab, AB/aB, Ab/ab, aB/ab, Ab/aB)$ are given respectively by:

$$\underset{\sim}{\rho}(1) = \left[\frac{1}{2}, \frac{1}{2}, 0, 0, \frac{1}{4}\left(1 + \frac{1-2p}{1+2p}\right), \frac{1}{4}\left(1 - \frac{1-2p}{1+2p}\right)\right],$$

$$\underset{\sim}{\rho}(2) = \left[\frac{1}{2}, 0, \frac{1}{2}, 0, \frac{1}{4}\left(1 - \frac{1-2p}{1+2p}\right), \frac{1}{4}\left(1 + \frac{1-2p}{1+2p}\right)\right],$$

$$\underset{\sim}{\rho}(3) = \left[0, \frac{1}{2}, 0, \frac{1}{2}, \frac{1}{4}\left(1 - \frac{1-2p}{1+2p}\right), \frac{1}{4}\left(1 + \frac{1-2p}{1+2p}\right)\right],$$

and

$$\underset{\sim}{\rho}(4) = \left[0, 0, \frac{1}{2}, \frac{1}{2}, \frac{1}{4}\left(1 + \frac{1-2p}{1+2p}\right), \frac{1}{4}\left(1 - \frac{1-2p}{1+2p}\right)\right].$$

(ii) The vectors of first time absorption probabilities into the 4 absorbing types $AB/AB, Ab/Ab, aB/aB, ab/ab$ given the transient types are given respectively by:

$$\underset{\sim}{g}_1(n) = \left(\frac{1}{2^{n+1}}, \frac{1}{2^{n+1}}, 0, 0, \delta_{1n}, \delta_{2n}\right)',$$

$$\underset{\sim}{g}_2(n) = \left(\frac{1}{2^{n+1}}, 0, \frac{1}{2^{n+1}}, 0, \delta_{2n}, \delta_{1n}\right)',$$

$$\underset{\sim}{g}_3(n) = \left(0, 0, \frac{1}{2^{n+1}}, \frac{1}{2^{n+1}}, \delta_{1n}, \delta_{2n}\right)',$$

$$\underset{\sim}{g}_4(n) = \left(0, 0, \frac{1}{2^{n+1}}, \frac{1}{2^{n+1}}, \delta_{1n}, \delta_{2n}\right)',$$

where

$$\delta_{1n} = \frac{1}{2^{n+1}} - \frac{1}{4}\left(\frac{1+2pq}{2}\right)\left[\frac{1}{2}(p^2 + q^2)\right]^{n-1} + \frac{1}{4}\left[\frac{1}{2}(q-p)\right]^n$$

and

$$\delta_{2n} = \frac{1}{2^{n+1}} - \frac{1}{4}\left(\frac{1+2pq}{2}\right)\left[\frac{1}{2}(p^2 + q^2)\right]^{n-1} - \frac{1}{4}\left[\frac{1}{2}(q-p)\right]^n.$$

(iii) The vectors of mean absorption times and variances of first time absorptions of the transient types are given respectively as:

$$\underset{\sim}{U}' = \left(2,\, 2,\, 2,\, 2,\, \frac{2(1+4pq)}{1+2pq},\, \frac{2(1+4pq)}{1+2pq}\right),$$

$$\underset{\sim}{V}' = (2,\, 2,\, v,\, v),$$

where $v = 12 - \frac{1}{(1+2pq)^2}\{2(3-2pq) + 4(14pq)^2\}$.

Example 2.15. The Wright model under mutation in population genetics. In Example 1.11, we have considered the Wright model for a single locus with two alleles A and a in population genetics. In this model, we now assume that there are no selection and no immigration and migration, but A mutates to a with rate α_1 and a to A with rate α_2. Then the transition probability from i A genes at generation n to j A genes at generation $n+1$ is

$$P_{ij} = \binom{2N}{j} p_i^j (1-p_i)^{2N-j},$$

where

$$p_i = \frac{i}{2N}(1-\alpha_1) + \left(1 - \frac{i}{2N}\right)\alpha_2 = \alpha_2 + \frac{i}{2N}(1-\alpha_1-\alpha_2),\ 0 \le \alpha_1, \alpha_2 \le 1.$$

In this model, if $0 < \alpha_1,\ \alpha_2 < 1$, then $P_{ij} > 0$ for all $i, j = 0, 1, 2, \ldots, 2N$. Thus, if $0 < \alpha_1,\ \alpha_2 < 1$, the chain is irreducible. On the other hand, if $\alpha_1 \ne 0,\ \alpha_1 < 1$ but $\alpha_2 = 0$, then 0 is an absorbing state and is the only persistent state, while all other states are transient; if $\alpha_2 \ne 0,\ \alpha_2 < 1$, but $\alpha_1 = 0$, then $\{2N\}$ is an absorbing state and is the only persistent state while all other states are transient. If both $\alpha_1 = \alpha_2 = 0$, then $\{0, 2N\}$ are the absorbing states and are the only persistent states while all other states are transient. To derive the absorption probabilities, we first derive the eigenvalues of the one-step transition matrix $P = (P_{ij})$.

(2.15.1) The eigenvalues of $P = (P_{ij})$. We will show that the eigenvalues of P are

$$\lambda_1 = 1,\ \text{and},\ \lambda_{k+1} = \frac{1}{(2N)^k}(1-\alpha_1-\alpha_2)^k \left\{\prod_{i=1}^{k}(2N-i+1)\right\},$$

for $k = 1, 2, \ldots, 2N$.

To prove this results, we will prove the following lemma.

Lemma 2.2. *Let $X(t)$ be the number of A alleles at generation t. If*

$$E\{X^j(t+1)|X(t)=i\} = \sum_{k=0}^{j} a_{kj}i^k, \quad j = 0, 1, 2, \ldots, 2N,$$

for some constants a_{kj}, then a_{jj} $(j = 0, 1, \ldots, 2N)$ are the eigenvalues of P.

Remark 2.1. Notice that the above condition is equivalent to stating that the jth conditional moment of $X(t+1)$ around 0 given $X(t) = i$ is a polynomial in i of degree j. The result of the lemma states that the coefficient of i^j in the polynomial $E\{X^j(t+1)|X(t)=i\}$ is the $(j+1)$th eigenvalue $(j = 0, \ldots, 2N)$ of P.

Proof of Lemma 2.2. Denote by $\underset{\sim}{P}_i'$ the ith row of P, $i = 1, 2, \ldots, 2N+1$. For $j = \{1, \ldots, 2N+1\}$, define the following $(2N+1) \times 1$ columns, $\underset{\sim}{r}_j = \{0^{j-1}, 1^{j-1}, \ldots, (2N)^{j-1}\}'$, $\underset{\sim}{v}_j = \{(j-1)^0, (j-1)^1, \ldots, (j-1)^{2N}\}'$ and $\underset{\sim}{a}_j = (a_{0,j-1}, a_{1,j-1}, \ldots, a_{j-1,j-1}, 0, \ldots, 0)'$. If the above condition is satisfied, then, with $O^0 = 1$, we have, for all $i = 1, 2, \ldots, 2N+1$ and $j = 1, 2, \ldots, 2N+1$:

$$E\{X^{j-1}(t+1)|X(t)=i-1\} = \sum_{k=0}^{2N} k^{j-1}P_{i-1,k} = \underset{\sim}{P}_i'\underset{\sim}{r}_j$$

$$= \sum_{k=0}^{j-1}(i-1)^k a_{k,j-1} = \underset{\sim}{v}_{i'}\underset{\sim}{a}_j.$$

Let R, H, A be $(2N+1) \times (2N+1)$ matrices with the jth column being given by $\underset{\sim}{r}_j$, $\underset{\sim}{v}_j$, $\underset{\sim}{a}_j$, respectively, $j = 1, \ldots, 2N+1$. Then, $H' = R$ and in matrix notation, the above equation reduces to: $PR = RA$. Thus $P = RAR^{-1}$, so that $|P - \lambda I| = |RAR^{-1} - \lambda I| = |A - \lambda I| = 0$. Thus, the eigenvalues of A are the eigenvalues of P. Since A is upper triangular, so $(a_{jj}, j = 0, 1, 2, \ldots, 2N)$, are the eigenvalues of P. $\qquad\square$

To derive the eigenvalues of P, we note that:

$$E\{X^0(t+1)|X(t)=i\} = 1,$$

$$E\{X(t+1)|X(t)=i\} = 2Np_i = (1 - \alpha_1 - \alpha_2)i + 2N\alpha_2,$$

and

$$E\{X^2(t+1)|X(t)=i\} = \sum_{j=0}^{2N} j(j-1)P_{ij} + \sum_{j=0}^{2N} jP_{ij}$$

$$= \left\{ p^2 \frac{\partial^2}{\partial p^2} \left[\sum_{j=0}^{2N} \binom{2N}{j} p^j q^{2N-j} \right] \right\}_{(p=p_i,\, q=1-p_i)}$$

$$+ \left\{ p \frac{\partial}{\partial p} \left[\sum_{j=0}^{2N} \binom{2N}{j} p^j q^{2N-j} \right] \right\}_{(p=p_i,\, q=1-p_i)}$$

$$= (2N)(2N-1)p_i^2 + 2Np_i$$

$$= (2N)(2N-1)\left[\frac{i}{2N}(1-\alpha_1-\alpha_2) + 2N\alpha_2 \right]^2$$

$$+ 2N\left[\frac{i}{2N}(1-\alpha_1-\alpha_2) + 2N\alpha_2 \right]$$

which is a polynomial in i of degree 2.

By mathematical induction, one can easily show that $E\{X^j(t+1)|X(t)=i\}$ is a polynomial in i of degree j, $j=0,1,2,\ldots,2N$. Furthermore, the coefficient of i^k in $E\{X^k(t+1)|X(t)=i\}$ is obtained from the coefficient of i^k of the following polynomial (polynomial in i):

$$\left\{ p^k \frac{\partial^k}{\partial p^k} \left[\sum_{j=0}^{2N} \binom{2N}{j} p^j q^{2N-j} \right] \right\}_{(p=p_i,\, q=1-p_i)}$$

$$= (2N)(2N-1)\cdots(2N-k+1)p_i^k$$

$$= \left\{ \prod_{i=1}^{k}(2N-i+1) \right\} \left[\frac{i}{2N}(1-\alpha_1-\alpha_2) + 2N\alpha_2 \right]^k,$$

for $k=1,2,3,\ldots,2N$.

Hence, the eigenvalues of P (the coefficient of i^k) are

$$\lambda_1 = 1, \quad \lambda_{k+1} = \frac{1}{(2N)^k}(1-\alpha_1-\alpha_2)^k \left\{ \prod_{i=1}^{k}(2N-i+1) \right\},$$

for $k=1,2,\ldots,2N$.

(2.15.2) The absorption probabilities. If $\alpha_i \neq 0$ for both $i = 1, 2$, the chain is irreducible and all states are persistent. To derive absorption probabilities, we thus consider the following three cases.

Case 1: If $\alpha_1 \neq 0, \alpha_1 < 1$ but $\alpha_2 = 0$, then 0 is an absorbing state and is the only persistent state, while all other states are transient. Since the chain is finite if $N < \infty$, the ultimate absorption probability of any transient state i $(i > 0)$ is one by results of Subsec. 2.5.1. Thus the probability is 1 that the A allele is eventually lost from the population.

In this case,

$$P = \begin{pmatrix} 1 & \underset{\sim}{0}' \\ \underset{\sim}{R}_1 & Q_1 \end{pmatrix}$$

and the eigenvalues of P are $\lambda_1 = 1$ and for $k = 1, \ldots, 2N, \lambda_{k+1} = \frac{1}{(2N)^k}(1 - \alpha_1)^k \prod_{i=1}^{k}(2N - i + 1), 0 \leq \lambda_{k+1} < 1$. Further, the eigenvalues of Q_1 are $\gamma_i = \lambda_{i+1}, i = 1, \ldots, 2N$. Since all eigenvalues are distinct and real, Q_1 can be expanded as a spectral expansion.

To find the absorption probabilities and the moments of first absorption times, put $E_i = \prod_{j \neq i} \frac{1}{\gamma_i - \gamma_j}(Q_1 - \lambda_j I_{2N}), i = 1, \ldots, 2N$. Then $E_i^2 = E_i, E_i E_j = 0$, if $i \neq j$ and $\sum_{i=1}^{2N} E_i = I_{2N}$. Since $\underset{\sim}{R}_1 = (I_{2N} - Q_1)\underset{\sim}{1}_{2N}$, we have:

(1) The vector $\underset{\sim}{F}(n)$ of absorption probabilities at or before n is:

$$\underset{\sim}{F}(n) = (I_{2N} - Q_1^n)\underset{\sim}{1}_{2N} = \sum_{i=1}^{2N}(1 - \gamma_i^n)E_i\underset{\sim}{1}_{2N}.$$

(2) The vector $\underset{\sim}{g}(n)$ of first time absorption probabilities at n is:

$$\underset{\sim}{g}(n) = Q_1^{n-1}(I_{2N} - Q_1)\underset{\sim}{1}_{2N} = \sum_{i=1}^{2N}\gamma_i^{n-1}(1 - \gamma_i)E_i\underset{\sim}{1}_{2N}, \; n = 1, \ldots, \infty.$$

(3) The vector $\underset{\sim}{U}$ of mean absorption times is:

$$\underset{\sim}{U} = (I_{2N} - Q_1)^{-1}\underset{\sim}{1}_{2N} = \sum_{i=1}^{2N}(1 - \gamma_i)^{-1}E_i\underset{\sim}{1}_{2N}.$$

(4) The vector $\underset{\sim}{V}$ of the variances of first absorption times is $\underset{\sim}{V} = \underset{\sim}{\eta} - \underset{\sim}{U}_{sq}$, where $\underset{\sim}{U}_{sq} = \{\nu_1^2, \ldots, \nu_{2N}^2\}'$ and

$$\underset{\sim}{\eta} = (I_{2N} + Q_1)(I_{2N} - Q_1)^{-2} \underset{\sim}{1}_{2N}$$

$$= \sum_{i=1}^{2N}(1 + \gamma_i)(1 - \gamma_i)^{-2} E_i \underset{\sim}{1}_{2N} . \tag{2.21}$$

Case 2: If $\alpha_2 \neq 0, \alpha_2 < 1$, but $\alpha_1 = 0$, then $\{2N\}$ is an absorbing state and is the only persistent states while all other states are transient. In this case, the eigenvalues of P are $\lambda_1 = 1$ and

$$\lambda_{k+1} = \frac{1}{(2N)^k}(1 - \alpha_2)^k \prod_{i=1}^{k}(2N - i + 1), k = 1, 2, \ldots, 2N ,$$

and $0 \leq \lambda_{k+1} < 1, k = 1, 2, \ldots, 2N$.

The one-step transition matrix is

$$P = \begin{pmatrix} Q_2 & \underset{\sim}{R}_2 \\ \underset{\sim}{0}' & 1 \end{pmatrix} ,$$

and the eignevalues of Q_2 are $\chi_i = \lambda_{i+1}, i = 1, \ldots, 2N$. In this case, the ultimate absorption probability of transient states into the absorbing state $2N$ is one. That is, the probability is 1 that the A allele is eventually fixed ("a" allele is lost in the population).

Put $E_i = \prod_{j \neq i} \frac{1}{\chi_i - \chi_j}(Q_2 - \chi_j I_{2N})$, $i = 1, \ldots, 2N$. Then $E_i^2 = E_i, E_i E_j = 0$, if $i \neq j$ and $\sum_{i=1}^{2N} E_i = I_{2N}$. Since $\underset{\sim}{R}_2 = (I_{2N} - Q_2)\underset{\sim}{1}_{2N}$, we have:

(1) The vector $\underset{\sim}{F}(n)$ of absorption probabilities at or before n is:

$$\underset{\sim}{F}(n) = (I_{2N} - Q_2^n)\underset{\sim}{1}_{2N} = \sum_{i=1}^{2N}(1 - \chi_i^n)E_i \underset{\sim}{1}_{2N} .$$

(2) The vector $\underset{\sim}{g}(n)$ of first time absorption probabilities at n is:

$$\underset{\sim}{h}(n) = Q_2^{n-1}(I_{2N} - Q_2)\underset{\sim}{1}_{2n} = \sum_{i=1}^{2N}\chi_i^{n-1}(1 - \chi_i)E_i \underset{\sim}{1}_{2N}, n = 1, \ldots, \infty .$$

(3) The vector $\underset{\sim}{U}$ of mean absorption times is:

$$\underset{\sim}{U} = (I_{2N} - Q_2)^{-1}\underset{\sim}{1}_{2N} = \sum_{i=1}^{2N}(1 - \chi_i)^{-1}E_i\underset{\sim}{1}_{2N}.$$

(4) The vector $\underset{\sim}{V}$ of the variances of first absorption times is $\underset{\sim}{V} = \underset{\sim}{\eta} - \underset{\sim}{U}_{sq}$, where $\underset{\sim}{U}_{sq} = \{\nu_1^2, \ldots, \nu_{2N}^2\}'$ and

$$\underset{\sim}{\eta} = (I_{2N} + Q_2)(I_{2N} - Q_2)^{-2}\underset{\sim}{1}_{2N}$$

$$= \sum_{i=1}^{2N}(1 + \chi_i)(1 - \chi_i)^{-2}E_i\underset{\sim}{1}_{2N}.$$

Case 3: If $\alpha_1 = \alpha_2 = 0$, then $\{0, 2N\}$ are absorbing states while the states $\{i = 1, \ldots, 2N-1\}$ are transient states. Hence the ultimate absorption probabilities of transient states are one. That is, gene A will either be lost or fixed in the population eventually as time progresses. This case corresponds to the case of "genetic drift" or "Wright drift" in population genetics.

In this case the one-step transition matrix is

$$P = \begin{pmatrix} 1 & 0 & 0 \\ 0 & 1 & 0 \\ \underset{\sim}{a}_1 & \underset{\sim}{a}_2 & Q \end{pmatrix}.$$

The matrix P have $2N - 1$ distinct eigenvalues $\lambda_1 = \lambda_2 = 1$, and

$$\lambda_{k+1} = \frac{1}{(2N)^k}\prod_{i=1}^{k}(2N - i + 1), \quad k = 2, \ldots, 2N.$$

The eigenvalue λ_1 has multiplicity 2 corresponding to the two absorbing states $\{0, 2N\}$ while all other eigenvalues have multiplicity 1 and have values between $0 < \lambda_i < 1$ $(i = 2, \ldots, 2N)$. It is easily observed that the eigenvalues of the $(2N - 1) \times (2N - 1)$ one-step transition matrix of transient states Q are $\sigma_i = \lambda_{i+2}$, $i = 1, \ldots, 2N - 1$.

Put $G_i = \prod_{j \neq i} \frac{1}{\sigma_i - \sigma_j}(Q - \sigma_j I_{2N})$, $i = 1, \ldots, 2N-1$. Then $G_i^2 = G_i$, $G_iG_j = 0$, if $i \neq j$ and $\sum_{i=1}^{2N-1} G_i = I_{2N-1}$. Let $i = 1$ correspond to the state 0 and

$i = 2$ correspond to the state $2N$. Since $\sum_{i=1}^{2} \underset{\sim}{a}_i = (I_{2N-1} - Q)\underset{\sim}{1}_{2N-1}$, we have:

(1) The vector $\underset{\sim}{F}_i(n)$ of absorption probabilities into the ith absorbing state at or before n is:

$$\underset{\sim}{F}_i(n) = (I_{2N-1} - Q^n)(I_{2N-1} - Q)^{-1}\underset{\sim}{a}_i$$

$$= \sum_{i=1}^{2N-1} (1 - \sigma_i^n)(1 - \sigma_i)^{-1}G_i\underset{\sim}{a}_i, \quad i = 1, 2.$$

The vector $\underset{\sim}{F}(n)$ of absorption probabilities into the absorbing states at or before n is:

$$\underset{\sim}{F}(n) = (I_{2N} - Q^n)\underset{\sim}{1}_{2N-1} = \sum_{i=1}^{2N-1} (1 - \sigma_i^n)G_i\underset{\sim}{1}_{2N-1}.$$

(2) The vector $\underset{\sim}{g}_i(n)$ of first time absorption probabilities into the ith absorbing state at n is:

$$\underset{\sim}{g}_i(n) = Q^{n-1}\underset{\sim}{a}_i = \sum_{j=1}^{2N-1} \sigma_j^{n-1}G_j\underset{\sim}{a}_i, \quad n = 1, \ldots, \infty.$$

The vector $\underset{\sim}{g}(n)$ of first time absorption probabilities at n is:

$$\underset{\sim}{g}(n) = Q^{n-1}(I_{2N-1} - Q)\underset{\sim}{1}_{2N-1} = \sum_{i=1}^{2N-1} \sigma_i^{n-1}(1 - \sigma_i)G_i\underset{\sim}{1}_{2N-1}, \quad n = 1, \ldots, \infty.$$

(3) The vector $\underset{\sim}{U}$ of mean absorption times is:

$$\underset{\sim}{U} = (I_{2N-1} - Q)^{-1}\underset{\sim}{1}_{2N-1} = \sum_{i=1}^{2N-1} (1 - \sigma_i)^{-1}G_i\underset{\sim}{1}_{2N-1}.$$

(4) The vector $\underset{\sim}{V}$ of the variances of first absorption times is $\underset{\sim}{V} = \underset{\sim}{\eta} - \underset{\sim}{U}_{sq}$, where $\underset{\sim}{U}_{sq} = \{\nu_1^2, \ldots, \nu_{2N-1}^2\}'$ and

$$\underset{\sim}{\eta} = (I_{2N-1} + Q)(I_{2N-1} - Q)^{-2}\underset{\sim}{1}_{2N-1}$$

$$= \sum_{i=1}^{2N-1} (1 + \sigma_i)(1 - \sigma_i)^{-2}G_i\underset{\sim}{1}_{2N-1}.$$

Example 2.16. The absorption probabilities of simple branching process. Let $\{X(t), t \in T = (0, 1, \ldots, \infty)\}$ be a Galton-Watson branching process with progeny distribution $\{p_j, j = 0, 1, \ldots, \infty\}$ as described in Example 2.10. (To avoid trivial cases, we assume $0 < p_0, p_1 < 1$ and $0 < p_0 + p_1 < 1$, unless otherwise stated.) In this section we derive some basic results for the absorption probabilities and mean absorption times of transient states if the ultimate absorption probability of transient states is one.

(2.16.1) Absorption probabilities. Let q_j be the ultimate absorption probability of the state j $(j > 0)$. Then, obviously, $q_j = q^j$, where $q = q_1$ is the ultimate extinction probability of $X(t)$ when $X(0) = 1$. When the branching process is the process describing the behavior of a mutant gene arising at time 0, q is the probability of ultimate extinction of the mutant. This problem is the well-known problem of the survival of a mutant in population genetics [14]. Although it is not necessary, for ease of illustration, in what follows we will often refer $X(t)$ as the number of mutants at generation t, unless otherwise stated.

Theorem 2.9. The survival probabilities of mutants. *Let* $x_n = \Pr\{X(n) = 0 | X(0) = 1\} = f_n(0)$, *where* $f_n(s)$ *is the pgf of* $X(n)$. *Then, we have*:

(1) *The limit* $\lim_{n\to\infty} x_n$ *exists and* $\lim_{n\to\infty} x_n = q$.

(2) q *satisfies the functional equation* $x = f(x)$, *where* $f(s) = f_1(s)$ *is the pgf of the progeny distribution.*

(3) q *is the smallest non-negative root of* $x = f(x)$.

Proof. To prove (1), notice first that x_n is the probability that the mutant is lost at or before generation n so that $0 \leq x_n \leq 1$. Hence, to show that $q = \lim_{n\to\infty} x_n$ exists, it suffices to show that q_n is a monotonic increasing function of n. We prove this by mathematical induction by first noting that $\frac{d}{ds} f(s) = f'(s) > 0$ for all $s > 0$ so that $f(s)$ is a monotonic increasing function of s if $s > 0$.

Now, $1 > f(0) = f_1(0) = x_1 = p_0 > 0$, so $f(p_0) = f[f(0)] = f_2(0) = x_2 \geq f(0) = f_1(0) = p_0 = x_1 > 0$; hence, $x_3 = f_3(0) = f[f_2(0)] \geq f[f(0)] = f_2(0) = x_2 \geq f(0) = x_1$. Suppose now $x_n = f_n(0) \geq x_{n-1} = f_{n-1}(0)$; then $x_{n+1} = f_{n+1}(0) = f[f_n(0)] \geq f[f_{n-1}(0)] = f_n(0) = x_n$. This shows that $x_n = f_n(0)$ is a monotonic increasing function of n. It follows that $\lim_{n\to\infty} x_n = q$ exists.

To prove that q satisfies the equation $x = f(x)$, notice that $x_{n+1} = f_{n+1}(0) = f[f_n(0)] = f(x_n) = \sum_{j=0}^{\infty} x_n^j \Pr\{X(1) = j \mid X(0) = 1\}$. Since $0 \le x_n \le 1$ and $\sum_{j=0}^{\infty} \Pr\{X(1) = j \mid X(0) = 1\} = 1$, so, by the Lebesque dominated convergence theorem (cf. Lemma 3.3),

$$q = \lim_{n \to \infty} x_{n+1} = \lim_{n \to \infty} \sum_{j=0}^{\infty} x_n^j \Pr\{X(1) = j \mid X(0) = 1\}$$

$$= \sum_{j=0}^{\infty} \{ \lim_{n \to \infty} x_n^j \} \Pr\{X(1) = j \mid X(0) = 1\}$$

$$= \sum_{j=0}^{\infty} q^j \Pr\{X(1) = j \mid X(0) = 1\} = f(q) \,.$$

This proves (2).

To prove (3), notice that $1 = f(1)$. Hence 1 is a solution of $x = f(x)$. Suppose now λ is another solution of $x = f(x)$. We proceed to show that $\lim_{n \to \infty} f_n(0) = q \le \lambda$ so that $\lim_{n \to \infty} x_n = q$ is the smallest non-negative root of $x = f(x)$. First, $x = 0$ is not a solution of $x = f(x)$ as $0 \ne f(0) = p_0$. Thus, $\lambda > 0$. Hence,

$$\lambda = f(\lambda) \ge f(0) = x_1 = p_0 \,;$$

$$\Rightarrow \lambda = f(\lambda) \ge f[f(0)] = f_2(0) = x_2 \,,$$

$$\Rightarrow \lambda = f(\lambda) \ge f[f_2(0)] = f_3(0) = x_3 \,.$$

Suppose now $\lambda = f(\lambda) \ge x_n = f_n(0)$, then $\lambda = f(\lambda) \ge f[f_n(0)] = f_{n+1}(0) = x_{n+1}$. Thus, $\lambda \ge x_n$ for all $n = 1, 2, \ldots$ so that $\lambda \ge \lim_{n \to \infty} x_n = q$. This proves (3). $\qquad\square$

Theorem 2.10. The fundamental theorem of simple branching process. *Let μ denote the mean number of the progeny distribution. Then $x = f(x)$ has an unique root η satisfying $0 < \eta < 1$ iff $f'(1) = \mu > 1$; or 1 is the only non-negative root of $x = f(x)$ iff $f'(1) = \mu \le 1$. In the case of the mutant gene, this is equivalent to stating that the probability is 1 that the mutant gene will eventually be lost (i.e. $r = 1$) iff the mean of the progeny distribution is less than or equal to 1.*

Proof. To prove Theorem 2.10, notice that under the assumptions $0 < p_0, p_1 < 1$ and $0 < p_0 + p_1 < 1$, $f''(s) = \frac{d^2}{ds^2} f(s) > 0$ for all $s > 0$. Thus, $f(s)$ is a convex function of s for $s > 0$. It follows that $s = f(s)$ intersects $y = f(s)$ in at most two points. Since $1 = f(1)$, there is at most one root η satisfying $0 < \eta < 1$. Furthermore, since $f''(s) > 0$ for $s > 0$, $f'(s)$ is a continuous and monotonic increasing function of s when $s > 0$.

(i) If there is a η such that $0 < \eta < 1$ and $\eta = f(\eta)$, we will show that $1 < \mu = f'(1)$. Now $\eta = f(\eta)$ leads to $1 - \eta = 1 - f(\eta)$ so that $\frac{1-f(\eta)}{(1-\eta)} = 1$. But, by the mean value theorem in Calculus, $1 - f(\eta) = f'(r)(1 - \eta)$ for some $0 < \eta \leq r \leq 1$ as $f(s)$ is convex, implying that $f'(r) = 1, 0 < r \leq 1$ and $r < 1$. Since $f'(s)$ is strictly monotonic increasing for $s > 0$, it follows that $f'(1) = \mu > f'(r) = 1$.

(ii) Conversely, suppose that $0 \leq f'(1) = \mu \leq 1$. Then, for any $s, 0 < s < 1$, we have $1 - f(s) = f'(r)(1 - s)$ for some $s \leq r < 1$. Since $f'(s)$ is strictly increasing for $s > 0$, so $f'(r) < f'(1) \leq 1$. It follows that $1 - f(s) < 1 - s$ or $s < f(s)$ for all $0 < s < 1$. Thus, 1 is the only non-negative root of $x = f(x)$. $\qquad\qquad\square$

As an example, consider a single locus with two alleles $A : a$ in a random mating diploid population with N individuals (The "a" gene is referred to as the mutant gene). Suppose that at the 0th generation, there is a mutant "a" entering into this population so that there is one individual with genotype Aa while all other individuals have genotype AA. Let the relative fitness of individuals with genotypes AA and Aa be given by 1 and $1 + s$ respectively. Then the frequency of "a" mutant is

$$P = \frac{\frac{1}{2N}(1+s)}{1 + \frac{s}{N}} = \frac{1}{2N}(1+s) + o\left(\frac{1}{N}\right),$$

and, in the next generation, the probability that there are j "a" genes is:

$$P_j = \binom{2N}{j}\left(\frac{1+s}{2N}\right)^j \left(1 - \frac{1+s}{2N}\right)^{2N-j}.$$

When N is large, $P_j \sim e^{-(1+s)}\frac{(1+s)^j}{j!}, j = 0, 1, 2 \cdots$.

Let $X(n)$ be the number of a allele at generation n. Under the assumption that there are no mutations from $A \to a$ or vice versa in future generations, $X(n)$ is a Galton-Watson branching process with progeny distribution $\{P_j, j =$

$0,\ldots,\}$. When N is sufficiently large, $1+s$ is the average number of progenies. Thus, in a sufficiently large population, the probability is 1 that the mutant "a" will eventually die out iff $s \le 0$, or iff there is disadvantageous selection for Aa individual. When $s > 0$, the survival probability is $\pi = 1 - \theta$, where θ is the smallest positive root of $\theta = e^{(1+s)(\theta-1)}$, as $e^{(1+s)(\theta-1)}$ is the probability generating function of

$$P_j = e^{-(1+s)}\frac{(1+s)^j}{j!},\ j = 0,1,2\cdots .$$

To find θ, notice that with $\pi = 1 - \theta$, $\theta = e^{(1+s)(\theta-1)}$, we have

$$1 - \pi = e^{-\pi(1+s)} = 1 - \pi(1+s) + \frac{1}{2}\pi^2(1+s)^2 + \frac{1}{6}\pi^3(1+s)^3 + \cdots .$$

If π and s are small, we may omit terms involving $\{\pi^r s^{n-r}, n \ge 3\}$, then $1 - \pi \cong 1 - \pi(1+s) + \frac{1}{2}\pi^2$ so that $\pi \cong 2s$. Better approximations may be obtained by expanding $f(\theta) = e^{(1+s)(\theta-1)}$ in Taylor series around $\theta = 1$ to give

$$\theta = f(\theta) = 1 + (\theta-1)f'(1) + \frac{1}{2}(\theta-1)^2 f'(1) + \frac{1}{6}(\theta-1)^3 f'''(1) + \cdots .$$

Thus, with $\pi = 1 - \theta$, $f'(1) = 1 + s$, $f''(1) = \sigma^2 - f'(1) + [f'(1)]^2 = \sigma^2 - (1+s) + (1+s)^2 = \sigma^2 + (1+s)s$; omitting $\{(\theta-1)^r, r \ge 3\}$, we have $-\pi \cong -\pi(1+s) + \frac{1}{2}\pi^2[\sigma^2 + s(1+s)]$. It follows that, $\pi \cong \frac{2s}{\sigma^2+s(1+s)} = \frac{2s}{\sigma^2(1+\frac{s(1+s)}{\sigma^2})} \cong \frac{2s}{\sigma^2}$ if we omit terms involving $\{s^r, r \ge 3\}$, where σ^2 is the variance of P_j and $\sigma^2 = (1+s)$.

(2.16.2) First absorption probabilities $g(n)$ and mean absorption time μ. In the Galton-Watson process, the set T of transient states is given by $T = \{1,2,3,\ldots\}$. Denote $w(n) = \Pr\{X(n) \in T | X(0) = 1\}$. Then $w(n) = 1 - x_n$, $w(n) \le w(n-1)$, and $g(n) = \Pr\{N_0 = n | X(0) = 1\} = x_n - x_{n-1} = w(n-1) - w(n)$, which is the probability that the chain enters the absorbing state 0 for the first time at n. This follows from the results that if $X(n) \in T$, then $X(n-1) \in T$ whereas given $X(n-1) \in T$, $X(n)$ may or may not be in T. Notice that $\sum_{n=1}^{\infty} g(n)$ is the ultimate absorption probability that the chain will eventually be absorbed into the state 0. By Theorem 2.11, this probability is 1 iff $\mu = f'(1) \le 1$. That is, the probability is 1 that the mutants will eventually die out iff the mean number of the progeny distribution is less than or equal to 1.

Assuming that the mean number μ of the progeny distribution is $\mu \leq 1$, then $g(n)$ forms a discrete probability density function and the mean absorption time μ (i.e. the expected number of generations for extinction given $X(0) = 1$) is:

$$\mu = \sum_{n=0}^{\infty} ng(n) = \sum_{n=0}^{\infty} n\{\omega(n-1) - \omega(n)\}$$

$$= \sum_{n=0}^{\infty} (n-1)\omega(n-1) + \sum_{n=0}^{\infty} \omega_1(n-1) - \sum_{n=0}^{\infty} n\omega(n)$$

$$= \sum_{n=0}^{\infty} \omega(n) \text{ as } \omega(-1) = 0 \,.$$

To derive $x_n = 1 - \omega(n)$, notice that $x_{n+1} = f_{n+1}(0) = f[f_n(0)] = f(x_n)$. When the progeny distribution is Poisson as above, then $x_{n+1} = e^{-(1+s)(1-x_n)}$. When $s = 0$, $\lim_{n\to\infty} x_n = q = 1$ and $x_{n+1} = e^{x_n-1}$. Putting $y_n = (1 - x_n)^{-1}$ (or $x_n = 1 - y_n^{-1}$, $y_n > 1$ as $1 - x_n < 1$), $y_{n+1} = (1 - e^{-y_n^{-1}})^{-1} = \{1 - (1 - y_n^{-1} + \frac{1}{2}y_n^{-2} - \frac{1}{6}y_n^{-3} + \theta_1 y_n^{-4})\}^{-1}$, where θ_1, depending on n, is less than some constant if y_n^{-1} is bounded. Or, $y_{n+1} = y_n(1 + \frac{1}{2}y_n^{-1} + \frac{1}{12}y_n^{-2} + \theta_2 y_n^{-3})$, where θ_2 is less than a constant. When n is large, the above indicates that y_n is dominated by $\frac{n}{2}$, $y_n \cong \frac{n}{2}$. Hence, $\sum_{m=1}^{n} \frac{1}{y_m} \cong 2\sum_{m=1}^{n} \frac{1}{m} \cong 2\log n$, when n is large. Thus, $y_n = \frac{n}{2} + \frac{1}{6}\log n + \theta_3$, where θ_3 is bounded. Hence, $x_n = 1 - \frac{1}{\frac{n}{2} + \frac{1}{6}\log n + \theta_3} = 1 - \frac{6}{3n + \log n + \theta_4}$, where θ_4 is bounded. Or, $\omega(n) = 1 - x_n = \frac{6}{3n + \log n + \theta_4}$ so that $\sum_{n=0}^{\infty} \omega(n) = \infty$. This shows that, although the probability is 1 that the mutant is eventually lost, the expected number of generations for extinction is ∞.

2.8. Finite Markov Chains

Consider a finite homogeneous Markov chains, $\{X(t), t \in T = (0, 1, 2, \ldots)\}$. Then, by Theorem 2.6, the chain must contain persistent states. Further, all persistent states are positive. If the chain contain transient states, then, the probability is 1 that the transient states will eventually be absorbed into persistent states. (In general, finite Markov chains may or may not contain transient states.)

2.8.1. *The canonical form of transition matrix*

Suppose that the chain contain r transient states and k closed sets of persistent states. Let P be the matrix of the 1-step transition probabilities of this chain and P_j the matrix of the 1-step transition probabilities of states in C_j. Then the following proposition gives the canonical form for P.

Proposition 2.1. *Let the first r states be transient states. Then, P can be expressed in the following canonical form:*

$$P = \begin{pmatrix} Q & R_1 & R_2 & \cdots & R_k \\ & P_1 & & & 0 \\ & & P_2 & & \\ & & & \ddots & \\ & 0 & & & P_k \end{pmatrix},$$

where $(R_1, R_2, \ldots, R_k) \neq 0$.

To prove the above proposition, first notice that the chain must contain persistent states. Thus, let j_1 be a given persistent state, that is, $f_{j_1 j_1} = 1$ so that $j_1 \leftrightarrow j_1$. Denote by C_1 the set of all states that j_1 leads to. Since $j_1 \in C_1$, so C_1 is not empty. By Theorem 2.7, for every $k \in C_1, k \leftrightarrow j_1$; hence, by using the Chapman-Kolmogorov equation, if $k \in C_1$ and if $\ell \in C_1$, then $k \leftrightarrow j_1, \ell \leftrightarrow j_1$ so that $k \leftrightarrow \ell$. This implies also that C_1 is closed. This follows since if $s \notin C_1$, then $k \nrightarrow s$ for any $k \in C_1$; for otherwise, $j_1 \leftrightarrow k, k \rightarrow s \Rightarrow j_1 \rightarrow s$ so that $s \in C_1$, a contradiction. Hence, C_1 is a non-empty closed irreducible set of persistent states, and for any $k \in C_1$ and $\ell \in C_1, k \leftrightarrow \ell$. If C_1 exhausts all persistent states, than all other states (if any) are transient states. Hence, with C_T denote the set of transient states, we have:

$$P = \begin{array}{c} \\ C_T \\ C_1 \end{array} \begin{array}{c} C_T \quad C_1 \\ \begin{pmatrix} Q & R \\ 0 & P_1 \end{pmatrix} \end{array}.$$

Since $C_1 \cup C_T = S$, the state space and $C_1 \cap C_T = \emptyset$; also $R \neq 0$ since, for otherwise, the chain can stay in transient states an infinite number of times. We notice that C_T may be empty, in which case, the chain is irreducible with $S = C_1$. Now, if C_1 does not exhaust all persistent states, then there exists an $j_2 \notin C_1$ and j_2 is a persistent state. Then, if we define C_2 as the set of all states

j_2 leading to, then C_2 is also a non-empty irreducible closed set of persistent states such that, $\ell \in C_2$ and $n \in C_2 \Rightarrow \ell \leftrightarrow n$. Moreover, $C_1 \cap C_2 = \emptyset$, for otherwise, $h \in C_1, h \in C_2$ so that $h \leftrightarrow j_1, h \leftrightarrow j_2 \Rightarrow j_1 \leftrightarrow j_2$ so that $j_2 \in C_1$, a contradiction. Continuing this process, since the chain is finite, we have, after a finite number of times, put the set of all persistent states into the disjoint union $C_1 \cup C_2 \cup \ldots \cup C_k$ of non-empty irreducible closed sets of persistent states. For any $k \in C_r$ and $\ell \in C_r, k \leftrightarrow \ell$, for all $r = 1, 2, \ldots, k$; and for $k \in C_r, \ell \in C_s, r \neq s, k \nrightarrow \ell$. Let C_T be the set of all transient state ($C_T = \emptyset$ if no transient states). Then $C_1 \cup C_2 \cup \cdots \cup C_k \cup C_T = S$, the state space, $C_T \cap C_r = \emptyset, r = 1, 2, \ldots, k, C_r \cap C_s = \emptyset, r \neq s$. Hence,

$$P = \begin{array}{c} \\ C_T \\ C_1 \\ C_2 \\ \\ C_k \end{array} \begin{array}{c} \begin{array}{ccccc} C_T & C_1 & & \cdots & C_k \end{array} \\ \left(\begin{array}{ccccc} Q & R_1 & R_2 & \cdots & R_k \\ 0 & P_1 & 0 & \cdots & 0 \\ 0 & 0 & P_2 & \cdots & 0 \\ \vdots & \ddots & & \vdots & \vdots \\ 0 & & & 0 & P_k \end{array} \right) \end{array},$$

where P_i is the transition probability matrix for states in C_i, R_i the transition probability matrix from states in C_T to states in C_i, and Q the transition matrix of transient states into transient states. We notice that, $R = (R_1, R_2, \ldots, R_k) \neq 0$, for otherwise C_T is closed so that the chain can stay in transient states an infinite number of times, violating the 0–1 law.

Using the above canonical form, it is easy to see that the matrix of the n-step transition probabilities is given by:

$$P(n) = P^n = \left(\begin{array}{ccccc} Q^n & H_{1n} & H_{2n} & \cdots & H_{kn} \\ 0 & P_1^n & 0 & \cdots & 0 \\ 0 & 0 & P_2^n & \cdots & 0 \\ \vdots & \ddots & & \vdots & \vdots \\ 0 & & \cdots & 0 & P_k^n \end{array} \right),$$

where

$$\left\{ \begin{array}{l} H_{i1} = R_i, \; i = 1, 2, \ldots, k \\[2mm] H_{in} = \displaystyle\sum_{s=0}^{n-1} Q^s R_i P_i^{n-1-s}, \; i = 1, 2, \ldots, k. \end{array} \right\}.$$

Since, by Theorem 2.4, if j is transient, then $\sum_n p_{ij}(n) < \infty$, so that $\lim_{n \to \infty} p_{ij}(n) = 0$. It follows that

$$\lim_{n \to \infty} Q^n = \lim_{n \to \infty} (p_{ij}(n)) = (\lim_{n \to \infty} p_{ij}(n)) = 0.$$

This implies the following two results which are basic for the analysis of homogeneous Markov chains.

(i) The eigenvalues of Q must have absolute value < 1. Furthermore, for every $i \in C_T$ and for every $j \in C_T$, $|p_{ij}(n)| \le cr^n$, where $0 \le r < 1$, and c a constant.

When Q has real eigenvalues $\{\lambda_i, i = 1, \ldots, r\}$, and can be expanded in spectral expansion, these results can easily be demonstrated. (The results hold regardless whether or not Q can be expanded in spectral expansion.)

For then

$$Q^n = \sum_{i=1}^{r} \lambda_i^n E_i,$$

where

$$|\lambda_1| \ge |\lambda_2| \ge \cdots \ge |\lambda_r|,$$

$\lambda_1, \lambda_2, \ldots, \lambda_r$ being the distinct eigenvalues of Q. $Q^n \to 0 \Rightarrow \lambda_i^n \to 0$ so that $|\lambda_i| < 1$. Further, if we let $e_i(u, v)$ be the (u, v)th element of E_i, then

$$p_{u,v}(n) = \sum_{i=1}^{r} \lambda_i^n e_i(u, v)$$

and

$$|p_{u,v}(n)| \le |\lambda_1|^n m_{u,v}, \text{ where } m_{u,v} = r \max_{1 \le i \le r} |e_i(u, v)|.$$

Putting $m = \max_{u \in C_T, v \in C_T} m_{u,v}$, then

$$|p_{ij}(n)| \le |\lambda_1|^n m, 0 \le |\lambda_1| < 1.$$

(ii) $(I - Q)^{-1}$ exists and $(I - Q)^{-1} = I + Q + Q^2 + \cdots$. This follows from the results: $Q^n \to 0$ as $n \to \infty$ and $(I - Q)(I + Q + Q^2 + \cdots + Q^n) = I - Q^{n+1}$. If Q has real eigenvalues $\{\lambda_i, i = 1, \ldots, r\}$ and can be expanded in terms of spectral expansion, then $Q = \sum_{i=1}^{r} \lambda_i E_i$, where $E_i = \prod_{j \ne i} \frac{1}{\lambda_i - \lambda_j}(Q - \lambda_j I)$. Notice that $E_i^2 = E_i, E_i E_j = 0$ for $i \ne j$ and $\sum_{i=1}^{r} E_i = I$. It follows that $(I - Q)^{-1} = \sum_{i=1}^{r} (\frac{1}{1 - \lambda_i}) E_i$.

2.8.2. *Absorption probabilities of transient states in finite Markov chains*

For homogeneous finite Markov chains, the absorption probabilities and the moments of first absorption times have been given in Sec. 2.5. However, these results are easily derived alternatively by using the above canonical form and by noting that the elements of rows of P^n sum up to 1. The latter condition is equivalent to that for any integer $m \geq 1$:

$$P_i^m \underset{\sim}{1}_{n_j} = \underset{\sim}{1}_{n_j}; \sum_{j=1}^{k} R_j \underset{\sim}{1}_{n_j} = (I_r - Q)\underset{\sim}{1}_r.$$

Because the methods are straightforward, we leave it as an exercise (Exercise 2.9).

2.9. Stochastic Difference Equation for Markov Chains With Discrete Time

When the chain is finite, in many cases one may use stochastic difference equations to represent and characterize the chain. When some data are available from the system, one may then derive a state space model for the system. As illustrated in Chap. 8, this will provide an avenue for validating the model and for making inferences about unknown parameters in Markov chains with discrete time. Notice also that in multi-dimensional Markov chains with discrete time, the traditional approaches often are very difficult and complicated, if not impossible. In these cases, the stochastic difference equation method appears to be an attractive alternative approach for solving many problems which prove to be very difficult from other approaches. In this section we illustrate how to develop stochastic difference equations for these Markov chains.

2.9.1. *Stochastic difference equations for finite Markov chains*

Consider a finite Markov chain $\{X(t), t \in T = (0, 1, \ldots, \infty)\}$ with state space $S = \{1, \ldots, k+1\}$. Denote the one step transition probabilities by $\{p_{ij}(t) = \Pr\{X(t+1) = j | X(t) = i\}\ i, j = 1, \ldots, k+1\}$. Let $X_i(t)$ $(i = 1, \ldots, k+1)$ be the number of individuals who are in state i at time t and $Z_{ij}(t)$ the number of individuals who are in state j at time $t + 1$ arising from individuals who are in state i at time t. Then, given $X_i(t)$, the probability distribution of $\{Z_{ij}(t), j = 1, \ldots, k\}$ follows a k-dimensional multinomial distribution with parameters $\{X_i(t), p_{ij}(t), j = 1, \ldots, k\}$. That is,

$$\{Z_{ij}(t+1), j = 1, \ldots, k\}|X_i(t) \sim ML\{X_i(t),\, p_{ij}(t),\, j = 1, \ldots, k\}. \quad (2.22)$$

Furthermore, conditional on $\{X_i(t), i = 1, \ldots, k+1\}$, $\{Z_{ij}(t), j = 1, \ldots, k\}$ is independently distributed of $\{Z_{rj}(t), j = 1, \ldots, k\}$ if $i \neq r$.

Using the above distribution results, we have, for $j = 1, \ldots, k + 1$:

$$X_j(t+1) = \sum_{i=1}^{k+1} Z_{ij}(t) = \sum_{i=1}^{k+1} X_i(t)p_{ij}(t) + \epsilon_j(t+1), \quad (2.23)$$

where $\epsilon_j(t+1) = \sum_{i=1}^{k+1}[Z_{ij}(t) - X_i(t)p_{ij}(t)]$ is the random noise for $X_j(t+1)$.

In Eq. (2.23), the random noise $\epsilon_j(t+1)$ is the sum of residues of random variables from its conditional mean values. Hence, $E[\epsilon_j(t+1)] = 0$ for all $j = 1, \ldots, k + 1$. Using the distribution result given in Eq. (2.22), one may readily derive the covariance $\text{Cov}\,(\epsilon_j(t+1), \epsilon_r(t+1))$ by noting the basic formulae

$$\text{Cov}\,(X, Y) = E\,\text{Cov}\,(X, Y|Z) + \text{Cov}\,[E(X|Z), E(Y|Z)]$$

for any three random variables (X, Y, Z). This gives (Exercise 2.10),

$$C_{jr}(t+1) = \text{Cov}\,[\epsilon_j(t+1), \epsilon_r(t+1)] = \sum_{i=1}^{k+1}[EX_i(t)]\{p_{ij}(t)[\delta_{jr} - p_{ir}(t)]\}.$$

Denote by $\underset{\sim}{X}(t) = \{X_1(t), \ldots, X_{k+1}(t)\}'$, $\underset{\sim}{\epsilon}(t) = \{\epsilon_1(t), \ldots, \epsilon_{k+1}(t)\}'$ and $F'(t) = (p_{ij}(t)) = P(t)$. Then, in matrix notation, Eq. (2.23) can be expressed as:

$$\underset{\sim}{X}(t+1) = F(t)\underset{\sim}{X}(t) + \underset{\sim}{\epsilon}(t+1). \quad (2.24)$$

In Eq. (2.24), the vector of random noises have expected value 0 and covariance matrix $V(t+1) = (C_{ij}(t+1))$. Let $\mu_i(t) = E[X_i(t)]$ and put $\underset{\sim}{\mu}(t) = \{\mu_1(t), \ldots, \mu_{k+1}(t)\}'$. Then, from Eq. (2.24),

$$\underset{\sim}{\mu}(t+1) = F(t)\underset{\sim}{\mu}(t).$$

For validating the model and for estimating the unknown parameters, suppose that some observed data are available on the number of states at times $t_j, j = 1, \ldots, n$. Let $Y_i(j)$ be the observed number of state i at time t_j, $i = 1, \ldots, k+1, j = 1, \ldots, n$. Then,

$$Y_i(j) = X_i(t_j) + e_i(j), i = 1, \ldots, k+1, j = 1, \ldots, n, \qquad (2.25)$$

where $e_i(j)$ is the measurement error for observing $Y_i(j)$.

One may assume that the $e_i(j)$'s have expected values $Ee_i(j) = 0$ and variance $\text{Var}\,[e_i(j)] = \sigma_j^2$ and are independently of one another for $i = 1, \ldots, k+1$ and $j = 1, \ldots, n$ and independently distributed of the random noises $\{\epsilon_r(t), r = 1, \ldots, k+1\}$.

Denote by $\underset{\sim}{Y}(j) = \{Y_1(j), \ldots, Y_{k+1}(j)\}'$ and $\underset{\sim}{e}(t) = \{e_1(t), \ldots, e_{k+1}(t)\}'$. Then, in matrix notation, Eq. (2.25) can be expressed as:

$$\underset{\sim}{Y}(j) = \underset{\sim}{X}(t_j) + \underset{\sim}{e}(j),\ j = 1, \ldots, n. \qquad (2.26)$$

Combining Eqs. (2.24) and (2.26), we have a linear state space model for the chain with stochastic system model given by Eq. (2.24) and with observation model given by Eq. (2.26). Using this state space model, one may then estimate the unknown parameters and validate the model. This will be illustrated in Chap. 9.

Example 2.17. Mixture of random mating and assortative mating. In Example 2.2, we have considered a large natural population involving one-locus with two alleles under random mating. Under this condition, the frequencies of the gene and the genotypes do not change from generation to generation. This steady state condition is the well-known Hardy-Weinberg law in population genetics. In natural populations, however, random mating is hardly the case; the real situation may be a mixture of several different types of mating types (see Subsec. 2.11.2). A frequent situation is the assortative mating by phenotypes. That is, individuals chose mating partners by phenotype, e.g. tall people chose tall people for mating. In this example, we consider the situation of mixture between random mating and assortative mating type. For a single locus with two alleles $A : a$, under assortative mating, there are then only three mating types: $AA \times AA, Aa \times Aa, aa \times aa$.

Let $\{X(t), t \in T = (0, 1, \ldots, \infty)\}$ be the Markov chain for the three genotypes $\{AA, Aa, aa\}$. Let $1 - \theta$ be the proportion of assortative mating type in each generation and $p(n)$ the frequency of A allele at generation n. Assume that there are no mutations, no selection and no immigration and migration. Then, $p(n) = p$ and, under mixture of random mating and assortative mating, the one-step transition matrix is given by (Exercise 2.11):

$$P = \theta \begin{pmatrix} p & 1-p & 0 \\ \dfrac{1}{2}p & \dfrac{1}{2} & \dfrac{1}{2}(1-p) \\ 0 & p & 1-p \end{pmatrix} + (1-\theta) \begin{pmatrix} 1 & 0 & 0 \\ \dfrac{1}{4} & \dfrac{1}{2} & \dfrac{1}{4} \\ 0 & 0 & 1 \end{pmatrix}.$$

Let $\underset{\sim}{X}'(t) = \{X_1(t), X_2(t), X_3(t)\}$ denote the frequencies of the three genotypes $\{AA, Aa, aa\}$ at generation t respectively and $\underset{\sim}{Y}'(j) = \{Y_1(j), Y_2(j), Y_3(j)\}$ the observed frequencies of the three genotypes $\{AA, Aa, aa\}$ at generation t_j $(j = 1, \ldots, n)$ respectively. Then we have, with $F = P'$:

$$\underset{\sim}{X}(t+1) = F\underset{\sim}{X}(t) + \underset{\sim}{\epsilon}(t+1). \tag{2.27}$$

and

$$\underset{\sim}{Y}(j) = \underset{\sim}{X}(t_j) + \underset{\sim}{e}(j), j = 1, \ldots, n, \tag{2.28}$$

where $\underset{\sim}{\epsilon}(t+1)$ and $\underset{\sim}{e}(j)$ are the vectors of random noises and of measurement errors as defined in Subsec. 2.9.1 respectively. Using this state space model, one may then use procedures given in Chap. 9 to estimate $\{\theta, p\}$ and to estimate and predict the state variables to validate the model.

2.9.2. *Markov chains in the HIV epidemic in homosexual or IV drug user populations*

For Markov chains of HIV epidemic considered in Example 2.6, the traditional approach as given above is very complicated and difficult. A feasible and rather simple alternative approach is by way of stochastic difference equations. This is the approach used by this author and his associates for modeling these models; see [7, 8, 10, 11, 15, 16]. In this section we illustrate how to develop stochastic difference equations for state variables in these models.

Example 2.18. The chain binomial model in the HIV epidemic. Consider the AIDS Markov chain $\{\underset{\sim}{Z}(t) = \{S(t), I(u,t), u = 0, \ldots, t, A(t)\}'$, $t \in T = (0, 1, \ldots, \infty)\}$ described in Example 2.6. For this chain, because the population size of non-AIDS people does not change significantly during a short period, we assume that for S people and $I(u)$ people, the number of immigrants and recruitment equal to the number of death and migration out of the population; thus as an approximation to the real world situation, we will ignore immigration and death of S people and $I(u)$ people. Then, S people would decrease only through HIV infection and $I(u)$ people would only decrease by developing AIDS symptoms to become clinical AIDS patients.

To develop stochastic difference equations for these state variables, let $F_S(t)$ be the total number of $S \to I$ during $[t, t+1)$, $F_I(u,t)$ the total number of $I(u) \to A$ during $[t, t+1)$. Let $p_S(t)$ be the probability of $S \to I$ during $[t, t+1)$ and $\gamma(u,t)$ the probability of $I(u) \to A$ during $[t, t+1)$ at time t. Then, we have:

$$S(t+1) = S(t) - F_S(t)\,, \tag{2.29}$$

$$I(0, t+1) = F_S(t)\,, \tag{2.30}$$

$$I(r+1, t+1) = I(r,t) - F_I(r,t)\,, \quad r = 0, 1, \ldots, t\,, \tag{2.31}$$

$$A(t+1) = A(t) + \sum_{r=0}^{t} F_I(r,t) \tag{2.32}$$

$$= A(t) + \sum_{r=0}^{t} [I(r,t) - I(r+1, t+1)]\,.$$

Assume that $p_S(t)$ and $\gamma(u,t)$ are deterministic (non-stochastic) functions. Then the probability distribution of $F_S(t)$ given $S(t)$ is binomial with parameters $\{S(t), p_S(t)\}$. That is, $F_S(t) \mid S(t) \sim B[S(t), p_S(t)]$. Similarly, $F_I(u,t) \mid I(u,t) \sim B[I(u,t), \gamma(u,t)]$.

Using these distribution results, one may subtract the conditional mean values from the random variables in Eqs. (2.29)–(2.32) respectively to obtain the following equivalent equations:

$$S(t+1) = S(t) - S(t)p_S(t) + \epsilon_S(t+1)\,, \tag{2.33}$$

$$I(0, t+1) = S(t)p_S(t) + \epsilon_0(t+1)\,, \tag{2.34}$$

$$I(r+1, t+1) = I(r,t) - I(r,t)\gamma(r,t) + \epsilon_{r+1}(t+1)\,, \quad r = 0, \ldots, t\,, \tag{2.35}$$

$$A(t+1) = A(t) + \sum_{r=0}^{t} I(r,t)\gamma(r,t) + \epsilon_A(t+1)\,. \tag{2.36}$$

In Eqs. (2.33)–(2.36), the random noises are given by:

$$\epsilon_S(t+1) = -[F_S(t) - S(t)p_S(t)]\,,\ \ \epsilon_0(t+1) = F_S(t) - S(t)p_S(t)\,,$$

$$\epsilon_{r+1}(t+1) = -[F_I(r,t) - I(r,t)\gamma(r,t)]\,,\ \ r = 0,\ldots,t\,,$$

$$\epsilon_A(t+1) = \sum_{r=0}^{t}[F(r,t) - I(r,t)\gamma(r,t)]\,.$$

Denote by $\underset{\sim}{X}(t) = \{S(t), I(u,t), u = 0, 1, \ldots, t\}'$. The above distribution results imply that the conditional probability $P\{\underset{\sim}{X}(t+1)|\underset{\sim}{X}(t)\}$ of $\underset{\sim}{X}(t+1)$ given $\underset{\sim}{X}(t)$ is

$$P\{\underset{\sim}{X}(t+1)|\underset{\sim}{X}(t)\} = \binom{S(t)}{I(0,t+1)} [p_S(t)]^{I(0,t+1)}[1 - p_S(t)]^{S(t+1)}$$

$$\times \prod_{r=0}^{t} \binom{I(r,t)}{I(r,t) - I(r+1,t+1)}$$

$$\times [\gamma(r,t)]^{I(r,t)-I(r+1,t+1)}[1 - \gamma(r,t)]^{I(r+1,t+1)}\,. \tag{2.37}$$

It follows that the probability density of $\mathbf{X} = \{\underset{\sim}{X}(1), \ldots, \underset{\sim}{X}(t_M)\}$ given $\underset{\sim}{X}(0)$ is

$$P\{\mathbf{X}|\underset{\sim}{X}(0)\} = \prod_{j=0}^{t_M-1} P\{\underset{\sim}{X}(j+1)|\underset{\sim}{X}(j)\}\,.$$

The above distribution results for $\{\underset{\sim}{X}(t), t = 1, \ldots, t_M\}$ have been referred to as the chain binomial distribution for the HIV epidemic [10, 11].

Let $Y(j)$ be the observed total number of new AIDS cases during $[t_{j-1}, t_j)$ $(j = 1, \ldots, n)$. Then

$$Y(j) = \sum_{i=t_{j-1}+1}^{t_j} A_I(i) + e(j),\ j = 1, \ldots, n \tag{2.38}$$

where $A_I(t)$ is the AIDS incidence during $[t, t+1)$ and $e(j)$ is the random measurement (reporting) error associated with observing $Y(j)$.

The above results give a state space model for the HIV epidemic for homosexual populations or populations of IV drug users first proposed by this author and his associates, see [10, 11, 15, 16]. For this state space model, the stochastic system model is given by Eqs. (2.33)–(2.36) and the observation model is given by Eq. (2.38).

Example 2.19. The staged chain multinomial model of HIV epidemic. In the studies of HIV epidemic, to account for effects of infection duration, the infective stages are usually partitioned into substage. For example, Longini *et al.* [17–19] and Satten and Longini [20–21] have partitioned the infective stage into 5 substage by the $CD4^{(+)}$ T cell counts per mm^3 of blood; see Fig. 1.1 in Example 1.12 for detail.

Complying with Fig. 1.1 of Example 1.12, in this example we consider a staging model for the homosexual population assuming that the infective stage has been partitioned into k sub-stages $(I_1, \ldots, I_k)$. Let $\{S(t), I(r,t), r = 1, \ldots, k\}$ denote the numbers of S people and I_r people at the tth month respectively and let $A(t)$ denote the total number of AIDS cases developed at the tth month. Under the assumption that the transition rates $\{\beta_r(s,t) = \beta_r(t), \gamma_r(s,t) = \gamma_r(t), \omega_r(s,t) = \omega_r(t)\}$ are independent of time s, $\underset{\sim}{X}(t) = \{S(t), I(r,t), r = 1, \ldots, k, A(t)\}'$ is a $(k+2)$-dimensional Markov chain with discrete time. For this chain, it is extremely difficult to develop analytical results. Thus, a feasible and reasonable approach is by way of stochastic difference equations.

To develop stochastic difference equations for these state variables, let $\{F_I(r,t), B_I(r,t), W_I(r,t), D_I(r,t)\}$ and $R_I(r,t)$ denote the numbers of $\{I_r \to I_{r+1}, I_r \to I_{r-1}, I_r \to A\}$, the number of death of I_r people, and the number of recruitment and immigration of I_r people during $[t, t+1)$, respectively. Let $F_S(t), D_S(t)$ and $R_S(t)$ be the number of $S \to I_1$, the number of death and retirement of S people and the number of immigrants and recruitment of S people during $[t, t+1)$, respectively. Assume that because of the awareness of AIDS, there are no immigration and recruitment of AIDS cases. Then we have the following stochastic equations for $S(t), I(r,t)$ and $A(t)$:

$$S(t+1) = S(t) + R_S(t) - F_S(t) - D_S(t); \tag{2.39}$$

$$I(1, t+1) = I(1, t) + R_I(1, t) + F_S(t) + B_I(2, t)$$
$$-F_I(1, t) - W_I(1, t) - D_I(1, t) ; \qquad (2.40)$$

$$I(r, t+1) = I(r, t) + R_I(r, t) + F_I(r-1, t)$$
$$+B_I(r+1, t) - F_I(r, t) - B_I(r, t)$$
$$-W_I(r, t) - D_I(r, t) \; r = 2, \ldots, k-1 ; \qquad (2.41)$$

$$I(k, t+1) = I(k, t) + R_I(k, t) + F_I(k-1, t)$$
$$-B_I(k, t) - F_I(k, t) - D_I(k, t) ; \qquad (2.42)$$

$$A(t+1) = A(t) + \sum_{r=1}^{k-1} W_I(r, t) + F_I(k, t) . \qquad (2.43)$$

Let $\mu_S(t)$ and $\mu_I(r, t)$ denote the probabilities of death of S people and I_r people during $[t, t+1)$ respectively. Let $p_S(t)$ be the probability of $S \to I_1$ during $[t, t+1)$. Then, given $\underset{\sim}{U}(t) = \{S(t), I(r, t), r = 1, \ldots, k\}$, the conditional distributions of the above random variables are given by:

$$\{F_S(t), D_S(t)\} | \underset{\sim}{U}(t) \sim ML\{S(t); p_S(t), \mu_S(t)\} ;$$

$$\{F_I(1, t), W_I(1, t), D_I(1, t)\} | \underset{\sim}{U}(t) \sim ML\{I(1, t); \gamma_1(t), \omega_1(t), \mu_I(1, t)\} ;$$

$$\{F_I(r, t), B_I(r, t), W_I(r, t), D_I(r, t)\} | \underset{\sim}{U}(t)$$

$$\sim ML\{I(r, t); \gamma_r(t), \beta_r(t), \omega_r(t), \mu_I(r, t)\} , \quad \text{for } (r = 2, \ldots, k-1) ;$$

$$\{F_I(k, t), B_I(k, t), D_I(k, t)\} | \underset{\sim}{U}(t) \sim ML\{I(k, t); \gamma_k(t), \beta_k(t), \mu_I(k, t)\} .$$

Assume further that the conditional distributions of $R_S(t)$ given $S(t)$ and of $R_I(r, t)$ given $I(r, t)$ are binomial with parameters $\{S(t), \nu_S(t)\}$ and $\{I(r, t), \nu_I(r, t)\}$ respectively, independently of other variables.

Define the random noises:

$$\epsilon_S(t+1) = [R_S(t) - S(t)\nu_S(t)] - [F_S(t) - S(t)p_S(t)]$$
$$-[D_S(t) - S(t)\mu_S(t)] ;$$

$$\epsilon_r(t+1) = [R_I(r, t) - I(r, t)\nu_I(r, t)] + [F_I(r-1, t) - I(r-1, t)\gamma_{r-1}(t)]$$

$$+(1 - \delta_{kr})[B_I(r+1,t) - I(r+1,t)\beta_{r+1}(t)]$$

$$-(1 - \delta_{kr})[F_I(r,t) - I(r,t)\gamma_r(t)] - (1 - \delta_{1r})[B_I(r,t)$$

$$-I(r,t)\beta_r(t)] - [W_I(r,t) - I(r,t)\omega_r(t)]$$

$$-[D_I(r,t) - I(r,t)\mu_r(t)], r = 1,\ldots,k\,;$$

$$\epsilon_A(t+1) = \sum_{r=1}^{k}[W_I(r,t) - I(r,t)\omega_r(t)]\,.$$

Then, Eqs. (2.39)–(2.43) are equivalent to the following stochastic difference equations:

$$S(t+1) = S(t)[1 + \nu_S(t) - p_S(t) - \mu_S(t)] + \epsilon_S(t+1)\,; \tag{2.44}$$

$$I(r,t+1) = I(r-1,t)\gamma_{r-1}(t) + (1 - \delta_{rk})I(r+1,t)\beta_{r+1}(t)$$

$$+I(r,t)[1 + \nu_I(r,t) - \gamma_r(t) - (1 - \delta_{1r})\beta_r(t)$$

$$-(1 - \delta_{rk})\omega_r(t) - \mu_I(r,t)] + \epsilon_r(t+1), \quad (r = 1, 2, \ldots, k)\,, \tag{2.45}$$

$$A(t+1) = A(t) + \sum_{r=1}^{k}I(r,t)\omega_r(t) + \epsilon_A(t+1)\,, \tag{2.46}$$

where $I(0,t) = S(t)$, $\gamma_0(t) = p_S(t)$, $\beta_1(t) = \beta_{k+1}(t) = 0$, $\omega_k(t) = \gamma_k(t)$ and δ_{ij} is the Kronecker's δ.

Because of the above distributions, the above model has also been referred to as chain multinomial model [7–9, 15].

2.10.   Complements and Exercises

Exercise 2.1.   For the self-fertilized two-loci linkage model as described in Example 2.4, show that the one-step transition matrix is as given in Example 2.4.

Exercise 2.2.   Consider a large natural population of diploid involving one-locus with three alleles A_1, A_2, A_3. Show that if the mating between individuals is random and if there are no selection, no mutations and no immigration and migration, then the Hardy-Weinberg holds (See Subsec. 2.11.1).

Exercise 2.3. **(Mixture of self-fertilization and random mating).** Consider a large natural population involving one-locus with two alleles $A : a$. Let $\{X(t), t \in T = (0, 1, \ldots, \infty)\}$ denote the Markov chain of the three genotypes $\{AA, Aa, aa\}$. Assume that the mating type is a mixture of random mating and self-fertilization. Let $1 - \theta$ be the proportion of random mating in each generation and let $p(n)$ be the frequency of A gene at generation n. Assume that there are no mutations, no selection and no immigration and migration. Show that $p(n) = p$ and the one-step transition matrix is given by:

$$P = (1 - \theta) \begin{pmatrix} p & 1-p & 0 \\ \dfrac{1}{2}p & \dfrac{1}{2} & \dfrac{1}{2}(1-p) \\ 0 & p & 1-p \end{pmatrix} + \theta \begin{pmatrix} 1 & 0 & 0 \\ \dfrac{1}{4} & \dfrac{1}{2} & \dfrac{1}{4} \\ 0 & 0 & 1 \end{pmatrix}.$$

Show that the eigenvalues of P are given by $\{\lambda_1 = 1, \lambda_2 = \frac{1}{2}\theta, \lambda_3 = \frac{1}{2}(1 + \theta)\}$. Obtain the spectral expansion $\{E_i, i = 1, 2, 3\}$ of P so that

$$P^n = \sum_{i=1}^{3} \lambda_i^n E_i \,.$$

Exercise 2.4. **(Markov chain under inbreeding in natural populations.)** Consider a large natural population involving one-locus with two alleles $A : a$. Let $\{X(t), t \in T = (0, 1, \ldots, \infty)\}$ denote the Markov chain of the three genotypes $\{AA, Aa, aa\}$. Assume that there are no mutations, no selection and no immigration and migration. Then under inbreeding with inbreeding coefficient F, the one-step transition matrix is given by:

$$P = (1 - F) \begin{pmatrix} p & 1-p & 0 \\ \dfrac{1}{2}p & \dfrac{1}{2} & \dfrac{1}{2}(1-p) \\ 0 & p & 1-p \end{pmatrix} + F \begin{pmatrix} 1 & 0 & 0 \\ \dfrac{1}{2} & 0 & \dfrac{1}{2} \\ 0 & 0 & 1 \end{pmatrix}.$$

Show that the eigenvalues of P are given by $\{\lambda_1 = 1, \lambda_2 = 0, \lambda_3 = \frac{1}{2}(1+F)\}$. Obtain the spectral expansion $\{E_i, i = 1, 2, 3\}$ of P so that

$$P^n = \sum_{i=1}^{3} \lambda_i^n E_i \,.$$

Exercise 2.5. Consider a Galton-Watson branching process $\{X(t), t \in T = (0, 1, \ldots,)\}$ with progeny distribution $\{p_j, j = 0, 1, \ldots, \}$. Suppose that the mean and the variance of the progeny distribution are given by $\{\mu, \sigma^2\}$ respectively. Show that the mean and variance of $X(n)$ given $X(0) = 1$ are given by $E[X(n)|X(0) = 1] = \mu^n$ and $\mathrm{Var}\,[X(n)|X(0) = 1] = \sigma^2 \mu^{n-1}(\sum_{i=0}^{n-1} \mu^i)$ respectively.

Exercise 2.6. Consider a Galton-Watson branching process $\{X(t), t \in T = (0, 1, \ldots)\}$ with progeny distribution $\{p_j(t), j = 0, 1, \ldots, \infty, t \geq 0\}$. Derive the one-step transition probabilities in each of the following.

(a) Assume that $p_j(t) = p_j$,
where

$$\left\{ p_j = \binom{\gamma + j - 1}{j} u^\gamma (1 - u)^j, \; j = 0, 1, \ldots, \infty, \; 0 \leq u \leq 1, \; \gamma > 0 \right\}.$$

(b) Assume that $p_j(t) = p_j(1)$ if $t \leq t_1$ and $p_j(t) = p_j(2)$ if $t > t_1$,
where

$$\left\{ p_j(i) = \binom{\gamma_i + j - 1}{j} u_i^{\gamma_i} (1 - u_i)^j, \; j = 0, 1, \ldots, \infty, \; 0 \leq u_i \leq 1, \; \gamma_i > 0 \right\}.$$

Exercise 2.7. **(Mating type involving sex-linked genes).** Consider a single locus with two alleles $A : a$ in sex chromosomes in human beings (see Remark 2.2.) Let the mating types $\{AA \times AY, aa \times aY\}$ be represented by the state e_1, the mating types $\{AA \times aY, aa \times AY\}$ by e_2 and the mating types $\{Aa \times aY, Aa \times AY\}$ by e_3. Let $X(t)$ be the Markov chain for the mating types under brother-sister mating in a large population. If there are no mutations, no immigration and migration, and no selection and if t represents generation, then $\{X(t), t \in T = (0, 1, \ldots,)\}$ is a homogeneous Markov chain with discrete time with state space $S = \{e_i, i = 1, 2, 3\}$.

(a) Show that e_1 is an absorbing state and the other two states are transient states. Show also that the matrix of the one-step transition probabilities is given by:

$$P = \begin{pmatrix} 1 & 0 & 0 \\ 0 & 0 & 1 \\ \dfrac{1}{4} & \dfrac{1}{4} & \dfrac{1}{2} \end{pmatrix}.$$

(b) Derive the eigenvalues, eigenvectors and the spectral expansion of the matrices P and P^n for positive integer n.

(c) Derive the probability densities of first absorption times of transient states.

(d) Derive the vectors of expected values and variances of the first absorption times.

Remark 2.2. (Sex-linked genes in human beings). In human beings, sex is determined by the sex chromosomes X and Y. The genotype of females is XX while that of males is XY. The Y chromosome has only a few genes so that if a gene is located in the sex chromosome, then the males are usually semizygote. That is, males have only one gene as they have only one X chromosome. Thus, for a single locus with two alleles $A : a$ in the sex chromosome, the genotypes of males are either AY or aY. (The genotypes of the females are AA, Aa, aa.) The mating types are:

$$AA \times AY, aa \times aY, AA \times aY, aa \times AY, Aa \times aY, Aa \times AY.$$

In the mating types, as a convention, the genotype on the left side denotes that of female and the genotype on the right side that of the male.

Exercise 2.8. (Parent-offspring mating types). Consider a single locus with two alleles A and a in a sufficient large diploid population. Assume that the matings always occur between the younger parents and the progenies. Then there are 9 mating types which we denote by the numbers $1, \ldots, 9$ respectively as follows (Remark 2.3):

Parent	$\times$	Offspring	State Number
AA	$\times$	AA	~ 1
aa	$\times$	aa	~ 2
AA	$\times$	Aa	~ 3
aa	$\times$	Aa	~ 4
Aa	$\times$	AA	~ 5
Aa	$\times$	aa	~ 6
Aa	$\times$	Aa	~ 7
aa	$\times$	AA	~ 8
AA	$\times$	aa	~ 9

Let $X(t), t \in T = \{0, 1, 2, \ldots\}$ denote the parent-offspring mating types at generation t. Then $\{X(t), t = 1, 2, \ldots\}$ is a Markov chain with state space $\Omega = \{1, \ldots, 9\}$. Assume that the fitness of the three genotypes are:

$$AA, \quad Aa, \quad aa$$
$$x \quad\quad 1 \quad\quad x, \quad x \geq 0.$$

Then the above chain is homogeneous.

(a) Show that the states $(1, 2)$ are absorbing states and all other states are transient states. Show also that the 1-step transition matrix is

$$P = \begin{pmatrix} P_1 & 0 & \underset{\sim}{0}' \\ 0 & P_2 & \underset{\sim}{0}' \\ \underset{\sim}{R}_1 & \underset{\sim}{R}_2 & Q \end{pmatrix},$$

where $P_1 = P_2 = 1$,

$$\underset{\sim}{R}_1 = \left\{0, 0, \frac{x}{1+x}, 0, 0, 0, 0\right\}', \quad \underset{\sim}{R}_2 = \left\{0, 0, 0, \frac{x}{1+x}, 0, 0, 0\right\}',$$

and

$$Q = \begin{pmatrix} 0 & 0 & \dfrac{x}{1+x} & 0 & \dfrac{1}{1+x} & 0 & 0 \\ 0 & 0 & 0 & \dfrac{x}{1+x} & \dfrac{1}{1+x} & 0 & 0 \\ \dfrac{1}{1+x} & 0 & 0 & 0 & 0 & 0 & 0 \\ 0 & \dfrac{1}{1+x} & 0 & 0 & 0 & 0 & 0 \\ 0 & 0 & \dfrac{x}{2(1+x)} & \dfrac{x}{2(1+x)} & \dfrac{1}{1+x} & 0 & 0 \\ 1 & 0 & 0 & 0 & 0 & 0 & 0 \\ 0 & 1 & 0 & 0 & 0 & 0 & 0 \end{pmatrix}.$$

(b) Obtain the eigenvalues and eigenvectors and the spectral expansion of matrices P and P^n.

(c) Derive the ultimate absorption probabilities into absorbing states.

(d) Derive the vector of probability densities of first absorption times.

(e) Derive the vectors of the means and the variances of first absorption times.

Remark 2.3. In this example, one may assume that the individuals have both sex organs as are the cases in some plants. In animal populations, one may assume a large population with equal number of males and females and that there are no differences in fitness between males and females.

Exercise 2.9. Consider a finite homogeneous Markov chain $\{X(n), n = 0, 1, \ldots,\}$. Assume that the chain contains k closed sets and also contain r transient states. Use the canonical form of the transition matrix as given in Sec. 2.8 to prove Eqs. (2.9)–(2.13) and (2.17)–(2.18).

Exercise 2.10. Derive the covariances between the random noises given in Eq. (2.23) in Subsec. 2.9.1.

Exercise 2.11. Let $\{X(t), t \in T = (0, 1, \ldots,)\}$ be the Markov chain given in Example (2.17). Assume that the population is very large and that there are no mutations, no selections and no immigration and migration. Show that the frequency $p(n)$ of the A allele at generation n is independent of n. Show also that the one-step transition matrix is as given in Example 2.17.

Exercise 2.12. (Multiple branching processes). Let $\{\underset{\sim}{X}(t) = [X_1(t), \ldots, X_k(t)]', t \in T = (0, 1, \ldots, \infty)\}$ be a k-dimensional Markov process with state space $S = \{\underset{\sim}{i} = (i_1, \ldots, i_k)', i_j = 0, 1, \ldots, \infty, j = 1, \ldots, k\}$. For each i $(i = 1, \ldots, k)$, let $\{q_i(\underset{\sim}{j}), \underset{\sim}{j} = (j_1, \ldots, j_k)' \in S\}$ denote a k-dimensional probability density defined on the space S.

Definition 2.5. $\{\underset{\sim}{X}(t), t \in T\}$ is called a *k-dimensional Galton-Watson branching process with progeny distributions* $\{q_i(\underset{\sim}{j}) = q_i(j_1, \ldots, j_k), \underset{\sim}{j} \in S, i = 1, \ldots, k\}$ iff

(a) $P\{X_i(0) = 1\} = 1 \quad$ *for some $i = 1, \ldots, k$*

and

(b) $\qquad P\{\underset{\sim}{X}(t+1) = \underset{\sim}{j} \,|\, \underset{\sim}{X}(t) = \underset{\sim}{i}\}$

$$= P\{X_r(t+1) = j_r, \, r = 1, \ldots, k | X_r(t) = i_r, \, r = 1, \ldots, k\}$$

$$= P\left\{ \sum_{u=1}^{k}\sum_{r=1}^{i_u} Z_1(r,u) = j_1, \ldots, \sum_{u=1}^{k}\sum_{r=1}^{i_u} Z_k(r,u) = j_k \right\}$$

$$= P\left\{ \sum_{u=1}^{k}\sum_{r=1}^{i_u} \underset{\sim}{Z}_u(r) = \underset{\sim}{j} \right\},$$

where $\underset{\sim}{Z}_i(r) = [Z_1(r,i), \ldots, Z_k(r,i)]' \sim \underset{\sim}{Z} = [Z_1(i), \ldots, Z_k(i)]'$ independently of one another for all $(i = 1, \ldots, k)$ and for all $(r = 1, \ldots, \infty)$ and where the $\underset{\sim}{Z}_i$ are random vectors independently distributed of one another with density $\{q_i(\underset{\sim}{j}) = q_i(j_1, \ldots, j_k), \underset{\sim}{j} \in S\}$. (Notice that $Z_j(r,i)$ is the number of Type-j progenies produced by the rth individual of Type-i parents.)

Let $g_i(\underset{\sim}{x}) = g_i(x_1, \ldots, x_k)$ $(i = 1, \ldots, k)$ denote the pgf of the ith progeny distribution (i.e. $q_i(\underset{\sim}{j}, \underset{\sim}{j} \in S)$. Let $f_i(\underset{\sim}{x};t)$ denote the pgf of $\underset{\sim}{X}(t)$ given $\underset{\sim}{X}(0) = \underset{\sim}{e}_i$, where $\underset{\sim}{e}_i$ is a $k \times 1$ column vector with 1 at the ith position and with 0 at other positions. Show that

$$\sum_{j_1=0}^{\infty} \cdots \sum_{j_k=0}^{\infty} \left(\prod_{r=1}^{k} x_r^{j_r} \right) P\{ \underset{\sim}{X}(t+1) = \underset{\sim}{j} \mid \underset{\sim}{X}(t) = \underset{\sim}{i} \} = \prod_{r=1}^{k} [g_r(\underset{\sim}{x})]^{i_r}.$$

Hence show that

$$f_i(\underset{\sim}{x};t+1) = g_i\{f_1(\underset{\sim}{x};t), \ldots, f_k(\underset{\sim}{x};t)\}.$$

Show also that

$$f_i(\underset{\sim}{x};t+1) = f_i\{g_1(\underset{\sim}{x}), \ldots, g_k(\underset{\sim}{x});t\}.$$

In the multiple branching process as defined in Definition 2.5. Let

$$m_{i,j}(n) = E[X_j(n) \mid \underset{\sim}{X}(0) = \underset{\sim}{e}_i] = \left(\frac{\partial}{\partial x_j} f_i(\underset{\sim}{x}) \right)_{(x_i=1,\ i=1,\ldots,k)}, \quad i,j = 1, \ldots, k.$$

Then $m_{i,j}(n)$ is the expected number of Type-j progenies at generation n from a Type-i parent at generation 0. Let $M(n) = [m_{i,j}(n)]$ be the $k \times k$ matrix with $m_{i,j}(n)$ as the (i,j)th element and put $M = M(1)$. Show that $M(n) = M^n$ for any positive integer n. The matrix M has been referred to as the matrix of expected progenies. Notice that the elements of M are non-negative.

The multiple branching process is called a positive regular if there exists a positive integer r such that all elements of M^r are positive. Such as matrix M is called a Perrron matrix. Denote by $g(\underset{\sim}{x}) = [g_1(\underset{\sim}{x}), \ldots, g_k(\underset{\sim}{x})]'$. The multiple branching process is called a singular process if $g(\underset{\sim}{x}) = A\underset{\sim}{x}$ for some matrix A. Notice that $g(\underset{\sim}{x}) = A\underset{\sim}{x}$ correspond to the case that each type has probability one of producing just one child.

The following results are useful for deriving basic results for multiple branching processes.

(1) If M is a Perron matrix, then M has a positive real eigenvalue λ_0 with multiplicity 1 whose corresponding normalized right eigenvector $(\underset{\sim}{u})$ and left eigenvector $(\underset{\sim}{v})$ have only real positive elements. Furthermore, the following results hold:

(a) If λ_i is any other eigenvalue of M, then $|\lambda_i| < \lambda_0$.
(b) There exists a matrix $M_2 = (c_{ij})$ such that

$$M^n = \lambda_0(\underset{\sim}{u}\,\underset{\sim}{v}') + M_2, \text{ with } |c_{ij}| < O(\alpha^n), i, j = 1, \ldots, k \text{ for some } 0 < \alpha < \lambda_0.$$

For proof, see Gantmacher [22], Chap. 13.

(2) For non-singular, positive-regular multiple branching processes, the following result is an extension of Theorem 2.11.

Theorem 2.11. *Let $\{\underset{\sim}{x}(t) = [X_1(t), \ldots, X_k(t)]', t \in T = (0, 1, \ldots, \infty)\}$ be a k-dimensional multiple branching process with state space $S = \{\underset{\sim}{i} = (i_1, \ldots, i_k)', i_j = 0, 1, \ldots, \infty, j = 1, \ldots, k\}$ and with progeny distributions $\{q_i(\underset{\sim}{j}), \underset{\sim}{j} = (j_1, \ldots, j_k)' \in S\}$. Let M be the matrix of expected number of progeny distributions per generation. Suppose that the process is non-singular and positive regular so that M has a real positive eigenvalue λ of multiplicity 1 satisfying $|\lambda_i| < \lambda$ for any other eigenvalue λ_i of M. Let $\underset{\sim}{q}$ denote the vector of extinction probabilities given $\{\underset{\sim}{X}(0) = \underset{\sim}{e}_i, i = 1, \ldots, k\}$. Then, $\underset{\sim}{q} = g(\underset{\sim}{q})$ and*

$$\underset{\sim}{q} = \underset{\sim}{1}_k, \text{ if } \lambda \le 1,$$

$$\underset{\sim}{0} \le \underset{\sim}{q} < \underset{\sim}{1}_k, \text{ if } \lambda > 1,$$

where $\underset{\sim}{g}(\underset{\sim}{x})$ is the vector of pgf's of the progeny distributions and $\underset{\sim}{e}_i$ the $k \times 1$ column vector with 1 at the ith position and with 0 at other positions.

For proof, see [23, p. 41].

2.11.   Appendix

2.11.1.   *The Hardy-Weinberg law in population genetics*

In natural populations, if the probability that each individual selects an individual of a particular type as its mating partner is given by the relative frequency of that particular type in the population, then the mating between individuals in the population is referred to as a random mating. The Hardy-Weinberg law in population genetics specifies that if the population size is very large and if the mating between individuals is random, then under some conditions, the frequencies of genes and genotypes do not change from generation to generation. The specific conditions under which this steady state condition holds are:

(1) The generation is discrete.
(2) There are no selection.
(3) There are no mutation.
(4) There are no immigration and no migration.

We now illustrate this law by considering a single locus in Subsec. 2.11.1.1 and two-linked loci in Subsec. 2.11.1.2 in diploid populations.

2.11.1.1.   *The Hardy-Weinberg law for a single locus in diploid populations*

Consider a single locus with two alleles (A and a) in a sufficiently large population of diploid. Suppose that the mating is random and that there are no selection, no mutation, no immigration and no migration. Then, after at most two generations, the frequency of A allele becomes p which is independent of time and the frequencies of the three genotypes $\{AA, Aa, aa\}$ become $\{p^2, 2pq, q^2\}$ where $q = 1 - p$, respectively. This steady-state result has been referred to as the Hardy-Weinberg law.

To prove the Hardy-Weinberg law, at generation t, let $\{n_m(1), n_m(2), n_m(3)\}$ denote the numbers of the three genotypes $\{AA, Aa, aa\}$ respectively for the males in the population; $\{n_f(1), n_f(2), n_f(3)\}$ the numbers of the three genotypes $\{AA, Aa, aa\}$ respectively for the females in the population. Denote by $\{u_m(i) = n_m(i)/[n_m(1) + n_m(2) + n_m(3)], u_f(i) = n_f(i)/[n_f(1) + n_f(2) + n_f(3)], i = 1, 2, 3; p_m = u_m(1) + \frac{1}{2}u_m(2), q_m = 1 - p_m, p_f = u_f(1) + \frac{1}{2}u_f(2), q_f = 1 - p_f$. Then, p_m and p_f are the frequency of A gene for the males and for the females respectively at generation t. Under random mating and assuming that there are no selection, no mutation, no immigration and no migration, the mating types (the first genotype denotes the mother), the frequencies of the mating types and the frequencies of the progenies are given by:

Mating types	Frequencies of mating types	Frequencies of different types in the daughters (or in the sons)
$AA \times AA$	$u_m(1)u_f(1)$	AA
$AA \times Aa$	$u_m(1)u_f(2)$	$\frac{1}{2}AA : \frac{1}{2}Aa$
$AA \times aa$	$u_m(1)u_f(3)$	Aa
$Aa \times AA$	$u_m(2)u_f(1)$	$\frac{1}{2}AA : \frac{1}{2}Aa$
$Aa \times Aa$	$u_m(2)u_f(2)$	$\frac{1}{4}AA : \frac{1}{2}Aa : \frac{1}{4}aa$
$Aa \times aa$	$u_m(2)u_f(3)$	$\frac{1}{2}Aa : \frac{1}{2}aa$
$aa \times AA$	$u_m(3)u_f(1)$	Aa
$aa \times Aa$	$u_m(3)u_f(2)$	$\frac{1}{2}Aa : \frac{1}{2}aa$
$aa \times aa$	$u_m(3)u_f(3)$	aa

Thus, for both the males and females, the frequencies of the three genotypes at generation $t + 1$ are:

$$(1) \ AA : u_m(1)u_f(1) + \frac{1}{2}u_m(1)u_f(2) + \frac{1}{2}u_m(2)u_f(1) + \frac{1}{4}u_m(2)u_f(2)$$

$$= \left\{ u_m(1) + \frac{1}{2}u_m(2) \right\} \left\{ u_f(1) + \frac{1}{2}u_f(2) \right\} = p_m p_f ;$$

$$(2) \ Aa : \frac{1}{2}u_m(1)u_f(2) + u_m(1)u_f(3) + \frac{1}{2}u_m(2)u_f(1) + \frac{1}{2}u_m(2)u_f(2)$$

$$+ \frac{1}{2}u_m(2)u_f(3) + u_m(3)u_f(1) + \frac{1}{2}u_m(3)u_f(2)$$

$$= u_m(1)q_f + \frac{1}{2}u_m(2)(p_f + q_f) + u_m(3)p_f$$

$$= \left[u_m(1) + \frac{1}{2}u_m(2)\right]q_f + \left[\frac{1}{2}u_m(2)u_m(3)\right]p_f$$

$$= p_m q_f + q_m p_f \ ;$$

$$(3) \ aa : u_m(3)u_f(3) + \frac{1}{2}u_m(3)u_f(2) + \frac{1}{2}u_m(2)u_f(3) + \frac{1}{4}u_m(2)u_f(2)$$

$$= \left\{u_m(3) + \frac{1}{2}u_m(2)\right\}\left\{u_f(3) + \frac{1}{2}u_f(2)\right\} = q_m q_f \ .$$

The frequency of the A allele at generation $t+1$ is $p = p_m p_f + \frac{1}{2}(p_m q_f + q_m p_f) = \frac{1}{2}(p_m + p_f)$ for both males and females. The frequencies of the three genotypes $\{AA, Aa, aa\}$ at generations $t + j$ $(j = 1, 2, \ldots,)$ are given respectively by $\{p^2, 2pq, q^2\}$, where $q = 1 - p$ for both males and females. The frequency of A allele is also p for both males and females at generations $t+j$ with $j = 1, 2, \ldots,$. Notice that if $u_m(i) = u_f(i), i = 1, 2, 3$, then $p_m = p_f = p$ and the steady-state condition is achieved in one generation. If $u_m(i) \neq u_f(i)$, then the steady-state condition is achieved in two generation with $p = \frac{1}{2}(p_m + p_f)$.

By a straightforward extension, the above results also extend to cases involving a single locus with more than two alleles; see Exercise 2.2. For example, suppose that the A locus has k alleles, $A_i, i = 1, 2, \ldots, k$ and that the following assumptions hold:

(1) The individuals are diploid and the population size is very large.
(2) The generation is discrete.
(3) The mating is random.
(4) There are no selection, no mutation, no immigration and no migration.

Then in at most two generations, the frequencies of the genotypes are p_i^2 for the genotype $A_i A_i$ and are $2p_i p_j$ for the genotype $A_i A_j, (i \neq j)$.

Example 2.20. **The ABO-Blood group in human beings.** In human beings, the blood group is controlled by three alleles $\{A, B, O\}$. The AB

type has genotype AB; the A-type has genotypes $\{AA, AO\}$; the B-type has genotypes $\{BB, BO\}$ and the O-type has genotype OO. Let $\{p, q, r\}$ ($r = 1 - p - q$) denote the frequencies of the $\{A, B, O\}$ alleles in the population respectively. Then, under Hardy-Weinberg equilibrium, the frequencies of the $\{AB$-type, A-type, B-type, O-type$\}$ are given by $\{2pq, p^2 + 2pr, q^2 + 2qr, r^2\}$ respectively; for more detail see Example 3.5.

2.11.1.2. *The Hardy-Weinberg law for linked loci in diploid populations*

To illustrate the Hardy-Weinberg equilibrium for linked loci, consider two linked-loci A and B in a very large diploid population. Assume that the A locus has k alleles $\{A_i, i = 1, \ldots, k\}$ and that the B locus has m alleles $\{B_j, j = 1, \ldots, m\}$. Assume further that the mating between individuals is random, the generation is discrete and that there are no selection, no mutation, no immigration and no migration. Then, as time progresses, with probability one the frequency of gamete with genotype A_iB_j will approach p_iq_j, where p_i and q_j are the frequencies of the alleles A_i and B_j in the population respectively; under random mating, the frequency of individuals with genotype A_iB_j/A_uB_v will approach $(2 - \delta_{iu})(2 - \delta_{jv})p_ip_uq_jq_v$, where δ_{ij} is the Kronecker's δ defined by $\delta_{ij} = 1$ if $i = j$ and $= 0$ if $i \neq j$. (As in the case with one locus, p_i is the average of the frequencies of the A_i allele of males and females at generation 0; q_i is the average of the frequencies of the B_j allele of males and females at generation 0).

To prove the above result, let $P_n(A_iB_j)$ denote the frequency of gamete with genotype A_iB_j at generation n; let θ_m and θ_f denote the crossing-over frequencies between the A locus and the B locus for males and for females respectively. Put $\theta = \frac{1}{2}(\theta_m + \theta_f)$. Then after two generations, we have the following relationship between the frequencies of gametes at generation $n - 1$ and generation n:

$$P_n(A_iB_j) = (1 - \theta)P_{n-1}(A_iB_j) + \theta p_iq_j,$$

for all $i = 1, \ldots, k; j = 1, \ldots, m$.

The above equation is derived by noting that the gametes at generation $n + 1$ are either products of a meiosis with a crossing over with probability θ or products of a meiosis without crossing-over with probability $1 - \theta$. If no crossing-over has occurred, then the frequency of gamete with genotype A_iB_j

at generation $n+1$ is the same as that at generation n; on the other hand, under crossing-over, the frequency of A_iB_j at generation $n+1$ is a product of random union of alleles A_i and B_j at generation n from the population. The above equation leads to

$$P_n(A_iB_j) - p_iq_j = (1-\theta)\{P_{n-1}(A_iB_j) - p_iq_j\} = (1-\theta)^{n-2}\{P_2(A_iB_j) - p_iq_j\},$$

for all $i = 1, \ldots, k$; $j = 1, \ldots, m$.

Since $1 - \theta < 1$, it follows that $\lim_{n\to\infty} P_n(A_iB_j) = p_iq_j$ for all $i = 1, \ldots, k, j = 1, \ldots, m$. This condition has been referred to as linkage equilibrium in the population.

The above results have also been extended to cases involving three or more linked loci by Bennett [24]. For example, if we have 4 linked loci, say A, B, C and D, then under linkage equilibrium, the frequency of gamete with genotype $A_iB_jC_uD_v$ is given by $p_iq_jr_uw_v$, where $\{p_i, q_j, r_u, w_v\}$ are the frequencies of alleles $\{A_i, B_j, C_u, D_v\}$ in the population respectively; for more detail, see [1].

The above Hardy-Weinberg law has also been extended to cases involving sex-linked genes and to polyploid populations; such extensions have been discussed in detail in [1].

2.11.2.   *The inbreeding mating systems*

In large natural populations, if there are no disturbances (mutation, selection, immigration and migration) and if the mating is random, then both the frequencies of the genes and the genotypes will not change from generation to generation. This steady state result has been referred to as the Hardy-Weinberg law; see Subsec. 2.11.1. If the mating is not random, then the frequencies of the genotypes will change over generations but the frequencies of genes remain stable and will not be affected. To illustrate, consider a single locus with k alleles $A_i, i = 1, \ldots, k$. Let F be the probability that the two genes in A_iA_i are exactly the same copies of the same gene in an ancestor. Then the frequencies of A_iA_i and A_iA_j $(i \neq j)$ are given respectively by $(1 - F)p_i^2 + Fp_i$ and $2(1 - F)p_ip_j$, where p_i is the frequency of the A_i allele in the population. In population genetics, F has been referred to as the inbreeding coefficient.

In wild natural populations, matings between individuals can hardly be expected to be random. Hence, in studying evolution theories, it is of considerable interests to study many other mating systems than random mating in the populations. Fisher [2] and Karlin [3, 4] have shown that many of these mating systems can be described by homogeneous Markov chains. In Examples 2.7 and 2.8, we have described the self-mating system and the full sib-mating system (brother-sister matings) respectively; other mating types which have been considered by Fisher and Karlin are the parent-offspring mating system, the half-sib mating system, the assortative mating system, first-cousin mating systems, second-cousin mating systems, as well as mixtures of these systems.

2.11.3. *Some mathematical methods for computing A^n, the nth power of a square matrix A*

Let A be a $p \times p$ matrix of real numbers. If there exists a number (in general, complex number) λ and a non-zero vector $\underset{\sim}{x}$ satisfying $A\underset{\sim}{x} = \lambda\underset{\sim}{x}$, then λ is called an eigenvalue of A and $\underset{\sim}{x}$ the eigenvector of A corresponding to λ. The restriction $\underset{\sim}{x} \neq \underset{\sim}{0}$ implies that the eigenvalues of A must satisfy $\det(A-\lambda I_p) = |A - \lambda I_p| = 0$. Now, $|A - \lambda I_p| = 0$ is a polynomial in λ of degree p; hence, $|A - \lambda I_p| = 0$ has p roots in the complex field. Let the distinct roots of A be $\lambda_1, \lambda_2, \ldots, \lambda_r$ repeated $k_1, k_2, \ldots, k_r$ times respectively ($k_1 + k_2 + \cdots + k_r = p$ as there are p roots). The matrix A is called *diagonable* iff, there exists a non-singular $p \times p$ matrix R such that:

$$R^{-1}AR = \Lambda = \begin{pmatrix} \lambda_1 I_{k_1} & & & \\ & \lambda_2 I_{k_2} & & 0 \\ & & \ddots & \\ 0 & & & \lambda_r I_{k_r} \end{pmatrix},$$

is a diagonal matrix with diagonal elements

$$\underbrace{\lambda_1, \ldots, \lambda_1,}_{k_1} \underbrace{\lambda_2, \ldots, \lambda_2,}_{k_2} \ldots, \underbrace{\lambda_r, \ldots, \lambda_r,}_{k_r} .$$

Partitioning R by $R = (\underset{k_1}{R_1}, \underset{k_2}{R_2}, \ldots \underset{k_r}{R_r})$ and R^{-1} by $R^{-1} = (\underset{k_1}{G'_1}, \underset{k_2}{G'_2}, \ldots, \underset{k_r}{G'_r})'$, then

$$R^{-1}AR = \Lambda \rightarrow AR = R\Lambda,\ R^{-1}A = \Lambda R^{-1},$$

and $A = R\Lambda R^{-1}$. These lead immediately to the following results:

1. $AR = R\Lambda \rightarrow AR_i = \lambda_i R_i$, $i = 1, 2 \cdots r$; and $R^{-1}A = \Lambda R^{-1} \rightarrow G_i A = \lambda_i G_i$, $i = 1, 2 \cdots r$. That is, the columns of R_i are independent right eigenvectors of the eigenvalue λ_i; the rows of G_i are independent left eigenvectors λ_i.

2. We have

$$A = R\Lambda R^{-1} = (R_1, R_2, \ldots, R_r)\Lambda(G_1', G_2', \ldots, G_r')' = \sum_{i=1}^{r} \lambda_i R_i G_i = \sum_{i=1}^{r} \lambda_i E_i,$$

where $E_i = R_i G_i$, $i = 1, 2, \ldots, r$. Next we show that

$$\sum_{i=1}^{r} E_i = I_p, \ E_i^2 = E_i, \text{ and } E_i E_j = 0$$

for all $i \neq j$ so that $A = \sum_{i=1}^{r} \lambda_i E_i$ is the spectral expansion of A. To prove these, note first that

$$\sum_{i=1}^{r} E_i = \sum_{i=1}^{r} R_i G_i = (R_1, R_2, \ldots, R_r)(G_1', G_2', \ldots, G_r')' = RR^{-1} = I_p;$$

$$I_p = R^{-1}R = (G_1', G_2', \ldots, G_r')'(R_1, R_2, \ldots, R_r)$$

$$= \begin{pmatrix} G_1 R_1 & G_1 R_2, & \cdots G_1 R_r \\ G_2 R_1 & G_2 R_2, & \cdots G_2 R_r \\ \vdots & \vdots & \vdots \\ G_r R_1 & G_r R_2, & \cdots G_r R_r \end{pmatrix}.$$

It follows that

$$G_i R_i = I_{k_i}, \ G_i R_j = 0, \quad \text{for } i \neq j.$$

Thus, $E_i^2 = E_i E_i = R_i G_i R_i G_i = R_i G_i = E_i$ and $E_i E_j = R_i G_i R_j G_j = 0$ for $i \neq j$.

If the eigenvalues λ_i $(i = 1, \ldots, r)$ of A are real numbers, then we have the following theorem for the spectral expansion of A.

Theorem 2.12.   Spectral expansion of square matrices. *If the $p \times p$ matrix A is diagonable with real eigenvalues, then the spectral expansion of A is* *given by $A = \sum_{i=1}^{r} \lambda_i E_i$, where $\lambda_1, \lambda_2, \ldots, \lambda_r$ are the real distinct eigenvalues of A and $E_i = \prod_{j \neq i} \frac{1}{(\lambda_i - \lambda_j)}(A - \lambda_j I_p)$.*

Proof.   A diagonable implies that there exists a non-singular matrix R such that:

$$
R^{-1}AR = \Lambda = \begin{pmatrix} \lambda_1 I_{k_1} & & & \\ & \lambda_2 I_{k_2} & & 0 \\ & & \ddots & \\ 0 & & & \lambda_r I_{k_r} \end{pmatrix},
$$

where $\lambda_1, \lambda_2, \ldots, \lambda_r$ are the real distinct eigenvalue of multiplicities $k_1, k_2, \ldots, k_r$ respectively. Further, λ_j real imply that R and R^{-1} are matrices of real numbers. Denote by D_j $(j = 1, \ldots, r)$ the $p \times p$ diagonal matrix with 1 at the $\sum_{i=1}^{j-1} k_i + 1, \ldots, \sum_{i=1}^{j-1} k_i + k_j$ diagonal positions but with 0 at all other positions. (The sum $\sum_{i=1}^{0} k_i$ is defined as 0 by convention). Then, $R^{-1}AR = \sum_{i=1}^{r} \lambda_i D_i; \sum_{i=1}^{r} D_i = I_p, D_i^2 = D_i$ and $D_i D_j = 0$ for $i \neq j$. Put $H_i = RD_i R^{-1}$, then $\sum_{i=1}^{r} H_i = I_p, H_i H_i = H_i, H_i H_j = 0$ for $i \neq j$. But,

$$
R^{-1} \left\{ \prod_{j \neq i} \frac{1}{(\lambda_i - \lambda_j)}(A - \lambda_j I_p) \right\} R = \left(\prod_{j \neq i} \frac{1}{(\lambda_i - \lambda_j)} \right) [R^{-1}(A - \lambda_1 I_p)R]
$$

$$
\times [R^{-1}(A - \lambda_2 I_p)R] \cdots [R^{-1}(A - \lambda_{i-1} I_p)R]
$$

$$
\times [R^{-1}(A - \lambda_{i+1} I_p)R] \cdots [R^{-1}(A - \lambda_r I_p)R] = D_i \,.
$$

Hence, $H_i = E_i$. $\qquad\qquad\qquad\qquad\qquad\qquad\qquad\qquad\qquad\qquad\qquad\qquad\square$

References

[1] J. F. Crow and M. Kimura, *An Introduction to Population Genetics Theory*, Harper and Row, New York (1970).

[2] R. A. Fisher, *The Theory of Inbreeding*, Second Edition, Oliver and Boyd, Edinburgh (1965).

[3] S. Karlin, *Equilibrium behavior of population genetic models with non-random mating, Part I, Preliminaries and special mating systems*, J. Appl. Prob. **5** (1968) 231–313.

[4] S. Karlin, *Equilibrium behavior of population genetic models with non-random mating, Part II, Pedigrees, homozygosity, and stochastic models*, J. Appl. Prob. **5** (1968) 487–566.

[5] W. Y. Tan, *Applications of some finite Markov chain theories to two locus selfing models with selection*, Biometrics **29** (1973) 331–346.

[6] D. T. Bishop, J. A. Williamson and M.H. Skolnick, *A model for restriction fragment length distributions*, Am. J. Hum. Genet. **35** (1983) 795–815.

[7] W. Y. Tan, *Chain multinomial models of HIV epidemic in homosexual population*, Math. Compt. Modelling **18** (1993) 29–72.

[8] W. Y. Tan, *On the chain multinomial model of HIV epidemic in homosexual populations and effects of randomness of risk factors*, in: *Mathematical population Dynamics* 3, eds. O. Arino, D. E. Axelrod and M. Kimmel, Wuerz Publishing Ltd., Winnepeg, Manitoba, Canada (1995).

[9] W. Y. Tan and R. H. Byers, *A stochastic model of HIV epidemic and the HIV infection distribution in a homosexual population*, Math Biosci. **113** (1993) 115–143.

[10] W. Y. Tan and Z. Z. Ye, *Estimation of HIV infection and HIV incubation via state space models.* Math. Biosci. **167** (2000) 31–50.

[11] W. Y. Tan and Z. Z. Ye, *Some state space models of HIV epidemic and applications for the estimation of HIV infection and HIV incubation*, Comm. Statist. (Theory and Methods) **29** (2000) 1059–1088.

[12] W. Y. Tan, S. R. Lee and S. C. Tang, *Characterization of HIV infection and seroconversion by a stochastic model of HIV epidemic*, Math. Biosci. **126** (1995) 81–123.

[13] W. Feller, *An Introduction to Probability Theory and Its Applications, Third Edition*, Wiley, New York (1968).

[14] R. A. Fisher, *The Genetic Theory of Natural Selection*, Second Edition, Dover Publication, New York (1958).

[15] W. Y. Tan and Z. Xiang, *State space models of the HIV epidemic in homosexual populations and some applications*, Math. Biosci. **152** (1998) 29–61.

[16] W. Y. Tan and Z. Xiang, *State space models for the HIV epidemic with variable infection in homosexual populations by state space models*, J. Statist. Planning Inference **78** (1999) 71–87.

[17] I. M. Longini, W. S. Clark, L. I. Gardner and J. F. Brundage, *The dynamics of CD4+ T–lymphocyte decline in HIV–infected individuals: A Markov modeling approach*, J. Aids **4** (1991) 1141–1147.

[18] I. M. Longini, R. H. Byers, N. A. Hessol and W. Y. Tan, *Estimation of the stage-specific numbers of HIV infections via a Markov model and backcalculation*, Statist. Med. **11** (1992) 831–843.

[19] I. M. Longini, W. S. Clark and J. Karon, *Effects of routine use of therapy in slowing the clinical course of human immunodeficiency virus (HIV) infection in a population based cohort*, Amer. J. Epidemiol. **137** (1993) 1229–1240.

[20] G. Satten and I. M. Longini, *Estimation of incidence of HIV infection using cross-sectional marker survey*, Biometrics **50** (1994) 675–688.

[21] G. Satten and I. M. Longini, *Markov chain with measurement error: Estimating the "True" course of marker of the progression of human immunodeficiency virus disease*, Appl. Statist. **45** (1996) 275–309.

[22] F. R. Gantmacher, *The Theory of Matrices*, Chelsea, New York (1959).

[23] T. Harris, *The Theory of Branching Processes*, Springer-Verlag, Berlin (1963).

[24] J. H. Bennett. *On the theory of random mating*, Ann. Eugen. **18** (1954) 311–317.

Chapter 3

Stationary Distributions and MCMC in Discrete Time Markov Chains

3.1. Introduction

In natural systems, most of the processes will eventually reach a steady-state condition as time progresses. This is true in Markov processes as well as in many non-Markovian processes. Thus, as time progresses, under some general conditions the homogeneous Markov chain with discrete time will eventually reach a steady-state condition under which the probability distribution of the chain is independent of time. In this chapter we will illustrate how and under what conditions the Markov chain $X(t)$ with discrete time will converge to some stationary distributions. Then we will illustrate how this theory can be used to generate some computational algorithms to solve many optimization problems in many applied fields including biomedical systems. This latter algorithm has been called the MCMC (Markov Chain Monte Carlo) method. First we give the following definition for stationary distributions of Markov chains (see Remark 3.1).

Definition 3.1.   Let $\{X(t), t = 0, 1, \ldots,\}$ be a homogeneous Markov chain with state space $S = \{0, 1, \ldots, \infty\}$. Then a probability density function π_k over S is called *a stationary distribution of the chain $X(t)$* iff

$$\pi_k = \sum_{j=0}^{\infty} \pi_j P_{jk}, \quad k = 0, 1, 2, \ldots,$$

where P_{jk} is the one-step transition probability.

Remark 3.1. Example 2.12 shows that homogeneous Markov chains with infinite states may not even have persistent states and hence may not have stationary distributions. However, for the definition of stationary distribution, the chain must be homogeneous.

Notice that if $\{\pi_j, j \in S\}$ is a stationary distribution of the Markov chain $X(t)$, then by application of the Chapman–Kolmogorov equation, it follows easily by mathematical induction that

$$\pi_k = \sum_{j=0}^{\infty} \pi_j P_{jk} \Rightarrow \pi_k = \sum_{j=0}^{\infty} \pi_j P_{jk}(n) \text{ for all } n \geq 1.$$

If the chain is finite with $m\,(m > 1)$ states and with 1-Step transition matrix $P = (P_{ij})$, then, with $\underset{\sim}{\pi}' = (\pi_1, \pi_2, \ldots, \pi_m)$, $\underset{\sim}{\pi}' = \underset{\sim}{\pi}'P(n) = \underset{\sim}{\pi}'P^n$ for all positive integer n.

In the next two sections, we will illustrate how the limiting results of the transition probabilities of the chain $X(t)$ characterize the stationary distributions of the chain $X(t)$. For deriving these results, we will need the following mathematical results which we present as Lemmas 3.1–3.3.

Lemma 3.1. **(Cesaro summability).** *If $a_n \to a$ as $n \to \infty$, then*

$$\frac{1}{N} \sum_{m=1}^{N} a_m \to a \text{ as } N \to \infty.$$

Note that the Converse result of this is not true in general.

For proof, see [1, p. 72].

Let $X(t)$, $t \in T = \{0, 1, 2, \ldots\}$ be a homogeneous Markov chain with state space $S = \{0, 1, 2, \ldots\}$.

Lemma 3.2. **(Fatou's lemma).** *If $f_j(n) \geq 0$, $P_{jk} \geq 0$ and if $\lim_{n\to\infty} f_j(n)$ exists for all j, then*

$$\lim_{n\to\infty} \sum_j f_j(n) P_{jk} \geq \sum_j \left(\lim_{n\to\infty} f_j(n) \right) P_{jk};$$

if $g(x) \geq 0$, $f(n,x) \geq 0$ and $\lim_{n\to\infty} f(n,x)$ exists for all x, then

$$\lim_{n\to\infty} \int f(n,x) g(x)\,dx \geq \int \left(\lim_{n\to\infty} f(n,x) \right) g(x)\,dx.$$

Lemma 3.3. **(Lebesque's dominated convergence theorem).** *If* $\lim_{n\to\infty} f_j(n)$ *exists for all* j, $P_{jk} \geq 0$, $|f_j(n)| \leq b_j$ *and if* $\sum_j b_j P_{jk} < \infty$, *then*

$$\lim_{n\to\infty} \left(\sum_j f_j(n) P_{jk} \right) = \sum_j \left(\lim_{n\to\infty} f_j(n) \right) P_{jk} \, ;$$

If $g(x) \geq 0, \lim_{n\to\infty} f(n,x)$ *exists for all* x, $|f(n,x)| \leq h(x)$ *and if* $\int h(x)g(x)\,dx < \infty$, *then*

$$\lim_{n\to\infty} \int f(n,x)g(x)\,dx = \int \left(\lim_{n\to\infty} f(n,x) \right) g(x)\,dx \,.$$

The proofs of Lemmas 3.2 and 3.3 can be found in most textbooks of real analysis, see for example, A. N. Kolmogorov and S. V. Fomin [2].

3.2. The Ergodic States and Some Limiting Theorems

Consider a homogeneous Markov chain $\{X(t), t = 0, 1, \ldots, \infty\}$ with state space $S = \{0, 1, \ldots, \infty\}$. If the state j is transient, then $\sum_{n=1}^{\infty} P_{ij}(n) < \infty$ so that $\lim_{n\to\infty} P_{ij}(n) = 0$ for all $i \in S$. Thus, by the Cesaro Lemma, $\frac{1}{N} \sum_{n=1}^{N} P_{ij}(n) \to 0$ as $N \to \infty$ if j is transient. If j is persistent, then the existence of $\lim_{n\to\infty} P_{ij}(n)$ depends on whether or not j is periodic.

Definition 3.2. The persistent state j is said to have *period* d_j $(d_j \geq 1)$ iff $P_{jj}(n) > 0$ and $n > 0$ imply $n = md_j$ for some $m > 0$. j is called aperiodic if $d_j = 1$.

Using this definition, it is seen that the greatest common divisor (g.c.d.) of n for which $P_{jj}(n) > 0$ is d_j; that is, for any finite set of integers $n_1, n_2, \ldots, n_r$ such that $P_{jj}(n_s) > 0$, $s = 1, 2, \ldots, r$, the g.c.d. of $\{n_1, n_2, \ldots, n_r\}$ is d_j. By results of the lemma in Sec. 3.8, there exists an integer $M > 0$ such that for all $m \geq M$, $P_{jj}(md_j) > 0$.

Theorem 3.1. *If* $j \in S$ *is persistent and aperiodic, then the limit* $\lim_{n\to\infty} P_{jj}(n)$ *exists and is given by*

$$\lim_{n\to\infty} P_{jj}(n) = \frac{1}{\mu_j} \,.$$

This limit is positive iff $\mu_j < \infty$ *or iff* j *is a positive state.*

Proof. To prove this result, notice that for $n = 2, \ldots,$

$$P_{ij}(n) = f_{ij}(1)P_{jj}(n-1) + \cdots + f_{ij}(n)P_{jj}(0) \,. \tag{3.1}$$

Let $Q_{ij}(s) = \sum_{n=0}^{\infty} s^n P_{ij}(n)$ and $F_{ij}(s) = \sum_{n=0}^{\infty} s^n f_{ij}(n)$.

On multiplying both sides of Eq. (3.1) by s^n and summing over n from 1 to ∞, the left side gives $Q_{ij}(s) - \delta_{ij}$ whereas the right side yields

$$\sum_{n=1}^{\infty} s^n \sum_{k=1}^{n} f_{ij}(k)P_{jj}(n-k) = \sum_{k=1}^{\infty} s^k f_{ij}(k) \sum_{n=k}^{\infty} s^{n-k} P_{jj}(n-k)$$

$$= F_{ij}(s)Q_{jj}(s) \,.$$

Hence, we have

$$Q_{ij}(s) - \delta_{ij} = F_{ij}(s)Q_{jj}(s) \,. \tag{3.2}$$

This gives

$$(1 - s)Q_{jj}(s) = \frac{1 - s}{1 - F_{jj}(s)} \,. \tag{3.3}$$

Notice that $F_{jj}(s)$ is the pgf of the first return time of the state j if j is persistent and aperiodic. Hence, if j is persistent and aperiodic, the mean return time μ_j of j is

$$\mu_j = \lim_{s \to 1} \frac{F_{jj}(s) - 1}{s - 1} = F'_{jj}(1) = \sum_{n=1}^{\infty} n f_{jj}(n) \,.$$

It follows that the right side of Eq. (3.3) is $1/\mu_j$. To show that the left side of Eq. (3.3) is $\lim_{n \to \infty} P_{jj}(n)$, notice that with $P_{ii}(-1) = 0$, we have $|P_{ii}(k) - P_{ii}(k-1)| \le 2$; hence, by the Lebesque's dominant convergence theorem (see Lemma 3.3), we obtain:

$$\lim_{s \to 1}(1 - s)Q_{ii}(s) = \lim_{s \to 1} \lim_{n \to \infty} \sum_{k=0}^{n} s^k \{P_{ii}(k) - P_{ii}(k-1)\}$$

$$= \lim_{n \to \infty} \lim_{s \to 1} \sum_{k=0}^{n} s^k \{P_{ii}(k) - P_{ii}(k-1)\}$$

$$= \lim_{n \to \infty} P_{ii}(n) \,.$$

Thus,

$$\lim_{n\to\infty} P_{ii}(n) = \frac{1}{\mu_i}\,.$$

If $i \neq j$, then by using Eq. (3.2), we have:

$$\lim_{s\to1}(1-s)Q_{ij}(s) = \lim_{s\to1} F_{ij}(s)\lim_{s\to1}(1-s)Q_{jj}(s) = \frac{f_{ij}}{\mu_j}\,.$$

With $P_{ij}(0) = 0$, if $i \neq j$, we obtain by using exactly the same approach as above:

$$\lim_{n\to\infty} P_{ij}(n) = \lim_{s\to1}(1-s)Q_{ij}(s)\,.$$

Hence, if j is persistent and aperiodic, then, for all i and j

$$\lim_{n\to\infty} P_{ij}(n) = \frac{f_{ij}}{\mu_j}\,.$$

The above proof is simple and straightforward but is not vigorous mathematically. More vigorous alternative proofs can be found in many texts; see for example, Karlin and Taylor [3].

By Theorem 2.7, if i is persistent and $i \to j$, then j is persistent and $j \to i$; furthermore, $f_{ij} = f_{ji} = 1$. Thus, if i and j are persistent and $i \to j$, then

$$\lim_{n\to\infty} P_{ij}(n) = \frac{1}{\mu_j}\,.$$

If j is persistent and periodic with period $d_j > 1$, then $P_{jj}(n) = 0$ if $n \neq rd_j$ for some integer r and the g.c.d. of m_i for $P_{jj}(m_i) > 0$ is d_j. The following theorem shows that although $\lim_{n\to\infty} P_{jj}(n)$ does not exist, $\lim_{n\to\infty} P_{jj}(nd_j)$ does exist, however.

Theorem 3.2. *If $j \in S$ is persistent and periodic with period $d_j > 1$, then* $\lim_{n\to\infty} P_{jj}(nd_j)$ *exists and is given by*

$$\lim_{n\to\infty} P_{jj}(nd_j) = \frac{d_j}{\mu_j}\,.$$

Proof. Defining $P_{jj}(nd_j) = P_{jj}^{(*)}(n)$ and $f_{jj}(nd_j) = f_{jj}^{(*)}(n)$, then we have:

$$P_{jj}^{(*)}(n) = \sum_{m=1}^{n} f_{jj}^{(*)}(m)P_{jj}^{(*)}(n-m)\,.$$

Since

$$\sum_{n=1}^{\infty} n f_{jj}^{(*)}(n) = \frac{1}{d_j} \sum_{n=1}^{\infty} (n d_j) f_{jj}(n d_j) = \frac{\mu_j}{d_j},$$

then we have, by Theorem 3.1:

$$\lim_{n\to\infty} P_{jj}(n d_j) = \lim_{n\to\infty} P_{jj}^{(*)}(n) = \frac{d_j}{\mu_j}. \qquad \square$$

By Theorem 3.2, if j is periodic with period $d_j > 1$ and if $\mu_j < \infty$, then the limit $\lim_{n\to\infty} P_{jj}(n)$ does not exist. This follows readily from the observation that the sub-sequence $\{P_{jj}(n d_j + 1), n = 1, \ldots, \infty\}$ converges to 0 whereas the sub-sequence $\{P_{jj}(n d_j), n = 1, \ldots, \infty\}$ converges to $d_j/\mu_j > 0$ by Theorem 3.2. The following theorem shows, however, that the Cesaro limit for j always exists regardless whether or not j is periodic.

Theorem 3.3.   *If $j \in S$ is persistent and periodic with period $d_j > 1$, then*

$$\lim_{n\to\infty} \frac{1}{N} \sum_{n=1}^{N} P_{jj}(n) = \frac{1}{\mu_j}.$$

The limit is positive iff $\mu_j < \infty$ iff j is positive.

Proof.   We have

$$\frac{1}{N} \sum_{n=1}^{N} P_{jj}(n) = \frac{1}{N} \sum_{m=1}^{M} P_{jj}(m d_j),$$

where $M d_j \leq N < (M+1) d_j$.

Noting that $\frac{M d_j}{N} \to 1$ and $\lim_{m\to\infty} P_{jj}^*(m) = P_{jj}(m d_j) = \frac{d_j}{\mu_j}$ by Theorem 3.2, we have:

$$\frac{1}{N} \sum_{n=1}^{N} P_{jj}(n) = \frac{M d_j}{N d_j} \times \frac{1}{M} \sum_{m=1}^{M} P_{jj}^*(m) \to \frac{1}{\mu_j}. \qquad \square$$

From Theorems 3.1–3.3, if j is persistent and positive, then the Cesaro limit is positive. If j is persistent, aperiodic and positive then the limit $\lim_{n\to\infty} P_{jj}(n)$ is positive. Persistent states which are aperiodic and positive are also called "ergodic states". For closed sets of persistent states, the properties "positiveness" and "periodicity" are shared by all states in the closed set. This is proved in the following theorem.

Theorem 3.4. *Consider a closed set C of persistent states. Then all states in C have the same period. Furthermore, either all states in C are positive or all states in C are null.*

Proof. Let $i \in C$ and $j \in C$. It then suffices to show:

(1) $d_i = d_j$, where d_i is the period of i,
and

(2) either both i and j are positive or both i and j are null.

Now $i \leftrightarrow j$ implies that there exist $m_0 > 0$ and $n_0 > 0$ such that

$$P_{ji}(m_0) > 0 \text{ and } P_{ij}(n_0) > 0.$$

By the Chapman–Kolmogorov equation, then,

$$P_{ii}(m_0 + n_0) \geq P_{ij}(n_0)P_{ji}(m_0) > 0$$

and

$$P_{jj}(m_0 + n_0) \geq P_{ji}(m_0)P_{ij}(n_0) > 0.$$

This implies that $m_0 + n_0$ are divisible by both d_i (period of i) and d_j (period of j).

Let k be a positive integer not divisible by d_i such that $P_{jj}(kd_j) > 0$. Then

$$P_{ii}(n_0 + m_0 + kd_j) \geq P_{ij}(n_0)P_{jj}(kd_j)P_{ji}(m_0) > 0.$$

Thus d_i divides $m_0 + n_0 + kd_j$ so that d_j is divisible by d_i. Similarly, by exactly the same approach, one can prove that d_i is also divisible by d_j. Hence $d_i = d_j = d$.

To prove that both i and j are positive or both i and j are null, notice that with $d_i = d_j = d$, $P_{ii}(m_0 + nd + m_0) \geq P_{ij}(m_0)P_{jj}(nd)P_{ji}(n_0) > 0$, and $P_{jj}(m_0 + nd + m_0) \geq P_{ji}(n_0)P_{ii}(nd)P_{ij}(m_0) > 0$. Hence j positive $\Rightarrow i$ positive; i null $\Rightarrow j$ null. $\qquad\square$

As an illustration, consider the random walk chain in Example 2.12. If $p = \frac{1}{2}$, then the chain is persistent and irreducible; since the state 0 has period 2, so all states have period 2. When $p = \frac{1}{2}$, the mean return time of 0 is

$$\nu_0 = \sum_{n=1}^{\infty}(2n)P_{00}(2n) \cong \sum_{n=1}^{\infty}(2n)/(\sqrt{n\pi}) = \infty.$$

Hence for the persistent random walk chain given in Example 2.12, the state 0 is null and so are all other states. On the other hand, all persistent states in Examples 2.1, 2.2 and 2.4 are aperiodic and have mean return time 1; hence all persistent states in these examples are ergodic states.

3.3. Stationary Distributions and Some Examples

With the above limiting results, we are now ready to derive the stationary distributions of homogeneous Markov chains.

Theorem 3.5. *In a homogeneous irreducible Markov chain $\{X(t), t \in T = (0, 1, 2, \ldots)\}$, if the states are persistent positive, then the stationary distribution exists and is uniquely given by the limits*

$$\lim_{N \to \infty} \frac{1}{N} \sum_{n=1}^{N} P_{jj}(n) = \frac{1}{\mu_j} = \pi_j > 0 \quad j = 0, 1, \ldots, \infty.$$

Proof. Since the chain is irreducible, we have, for any two states j and k, $j \leftrightarrow k$; further, since the states are persistent, $f_{jk} = 1$. Now, by Theorem 3.3,

$$\lim_{N \to \infty} \frac{1}{N} \sum_{n=1}^{N} P_{jk}(n) = \frac{f_{jk}}{\mu_k} = \frac{1}{\mu_k} > 0$$

as the states are positive.

But, using Chapman–Kolmogorov equation, $P_{kk}(n+1) = \sum_j P_{kj}(n) P_{jk}$ so that

$$\frac{N+1}{N} \left[\frac{1}{N+1} \sum_{n=1}^{N+1} P_{kk}(n) - \frac{P_{kk}}{N+1} \right] = \frac{1}{N} \sum_{n=1}^{N} P_{kk}(n+1)$$

$$= \sum_j \left[\frac{1}{N} \sum_{n=1}^{N} P_{kj}(n) \right] P_{jk}.$$

Letting $N \to \infty$, we obtain $\pi_k \geq \sum_j \pi_j P_{jk}$ by Fatou's lemma. Summing over k,

$$\sum_k \pi_k \geq \sum_j \pi_j \sum_k P_{jk} = \sum_j \pi_j.$$

Thus $\pi_k = \sum_j \pi_j P_{jk}$.

Now $1 = \sum_k P_{jk}(n)$ for all $n \geq 1$ so that

$$1 = \sum_k \frac{1}{N} \sum_{n=1}^{N} P_{jk}(n)$$

for all N.

By Fatous Lemma,

$$1 \geq \sum_k \pi_k > 0 \,.$$

Further,

$$\pi_k = \sum_j \pi_j P_{jk} \Rightarrow \pi_k = \sum_j \pi_j P_{jk}(n)$$

for all $n \geq 1$. This implies that

$$\pi_k = \sum_j \pi_j \frac{1}{N} \sum_{n=1}^{N} P_{jk}(n) \,.$$

Since $\lim_{N \to \infty} \frac{1}{N} \sum_{n=1}^{N} P_{jk}(n)$ exists, $1 \geq \frac{1}{N} \sum_{n=1}^{N} P_{jk}(n) \geq 0$ and $0 < \sum_j \pi_j \leq 1$, by Lebesque's dominated convergence theorem,

$$\pi_k = \lim_{N \to \infty} \sum_j \pi_j \frac{1}{N} \sum_{n=1}^{N} P_{jk}(n) = \left(\sum_j \pi_j \right) \pi_k \,.$$

Thus, $\sum_j \pi_j = 1$, $1 > \pi_j > 0$; or, $(\pi_1, \pi_2, \ldots)$ is a stationary distribution of the chain. Suppose $(\mu_1, \mu_2, \ldots)$ is another stationary distribution, i.e., $\mu_k = \sum_j \mu_j P_{jk}$; then $\mu_k = \sum_j \mu_j P_{jk}(n)$ for all $n \geq 1$ so that

$$\mu_k = \sum_j \mu_j \frac{1}{N} \sum_{n=1}^{N} P_{jk}(n)$$

By Lebesque's dominated convergence theorem, we then have

$$\mu_k = \left(\sum_j \mu_j \right) \pi_k = \pi_k \,, \quad k = 0, 1, 2, \ldots \qquad \square$$

Corollary 3.1. *Let $\{X(n), n = 1, \ldots, \infty\}$ be a homogeneous Markov chain with transition probabilities $\{P_{ij}(n), i, j = 0, 1, \ldots, \infty\}$. If the chain is irreducible, aperiodic and positive, then $\lim_{n \to \infty} P_{ij}(n) = \pi_j > 0$ exists and the stationary distribution is uniquely given by $\pi_j, j = 0, 1, \ldots, \infty$.*

The proof is trivial and is left as an exercise; see Exercise 3.1.

Let $\{X(t), t \in T = (0, 1, 2, \ldots)\}$, be a homogeneous Markov chain. If the chain contains persistent states, then, as shown previously, the persistent states can be put into a disjoint union of irreducible closed sets of persistent states. Suppose that this chain contains persistent positive states. Then, there exists an irreducible closed set C_k of positive persistent states. For this closed set, we have, using Theorem 3.5, a stationary distribution, say $\underset{\sim}{\pi}_k^{(\star)} = (\pi_{1k}^{(\star)}, \pi_{2k}^{(\star)}, \ldots)$. It is then easily seen that $(\underset{\sim}{Q}', \underset{\sim}{\pi}_k^{(\star)}, \underset{\sim}{Q}', \underset{\sim}{Q}')$ is a stationary distribution of the chain. Suppose we have another irreducible closed set C_l of positive persistent states, $(l > k)$ then $(\underset{\sim}{Q}', \underset{\sim}{Q}', \underset{\sim}{\pi}_l^{(\star)}, \underset{\sim}{Q}')$ is another stationary distribution. Obviously, for any $0 < c < 1$,

$$c(\underset{\sim}{Q}', \underset{\sim}{\pi}_k^{(\star)}, \underset{\sim}{Q}', \underset{\sim}{Q}') + (1 - c)(\underset{\sim}{Q}', \underset{\sim}{Q}', \underset{\sim}{\pi}_l^{(\star)}, \underset{\sim}{Q}')$$

is also a stationary distribution of the chain. Hence, if the chain contains more than one irreducible closed set of positive persistent states, then the chain contains an infinite number of stationary distributions. To summarize, we have:

(a) If the chain does not contain positive persistent states, then the stationary distribution does not exist. Notice that this is possible only if the state space of the chain contains an infinite number of states.

(b) If the chain contains only one irreducible closed set of positive persistent states, the stationary distribution exists and is unique.

(c) If the chain contains more than one irreducible closed set of persistent states, the stationary distributions exist and is infinitely many.

As a special case of the above result, we see that, for finite homogeneous Markov chain, the stationary distributions always exist. The stationary distribution is unique if and only if the chain contains only one irreducible closed set. All these results follow from the fact that finite homogeneous Markov chain must contain persistent states and all persistent states must be positive; see Exercise 3.2.

Example 3.1. The base substitution model in human DNA dimers.
In Example 2.5, we have considered the DNA base substitution model in humans considered by Bishop *et al.* [4]. This is an irreducible, aperiodic and homogeneous Markov chain with discrete time and with state space $S = \{A, C, G, T, AG, AGC, AGCT\}$. Since the rows of the one-step transition matrix P sum up to 1, 1 is an eigenvalue with multiplicity 1 and a right eigenvector of 1 is $\underset{\sim}{1}_7$, a 7×1 column of 1"s. Let $\underset{\sim}{\pi} = \{\pi_1, \ldots, \pi_7\}'$ be the left eigenvector of 1. Then $\underset{\sim}{\pi}$ is the stationary distribution and by Corollary of Theorem 3.5, $\lim_{n \to \infty} P(n) = \lim_{n \to \infty} P^n = \underset{\sim}{1}_7 \underset{\sim}{\pi}'$. That is, $\underset{\sim}{\pi}'P = \underset{\sim}{\pi}'$.

To derive $\underset{\sim}{\pi}$, partition $\underset{\sim}{\pi}' = (\underset{\sim}{x}', \underset{\sim}{y}')$, where $\{\underset{\sim}{x}' = (\pi_1, \ldots, \pi_4), \underset{\sim}{y}' = (\pi_5, \pi_6, \pi_7)\}$ and partition P by

$$P = \begin{pmatrix} P_{11} & P_{12} \\ P_{21} & P_{22} \end{pmatrix}, \quad \text{where} \quad P_{11} = \begin{pmatrix} 0.32 & 0.18 & 0 & 0.27 \\ 0.37 & 0.23 & 0.05 & 0.35 \\ 0.30 & 0.21 & 0.25 & 0.24 \\ 0.23 & 0.19 & 0.25 & 0.33 \end{pmatrix},$$

$$P_{12} = \begin{pmatrix} 0.23 & 0 & 0 \\ 0 & 0 & 0 \\ 0 & 0 & 0 \\ 0 & 0 & 0 \end{pmatrix},$$

$$P_{21} = \begin{pmatrix} 0.30 & 0 & 0.25 & 0.24 \\ 0.37 & 0.23 & 0.05 & 0 \\ 0.23 & 0.19 & 0.25 & 0.33 \end{pmatrix} \quad \text{and} \quad P_{22} = \begin{pmatrix} 0 & 0.21 & 0 \\ 0 & 0 & 0.35 \\ 0 & 0 & 0 \end{pmatrix}.$$

Then $\underset{\sim}{x}'P_{12} + \underset{\sim}{y}'P_{22} = \underset{\sim}{y}'$ which gives

$$\underset{\sim}{y}' = (\pi_5, \pi_6, \pi_7) = \pi_1(0.23, \ 0.0483, \ 0.016905) = \pi_1 \underset{\sim}{a}'.$$

Further, the constraint $\underset{\sim}{1}'_7 \underset{\sim}{\pi} = 1$ gives

$$s\pi_1 + \pi_2 + \pi_3 + \pi_4 = 1, \quad \text{where } s = 1.289205.$$

Let $\underset{\sim}{b}_i$ be the ith column of $I_4 - P'_{11}$ and put

$$\underset{\sim}{c}_1 = \underset{\sim}{b}_1 - P'_{21}\underset{\sim}{a}, \ \underset{\sim}{c}_i = \underset{\sim}{b}_i, \quad i = 2,3,4, \quad \underset{\sim}{c}_5 = (s,1,1,1)'.$$

Let $\underset{\sim}{e}' = (0,0,0,0,1)$. Let C_1 be the 4×4 matrix with the ith column being given by $\underset{\sim}{c}_i$ and C the 5×4 matrix with the first 4 rows being given by the rows of C_1 and the 5th row being given by $\underset{\sim}{c}'_5$. On substituting these results into the equation $\underset{\sim}{\pi}'P = \underset{\sim}{\pi}'$, we obtain, with $\underset{\sim}{0}$ denoting a 4×1 column of 0's:

$$\underset{\sim}{x}'P_{11} + \pi_1 \underset{\sim}{a}'P_{21} = \underset{\sim}{x}' \text{ so that } (I_4 - P_{11})'\underset{\sim}{x} - \pi_1 P'_{21}\underset{\sim}{a} = C_1\underset{\sim}{x} = \underset{\sim}{0}\,;$$

$$C\underset{\sim}{x} = \underset{\sim}{e} \text{ so that } \underset{\sim}{x} = (C'C)^{-1}C'\underset{\sim}{e}\,.$$

This gives the stationary distribution as

$$\underset{\sim}{\pi}' = (\underset{\sim}{x}', \pi_1\underset{\sim}{a}') = (0.2991,\ 0.1846,\ 0.1354,\ 0.2906,\ 0.0688,\ 0.0144,\ 0.0051)\,.$$

Example 3.2.   The frequencies of genotypes under equilibrium conditions in natural populations. For the Markov chain of genotype frequencies described in Example 2.2, the eigenvalues of the one-step transition probability P are given by $\{\lambda_1 = 1, \lambda_2 = \frac{1}{2}, \lambda_3 = 0\}$. Then, with $\underset{\sim}{1}_3$ denoting a 3×1 column of 1's and with $\underset{\sim}{u}' = (p^2, 2pq, q^2)$, $q = 1 - p$, the spectral matrices are given by:

$$E_1 = \prod_{j=2}^{3} \frac{1}{\lambda_1 - \lambda_j}(P - \lambda_j I_3) = \begin{pmatrix} p^2 & 2pq & q^2 \\ p^2 & 2pq & q^2 \\ p^2 & 2pq & q^2 \end{pmatrix} = \underset{\sim}{1}_3\underset{\sim}{u}'\,,$$

$$E_2 = \prod_{j\neq 2} \frac{1}{\lambda_2 - \lambda_j}(P - \lambda_j I_3) = \begin{pmatrix} 2pq & 2q(q-p) & -2q^2 \\ p(q-p) & 1-4pq & -q(q-p) \\ -2p^2 & -2p(q-p) & 2pq \end{pmatrix}\,,$$

$$E_3 = \prod_{j=1}^{2} \frac{1}{\lambda_3 - \lambda_j}(P - \lambda_j I_3) = \begin{pmatrix} q^2 & -2q^2 & q^2 \\ -pq & 2pq & -pq \\ p^2 & -2p^2 & p^2 \end{pmatrix}\,.$$

Hence the spectral expansion of $P(n) = P^n$ is given by:

$$P(n) = P^n = \sum_{i=1}^{3} \lambda_i^n E_i = E_1 + \frac{1}{2^n} E_2 \,.$$

It follows that $\lim_{n \to \infty} P(n) = E_1 = \underset{\sim}{1}_3 \underset{\sim}{u}'$. That is, the stationary distribution of the frequencies of the three genotypes $\{AA, Aa, aa\}$ are given by $(p^2, 2pq, q^2)$ respectively. This is equivalent to saying that the Hardy–Weinberg law is in fact the stationary distribution of the genotypes.

Example 3.3. The Wright model under mutation in population genetics. Consider the Wright model under mutation given by Example 2.11 for a single locus with two alleles A and a. This is a Markov chain with the one-step transition probabilities given by:

$$P_{ij} = \binom{2N}{j} p_i^j (1 - p_i)^{2N-j} \,,$$

where

$$p_i = \frac{i}{2N}(1 - \alpha_1) + \left(1 - \frac{i}{2N}\right) \alpha_2 = \alpha_2 + \frac{i}{2N}(1 - \alpha_1 - \alpha_2), \ 0 \le \alpha_1, \alpha_2 \le 1 \,.$$

This chain is homogenous if the α_i, $(i = 1, 2)$ are independent of time t. Also, as proved in Example 2.15, the eigenvalues of the one-step transition matrix $P = (P_{ij})$ are given by:

$$\lambda_1 = 1, \text{ and } \lambda_{k+1} = \frac{1}{(2N)^k}(1 - \alpha_1 - \alpha_2)^k \left\{ \prod_{i=1}^{k}(2N - i + 1) \right\} \,,$$

for $k = 1, 2, \ldots, 2N$.

Assuming that the α_i's are independent of time t so that the chain is homogeneous, we will derive the stationary distribution in two cases.

Case 1: If $\alpha_i \ne 0$, for $i = 1, 2$ and if $\alpha_1 + \alpha_2 < 1$, then the chain is irreducible and aperiodic; further, all eigenvalues of P are distinct. (The largest eigenvalue is $\alpha_1 = 1$ and all other eigenvalues are positive but less than 1).

Let $\underset{\sim}{x}_j$ and $\underset{\sim}{y}_j$ be the right eigenvector and the left eigenvector of λ_j satisfying $\underset{\sim}{x}_j' \underset{\sim}{y}_j = 1$, respectively. Then, since the rows of P sum to one,

$\underset{\sim}{x}'_1 = \underset{\sim}{1}_{2N+1}$, a $(2N+1) \times 1$ column of 1's and

$$P(n) = P^n = \underset{\sim}{1}_{2N+1}\,\underset{\sim}{y}'_1 + \sum_{i=2}^{2N+1} \lambda_i^n E_i\,,$$

where $E_i = \underset{\sim}{x}_i\,\underset{\sim}{y}'_i$, $i = 1,\ldots,2N+1$.

Hence, $\underset{\sim}{y}_1$ is the stationary distribution of the above Markov chain if $\alpha_i \neq 0$ and if α_i are independent of time t. Although for given $\{N, \alpha_i, i = 1, 2\}$, numerical solution of $\underset{\sim}{y}'_1 P = \underset{\sim}{y}'_1$ is possible, exact analytical solution is not yet available; in Chap. 6, we will approximate this stationary distribution for large N by a diffusion process.

Case 2: If $\alpha_1 = \alpha_2 = 0$, then 0 and $2N$ are absorbing states while all other states are transient. In this case $P_{00} = P_{2N,2N} = 1$ and there are $2N$ distinct eigenvalues given by:

$$\lambda_1 = \lambda_2 = 1, \text{ and } \lambda_{k+2} = \frac{1}{(2N)^k} \prod_{i=1}^{k}(2N - i + 1), \quad k = 1,\ldots,2N - 1\,.$$

The eigenvalue 1 has multiplicity 2 while all other eigenvalues have multiplicity 1 and have values between $0 < \lambda_i < 1$ $(i = 3,\ldots,2N+1)$. It is easily observed that $\underset{\sim}{y}_1 = (1, 0, \ldots, 0)'$ and $\underset{\sim}{y}_2 = (0, 0, \ldots, 0, 1)'$ are two independent left eigenvectors of $\lambda_1 = \lambda_2 = 1$. Hence, for any real $0 < u < 1$, $u\underset{\sim}{y}_1 + (1-u)\underset{\sim}{y}_2$ is a stationary distribution of the above chain.

This case corresponds to the case of "genetic drift" or "Wright drift" in population genetics.

3.4. Applications of Stationary Distributions and Some MCMC Methods

In the past 15 years, important breakthroughs have been made in computational algorithms by the application of the stationary distribution of discrete time Markov chains. The basic idea is that if one wishes to generate random samples from some unknown probability distribution, one needs only to construct an irreducible aperiodic Markov chain with discrete time so that the stationary distribution of the chain coincides with the probability distribution in question. Thus, even though the probability distribution is unknown, it is still possible to generate random samples from this probability distribution if an irreducible Markov chain can be constructed. This type of methods has been called the Markov Chain Monte Carlo Method (MCMC). In this section, we will describe three such methods and illustrate its applications by some examples from genetics. In Chap. 9 we will illustrate how to apply these methods to solve some difficult estimation problems in AIDS and carcinogenesis. For more general theories and methods, we refer the readers to text books [5, 6].

3.4.1. *The Gibbs sampling method*

To illustrate, consider k random variables, say, $X_1, \ldots, X_k$. Suppose that the conditional distribution $p(x_i|x_1, \ldots, x_{i-1}, x_{i+1}, \ldots, x_k) = p(x_i|\underset{\sim}{x}_{(i)})$ of X_i given all other random variables is available for all $i = 1, \ldots, k$. (We denote by $\underset{\sim}{x}_{(i)}$ the sample with x_i being deleted.) Then the Gibbs sampling procedure starts with an arbitrary initial point $\underset{\sim}{x}^{(0)} = (x_1^{(0)}, x_2^{(0)}, \ldots, x_k^{(0)})$ and generates a series of random points $\underset{\sim}{x}^{(1)}, \underset{\sim}{x}^{(2)}, \underset{\sim}{x}^{(3)}, \ldots$, where $\underset{\sim}{x}^{(m+1)}$ is derived from $\underset{\sim}{x}^{(m)}$ in the following way.

(1) $x_1^{(m+1)}$ is drawn randomly from $p(x_1|x_2^{(m)}, x_3^{(m)}, \ldots, x_k^{(m)})$;

(2) $x_2^{(m+1)}$ is drawn randomly from $p(x_2|x_1^{(m+1)}, x_3^{(m)}, \ldots, x_k^{(m)})$;

$$\cdots \quad \cdots \quad \cdots$$

(3) $x_k^{(m+1)}$ is drawn randomly from $p(x_k|x_1^{(m+1)}, x_2^{(m+1)}, \ldots, x_{k-1}^{(m+1)})$.

Perform the above random drawings independently and repeatedly until convergence. When convergence is reached at $m = N$, the sample point $x_i^{(N)}$ is a random sample of size 1 from the marginal distribution $p(x_i)$ of X_i, $i = 1, \ldots, k$. That is, the theory of Gibb sampler indicates that one can generate random samples from the marginal distribution of X_i by generating random samples from the conditional distributions $p(x_i|\underset{\sim}{x}_{(i)})$ through an iterative procedure. The proof of this theory is based on the limiting stationary distribution of an irreducible and aperiodic Markov chain and hence has also been referred to as a Markov Chain Monte Carlo Method (MCMC). The multi-level Gibbs sampler is an extension of the Gibbs sampler method in that each x_i is a vector of random variables. This method has been used extensively by this author in AIDS research; see Chap. 9 for details.

To prove the above result and the convergence, it suffices to consider two discrete random variables X and Y. For illustration, assume that the sample space of X and Y are given respectively by $S_x = \{a_1, \ldots, a_m\}$ and $S_y = \{b_1, \ldots, b_n\}$. Then,

$$P(X = a_i | X = a_j) = \sum_{r=1}^{n} P(X = a_i | Y = b_r) P(Y = b_r | X = a_j),$$

and

$$P(Y = b_i | Y = b_j) = \sum_{r=1}^{m} P(Y = b_i | X = a_r) P(X = a_r | Y = b_j).$$

Let $A(x|x)$ be the $m \times m$ matrix with the (i,j)th element given by $P(X = a_i | X = a_j)$, $A(y|y)$ the $n \times n$ matrix with the (i,j)th element given by $P(Y = b_i | Y = b_j)$, $A(x|y)$ be the $m \times n$ matrix with the (i,j)th element given by $P(X = a_i | Y = b_j)$ and $A(y|x)$ be the $n \times m$ matrix with the (i,j)th element given by $P(Y = b_i | X = a_j)$. Then, $A(x|x) = A(x|y)A(y|x)$ and $A(y|y) = A(y|x)A(x|y)$. Consider a discrete time Markov chain C_x with state space S_x and with one-step transition matrix $P_x = A'(x|x)$. Then, since all elements of $A(x|x)$ are positive, the chain is irreducible and aperiodic. Thus, starting with any distribution of X, the chain will converge to a stationary distribution $\underset{\sim}{g}_x = \{g_x(1), \ldots, g_x(m)\}'$ which satisfies the condition $\underset{\sim}{g}_x = A(x|x)\underset{\sim}{g}_x$. This stationary distribution is unique as the chain is irreducible. We next show that $g_x(i)$ is in fact given by the marginal distribution of X.

To prove this, let $f_x(i) = P(X = a_i)$ $(i = 1, \ldots, m)$ and $f_y(j) = P(Y = b_j)$ $(j = 1, \ldots, n)$. Then the joint density of (X, Y) is

$$P(X = a_i, Y = b_j) = f(i,j) = P(X = a_i)P(Y = b_j | X = a_i)$$

$$= f_x(i)P(Y = b_j | X = a_i)$$

$$= f_y(j)P(X = a_i | Y = b_j).$$

Hence the marginal pdf of X is $f_x(i) = \sum_{j=1}^{n} f_y(j)P(X = a_i | Y = b_j)$ and the marginal pdf of Y is $f_y(j) = \sum_{i=1}^{m} f_x(i)P(Y = b_j | X = a_i)$. It follows that

$$\sum_{j=1}^{m} P(X = a_i | X = a_j) f_x(j)$$

$$= \sum_{j=1}^{m} \left\{ \sum_{r=1}^{n} P(X = a_i | Y = b_r) P(Y = b_r | X = a_j) \right\} f_x(j)$$

$$= \sum_{r=1}^{n} P(X = a_i | Y = b_r) \left\{ \sum_{j=1}^{m} f_x(j) P(Y = b_r | X = a_j) \right\}$$

$$= \sum_{r=1}^{n} P(X = a_i | Y = b_r) f_y(r) = f_x(i).$$

Put $\underset{\sim}{f}_x = \{f_x(1), \ldots, f_x(m)\}'$. The above is equivalent to $A(x|x) \underset{\sim}{f}_x = \underset{\sim}{f}_x$ so that the density $\underset{\sim}{f}_x$ of X is indeed the stationary distribution of the above Markov chain. Hence $\underset{\sim}{f}_x = \underset{\sim}{g}_x$ as the stationary distribution is unique. Similarly, one consider a discrete time Markov chain C_y with state space S_y and with one-step transition matrix $P_y = A'(y|y)$. Then, this chain is irreducible and aperiodic. Further, it can similarly be shown that the density function $f_y(j)$ $(j = 1, \ldots, n)$ of Y is the stationary distribution of C_y.

The above idea can readily be extended to continuous random variables. If the sample space of X and Y are finite intervals, this can easily be shown by using embedded Markov chains; see Chap. 4 for general theories. However, for continuous random variables, if the sample space is not bounded, then there is a possibility that the chain may not converge to a density function. This has been called the improper posterior in the literature. An example of this type has been given by Casella and George [7]. Hence, to use the Gibbs sampler method for continuous random variables, it is important to monitor the convergence. Methods to monitor the convergence have been described in [8–12].

To implement the above Gibbs sampling procedure, one needs to generate a random sample of size 1 from a conditional density. In many practical problems, it is often very difficult to draw a sample either because the density is not completely specified or because the density is very complicated. To get around this difficulty, in the next two sections we describe two indirect methods. In Chap. 9, we will use these methods to implement the Gibbs sampling procedures.

3.4.2. *The weighted bootstrap method for generating random samples*

Suppose that $g(x)$ is a pdf and $f(x) \geq 0$ for all x ($f(x)$ may not be a density). Suppose further that both $g(x)$ and $f(x)$ are computable for all x. The weighted bootstrap method is a method to generate a random sample from the density $h(x) = \frac{f(x)}{\int_{-\infty}^{\infty} f(x)}$ through generating random samples from the density $g(x)$. This method was proposed by Smith and Gelfand [13] and is given by the following procedures:

(i) Generate a random sample $(x_1, \ldots, x_n)$ from $g(x)$.

(ii) Compute $\omega_i = f(x_i)/g(x_i)$ and $q_i = \omega_i / \sum_{j=1}^{n} \omega_j$ for $i = 1, \ldots, n$.

(iii) Let Y denote the random variable with sample space $\Pi = \{x_1, \ldots, x_n\}$ and with probability density $\{q_i, i = 1, \ldots, n\}$ so that $P(Y = x_i) = q_i$. Then draw a random sample $\{y_1, \ldots, y_N\}$ from Y. That is, draw a random sample $\{y_1, \ldots, y_N\}$ of size N with replacement from Π with probabilities $\{q_i, i = 1, \ldots, n\}$. ($N$ need not be the same as n). If n is sufficiently large, $\{y_1, \ldots, y_N\}$ is approximately a random sample of size N from $h(x)$.

To prove the above algorithm, observe that

$$P(Y \leq y) = \sum_{i=1}^{n} q_i I_{(-\infty,\, y]}(x_i) = \frac{1}{n} \sum_{i=1}^{n} \omega_i I_{(-\infty,\, y]}(x_i) \bigg/ \left\{ \frac{1}{n} \sum_{i=1}^{n} \omega_i \right\}$$

where $I_A(x)$ is the indicator function of the set A.

Since $\{x_1, \ldots, x_n\}$ is a random sample from $g(x)$, if n is sufficiently large,

$$\frac{1}{n} \sum_{i=1}^{n} \omega_i I_{(-\infty,\, y]}(x_i) \cong E\{\omega(X) I_{(-\infty,\, y]}(X)\}$$

$$= \int_{-\infty}^{\infty} \frac{f(x)}{g(x)} I_{(-\infty,\, y]}(x) g(x) dx$$

$$= \int_{-\infty}^{\infty} f(x) I_{(-\infty,\, y]}(x) dx = \int_{-\infty}^{y} f(x) dx\,.$$

Similarly, if n is sufficiently large,

$$\frac{1}{n}\sum_{i=1}^{n}\omega_i \cong E\{\omega(X)\} = \int_{-\infty}^{\infty}\frac{f(x)}{g(x)}g(x)dx = \int_{-\infty}^{\infty}f(x)dx\,.$$

Hence, for large n,

$$P(Y \le y) \cong \frac{\int_{-\infty}^{y}f(x)dx}{\int_{-\infty}^{\infty}f(x)dx} = \int_{-\infty}^{y}h(x)dx\,.$$

Tan and Ye [14, 15] have used the above method to estimate unknown parameters and state variables in an AIDS epidemic model involving homosexual and IV drug populations. The method will be illustrated in Chap. 9.

3.4.3. *The Metropolis–Hastings algorithm*

This is an algorithm to generate data from a density $\{\pi_i, i = 0, 1, \ldots\}$ without fully specifying the form of π_i. Specifically, one assumes that $\pi_i \propto f(i)$ with $f(i)$ computable; however, the normalizing constant is very complicated so that π_i is not completely specified. The problem is how to generate data from π_i under these conditions? The main idea of the Metropolis–Hastings algorithm is to construct a Markov chain with π_i as the stationary distribution. We will illustrate this by first proving the following theorem. This theorem is the basis of the Metropolis–Hastings algorithm.

Theorem 3.6.   *Let $p(i, j)$ be the one-step transition probability from state i to state j of a homogeneous Markov chain with state space $S = \{0, 1, \ldots, \infty\}$ and parameter space $T = \{0, 1, \ldots, \infty\}$. Suppose that this chain is irreducible and aperiodic. Assume that the density $\{\pi_i, i = 0, 1, \ldots, \infty\}$ satisfies the following conditions.*

$$\pi_i\, p(i, j) = \pi_j\, p(j, i) \text{ for all } i \text{ and } j\,.$$

Then $\{\pi_i, i = 0, 1, \ldots, \}$ is the stationary distribution of this chain.

Proof.   This theorem follows easily from the observation that

$$\sum_{i=0}^{\infty}\pi_i\, p(i, j) = \sum_{i=0}^{\infty}\pi_j\, p(j, i) = \pi_j\sum_{i=0}^{\infty}p(j, i) = \pi_j \text{ for all } j\,. \qquad \square$$

To construct a Markov chain with $\{\pi_i, i = 0, 1, \ldots\}$ as the stationary distribution, let $q(i, j)$ be the one-step transition probabilities of an irreducible and aperiodic Markov chain. Let $\alpha(i, j)$ be positive quantities satisfying $\{0 < \alpha(i, j) \le 1$ for all $i, j = 0, 1, \ldots, \}$. Then we have a new Markov chain with transition probabilities $p(i, j)$ given by:

$$
p(i, j) = \begin{cases} q(i, j)\alpha(i, j), & \text{if } i \ne j; \\ 1 - \sum_{j \ne i} q(i, j)\alpha(i, j), & \text{if } i = j. \end{cases}
$$

We will chose $\alpha(i, j)$ such that $\pi_i\, p(i, j) = \pi_j\, p(j, i)$. Then, by Theorem 3.6, $\{\pi_i, i = 0, 1, \ldots, \}$ is the stationary distribution of this chain. Assuming that for all i, j, $\pi_i\, q(i, j) > 0$ and $\pi_j\, q(j, i) > 0$, Hastings [16] proposed choosing $\alpha(i, j)$ by the following forms:

$$
\alpha(i, j) = s_{ij} \left\{ 1 + \frac{\pi_i\, q(i, j)}{\pi_j\, q(j, i)} \right\}^{-1},
$$

where the s_{ij}'s are given by

$$
s_{ij} = \begin{cases} 1 + \dfrac{\pi_i\, q(i, j)}{\pi_j\, q(j, i)}, & \text{if } \dfrac{\pi_i\, q(i, j)}{\pi_j\, q(j, i)} \le 1, \\[2ex] 1 + \dfrac{\pi_j\, q(j, i)}{\pi_i\, q(i, j)}, & \text{if } \dfrac{\pi_i\, q(i, j)}{\pi_j\, q(j, i)} > 1. \end{cases}
$$

From these definitions, obviously $s_{ij} = s_{ji}$. Further,

$$
\begin{aligned}
\pi_i\, p(i, j) &= \pi_i q(i, j) s_{ij} \{ 1 + \pi_i\, q(i, j)[\pi_j\, q(j, i)]^{-1} \}^{-1} \\
&= \pi_j q(j, i) s_{ji} \pi_i q(i, j) \{ \pi_i\, q(i, j) + \pi_j\, q(j, i) \}^{-1} \\
&= \pi_j q(j, i) s_{ji} \{ 1 + \pi_j\, q(j, i)[\pi_i\, q(i, j)]^{-1} \}^{-1} \\
&= \pi_j\, q(j, i)\alpha(j, i) = \pi_j\, p(j, i),
\end{aligned}
$$

for all $i, j = 0, 1, \ldots$.

This shows that π_i is indeed the stationary distribution of the newly constructed Markov chain. In this construction, we have further that $\alpha(i, j) = 1$

if $\frac{\pi_i\, q(i,j)}{\pi_j\, q(j,i)} \leq 1$ and

$$\alpha(i,j) = \left\{1 + \frac{\pi_j\, q(j,i)}{\pi_i\, q(i,j)}\right\}\left\{1 + \frac{\pi_i\, q(i,j)}{\pi_j\, q(j,i)}\right\}^{-1}$$

$$= \frac{\pi_j\, q(j,i)}{\pi_i\, q(i,j)} < 1, \quad \text{if } \frac{\pi_i\, q(i,j)}{\pi_j\, q(j,i)} > 1.$$

Since $\pi_i\, q(i,j)\{\pi_j\, q(j,i)\}^{-1} = \{f(i)q(i,j)\}\{f(j)q(j,i)\}^{-1}$, the above construction leads to the following algorithm known as the Metropolis–Hastings algorithm:

(1) Given i, generate j by using the transition probabilities $q(i,j)$.

(2) Compute the ratio $\lambda = \{f(i)q(i,j)\}\{f(j)q(j,i)\}^{-1}$. If this ratio is less than or equal to 1, keep j.

(3) If the ratio in Step (2) is greater than 1, then keep j with probability $\alpha(i,j) = \{f(j)q(j,i)\}\{f(i)q(i,j)\}^{-1}$ and keep i with probability $1 - \alpha(i,j)$ and go back to Step (1).

To implement Step (3), one generate an uniform variable U from the $U(0,1)$ distribution. If $U \leq \alpha(i,j)$, keep j; otherwise, keep i and go back to Step (1) and repeat the process.

In the literature, the distribution $q(i,j)$ has been referred to as the proposal distribution whereas the probabilities $\alpha(i,j)$ the acceptance probabilities.

The above algorithm extends readily to continuous density $g(x) \propto f(x)$ by noting that if $g(x)p(x,y) = g(y)p(y,x)$, then

$$\int g(x)p(x,y)dx = g(y) \int p(y,x)dx = g(y)\,.$$

In this case, assuming that the support of $g(y)$ and $q(x,y)$ are the same as S, the algorithm becomes (See Remark 3.2):

(1) Given $x \in S$, generate $y \in S$ by using the transition probabilities $q(x,y)$.

(2) Assuming $\{f(x)q(x,y) > 0,\ f(y)q(y,x) > 0\}$ for all $x \in S$ and $y \in S$, compute the ratio $\lambda = \{f(x)q(x,y)\}\{f(y)q(y,x)\}^{-1}$. Then keep y with probability $\alpha(x,y) = \min\{1, \frac{f(y)q(y,x)}{f(x)q(x,y)}\}$.

Remark 3.2. For the validity of the algorithm, it is necessary to assume that the support of $g(x)$ is the same as the support of $q(x,y)$. The result is not valid if this assumption fails; see Tanner [17].

Example 3.4.   **The Linkage Problem.** As illustrated in Sec. 3.5, to estimate the linkage fraction between two genes, one would usually need to generate samples from the density of the form:

$$f(\phi) \propto g(\phi) = \phi^{a-1}(1-\phi)^{b-1}(2-\phi)^m, \ 0 \le \phi \le 1.$$

Given the above density $f(\phi)$, $E(\phi)$ and $E(\phi^2)$ are easily derived as

$$\nu_1 = E(\phi) = \sum_{i=0}^{m} \omega_i \frac{a}{a+b+i},$$

where $\omega_i = c_i / \sum_{j=0}^{m} c_j$ with

$$c_j = \prod_{i=1}^{j} \left[\frac{(m-i)(b+j-i)}{i(a+b+j-i)} \right] . \quad \left(\prod_{i=1}^{0} \text{ is defined as 1.} \right)$$

and

$$\nu_2 = E(\phi^2) = \sum_{i=0}^{m} \omega_i \frac{(a)(a+1)}{(a+b+i)(a+b+i+1)} .$$

To generate data from $f(\phi)$, notice that $f(\phi)$ can be closely approximated by the density of a beta distribution with the two parameters (d_1, d_2) being derived by equating the first two moments of this density to $\{\nu_1, \nu_2\}$ respectively. That is, we put:

$$\frac{d_1}{d_1 + d_2} = \nu_1 \text{ and } \frac{d_1(d_1+1)}{(d_1+d_2)(d_1+d_2+1)} = \nu_2 .$$

This gives

$$d_1 = \frac{\nu_1(\nu_2 - \nu_1)}{\nu_1^2 - \nu_2} \text{ and } d_2 = \frac{d_1(1-\nu_1)}{\nu_1} .$$

To speed up convergence, we use this beta distribution to generate an initial value of ϕ. Then, given ϕ_1, we will use as proposal density $q(\phi_1, \phi)$, where

$$q(\phi_1, \phi) = \frac{1}{B(k_1, d_2)} \phi^{k_1 - 1}(1-\phi)^{d_2 - 1}$$

where $k_1 = d_2\phi_1/(1-\phi_1)$.

For the above linkage problem, the Metropolis–Hastings algorithm is then given as follows:

(1) Using the above approximated beta-density to generate the initial starting value, say ϕ_1 to start the process.

(2) Given ϕ_1, use the above proposal density $q(\phi_1, \phi)$ to generate ϕ_2.

(3) Compute the ratio $\lambda = \{f(\phi_1)q(\phi_1, \phi_2)\}\{f(\phi_2)q(\phi_2, \phi_1)\}^{-1}$. If this ratio is less than or equal to 1, keep ϕ_2.

(4) If the ratio in Step (2) is greater than 1, then keep ϕ_2 with probability $\alpha(\phi_1, \phi_2) = \{f(\phi_2)q(\phi_2, \phi_1)\}\{f(\phi_1)q(\phi_1, \phi_2)\}^{-1}$ and keep ϕ_1 with probability $1 - \alpha(i, j)$ and go back to Step (1).

Notice that in implementing the Metropolis–Hastings algorithm, one would need to specify $q(x, y)$; furthermore, $f(x)$ must be computable. Different choices of $q(x, y)$ and the pro's and con's have been discussed by Tierney [18] and by Chip and Greenberg [19]. Intuitively, to speed up the convergence, it is advantageous to choose $q(x, y)$ as close as possible to the stationary density $g(x)$ (in discrete cases, π_i). Thus, we propose the following approach which will be most convenient especially in handling random vectors of variables.

To illustrate, suppose that we wish to generate $\underset{\sim}{\theta}$ from $P(\underset{\sim}{\theta}|Y)$, where $\underset{\sim}{\theta}$ is a vector of variables. Suppose that $P(\underset{\sim}{\theta}|Y)$ is not completely specified but $P(\underset{\sim}{\theta}|Y) \propto h(\underset{\sim}{\theta}) = \exp\{g(\underset{\sim}{\theta})\}$ with $h(\underset{\sim}{\theta})$ computable. Suppose further that $g(\underset{\sim}{\theta})$ is a concave function of $\underset{\sim}{\theta}$ defined over Ω. (This implies that the matrix V of second derivatives of $g(\underset{\sim}{\theta})$ is negative definite). Then one may proceed as follows:

(1) Given $\underset{\sim}{\theta}_0$, approximate $g(\underset{\sim}{\theta})$ by a Taylor series expansion up to second order to give:

$$g(\underset{\sim}{\theta}) \simeq g(\underset{\sim}{\theta}_0) + (\underset{\sim}{\theta} - \underset{\sim}{\theta}_0)'\hat{\underset{\sim}{u}}_0 - \frac{1}{2}(\underset{\sim}{\theta} - \underset{\sim}{\theta}_0)'\hat{V}(\underset{\sim}{\theta} - \underset{\sim}{\theta}_0).$$

where $\hat{\underset{\sim}{u}}_0 = (\frac{\partial}{\partial \underset{\sim}{\theta}} g(\underset{\sim}{\theta}))_{\underset{\sim}{\theta} = \underset{\sim}{\theta}_0}$ and $\hat{V} = (\hat{V}(r, s))$ with $\hat{V}(r, s) = -(\frac{\partial^2}{\partial \theta_r \partial \theta_s} g(\underset{\sim}{\theta}))_{\underset{\sim}{\theta} = \underset{\sim}{\theta}_0}$.

(Notice that since $g(\underset{\sim}{\theta})$ is concave, $\hat{V}$ is positive definite or semi-positive definite if $\underset{\sim}{\theta}_0 \in \Omega$).

Then

$$\exp(g(\underset{\sim}{\theta})) \propto \exp\left\{ -\frac{1}{2}(\underset{\sim}{\theta} - \hat{\underset{\sim}{u}})'\hat{V}(\underset{\sim}{\theta} - \hat{\underset{\sim}{u}}) \right\}.$$

where $\hat{\underset{\sim}{u}} = \underset{\sim}{\theta}_0 + (\hat{V} + \delta I)^{-1}\hat{\underset{\sim}{u}}_0$ with $\delta \geq 0$. (If $\hat{V}$ is nonsingular, take $\delta = 0$; otherwise, let δ be a small positive number such as 10^{-5}). Let $q(\underset{\sim}{\theta}|\underset{\sim}{\theta}_0)$ denote the multivariate normal density of $N\{\hat{\underset{\sim}{u}}, (\hat{V} + \delta I)^{-1}\}$.

(2) Generate $\underset{\sim}{\theta}_1$ from $q(\underset{\sim}{\theta}|\underset{\sim}{\theta}_0)$ and compute

$$\alpha = \begin{cases} \min\left\{ 1, \frac{h(\underset{\sim}{\theta}_1)q(\underset{\sim}{\theta}_0|\underset{\sim}{\theta}_1)}{h(\underset{\sim}{\theta}_0)q(\underset{\sim}{\theta}_1|\underset{\sim}{\theta}_0)} \right\}, & \text{if } h(\underset{\sim}{\theta}_0)q(\underset{\sim}{\theta}_1|\underset{\sim}{\theta}_0) > 0, \\ 1, & \text{if } h(\underset{\sim}{\theta}_0)q(\underset{\sim}{\theta}_1|\underset{\sim}{\theta}_0) = 0. \end{cases}$$

(3) Keep $\underset{\sim}{\theta}_1$ with probability α.

(4) To implement the above algorithm when $\alpha < 1$, generate an uniform variable U from $U(0,1)$ and keep $\underset{\sim}{\theta}_1$ if $U \leq \alpha$; keep $\underset{\sim}{\theta}_0$ and repeat the process if $U > \alpha$.

3.5.   Some Illustrative Examples

To illustrate the above MCMC methods, in this section we give some specific examples.

Example 3.5.   Estimation of frequencies of genes in the ABO blood groups via the Gibbs sampling method. In human beings, there are four different types of blood: The AB-type, the A-type, the B-type and the O-type. These blood types are controlled by three alleles $\{A, B, O\}$. The genotypes of these blood types are given respectively by:

Blood Types	AB-type	A-type	B-type	O-type
Genotypes	AB	AA, AO	BB, BO	OO

Let $\underset{\sim}{\theta} = (p, q)$ and r $(r = 1-p-q)$ denote the frequencies of the alleles (A, B) and O in the population respectively. Then, under steady-state conditions, the frequencies of the four blood types in the population are given respectively by:

Blood Type	AB-type	A-type	B-type	O-type
Frequency	$2pq$	$p^2 + 2pr$	$q^2 + 2qr$	r^2

Suppose that a random sample of size n is taken from the population. Among this sample of n individuals, let the numbers of individuals with blood types $\{AB, A, B, O\}$ be denoted by $\underset{\sim}{y}' = \{y_1, y_2, y_3, y_4\}$ $(\sum_{i=1}^{4} y_i = n)$ respectively. Then the probability distribution of $\underset{\sim}{y}$ is multinomial with density

$$P\{\underset{\sim}{y}|\underset{\sim}{\theta}\} = C_0(\underset{\sim}{y})(2pq)^{y_1}(p^2 + 2pr)^{y_2}(q^2 + 2qr)^{y_3}(r^2)^{y_4} , \qquad (3.4)$$

where $C_0(\underset{\sim}{y}) = \dfrac{n!}{\prod_{j=1}^{4} y_j!}$ and y_j are non-negative integers satisfying $\sum_{i=1}^{4} y_i = n$.

Let Z_1 be the number of individuals with genotype AA in the population and Z_2 be the number of individuals with genotype BB in the population. Then $Z_i \in \{0, 1, \ldots, y_{i+1}\}$ $i = 1, 2$ and the joint density of $\underset{\sim}{y}$ and $\underset{\sim}{Z} = (Z_1, Z_2)'$ is

$$P\{\underset{\sim}{y}, \underset{\sim}{Z}|\underset{\sim}{\theta}\} = C_1(\underset{\sim}{y})(2pq)^{y_1}(p^2)^{Z_1}(2pr)^{y_2 - Z_1}(q^2)^{Z_2}(2qr)^{y_3 - Z_2}(r^2)^{y_4} ,$$

where

$$C_1(\underset{\sim}{y}) = C_0(\underset{\sim}{y}) \prod_{j=1}^{2} \binom{y_{j+1}}{Z_j} ,$$

and

$$P\{\underset{\sim}{y}|\underset{\sim}{\theta}\} = \sum_{Z_1=0}^{y_2} \sum_{Z_2=0}^{y_3} P\{\underset{\sim}{y}, \underset{\sim}{Z}|\underset{\sim}{\theta}\} .$$

The conditional density of $\underset{\sim}{Z}$ given $\underset{\sim}{y}$ is

$$P\{\underset{\sim}{Z}|\underset{\sim}{y}, \underset{\sim}{\theta}\} = \left[\prod_{j=1}^{2}\binom{y_{j+1}}{Z_j}\right]\left(\frac{p^2}{p^2 + 2pr}\right)^{Z_1}\left(\frac{2pr}{p^2 + 2pr}\right)^{y_2 - Z_1}$$

$$\times \left(\frac{q^2}{q^2 + 2qr}\right)^{Z_2}\left(\frac{2qr}{q^2 + 2qr}\right)^{y_3 - Z_2}$$

where $Z_1 = 0, 1, \ldots, y_2$, $Z_2 = 0, 1, \ldots, y_3$.

Let $P(\underset{\sim}{\theta})$ be the prior distribution of θ. Then the conditional density $P(\underset{\sim}{\theta}|\underset{\sim}{y}, \underset{\sim}{Z})$ of θ given $(\underset{\sim}{y}, \underset{\sim}{Z})$ is

$$P(\underset{\sim}{\theta}|\underset{\sim}{y}, \underset{\sim}{Z}) \propto P(\underset{\sim}{\theta})P\{\underset{\sim}{y}, \underset{\sim}{Z}|\underset{\sim}{\theta}\}$$

$$\propto P(\underset{\sim}{\theta})(pq)^{y_1}p^{2Z_1}(pr)^{y_2-Z_1}q^{2Z_2}(qr)^{y_3-Z_2}r^{2y_4}$$

$$\propto P(\underset{\sim}{\theta})p^{m_1}q^{m_2}(1-p-q)^{m_3},$$

where $m_1 = y_1 + y_2 + Z_1$, $m_2 = y_1 + y_3 + Z_2$ and $m_3 = y_2 + y_3 - Z_1 - Z_2 + 2y_4$.

To implement the Gibbs sampling method, a natural conjugate prior distribution is usually taken for $P(\underset{\sim}{\theta})$ so that $P(\underset{\sim}{\theta}) \propto p^{a_1-1}q^{a_2-1}(1-p-q)^{a_3-1}$, where the hyper-parameters a_i are determined by some previous studies. (This is the empirical Bayesian method; see [20]). In the event that there is no prior information or previous studies, one may take ($a_i = 1$, $i = 1, 2, 3$) to reflect the situation that the prior information is vague and imprecise. The latter prior is referred to as the noninformative uniform prior.

Given the above conditional distributions, notice that $P\{\underset{\sim}{Z}|\underset{\sim}{y}, \underset{\sim}{\theta}\}$ is the product of two binomial densities $\{P_i(Z_i), i = 1, 2\}$, where $P_1(Z_1)$ is the density of $B(y_2, p^2/\{p^2 + 2pr\})$ and $P_2(Z_2)$ is the density of $B(y_3, q^2/\{q^2 + 2qr\})$. Similarly, one may also note that $P(\underset{\sim}{\theta}|\underset{\sim}{y}, \underset{\sim}{Z})$ is the density of a bivariate Beta-distribution with $B(m_1, m_2, m_3)$ with parameters $\{m_i, i = 1, 2, 3\}$. Hence the algorithm of the Gibbs sampler is given by the following procedures:

(1) Given $\{\underset{\sim}{y}, \underset{\sim}{\theta}\}$, generate $Z_1^{(*)}$ from the binomial distribution $Z_1 \sim B(y_2, p^2/\{p^2 + 2pr\})$; generate $Z_2^{(*)}$ from the binomial distribution $Z_2 \sim B(y_3, q^2/\{q^2 + 2qr\})$.

(2) Given $\underset{\sim}{y}$ and with $\underset{\sim}{Z} = \underset{\sim}{Z}^{(*)} = \{Z_1^{(*)}, Z_2^{(*)}\}'$, generate $\underset{\sim}{\theta}^{(*)} = \{p^{(*)}, q^{(*)}\}'$ from the bivariate Beta distribution $(p, q) \sim B(m_1 + a_1, m_2 + a_2, m_3 + a_3)$.

(3) With $\underset{\sim}{\theta} = \underset{\sim}{\theta}^{(*)}$, go to Step (1) and repeat the above (1)–(2) loop until convergence.

At convergence, the above Gibbs sampling method then gives a random sample $\underset{\sim}{Z}$ of size 1 from $P(\underset{\sim}{Z}|\underset{\sim}{y})$ and a random sample $\underset{\sim}{\theta}$ from $P(\underset{\sim}{\theta}|\underset{\sim}{y})$. The convergence is guaranteed by the basic theory of homogeneous Markov chains. To estimate the parameters $\underset{\sim}{\theta}$, one then generate a random sample from $P(\theta|\underset{\sim}{y})$ of size n. The sample means (the posterior means) and the sample variances (the posterior variances) and the sample covariances (the posterior covariances) from this sample may then be used as the estimates of the parameters and the estimates of the variances and covariances of these estimates.

To illustrate the above procedure, we use the data from Li [21, p. 48] which provided survey results of blood types of 6,000 people in Kunming, Yunnan, China. This data is given as follows:

Blood Types	AB-type	A-type	B-type	O-type
Observed Number	607	1920	1627	1846

Applying the above procedures to this data set, one can readily estimate the parameters $\{p, q, r = 1 - p - q\}$. Plotted in Fig. 3.1 are the estimates of these parameters by the Bayesian Gibbs sampling method with uniform prior for the parameters. From Fig. 3.1, it is clear that after eight iterations, the results

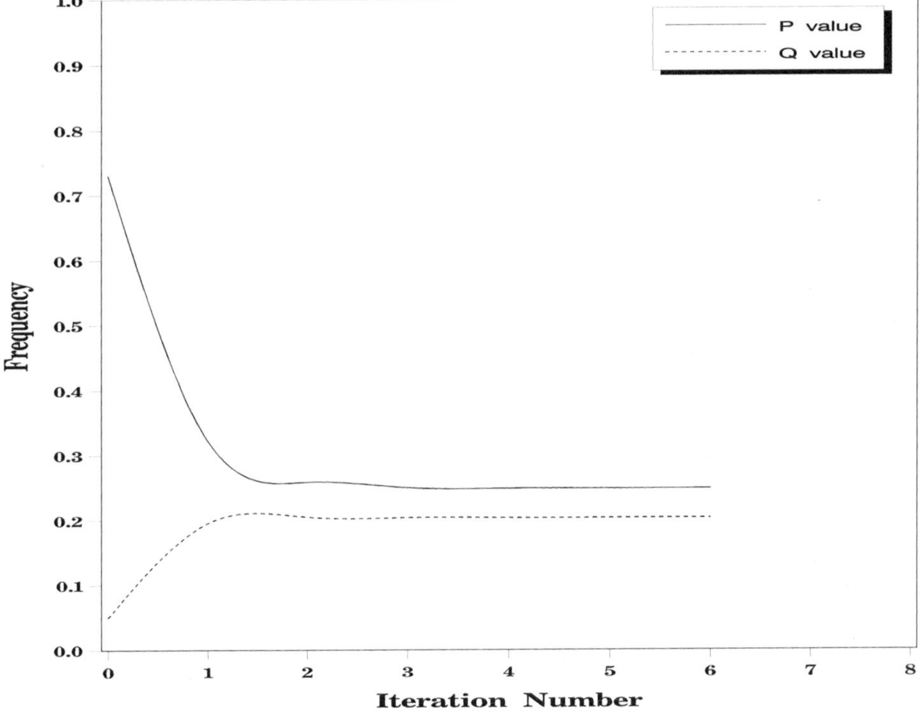

Fig. 3.1. Plots showing estimates of the frequencies (p, q) of blood type genes by the Gibbs sampling method.

converge to $\{p = 0.2356, q = 0.2059\}$. Based on 100 repeated sampling, the standard errors of these estimates are given by 0.0041 and 0.0039 respectively.

Remark 3.3.   The EM-algorithm. From the classical sampling theory approach, the EM-algorithm has been proposed to estimate the frequencies of genes in the ABO-blood groups; see [22]. The EM-algorithm is an algorithm to compute the MLE (the maximum likelihood estimates) of the unknown parameters. This method is an iterative method which loops between the E-step and the M-step until convergence. For the above problem, the E-step is to computer the expected value of $\underset{\sim}{Z}$ from the conditional density $P\{\underset{\sim}{Z}|\underset{\sim}{y},\,\underset{\sim}{\theta}\}$ to give:

$$\hat{Z}_1 = E\{Z_1|\underset{\sim}{y},\,\underset{\sim}{\theta}\} = y_2 \frac{p^2}{p^2 + 2pr},$$

and

$$\hat{Z}_2 = E\{Z_2|\underset{\sim}{y},\,\underset{\sim}{\theta}\} = y_3 \frac{q^2}{q^2 + 2qr}.$$

With $\{Z_1 = \hat{Z}_1,\, Z_2 = \hat{Z}_2\}$, the M-step is to derive the MLE of $\underset{\sim}{\theta}' = (p, q)$ by maximizing the function $H(p, q) = p^{m_1} q^{m_2} (1 - p - q)^{m_3}$ to give

$$\hat{p} = \frac{\hat{m}_1}{\sum_{i=1}^{3} \hat{m}_i} \quad \text{and} \quad \hat{q} = \frac{\hat{m}_2}{\sum_{i=1}^{3} \hat{m}_i},$$

where the $\hat{m}_i$'s are computed from m_i by substituting $\hat{Z}_i$ for Z_i.

To compare the EM–algorithm with the Gibbs sampling method, notice that computing the $\{\hat{Z}_i, i = 1, 2\}$ is equivalent to generating a large sample from $P\{\underset{\sim}{Z}|\underset{\sim}{y},\,\underset{\sim}{\theta}\}$ and then computing the sample means; similarly, computing the MLE of $\underset{\sim}{\theta}$ by maximizing the function $H(p, q)$ is equivalent numerically to generating a large sample from the Beta-distribution $B(m_1 + 1, m_2 + 1, m_3 + 1)$ and then computing the sample means. Hence, for the above problem, the EM-algorithm is numerically equivalent to the Gibbs sampling method only under the noninformative uniform prior. Comparing the above two approaches, notice that there are significant differences between them both in concept and in results:

(1) The EM-algorithm is the sampling theory approach while the Gibbs sampling method is based on the Bayesian approach. Hence in the

EM-algorithm, the parameters are unknown constants whereas the sample data are random variables; on the other hand, in the above Gibbs sampling method, the parameters are random variables whereas the sample data are given constants.

(2) The probability concept between the two approaches are quite different. In the EM-method, the probability is defined in terms of relative frequency whereas in the above Gibbs sampling method, the probability is subjective and is based on personnel degree of beliefs; see [23].

(3) In the EM-method, prior information about the parameters are ignored whereas in the above Gibbs sampling method, prior information about the parameters are taken into account through the prior distribution. Hence the two method give identical numerical results only if one uses the noninformative uniform prior.

(4) In the above Gibbs sampling method, the sample posterior variances and covariances can be used as estimates of the variances and covariances of the estimates. In the EM-algorithm, one would need to compute the estimates of the variances and covariances of the estimates given the MLE. Given the MLE, general formula for estimating the asymptotic variances and covariances have been provided by Louis [24]; see also [17, pp. 47–52].

3.6. Estimation of Linkage Fraction by Gibbs Sampling Method

In living beings, all characters are controlled by genes, each of which is a segment of DNA in a chromosome. Since the number of genes are very large whereas the number of chromosomes for each species is finite, one may expect that many genes are located in the same chromosome. (For example, in human beings, there are more than 100,000 genes identified yet there are only 23 pairs of chromosomes). Thus, many of the genes are linked and they tend to segregate together into the same cell during cell division. To identify association between different characters, it is important in many cases to estimate how far genes are apart in the same chromosome. This is the problem of estimating the linkage proportion or recombination fraction for linked loci. For this type of problem, the Gibbs sampling method is particularly useful, especially in analyzing human pedigree data; see [25, 26]. By using some simple examples, in this section, we will illustrate how to use the Gibbs sampling method to estimate the recombination fraction between two loci in the same chromosome.

To begin with, consider two loci in the same chromosome. Suppose that the first locus has alleles A and a and that the second locus has alleles B and b. Denote by AB/ab the genotype of a diploid individual with phenotype $AaBb$, where the genes A and B are in the same chromosome and where the genes a and b are in the other homologous chromosome. (In the genetic literature, the genotype AB/ab has been referred to as the coupling phase whereas the genotype Ab/aB has been referred to as the repulsion phase). Then, for individuals with genotype AB/ab, most of the gametes (haploid) produced have genotypes AB and ab because genes in the same chromosome tend to segregate together; however, a small fraction of the gametes of AB/ab individuals have genotypes Ab and aB due to crossing-over between the two homologous chromosomes during cell division. Let $\phi\,(1 \geq \phi \geq 0)$ be the frequency of crossing-over between the two loci. Since there are four chromatids in each pair of homologous chromosomes during meiosis and crossing-over occurs only between any two of the chromatids, each crossing over will give on the average an equal number of the $\{AB, Ab, aB, ab\}$ gametes; on the other hand, when there are no crossing over, each meiosis will give only AB and ab gametes with frequency $1/2$ for each. Hence, the proportion of AB or ab gametes is $\frac{1}{4}\phi + \frac{1}{2}(1 - \phi) = \frac{1}{2}(1 - \phi/2)$ whereas the proportion of Ab or aB gametes is $\frac{1}{4}\phi = \frac{1}{2}(\phi/2)$. Denote by $\theta = \phi/2$. Then $\frac{1}{2} \geq \theta \geq 0$ and the expected proportion of the four types of gametes from AB/ab individuals are:

Gamete Genotype	AB	Ab	aB	ab
Proportion	$\frac{1}{2}(1-\theta)$	$\frac{1}{2}\theta$	$\frac{1}{2}\theta$	$\frac{1}{2}(1-\theta)$
	$\frac{1}{4}(2-\phi)$	$\frac{1}{4}\phi$	$\frac{1}{4}\phi$	$\frac{1}{4}(2-\phi)$

Similarly, the expected proportion of the four types of gametes from Ab/aB individuals are:

Gamete	AB	Ab	aB	ab
Proportion	$\frac{1}{2}\theta$	$\frac{1}{2}(1-\theta)$	$\frac{1}{2}(1-\theta)$	$\frac{1}{2}\theta$
	$\frac{1}{4}\phi$	$\frac{1}{4}(2-\phi)$	$\frac{1}{4}(2-\phi)$	$\frac{1}{4}\phi$

In the above, θ has been referred to as the recombination fraction and $1-\theta$ the linkage fraction. The constraint $\frac{1}{2} \geq \theta \geq 0$ implies that $1-\theta \geq \theta$ so that the linkage fraction is always greater than or equal to the recombination fraction. The case with $\theta = \frac{1}{2}$ is equivalent to independent segregation in which case the four types of gametes are equally probable (i.e. each with frequency $\frac{1}{4}$).

With the above background, in the next two examples we will illustrate how the Gibbs sampling method can be used to estimate the recombination fraction θ. Because of the constraint $\frac{1}{2} \geq \theta \geq 0$, it is convenient to work with $\phi = 2\theta$ as the constraint of ϕ is $1 \geq \phi \geq 0$. Notice that given the estimate $\hat{\phi}$, the estimate of θ is $\hat{\theta} = \frac{1}{2}\hat{\phi}$ and $\mathrm{Var}\{\hat{\theta}\} = \frac{1}{4}\mathrm{Var}\{\hat{\phi}\}$.

Example 3.6. Estimation of recombination proportion between two linked loci in self-fertilized populations by Gibbs sampling method. In this example we consider two linked loci each with two alleles (say $A : a$ and $B : b$) in a self-fertilized population such as rice or wheat. (This example is commonly used in most of the statistic texts such as Rao [27]).

To start with, consider a crossing $AABB \times aabb$ between two pure-lines with genotypes $AABB$ and $aabb$ respectively. Then, in F_1 (i.e. the first generation), all individuals have genotype AB/ab (coupling phase). Hence, in F_2 (i.e. the second generation), there are 10 genotypes

$$\{AB/AB,\ AB/Ab,\ AB/aB,\ AB/ab,\ Ab/Ab,$$

$$Ab/aB,\ Ab/ab,\ aB/aB,\ aB/ab,\ ab/ab\}$$

with frequencies

$$\left\{\frac{1}{4}(1-\theta)^2,\ \frac{1}{2}\theta(1-\theta),\ \frac{1}{2}\theta(1-\theta),\ \frac{1}{2}(1-\theta)^2,\ \frac{1}{4}\theta^2,\right.$$

$$\left.\frac{1}{2}\theta^2,\ \frac{1}{2}\theta(1-\theta),\ \frac{1}{4}\theta^2,\ \frac{1}{2}\theta(1-\theta),\ \frac{1}{4}(1-\theta)^2\right\}$$

respectively. If the allele A is dominant over the allele a and B is dominant over b, then there are only four phenotypes which are denoted by $\{A\text{--}B\text{--},$ $A\text{--}bb,\ aaB\text{--},\ aabb\}$. Notice that individuals having any of the genotypes $\{AB/AB, AB/Ab, AB/aB, AB/ab, Ab/aB\}$ have the same phenotype $A\text{--}B\text{--}$; individuals having any of the genotypes $\{Ab/Ab, Ab/ab\}$ have the same phenotype $A\text{--}bb$; individuals having any of the genotypes $\{aB/aB, aB/ab\}$ have the same phenotype $aaB\text{--}$ and individuals with the genotype ab/ab have the phenotype *aabb*. Hence, among the progenies of a $AB/ab \times AB/ab$ mating, the frequencies of the A–B–phenotype, the A–bb phenotype, the aaB–phenotype and the *aabb* phenotype are given respectively by:

(1) **A–B–Phenotype**

$$\frac{1}{4}(1-\theta)^2 + \frac{1}{2}\theta(1-\theta) + \frac{1}{2}\theta(1-\theta) + \frac{1}{2}(1-\theta)^2 + \frac{1}{2}\theta^2$$

$$= \frac{1}{4}[(1-\theta)^2 + 2] = \frac{1}{4}(3 - 2\theta + \theta^2) = \frac{1}{16}(12 - 4\phi + \phi^2).$$

(2) **A–bb Phenotype**

$$\frac{1}{4}\theta^2 + \frac{1}{2}\theta(1-\theta) = \frac{1}{4}\theta[2(1-\theta) + \theta] = \frac{1}{4}\theta(2-\theta) = \frac{1}{16}\phi(4-\phi).$$

(3) **aaB–Phenotype**

$$\frac{1}{4}\theta^2 + \frac{1}{2}\theta(1-\theta) = \frac{1}{4}\theta[2(1-\theta) + \theta] = \frac{1}{4}\theta(2-\theta) = \frac{1}{16}\phi(4-\phi).$$

(4) **aabb Phenotype**

$$\frac{1}{4}(1-\theta)^2 = \frac{1}{16}(2-\phi)^2.$$

Suppose that in F_2, the observed number of individuals with phenotypes $\{A\text{–}B\text{–}, A\text{–}bb, aaB\text{–}, aabb\}$ are given respectively by $\underset{\sim}{y}' = \{y_1, y_{21}, y_{22}, y_3\}$, where $y_2 = \sum_{j=1}^{2} y_{2j}$ and $\sum_{i=1}^{3} y_i = n$. Then the probability distribution for these observed data is multinomial with density:

$$P\{\underset{\sim}{y}|\theta\} = C(n; y_i, i=1,2,3)\frac{1}{4^n}(3 - 2\theta + \theta^2)^{y_1}[\theta(2-\theta)]^{y_2}(1-\theta)^{2y_3}, \quad (3.5)$$

where

$$C(n; y_i, i=1,2,3) = \frac{n!}{\{\prod_{j=1}^{3} y_j!\}}$$

and y_j are non-negative integers satisfying $\sum_{i=1}^{3} y_i = n$.

To estimate θ using data $\underset{\sim}{y}$ as above, the classical sampling theory approach is to derive estimates of θ by maximizing $P\{\underset{\sim}{y}|\theta\}$ under the constraint $\frac{1}{2} \geq \theta \geq 0$. Because with positive probability, the MLE without the constraint may give estimates with value greater $\frac{1}{2}$, the classical method without constraint is not satisfactory. In the sampling theory approach, to date, efficient statistical procedures for estimating θ under the constraint remain to be developed. Because of these difficulties, Tan [28] has proposed the Bayesian method. Through Monte Carlo studies, he has shown that the Bayesian method was considerably more efficient than the classical MLE approach. Results of Monte Carlo studies by Tan [28] showed that in almost all cases, the Bayesian method gave estimates much closer to the true values than the MLE without the constraint. Under uniform noninformative prior, the Bayesian estimates are numerically equal to the MLE under the constraint which can be derived by the EM-algorithm. As an illustration, we now illustrate how to use the Gibbs sampling method to derive the Bayesian estimate of θ.

To derive the Bayesian estimate of θ, denote by Z_1 the number of individuals with genotypes AB/AB or AB/ab, Z_2 the number of individuals with genotypes AB/Ab or AB/aB, respectively, among the y_1 individuals with phenotype $A\!-\!B\!-$. Let W denote the number of individuals with genotypes Ab/Ab or aB/aB among the y_2 individuals with phenotypes $A\!-\!bb$ or $aaB\!-$. Then $Z_3 = y_1 - \sum_{i=1}^{2} Z_i$ is the number of individuals with genotypes Ab/aB among the y_1 individuals with phenotype $A\!-\!B\!-$ and $y_2 - W$ the number of individuals with genotypes Ab/ab or aB/ab among the y_2 individuals with phenotypes $A\!-\!bb$ or $aaB\!-$. The conditional density of $\underset{\sim}{Z} = \{Z_1, Z_2\}$ given $\underset{\sim}{y}$ and θ is that of a 2-dimensional multinomial distribution with parameters $\{y_1, \frac{3(1-\theta)^2}{3-2\theta+\theta^2}, \frac{4\theta(1-\theta)}{3-2\theta+\theta^2}\}$. That is,

$$\underset{\sim}{Z}\,|\{\underset{\sim}{y},\,\theta\} \sim ML\left\{ y_1;\ \frac{3(1-\theta)^2}{3-2\theta+\theta^2},\ \frac{4\theta(1-\theta)}{3-2\theta+\theta^2} \right\}; \qquad (3.6)$$

the conditional distribution of W given $\underset{\sim}{y}$ and θ is binomial with parameters $\{y_2, \frac{\theta}{2-\theta}\}$. That is,

$$W|\{\underset{\sim}{y},\,\theta\} \sim B\left\{ y_2,\ \frac{\theta^2}{\theta(2-\theta)} \right\}, \qquad (3.7)$$

Further, given $\{\underset{\sim}{y},\,\theta\}$, $\underset{\sim}{Z}$ and W are independently distributed of one another. Denote by $\underset{\sim}{X}' = \{\underset{\sim}{Z}',\,W\}$. Then the joint probability density function of $\{\underset{\sim}{y}, \underset{\sim}{X}\}$ given θ is

$$P\{\underset{\sim}{y}, \underset{\sim}{X}|\theta\} = P\{\underset{\sim}{y}|\theta\}P\{\underset{\sim}{Z}|y_1,\theta\}P\{W|y_2,\theta\}$$

$$= C_1(1-\theta)^{2Z_1}[\theta(1-\theta)]^{Z_2}\theta^{2Z_3}$$

$$\times (1-\theta)^{2y_3}\{\theta^{2W}[\theta(1-\theta)]^{y_2-W}\}$$

where C_1 is a function of $(\underset{\sim}{y}, \underset{\sim}{Z})$ but is independent of θ. C_1 is in fact given by

$$C_1 = 3^{Z_1}2^{2Z_2+Z_3+y_2-W-2n}C(n;y_i,i=1,2,3)C(y_1;Z_j,j=1,2,3)\binom{y_2}{W},$$

where $C(m;k_i,i=1,\ldots,r) = m!/\{\prod_{i=1}^{r}k_i!\}$.

Let $P(\theta)$ be the prior distribution of θ. Then the joint density of $\{\theta, \underset{\sim}{y}, \underset{\sim}{X}\}$ is $P(\theta)P\{\underset{\sim}{y}, \underset{\sim}{X}|\theta\}$ and the posterior distribution of θ given $\{\underset{\sim}{y}, \underset{\sim}{X}\}$ is

$$P\{\theta|\underset{\sim}{y}, \underset{\sim}{X}\} \propto P(\theta)(1-\theta)^{2Z_1}[\theta(1-\theta)]^{Z_2}(\theta)^{2Z_3}$$

$$\times (1-\theta)^{2y_3}\{\theta^{2W}[\theta(1-\theta)]^{y_2-W}\}$$

$$= P(\theta)\theta^{m_1}(1-\theta)^{m_2}, \tag{3.8}$$

where $m_1 = m_1(\underset{\sim}{y}, \underset{\sim}{X}) = 2Z_3 + Z_2 + y_2 + W$ and $m_2 = m_2(\underset{\sim}{y}, \underset{\sim}{X}) = 2(Z_1 + y_3) + Z_2 + (y_2 - W)$.

Because $\frac{1}{2} \geq \theta \geq 0$, a natural conjugate prior of θ is

$$P(\theta) \propto (2\theta)^{a_1-1}(1-2\theta)^{a_2-1},$$

where $a_i > 0$, $i = 1, 2$; see [25].

Using this prior, then, with $\phi = 2\theta$, the posterior distribution of ϕ is

$$P\{\phi|\underset{\sim}{y}, \underset{\sim}{X}\} \propto \phi^{a_1+m_1-1}(1-\phi)^{a_2-1}(2-\phi)^{m_2}, \ 0 \leq \phi \leq 1. \tag{3.9}$$

Using the density in (3.9) and noting $(2-\phi)^{m_2} = \sum_{i=0}^{m_2}\binom{m_2}{i}(1-\phi)^i$, we obtain after simplification the conditional expected value of ϕ given $\{\underset{\sim}{y}, \underset{\sim}{X}\}$ as

$$E\{\phi|\underset{\sim}{y}, \underset{\sim}{X}\} = \sum_{i=0}^{m_2}\omega_i\frac{m_1+a_1}{m_1+a_1+a_2+i},$$

where $\omega_i = b_i / \sum_{j=0}^{m_2} b_j = c_i / \sum_{j=0}^{m_2} c_j$ with

$$b_j = \binom{m_2}{j} \frac{\Gamma(a_2 + j)}{\Gamma(m_1 + a_1 + a + 2 + j)},$$

and

$$c_j = \prod_{i=1}^{j} \left[\frac{(m_2 - i)(a_2 + j - i)}{i(m_1 + a_1 + a_2 + j - i)} \right] \cdot \quad \left(\prod_{i=1}^{0} \text{ is defined as 1.} \right)$$

Given the above results, the algorithm of the Gibbs sampler for estimating $\{ \underset{\sim}{Z}, \phi = 2\theta \}$ is given by the following procedures:

(1) Given $\{ \underset{\sim}{y}, \phi = 2\theta \}$, generate $\underset{\sim}{Z}^{(*)}$ from the multinomial distribution

$$\underset{\sim}{Z} \sim \text{ML} \left\{ y_1; \frac{3(2 - \phi)^2}{12 - 4\phi + \phi^2}, \frac{4\phi(2 - \phi)}{12 - 4\phi + \phi^2} \right\};$$

given $\{ \underset{\sim}{y}, \phi = 2\theta \}$, generate $W^{(*)}$ from the binomial distribution

$$W \sim B \left(y_2, \frac{\phi}{4 - \phi} \right).$$

(2) Given $\underset{\sim}{y}$ and with $\underset{\sim}{X} = \underset{\sim}{X}^{(*)} = \{ \underset{\sim}{Z}^{(*)'}, W^{(*)} \}'$, generate $\phi^{(*)}$ from the density $P\{ \phi | \underset{\sim}{y}, \underset{\sim}{X}^{(*)} \}$.

(3) With $\phi = \phi^{(*)}$, go to Step (1) and repeat the above Step (1)–(2) loop until convergence.

At convergence, the above Gibbs sampling method then gives a random sample $\underset{\sim}{X}$ of size 1 from $P(\underset{\sim}{X} | \underset{\sim}{y})$ and a random sample ϕ from $P(\phi | \underset{\sim}{y})$. The convergence is guaranteed by the basic theory of homogeneous Markov chains. To estimate the parameter ϕ, one then generate a random sample from $P(\phi | \underset{\sim}{y})$ of size n. The sample means (the posterior means) and the sample variance (the posterior variance) of $\phi = 2\theta$ from this sample may then be used as the estimate of $\phi = 2\theta$ and the estimate of the variance of this estimate.

To implement the above algorithm, as illustrated in Example 3.4, one may use the Metropolis–Hasting algorithm.

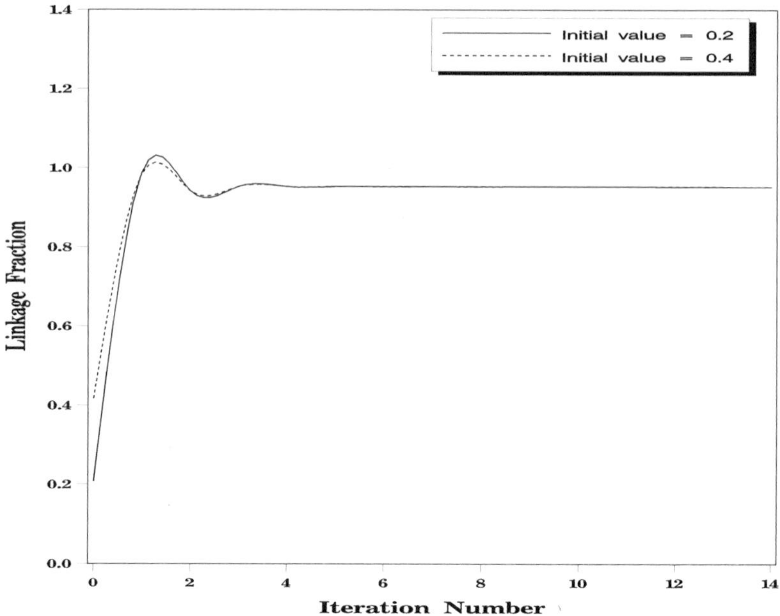

Fig. 3.2.   Plots showing estimates of linkage fraction by the Gibbs sampling method.

As an example, consider the data given in Rao [27, p. 369] which give the observed number of different phenotypes in F_2 under coupling phase.

Phenotype in F_2	A–B–	A–bb	aaB–	$aabb$
Observed Number	125	18	20	34

Applying the above procedures to this data set, one can readily estimate the linkage fraction θ. Plotted in Fig. 3.2 are the estimates of θ by the Bayesian Gibbs sampling method with uniform prior. From Fig. 3.2, it is clear that after a few iterations, the results converge to $\phi = 2\theta = 0.9525$ (or $\theta = 0.4762$). Based on 100 repeated sampling, the standard errors of this estimate is 0.0028.

Remark 3.4.   The EM-algorithm. Under uniform noninformative prior, the Bayesian estimate is numerically equal to the MLE under the constraint. Hence, putting $a_i = 1, (i = 1, 2)$ and using the conditional expected values as the generated number in each of the Steps (1)–(2) above, the above Gibbs sampling method is numerically identical to the EM-algorithm to derive the

MLE of ϕ and hence θ under the constraint. Notice that the EM-algorithm in this case is much more easier and converges much faster than the GIbbs sampling method. However, the EM-algorithm has ignored the prior information as it is a sampling theory approach whereas the Gibbs sampling method is a Bayesian approach. Further, the probability concept is very different between the two approaches. Because computing the posterior means are equivalent to generating a large sample and then computing the sample mean, for speeding up the convergence and for ease of computation one may use the posterior mean as the generated sample in Steps (1)–(2) of the above algorithm.

Example 3.7. Estimation of recombination proportion between two linked loci in human populations by Gibbs sampling method. In this example we illustrate how to estimate the recombination fraction between two linked loci each with two alleles (say $A : a$ and $B : b$) in a human population. Again we assume that A is dominant over a and B is dominant over b.

Suppose that we have data from k independent families in each of which the phenotype of one parent is A–B– and the phenotype of the other parent is $aabb$. That is, the mating in each family is A–B– $\times$ $aabb$. Assume that the jth family has $n_j (n_j > 1)$ progenies and in this family, denote by $\underset{\sim}{y}'_j = \{y_j(1), y_j(2,1), y_j(2,2), y_j(3)\}$ the observed numbers of progenies with phenotypes $\{A$–B–, A–bb, aaB–, $aabb\}$ respectively. To rule out the possible genotypes $\{AB/AB, AB/Ab.AB/aB\}$ for the parent with phenotype A–B–, we restrict ourselves to the situation that in each family, either the phenotype $aabb$ or at least three different phenotypes have been observed. Then the genotype of the parent with phenotype A–B– is either AB/ab or Ab/aB. (Notice also that in the mating A–B– $\times aabb$, the genotypes $\{AB/AB, AB/Ab.AB/aB\}$ for the parent with phenotype A–B–provide no information about the recombination fraction θ.) Further, under linkage equilibrium, the frequencies of the AB/ab and aB/Ab genotypes among parent with phenotype A–B– are $\frac{1}{2}$; see [26]. It follows that the probability density for observing $\underset{\sim}{y}_j$ given θ for the jth family is

$$P\{\underset{\sim}{y}_j|\theta\} = \frac{1}{2}C_0\{\theta^{y_j(2)}(1-\theta)^{y_j(1)+y_j(3)} + \theta^{y_j(1)+y_j(3)}(1-\theta)^{y_j(2)}\},$$

where $C_0 = (n_j!)/\{4^{n_j}\prod_{i=1}^{3}y_j(i)!\}$ with $y_j(2) = \sum_{r=1}^{2}y_j(2,r)$.

The probability density for observing $\boldsymbol{Y} = \{\underset{\sim}{y}_j, j = 1,\ldots,k\}$ given θ for all families is $P(\boldsymbol{Y}|\theta) = \prod_{j=1}^{k}P\{\underset{\sim}{y}_j|\theta\}$.

For applying the Gibbs sampling method, in the jth family, let Z_j be defined by $Z_j = 1$ if the genotype of the parent with phenotype $A\text{–}B\text{–}$ is AB/ab and $Z_j = 0$ if otherwise. Then the Z_j's are independently and identically distributed as binomial with parameters $\{1, \frac{1}{2}\}$. That is, for $j = 1, \ldots, k$, $P(Z_j = 1) = P(Z_j = 0) = \frac{1}{2}$ independently. The joint density of $\{Z_j, \underset{\sim}{y}_j\}$ given θ is

$$P\{Z_j, \underset{\sim}{y}_j | \theta\} = \frac{1}{2} C_0 \theta^{N_1(j)} (1 - \theta)^{N_2(j)},$$

where $N_1(j) = Z_j y_j(2) + (1 - Z_j)[y_j(1) + y_j(3)]$ and $N_2(j) = Z_j[y_j(1) + y_j(3)] + (1 - Z_j) y_j(2)$.

Hence the conditional density of Z_j given $\underset{\sim}{y}_j$ and θ is

$$P\{Z_j | \underset{\sim}{y}_j, \theta\} = P\{Z_j, \underset{\sim}{y}_j | \theta\} / P\{\underset{\sim}{y}_j | \theta\} = \psi_j^{Z_j} (1 - \psi_j)^{1 - Z_j}$$

$$\psi_j = \{\theta^{y_j(2)} (1 - \theta)^{y_j(1) + y_j(3)}\} \{\theta^{y_j(2)} (1 - \theta)^{y_j(1) + y_j(3)}$$

$$+ \theta^{y_j(1) + y_j(3)} (1 - \theta)^{y_j(2)}\}^{-1}.$$

The conditional density of $\underset{\sim}{Z}' = \{Z_j, j = 1, \ldots, k\}$ given $\boldsymbol{Y}$ and θ is

$$P\{\underset{\sim}{Z} | \boldsymbol{Y}, \theta\} = \prod_{j=1}^{k} P\{Z_j | \underset{\sim}{y}_j, \theta\} = \prod_{j=1}^{k} \psi_j^{Z_j} (1 - \psi_j)^{1 - Z_j}.$$

Let $P(\theta)$ be the prior distribution of θ. Then the posterior distribution of θ given $\{\underset{\sim}{Z}, \boldsymbol{Y}\}$ is

$$P\{\theta | \underset{\sim}{Z}, \boldsymbol{Y}\} \propto P(\theta) \theta^{N_1} (1 - \theta)^{N_2},$$

where $N_i = \sum_{j=1}^{k} N_i(j), i = 1, 2$.

Because $\frac{1}{2} \le \theta \le 0$, a natural conjugate prior of θ is

$$P(\theta) \propto (2\theta)^{a_1 - 1} (1 - 2\theta)^{a_2 - 1},$$

where $a_i > 0, i = 1, 2$; see [25].

Using this prior, then, with $\phi = 2\theta$, the posterior distribution of ϕ is

$$P\{\phi | \boldsymbol{Y}, \underset{\sim}{Z}\} \propto \phi^{a_1 + N_1 - 1} (1 - \phi)^{a_2 - 1} (2 - \phi)^{N_2}.$$

The conditional expected value of ϕ given $\{\boldsymbol{Y}, \underset{\sim}{Z}\}$ is

$$E\{\phi|\boldsymbol{Y}, \underset{\sim}{Z}\} = \sum_{i=0}^{N_2} \lambda_i \frac{N_1 + a_1}{N_1 + a_1 + a_2 + i}\,,$$

where $\lambda_i = u_i/\{\sum_{j=0}^{N_2} u_j\} = v_i/\{\sum_{j=0}^{N_2} v_j\}$ with

$$u_j = \binom{N_2}{j} \frac{\Gamma(a_2 + j)}{\Gamma(N_1 + a_1 + a + 2 + j)}\,,$$

and

$$v_j = \prod_{i=1}^{j} \left[\frac{(N_2 - i)(a_2 + j - i)}{i(N_1 + a_1 + a_2 + j - i)} \right] \cdot \left(\prod_{i=1}^{0} \text{ is defined as 1.} \right)$$

Given the above results, the algorithm of the Gibbs sampler for estimating $\{\underset{\sim}{Z}, \phi = 2\theta\}$ is given by the following procedures:

(1) Given $\{\boldsymbol{Y}, \phi = 2\theta\}$, generate $Z_i^{(*)}$ from the binomial distribution

$$Z_i \sim B\{1; \psi_i\}, \quad i = 1, \ldots, k\,.$$

(2) Given $\boldsymbol{Y}$ and with $\underset{\sim}{Z} = \underset{\sim}{Z}^{(*)} = \{Z_i^{(*)}, i = 1, \ldots, k\}'$, generate $\phi^{(*)}$ from the density $P\{\phi|\boldsymbol{Y}, \underset{\sim}{Z}^{(*)}\}$.

(3) With $\phi = \phi^{(*)}$, go to Step (1) and repeat the above Steps (1)–(2) loop until convergence.

At convergence, the above Gibbs sampling method then gives a random sample $\underset{\sim}{Z}$ of size 1 from $P(\underset{\sim}{Z}|\boldsymbol{Y})$ and a random sample ϕ from $P(\phi|\boldsymbol{Y})$. The convergence is guaranteed by the basic theory of homogeneous Markov chains. To estimate the parameter ϕ, as in Example 3.6, one generates a random sample from $P(\phi|\boldsymbol{Y})$ of size n. The sample means (the posterior means) and the sample variance (the posterior variance) of $\phi = 2\theta$ from this sample may then be used as the estimate of $\phi = 2\theta$ and the estimate of the variance of this estimate.

To implement the above algorithm, as in Example 3.6, one may use the posterior mean as the generated sample in Step (2) of the above algorithm.

In the above, we have considered only a special case involving an *aabb* parent. Many other cases have been considered in [26]. Also, we are only involving qualitative traits. The Gibbs sampling method will be especially useful in estimating the linkage recombination involving quantitative traits.

Some preliminary studied have been made in [25, 29]. There are many interesting problems which remain to be solved, however; and the Gibbs sampling methods can provide a solution to these problems. We will not go any further here, however.

3.7. Complements and Exercises

Exercise 3.1. Prove the Corollary 3.1.

Exercise 3.2. Let $\{X(t), t \in T = (0, 1, \ldots)\}$ be a finite homogeneous Markov chain. If the chain is irreducible, show that all states are persistent and positive. Hence, for homogeneous finite Markov chains with discrete time, the stationary distribution exists and is unique.

Exercise 3.3. Consider the finite Markov chain for mixtures of selfing and random mating as described in Exercise 2.2. Show that the chain is irreducible. Derive the stationary distribution of this chain.

Exercise 3.4. Consider the finite Markov chain for inbreeding as described in Exercise 2.3. Show that the chain is irreducible. Derive the stationary distribution of this chain.

Exercise 3.5. **(Estimation of Inbreeding Coefficient).** Given below are data on haptoglobin genotypes from 1,948 people from northeast Brazil cited by Yasuda [30]. Here, there are three alleles $\{G_i, i = 1, 2, 3\}$. Let $\{p_i, i = 1, 2, 3\}$ be the frequencies of the genes. Under Hardy–Weinberg condition, the frequencies of the genotypes G_i/G_i and G_i/G_j $(i \neq j)$ are then given by $Fp_i + (1 - F)p_i^2$ and $(1 - F)2p_ip_j$ respectively, where F is the inbreeding coefficient.

Genotype	Observed Number	Genotype Frequency
G_1/G_1	108	$Fp_1 + (1 - F)p_1^2$
G_1/G_2	196	$(1 - F)2p_1p_2$
G_1/G_3	429	$(1 - F)2p_1p_3$
G_2/G_2	143	$Fp_2 + (1 - F)p_2^2$
G_2/G_3	513	$(1 - F)2p_2p_3$
G_3/G_3	559	$Fp_3 + (1 - F)p_3^2$

By introducing dummy variables for the genotypes G_i/G_i, develop an EM-algorithm for deriving the MLE of $\{p_i, i = 1, 2, F\}$. Assuming uniform prior, derive a Bayesian Gibbs sampling procedure for estimating these parameters.

Exercise 3.6. **(Estimation of Linkage in Repulsion Case).** Consider two linked loci in a self-fertilized population each with two alleles, say $A : a$ and $B : b$. Suppose that a repulsion crossing $AAbb \times aaBB$ is made at time 0. Assume that there are no interactions between the two loci and that A is dominant over a and B dominant over b. Derive the expected frequencies of the four phenotypes $\{A\text{–}B\text{–}, A\text{–}bb, aaB\text{–}, aabb\}$. Assuming that the observed numbers of the four phenotypes are given by $\{n_{11}, n_{12}, n_{21}, n_{22}\}$ respectively, derive an EM-algorithm to estimate the linkage recombination θ between the A-locus and the B-locus. Let the prior distribution of θ be given by $P(\theta) \propto \theta^{a-1}(1-\theta)^{b-1}, a \geq 1, b \geq 1$, derive a Bayesian Gibbs sampling procedure to estimate θ.

Exercise 3.7. **(Estimation of linkage through backcrossing between *aabb* and *Ab/aB* or *AB/ab*).** Consider two linked loci in a human population each with two alleles, say $A : a$ and $B : b$. Assume that A is not dominant over a and that B is not dominant over b. Take a random sample of size n from the population with phenotype $AaBb$ and mate each individual in the sample with a person with genotype $aabb$. (This is called the *backcrossing*.)

(a) Assuming that the frequencies of AB/ab and Ab/aB are equal in the population with phenotype $AaBb$, derive the expected frequencies of the observed phenotypes.

(b) By introducing dummy variables, derive an EM algorithm for estimating the MLE of the recombination fraction θ between the two loci.

(c) Assuming a beta prior for θ, derive the Bayesian Gibbs sampling procedures to estimate θ.

3.8. Appendix: A Lemma for Finite Markov Chains

Lemma 3.4. *Let $P(x)$ be a real-valued function defined over the set of integers. Let $I_{(+)}$ be a subset of positive integers defined by: $n \in I_{(+)}$ iff (if* *and only if)* $P(n) > 0$. *Suppose that the following conditions hold*:

(a) $I_{(+)}$ *is not empty. That is, there is a positive integer* m *such that* $P(m) > 0$.

(b) *The g.c.d. (greatest common divisor) of elements of* $I_{(+)}$ *is 1.*

(c) *If* $a \in I_{(+)}$ *and* $b \in I_{(+)}$, *then* $a + b \in I_{(+)}$.

Then, there exists a positive integer N_0 *such that for all integer* n *satisfying* $n > N_0$, $n \in I_{(+)}$. *That is, for all integer* n *such that* $n > N_0$, $P(n) > 0$.

Proof.   To prove the Lemma, we first prove the following three results:

(1) First we prove that there exists a positive integer a in $I_{(+)}$ with $a > 1$. To prove this, notice that by assumption (a), there exists a positive integer b such that $b \in I_{(+)}$; then by condition (c), $mb \in I_{(+)}$ for all positive integers $m = 1, 2, 3, \ldots$; hence $I_{(+)}$ contains infinite many elements and there exists an element a in $I_{(+)}$ with $a > 1$.

(2) Next we notice that for every positive integers n and a, if $a > 1$ and if n is not divisible by a, then n can be expressed by $n = r_0 a + r$, where r_0 is a non-negative integer and r is a positive integer satisfying $0 < r < a$; for otherwise, n is divisible by a, contradicting the assumption.

(3) Third, we will show that there exist two consecutive integers in $I_{(+)}$, say N_1 and $N_2 (N_2 > N_1)$, such that $N_2 = N_1 + 1$. To prove this, suppose that the maximum difference between consecutive numbers in $I_{(+)}$ is k $(k \geq 1)$. That is, there exist $n_1 \in I_{(+)}$ and $n_2 = n_1 + k \in I_{(+)}$ and n_1 and $n_2 (n_2 > n_1)$ are consecutive numbers in $I_{(+)}$. We will show that $k = 1$ by using argument of contradiction.

Suppose that $k > 1$. Then there exists a positive integer n in $I_{(+)}$ such that n is not divisible by k. (Notice that such an n always exists; for otherwise, k is the g.c.d. of elements of $I_{(+)}$ but $k > 1$.). Then by result (2), n can be expressed by $n = m_1 k + m_2$ where m_1 is a non-negative integer and m_2 is a positive integer satisfying $0 < m_2 < k$. By condition (c), $d = (m_1 + 1)(n_1 + k) \in I_{(+)}$ and $e = n + (m_1 + 1)n_1 \in I_{(+)}$. But, $d - e = (m_1 + 1)k - m_1 k - m_2 = k - m_2 > 0$ which is less than k contradicting that k is the maximum difference between consecutive numbers. Hence $k = 1$.

With results from (1)–(3), we are now ready to prove the final result of the lemma. By result (3), there exists an N_1 in $I_{(+)}$ such that $N_1 + 1 \in I_{(+)}$. By Condition (c), obviously, one may assume $N_1 > 1$. Let $N_0 = N_1^2$ and

$n - N_0 = j$. If j is a positive integer, then by result (2), $j = r_1 N_1 + r_2$ where r_1 is a non-negative integer and r_2 a positive integer satisfying $0 < r_2 < N_1$. It follows that

$$n = N_0 + j = N_1^2 + j = N_1^2 + r_1 N_1 - r_2 N_1 + r_2 + r_2 N_1$$

$$= N_1(N_1 - r_2 + r_1) + r_2(N_1 + 1).$$

Since for all positive integer j, $N_1(N_1 - r_2 + r_1) \in I_{(+)}$ and $r_2(N_1 + 1) \in I_{(+)}$, so, $n = N_0 + j \in I_{(+)}$ for all $j = 1, 2, \ldots$. $\qquad\square$

References

[1] K. Knopp, *Theory and Applications of Infinite Series*, Fourth edition (Translated from Germany by R.C.H. Young), Blackie and Sons, London (1957).

[2] A. N. Kolmogorov and S. V. Fomin, *Elements of the Theory of Functions and Functional Analysis*, Vol. 2. (Translated from Russian by H. Kamel and H. Comm.), Graylock Press, Albany, New York (1961).

[3] S. Karlin and H. M. Taylor, *A First Course in Stochastic Processes*, Second Edition, Academic Press, New York (1975).

[4] D. T. Bishop, J. A. Williamson and M. H. Skolnick, *A model for restriction fragment length distributions*, Am. J. Hum. Genet. **35** (1983) 795–815.

[5] C. P. Robert and G. Casella, *Monte Carlo Statistical Methods*, Springer-Verlag, Berlin (1999).

[6] M.-H. Chen, Q.-M. Shao and J. G. Ibrahim, *Monte Carlo Methods in Bayesian Computation*, Springer-Verlag, Berlin (2000).

[7] G. Casella and E. George, *Explaining the Gibbs ampler*, American Statistician **46** (1992) 167–174.

[8] J. Besag, P. Green, D. Higdon and K. Mengersen, *Bayesian computation and stochastic systems (with discussion)*, Statistical Science **10** (1995) 3–66.

[9] M. K. Cowles and B. P. Carlin, *Markov chain Monte Carlo convergence diagnostics: A comparative review*, J. Amer. Statist, Association **91** (1996) 883–904.

[10] A. Gelman, *Inference and monitoring convergence*, in *Markov Chain Monte Carlo in Practice*, eds. W. R. Gilks, S. Richardson and D. J. Spiegelhalter, Chapman and Hall, London (1996) 131–143.

[11] R. E. Kass, P. R. Carlin, A. Gelman and R. M. Neal, *Markov chain Monte Carlo in practice: A roundtable discussion*, American Statistician **52** (1998) 93–100.

[12] G. O. Roberts, *Convergence control methods for Markov chain Monte Carlo algorithms*, Statistical Science **10** (1995) 231–253.

[13] A. F. M. Smith and A. E. Gelfand, *Bayesian statistics without tears: A sampling-resampling perspective*, American Statistician **46** (1992) 84–88.

[14] W. Y. Tan and Z. Z. Ye, *Estimation of HIV infection and HIV incubation via state space models*, Math. Biosciences **167** (2000) 31–50.

[15] W. Y. Tan and Z. Z. Ye, *Some state space models of HIV epidemic and applications for the estimation of HIV infection and HIV incubation*, Comm. Statistics (Theory and Methods) **29** (2000) 1059–1088.

[16] W. K. Hastings, *Monte Carlo sampling methods using Markov chains and their application*, Biometrika **57** (1970) 97–109.

[17] M. A. Tanner, *Tools for Statistical Inference*, Second Edition, Springer-Verlag, Berlin (1993).

[18] L. Tierney, *Markov Chains for exploring posterior distributions (with discussion)*, Annals of Statistics **22** (1994) 1701–1762.

[19] S. Chib and E. Greenberg, *Understanding the Metropolis–Hastings algorithm*, American Statistician **49** (1995) 327–335.

[20] B. P. Carlin and T. A. Louis, *Bayes and Empirical Bayes Methods for Data Analysis*, Chapmand and Hall, London (1996).

[21] C. C. Li, *Population Genetics*, University of Chicago Press, Chicago (1955).

[22] A. P. Dempster, N. Laird and D. B. Rubin, *Maximum likelihood from incomplete data via the EM algorithm*, J. Royal Statist. Soc. B **39** (1977) 1–38.

[23] G. E. P. Box and G. C. Tiao, *Bayesian Inference in Statistical Analysis*, Addison-Wesley, Reading, MA. (1973).

[24] T. A. Louis, *Finding the observed information using the EM algorithm*, J. Royal Statist. Soc. B **44** (1982) 98–130.

[25] D. Thomas and V. Cortessis, *A Gibbs sampling approach to linkage analysis*, Human Heredity **42** (1992) 63–76.

[26] J. Ott, *Analysis of Human Genetic Linkage*, Johns Hopkins University Press, Baltimore (1985).

[27] C. R. Rao, *Linear Statistical Inference and Its Applications*, Second Edition, Wiley, New York (1973).

[28] W. Y. Tan, *Comparative studies on the estimation of linkage by Bayesian method and maximum likelihood method*, Comm. in Statistics **B9** (1980) 19–41.

[29] S. W. Guo and E. A. Thompson, *A Monte Carlo method for combined segregation and linkage analysis*, Am. J. Hum. Genet. **51** (1992) 1111–1126.

[30] N. Yasuda, *Estimation of the inbreeding coefficient from phenotype frequencies by a method of maximum likelihood scoring*, Biometrics **24** (1968) 915–934.

Chapter 4

Continuous-Time Markov Chain Models in Genetics, Cancers and AIDS

In the previous two chapters, we have discussed a class of stochastic models which are Markov chains with discrete time. We have developed some basic theories and have demonstrated some applications of this class of stochastic models to some genetic problems and to some biomedical problems. In this chapter, we will extend these results into Markov chains with continuous time. As it turns out, this class of stochastic models is very rich and has a wide range of applications to many biomedical problems as well as ecological systems. It includes stochastic birth-death processes, many genetic processes, filtered Poisson processes, HIV epidemiology models as well as many cancer models.

4.1. Introduction

Let $\{X(t), t \in T\}$ be a Markov chain with $T = \{t \geq 0\}$ and with state space $S = \{0, 1, \ldots, \infty\}$. Denote the transition probabilities of this Markov chain by $p_{ij}(s, t) = \Pr\{X(t) = j | X(s) = i\}$ for $t \geq s$ and for $i, j = 0, 1, \ldots, \infty$. Then, as in Markov chains with discrete time, the chain $\{X(t), t \geq 0\}$ is defined as a homogeneous chain if $p_{ij}(s, t) = p_{ij}(t - s) = P\{X(t - s) = j | X(0) = i\}$. Furthermore, as in Markov chains with discrete time, the following two results are immediate:

(1) $\lim_{s \to t} p_{ij}(s, t) = \lim_{t \to s} p_{ij}(s, t) = \delta_{ij}$, where δ_{ij} is the kronecker's δ defined by $\delta_{ij} = 1$ if $i = j$ and $\delta_{ij} = 0$ if $i \neq j$. If the chain is homogeneous, then $\lim_{\Delta t \to 0} p_{ij}(\Delta t) = \delta_{ij}$.

(2) For any times $s \leq r \leq t$ and $\{0 \leq i, j \leq \infty\}$, the following Chapman–Kolmogorov equation holds:

$$p_{ij}(s,t) = \sum_{k=0}^{\infty} p_{ik}(s,r)p_{kj}(r,t)\,.$$

If the chain is finite, then, with the transition matrix being denoted by $P(s,t) = (p_{ij}(s,t))$, the above Chapman–Kolmogorov equation can be expressed in matrix notation as:

$$P(s,t) = P(s,r)P(r,t) \text{ for any } r \text{ satisfying } s \leq r \leq t\,.$$

It follows that if the chain is finite and homogeneous, then with $t = n\Delta t$,

$$P(t) = P(\Delta t)^n = \lim_{n \to \infty} P(t/n)^n.$$

Using $p_{ij}(\Delta t)$ as analog of the one-step transition probabilities in Markov chains with discrete time, one may then construct a Markov chain with discrete time with Δt corresponding to one time unit. This Markov chain will be referred to as an embedded Markov chain for the original chain $X(t)$; see Remark 4.1.

As in Markov chain with discrete time, one may also define $i \to j$ for $i \in S$ and $j \in S$ and define closed sets in the state space S. Thus, if $\{X(t), t \in T = [0, \infty)\}$ is a homogeneous Markov chain with continuous parameter space and with state space $S = \{0, 1, \ldots, \infty\}$, then $i \to j$ for $i \in S$ and $j \in S$ iff there exists a time $t > 0$ in T such that $p_{ij}(t) > 0$; we define the subset C in S (i.e. $C \subset S$) as a closed set iff for every $i \in C$ and for every $j \in C$, $i \leftrightarrow j$. The chain $X(t)$ is said to be irreducible iff the state space S does not contain proper closed subsets, or, iff for every $i \in S$ and for every $j \in S$, $i \leftrightarrow j$.

As in Markov chains with discrete time, the states in Markov chains with continuous time can also be classified as persistent states (or recurrent states) and transient states (or non-recurrent states). To be specific, let T_{ij} be the first passage time to the state j from the state i at time 0 and denote by $P\{T_{ij} \in [t, t + \Delta t)|X(0) = i\} \cong f_{ij}(t)\Delta t$. Then, the state i is classified as a persistent state (or recurrent state) iff $\int_0^\infty f_{ii}(t)dt = 1$. ($i$ is classified as a transient state (or non-recurrent state) if i is not persistent.)

If i is persistent, then $f_{ii}(t)$ is the pdf of the first return time T_{ii} of the state i. The mean return time of the persistent state i is $\nu_i = \int_0^\infty t f_{ii}(t)dt$. As in Markov chains with discrete time, the persistent state i is called a positive state if the mean return time ν_i of i is finite, i.e. $\nu_i < \infty$; the persistent state i is called a null state if the mean return time ν_i of i is infinite, i.e. $\nu_i = \infty$. Unlike Markov chains with discrete time, however, the problem of "periodicity" does not exist in Markov chains with continuous time. That is, in the embedded Markov chains, all persistent states are aperiodic.

Remark 4.1. In [1], any Markov chain derived from the original chain is called an imbedded chain. We will follow Karlin [2] to define the discrete-time Markov chain with Δt as a fixed time unit as an embedded chain to differ from the imbedded chain as defined in [1].

4.2. The Infinitesimal Generators and an Embedded Markov Chain

In Markov chains with discrete time, the chain is characterized and specified by the one-step transition probabilities. In Markov chains with continuous time, the role of these probabilities are played by the transition rates (or the incidence functions or the infinitesimal generator) $\alpha_{ij}(t)$, where for $i \neq j$,

$$\alpha_{ij}(t) = \lim_{\Delta t \to 0} \frac{1}{\Delta t} P\{X(t + \Delta t) = j | X(t) = i\},$$

and $\alpha_{ii}(t) = \sum_{j \neq i} \alpha_{ij}(t)$. Or, equivalently, for $i \neq j$,

$$P\{X(t + \Delta t) = j | X(t) = i\} = \alpha_{ij}(t)\Delta t + o(\Delta t),$$

where $\lim_{\Delta t \to 0} \frac{o(\Delta t)}{\Delta t} = 0$; and

$$P\{X(t + \Delta t) = i | X(t) = i\} = 1 - \sum_{j \neq i} P\{X(t + \Delta t) = j | X(t) = i\}$$

$$= 1 - \alpha_{ii}(t)\Delta t + o(\Delta t). \tag{4.1}$$

In the literature, $\alpha_{ij}(t)$ have also been referred to as infinitesimal generators (or incidence functions or hazard functions). (In what follows, we assume that the $\alpha_{ij}(t)$'s are continuous functions of t unless otherwise stated.)

Given $\Delta t > 0$, denote by $p_{ij}^{(*)}(m, n) = P\{X(n\Delta t) = j | X(m\Delta t) = i\}$ for all integers $n \geq m \geq 0$. Then one may construct a new Markov chain $\{Y(t), t \in T = (0, 1, \ldots, \infty)\}$ with discrete time and with transition probabilities given by $P\{Y(n) = j | Y(m) = i\} = P_{ij}^{(*)}(m, n)$. The state space of this chain is $S = \{0, 1, \ldots, \infty\}$ and the one-step transition probabilities are $P_{ij}^{(*)}(m, m + 1) = \alpha_{ij}(m\Delta t)\Delta t$. This Markov chain $Y(t)$ has discrete time $T = \{0, 1, \ldots, \infty\}$. In this book, this chain is referred to as an embedded Markov chain embedded in the Markov chain $X(t)$; see Remark 4.1. In this embedded Markov chain, one time unit corresponds to Δt in the continuous time scale.

By using the above embedded Markov chains, one can extend results of Markov chains with discrete time to those of Markov chains with continuous time through the following procedures:

(1) For given $\Delta t > 0$, construct an embedded Markov chain.

(2) Write down results of Markov chain with discrete time using this embedded Markov chain.

(3) Derive results of Markov chain with continuous time by letting $\Delta t \to 0$.

By using this approach, for Markov chains with continuous time, the following results are immediate:

(1) Starting with any transient state, if the set C_T of transient states is finite, then with probability one the chain will eventually be absorbed into a persistent state. It follows that if C_T is finite, then the chain must contain persistent states.

(2) With probability one, persistent states will return to itself an infinitely many times; transient states will return to itself only a finite number of times.

(3) Persistent states go only to persistent states. That is, if i is persistent and if $i \to j$, then $j \to i$ and j is persistent; furthermore,

$$\int_0^\infty f_{ij}(t)dt = \int_0^\infty f_{ji}(t)dt = 1\,.$$

(4) For finite Markov chains, not all states are transient. That is, there exist at least one persistent state.

(5) For finite Markov chains, all persistent states are positive.

(6) If C is a closed set of persistent states, then either all states are positive or all states are null. If the closed set is finite, then it must be a closed set of persistent states and all states in C are positive.

Example 4.1. Stochastic birth-death processes. A Markov chain $\{X(t), t \in T = [0, \infty)\}$ with continuous time and with state space $S = \{0, 1, \ldots, \}$ is called a stochastic birth and death process with birth rate $b_i(t) = \alpha_{i,i+1}(t)$ and death rate $d_i(t) = \alpha_{i,i-1}(t)$ iff the transition rates satisfy the following conditions:

(1) $\alpha_{i,i+1}(t) = b_i(t) \geq 0$,
(2) $\alpha_{i,i-1}(t) = d_i(t) \geq 0$, and
(3) $\alpha_{ij}(t) = 0$ for all $|i - j| \geq 2$.

The birth-death process is called the *Feller–Arley birth-death process with birth rate $b_i(t)$ and death rate $d_i(t)$* if $b_i(t) = ib(t)$ and $d_i(t) = id(t)$, the stochastic Gompertz birth-death process if $b_i(t) = ib(t)e^{-\lambda t}$ and $d_i(t) = id(t)e^{-\lambda t}$ with $\lambda > 0$ [3], and the stochastic logistic birth-death process if $b_i(t) = ib(t)\{1-i/M\}$, $d_i(t) = id(t)\{1-i/M\}$ and $S = \{0, 1, \ldots, M\}$, where M denotes the maximum population size [4, 5]. A Markov process $\{X(t), t \in T = [0, \infty)\}$ with state space $S = (0, 1, \ldots, \infty)$ is called a birth-death process with immigration iff (1) $X(t)$ is a birth-death process, and (2) the birth rate and death rate are given by $\{b_i(t) + \alpha(t)$ $(\alpha(t) > 0, b_i(t) \geq 0)$ and $d_i(t)$ $(d_i(t) \geq 0)$ respectively.

Stochastic birth-death processes are the most widely used processes in natural sciences including biology, medical sciences, businesses, social sciences and engineering. It has been used in cancer models [3, 4, 6] and in AIDS models [7, Chaps. 7 and 8]. It has been shown by [4, 6] and Tan and Piantadosi [5] that the cell proliferation and differentiation of normal stem cells and cancer cells of female breasts are best described by stochastic logistic birth-death processes; similarly, the growth and relapse of cancer tumors are best described by stochastic Gompertz birth-death processes; see [3, 6]. In [7, Chaps. 7 and 8] and [8], stochastic logistic birth-death processes have been used to model proliferation of CD4 T cells by infection by antigens and HIV.

Example 4.2.   The Moran genetic model as a finite stochastic birth-death process. The model was first proposed by Moran [9]. It considers a haploid population of fixed population size M, consisting of two types of alleles A_1 and A_2 together with the following basic assumptions:

(i) During $[t, t + \Delta t)$, the probability of having more than one death is $o(\Delta t)$.

(ii) The pdf of the life time distribution of each A_j allele is $\lambda_j e^{-\lambda_j t}, t \geq 0$, where $\lambda_j > 0$ is independent of t. (Notice that if there is selection among the A_1 and A_2 alleles, then $\lambda_1 \neq \lambda_2$; if there is no selection among the two types of alleles, then $\lambda_1 = \lambda_2 = \lambda$.)

(iii) Whenever a death occurs among the alleles during $[t, t + \Delta t)$, it is replaced immediately by an allele which is A_1 or A_2 allele with respective probabilities $p_t^{(*)} = \frac{X(t)}{M}(1 - \alpha_1) + (1 - \frac{X(t)}{M})\alpha_2$ and $q_t^{(*)} = 1 - p_t^{(*)}$, where $X(t)$ is the number of A_1 allele at time t and where α_1 ($1 \geq \alpha_1 \geq 0$) and α_2 ($1 \geq \alpha_2 \geq 0$) are the mutation rates from A_1 to A_2 and from A_2 to A_1 respectively. (Notice that because the population size is fixed at M, if $X(t)$ is the number of A_1 allele in the population at time t, then the number of A_2 allele at time t is $M - X(t)$.)

We now proceed to show that the above process $X(t)$ with $t \in T = [0, \infty)$ and with state space $S = \{0, 1, 2, \ldots, M\}$ is a finite homogeneous birth-death process with birth rate b_j and death rate d_j being given respectively by

$$b_j = \lambda_2 p_j (M - j) \text{ and } d_j = \lambda_1 j q_j, \text{ where}$$

$$p_j = \frac{j}{M}(1 - \alpha_1) + \left(1 - \frac{j}{M}\right)\alpha_2 \text{ and } q_j = 1 - p_j, j = 0, 1, 2, \ldots, M.$$

To prove the above claim, let T_j denote the survival time of an A_j allele. Then, by (ii) given above, the probability density of T_j is

$$f_j(t) = \lambda_j e^{-\lambda_j t}, t \geq 0 \text{ so that } \Pr\{T_j \geq t\} = \int_t^{\infty} \lambda_j e^{-\lambda_j z} dz = e^{-\lambda_j t}.$$

Hence the conditional probability that an A_j allele dies during $[t, t + \Delta t)$ given an A_j allele at time t is

$$\Pr\{T_j \geq t | T_j \geq t\} - \Pr\{T_j \geq t + \Delta t | T_j \geq t\}$$

$$= 1 - \Pr\{T_j \geq t + \Delta t | T_j \geq t\}$$

$$= 1 - \frac{\Pr\{T_j \geq t + \Delta t\}}{\Pr\{T_j \geq t\}} = 1 - \frac{e^{-\lambda_j(t+\Delta t)}}{e^{-\lambda_j t}}$$

$$= 1 - e^{-\lambda_j \Delta t} = \lambda_j \Delta t + o(\Delta t), \qquad j = 1, 2.$$

Thus, the conditional probability that an A_1 allele dies during $[t, t + \Delta t)$ given that there are k A_1 alleles at time t is

$$\Pr\{\text{An } A_1 \text{ allele dies during } [t, t + \Delta t) | X(t) = k\} = k\lambda_1 \Delta t + o(\Delta t).$$

Similarly, the conditional probability that an A_2 allele dies during $[t, t+\Delta t)$ given that there are $M - k$ A_2 alleles at time t is

$$\Pr\{\text{An } A_2 \text{ allele dies during } [t, t + \Delta t) | X(t) = k\} = (M - k)\lambda_2 \Delta t + o(\Delta t) \,.$$

Since the conditional probability that an A_2 allele dies and is replaced by an A_1 allele during $[t, t + \Delta t)$ given that there are j A_1 allele is $P_{j,j+1}(\Delta t) = \Pr\{X(t + \Delta t) = j + 1 | X(t) = j\}$, we have:

$$P_{j,j+1}(\Delta t) = \Pr\{X(t + \Delta t) = j + 1 | X(t) = j\} = (M - j)\lambda_2 p_j \Delta t + o(\Delta t) \,.$$

Similarly, since the conditional probability that an A_1 allele dies and is replaced by an A_2 allele during $[t, t + \Delta t)$ given that there are j A_1 allele is $P_{j,j-1}(\Delta t) = \Pr\{X(t + \Delta t) = j - 1 | X(t) = j\}$, we have:

$$P_{j,j-1}(\Delta t) = \Pr\{X(t + \Delta t) = j - 1 | X(t) = j\} = j\lambda_1 q_j \Delta t + o(\Delta t) \,;$$

and by assumption (i) above,

$$P_{j,k}(\Delta t) = \Pr\{X(t + \Delta t) = k | X(t) = j\} = o(\Delta t) \text{ if } |k - j| \geq 2 \,.$$

This shows that the Moran's genetic model is indeed a finite homogeneous birth-death process with birth rate $b_j = (M - j)\lambda_2 p_j$ and death rate $d_j = j\lambda_1 q_j$ and with state space $S = \{0, 1, 2, \ldots, M\}$. This is a finite Markov chain with continuous time. In this chain, if $\alpha_i = 0, i = 1, 2$, then the states 0 and M are absorbing states (persistent states) and all other states (i.e. $1, 2, \ldots, M-1$) are transient states and the chain contains two proper subsets. If $\alpha_i > 0, i = 1, 2$, then all states are persistent and the chain is irreducible.

Example 4.3.   The nucleotide substitution model as a finite Markov chain with continuous time. In molecular evolution, Kimura [10] showed that the nucleotide substitution in Eukaryotes were best described by Markov chains with continuous time. In these cases, the four DNA bases $\{A, T, C, G\}$ are generated by a Markov chain with continuous time with transition rates $\{\alpha, \beta, \gamma, \delta, \epsilon, \lambda, \kappa, \sigma\}$ as described by Fig. 4.1. This is a homogeneous Markov chain with state space $S = \{A, C, T, G\}$.

Example 4.4.   The AIDS epidemic as Markov chains with continuous time. Markov models with continuous time have been used extensively in the literature to study the dynamic of the HIV epidemic [7, 11–15].

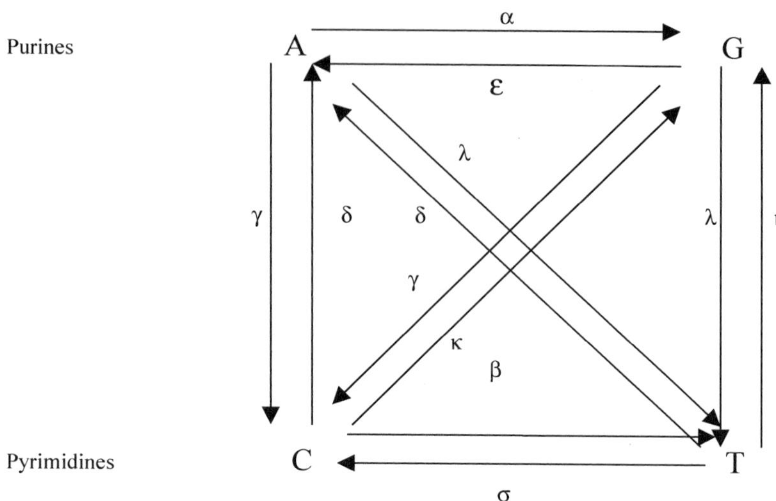

Fig. 4.1. Plots showing transitions in the nucleotide substitution model in molecular evolution.

To illustrate, consider a large population consisting of HIV infected individuals and AIDS cases with transitions described by Fig. 1.1.

If we assume that the transition rates $\beta_i(s,t) = \beta_i(t), \gamma_i(s,t) = \gamma_i(t)$ and $\omega_i(s,t) = \omega_i(t)$ are independent of the initiation time s, then we have a Markov chain with state space $\Omega = \{I_i, i = 1, \ldots, k, A = I_{k+1}\}$, where A denotes the AIDS stage and I_i the ith infective stage. When time is treated as a continuous variables, this is a finite Markov chain with continuous time, the special cases of which have been considered by Longini and his associates [11–15]. In this chain, A is the absorbing state and all other states I_i ($i = 1, \ldots, k$ are transient states. If $\{\beta_i(t) = \beta_i, \gamma_i(t) = \gamma_i, \omega_i(t) = \omega_i\}$ ($i = 1, \ldots, k$) are independent of time t, then the process is also time-homogeneous. Notice also that if $\beta_i = 0$ for $i = 1, \ldots, k$ and if $\omega_j = 0$ for $j = 1, \ldots, k - 1$, then the model reduces to the model considered by Longini *et al.* [11, 12].

Example 4.5. The drug resistant tumor cells in chemotherapy as Markov chains with continuous time. In Example 1.4, we have considered a drug-resistant model for cancer chemotherapy. In this model, there are two types of cancer tumor cells: The sensitive cancer tumor cells (X_1) and the resistant cancer tumor cells (X_2). Assuming that the X_i cancer tumor cells follow stochastic birth and death processes for cell proliferation and differentiation and that the resistant cancer tumor cells arise from sensitive cancer tumor cells by mutation. Let $X_i(t)$ be the number of X_i cancer tumor cells at time t and let 0 be the time starting treatment. Then $\{[X_1(t), X_2(t)], t > 0\}$ is a 2-dimensional Markov chain with continuous parameter space $T = \{t > 0\}$ and with state space $S = \{(i,j), i, j = 0, 1, \ldots, \}$. This is the stochastic process first studied by Tan and Brown [16].

Example 4.6. The Chu's multi-event model of carcinogenesis as Markov chains with continuous time. Many biological studies have shown that carcinogenesis is a multi-stage random process with intermediate cells subjected to stochastic cell proliferation and differentiation and with the number of stages greater than 2. Specific examples include colon cancer which has been shown to involve at lease one dominantly acting oncogene (ras) and several antioncogenes in chromosomes 5q, 17p and 18q [17, 18]; other examples involving more than 2 stages have been given in [6, 19–22]. This has led Chu [23], Chu *et al.* [24], Little and coworkers [19–22] to extend the MVK two-stage model into a multistage model with the normal stem cells and the initiated cells subjecting to stochastic proliferation (birth) and differentiation (death). When the number of stages for initiated cells equals to 2, this model reduces to the MVK two-stage model. Chu [23] has called his model the multi-event model to distinguish it from the classical Armitage-Doll model [6, Chap. 1]. Let $N(t)$ denote the number of normal stem cells at time t, $I_i(t)$ $(i = 1, \ldots, k)$ the number of the i-stage initiated cells at time t and $T(t)$ the number of malignant cancer tumors at time t. As in the MVK two-stage model, the above multi-event model also assumes that with probability one each of the I_k cells grows instantaneously into a malignant cancer tumor so that the I_k cells may be considered as cancer tumors. In this case, the process $\{N(t), I_i(t), i = 1, \ldots, k-1, I_k(t) = T(t)\}$ is a $(k+1)$-dimensional Markov chain with continuous time $T = [0, \infty)$. The state space is $S = \{(i_0, i_1, \ldots, i_k), \text{with } i_j, j = 0, 1, \ldots, k \text{ being non-negative integers}\}$. Notice, however, that in many practical situations the assumption $I_k \cong T$ may not hold [25, 26]; therefore, as noted in Example 1.13, $T(t)$ is not Markov. In any case, the process $\{N(t), I_i(t), i = 1, \ldots, k-1\}$ is always a Markov chain with continuous time.

Example 4.7. A continuous time Galton–Watson branching process. Consider a large haploid population (i.e. the individuals have only one set of homologous chromosomes in the genome.). Assume that at time 0 a mutant is introduced into the population. Let $X(t)$ be the number of mutants at time t. Then $X(t)$ is a homogeneous Markov chain with parameter space $T = [0, \infty)$ and with state space $S = \{0, 1, \ldots, \infty\}$ if the following conditions hold:

(i) After time 0, no new mutants are introduced into the population.

(ii) The density of the life time distribution of each mutant is exponential with density $h_M(t) = \lambda\, e^{-\lambda t}, \lambda > 0, t \in T$.

(iii) Whenever a mutant dies with probability $p_j (j = 0, 1, \ldots, \infty)$ such that $\sum_{j=0}^{\infty} p_j = 1$, it leaves beyond j mutants. (To avoid trivial situations, we will assume $0 < p_0 < 1$.)

(vi) Each mutant proceeds according to the above rules and probability laws independently of other mutants.

In the above Markov chain, the state 0 is persistent and all other states (i.e. $1, 2, \ldots, \infty$) are transient.

To find the transition rates α_{ij}, let T_M be the survival time of the mutant. Since the probability that a mutant at time 0 will survive at least t period is $P(T_M \geq t) = \int_t^{\infty} h_M(x)dx = e^{-\lambda t}$, the conditional probability that a mutant at time 0 will die during $[t, t + \Delta t)$ given that it has survived t period is

$$\frac{1}{P(T_M \geq t)}\{P(T_M \geq t) - P(T_M \geq t + \Delta t)\} = \lambda\Delta t + o(\Delta t)\,.$$

It follows that for $j \neq 1$,

$$\Pr\{X(t + \Delta t) = j | X(t) = 1\} = \lambda p_j \Delta t + o(\Delta t)\,.$$

If $j = 1$, then, since the mutant remains one mutant if it does not die,

$$\Pr\{X(t + \Delta t) = 1 | X(t) = 1\} = 1 - \lambda\Delta t + \lambda p_1 \Delta t + o(\Delta t)$$

$$= 1 - (1 - p_1)\lambda\Delta t + o(\Delta t)\,.$$

(In the above equation, the first term $1 - \lambda\Delta t$ is the probability that the mutant survives the time period $[t, t + \Delta t)$.)

It follows that $\alpha_{1j} = p_j\lambda$ for $j \neq 1$ and $\alpha_{11} = (1 - p_1)\lambda$. Let $q_j = \alpha_{1j}\Delta t$ if $j \neq 1$ and $q_1 = 1 - \alpha_{11}\Delta t$. Let $Y(n)$ be the Galton–Watson branching process with progeny distribution $\{q_j, j = 0, 1, \ldots, \infty\}$. Then $\sum_{j=0}^{\infty} q_j = 1$

and $Y(n)$ is an embedded Markov chain of $X(t)$. The pgf of $\{q_j, j \in S\}$ is

$$h(s) = \sum_{j=0}^{\infty} s^j q_j = s + [g(s) - s]\lambda\Delta t = s + u(s)\Delta t\,,$$

where $g(s)$ is the pgf of $p_j, j \in S$ and $u(s) = [g(s) - s]\lambda$.

Since each mutant proceeds according to the same rule and the same probability laws independently of one another, given i mutants at time t, the pgf of the number of mutants at time $t + \Delta t$ is $[h(s)]^i$. Now,

$$[h(s)]^i = \{s + u(s)\Delta t\}^i = s^i + is^{i-1}u(s)\Delta t + o(\Delta t)$$

$$= s^i[1 - i(1 - p_1)\lambda\Delta t] + is^{i-1}p_0\lambda\Delta t$$

$$+ i\sum_{j=2}^{\infty} s^{i+j-1}p_j\lambda\Delta t + o(\Delta t)\,.$$

It follows that

$$\Pr\{X(t + \Delta t) = j | X(t) = i\} = \begin{cases} o(\Delta t)\,, & \text{if } j < i - 1\,; \\ ip_0\lambda\Delta t + o(\Delta t)\,, & \text{if } j = i - 1\,; \\ 1 - i(1 - p_1)\lambda\Delta t + o(\Delta t)\,, & \text{if } j = i\,; \\ ip_{j-i+1}\lambda\Delta t + o(\Delta t)\,, & \text{if } j > i\,. \end{cases}$$

Hence, for $i > 1$, $\alpha_{ij} = 0$ if $j < i - 1$; $\alpha_{ij} = ip_0\lambda$ if $j = i - 1$; $\alpha_{ii} = i(1 - p_1)\lambda$ if $j = i$; and $\alpha_{ij} = ip_{j-i+1}\lambda$ if $j > i$.

4.3. The Transition Probabilities and Kolmogorov Equations

In Markov chains with discrete time, one derives the general transition probabilities by giving the one-step transition probabilities and then applying the Chapman–Kolmogorov equation. Extending this to continuous time, one then derives the general transition probabilities by giving the transition rates and then applying the Kolmogorov forward and/or backward equations. These are systems of differential and difference equations. The differential operator applies to time as time is continuous whereas the difference operator applies to the state variables as the state variables are discrete.

There are two types of systems of equations. One is referred to as the Kolmogorov forward equations whereas the other as the Kolmogorov backward equations. The Kolmogorov forward equations are given by:

$$\frac{d}{dt}p_{ij}(s,t) = \sum_{k \neq j} p_{ik}(s,t)\alpha_{kj}(t) - p_{ij}(s,t)\alpha_{jj}(t)\,, \qquad (4.2)$$

where $i,j = 0,1,\ldots,\infty$ and $p_{ij}(s,s) = \delta_{ij}$.

The Kolmogorov backward equations are given by:

$$-\frac{d}{ds}p_{ij}(s,t) = \sum_{k \neq i} \alpha_{ik}(s)p_{kj}(s,t) - \alpha_{ii}(s)p_{ij}(s,t)\,, \qquad (4.3)$$

where $i,j = 0,1,\ldots,\infty$ and $p_{ij}(s,s) = \delta_{ij}$.

Notice that in the Kolmogorov forward equations, the derivatives are taken with respect to the forward time t whereas in the Kolmogorov backward equations, the derivatives are taken with respect to the backward time s. If the $\alpha_{ij}(t) = \alpha_{ij}$'s are independent of t, then the chain is homogeneous so that $p_{ij}(s,t) = p_{ij}(t-s)$. In these cases, the above Kolmogorov forward and backward equations reduce respectively to:

(1) The Kolmogorov Forward Equations:

$$\frac{d}{dt}p_{ij}(t) = \sum_{k \neq j} p_{ik}(t)\alpha_{kj} - p_{ij}(t)\alpha_{jj}\,,$$

where $i,j = 0,1,\ldots,\infty$ and $p_{ij}(0) = \delta_{ij}$.

(2) The Kolmogorov Backward Equations:

$$\frac{d}{dt}p_{ij}(t) = \sum_{k \neq i} \alpha_{ik}p_{kj}(t) - \alpha_{ii}p_{ij}(t)\,,$$

where $i,j = 0,1,\ldots,\infty$ and $p_{ij}(0) = \delta_{ij}$.

To prove Eqs. (4.3), note that by the Chapman–Kolmogorov equation, we have for $t > s$ and for $\Delta s > 0$:

$$p_{ij}(s - \Delta s, t) = \sum_{k=0}^{\infty} p_{ik}(s - \Delta s, s)p_{kj}(s,t)$$

$$= \sum_{k \neq i}^{\infty} \alpha_{ik}(s - \Delta s)p_{kj}(s,t)\Delta s$$

$$+ \{1 - \alpha_{ii}(s - \Delta s)\Delta s\}p_{ij}(s,t)\Delta s + o(\Delta s)\,.$$

Subtracting $p_{ij}(s,t)$ from both sides of the above equation and dividing by Δs, we obtain:

$$\frac{1}{\Delta s}\{p_{ij}(s - \Delta s, t) - p_{ij}(s,t)\} = \sum_{k \neq i}^{\infty} \alpha_{ik}(s - \Delta s)p_{kj}(s,t)$$
$$- \alpha_{ii}(s - \Delta s)p_{ij}(s,t) + \frac{o(\Delta s)}{\Delta s}.$$

By letting $\Delta s \to 0$, one then drives the Kolmogorov backward equations as given by (4.3). By using exactly the same approach, one may derives the Kolmogorov forward equations as given by (4.2); see Exercise 4.1.

From above, theoretically, given the transition rates, one may derive the general transition probabilities by solving the Kolmogorov forward equations or by solving the Kolmogorov backward equations. (In practice, the solution of the above system equations may be very difficult if not impossible.) We now illustrate this by some examples.

Example 4.8. The Feller–Arley stochastic birth-death process. Let $X(t)$ be a Feller–Arley birth-death process with birth rate $b_i(t) = ib(t)$ and death rate $d_i(t) = id(t)$. Then $\alpha_{i,i+1}(t) = ib(t)$, $\alpha_{i,i-1}(t) = id(t)$, and $\alpha_{ij}(t) = 0$ if $|i - j| > 2$. Hence the Kolmogorov forward equations become:

$$\frac{d}{dt}p_{ij}(s,t) = p_{i,j-1}(s,t)(j - 1)b(t) + p_{i,j+1}(s,t)(j + 1)d(t)$$
$$- p_{ij}(t)j[b(t) + d(t)], \tag{4.4}$$

where $i, j = 0, 1, \ldots, \infty$ and $p_{ij}(s,s) = \delta_{ij}$.

We now use the following basic steps to solve the above system of equations:

(1) Transform the system of equations into a partial differential equation (PDE) for the pfg of $p_{ij}(s,t)$:

$$Q_i(u; s, t) = \sum_{k=0}^{\infty} u^k p_{ik}(s,t).$$

(2) Solve the PDE of $Q_i(u; s, t)$.

(3) Obtain $p_{ij}(s,t)$ by taking derivatives repeatedly with respect to u to obtain $p_{ij}(s,t)$ as

$$p_{ij}(s,t) = \frac{1}{j!}\left(\frac{d^j}{du^j}Q_i(u; s, t)\right)_{u=0}.$$

To solve Eq. (4.4), multiplying both sides of (4.4) by u^j and summing over j from $j = 0$ to $j = \infty$, the left side is $\frac{\partial}{\partial t}Q_i(u; s, t)$; the right side is given by $g(u, t)\frac{\partial}{\partial u}Q_i(u; s, t)$ where $g(u, t) = (u - 1)[ub(t) - d(t)] = (u - 1)[(u - 1)b(t) + \gamma(t)]$ with $\gamma(t) = b(t) - d(t)$, by using the results given below:

$$\sum_{j=0}^{\infty} u^j(j - 1)p_{i,j-1}(s, t) = u^2 \sum_{j=1}^{\infty} u^{j-2}(j - 1)p_{i,j-1}(s, t)$$

$$= u^2 \sum_{j=0}^{\infty} \frac{\partial}{\partial u} u^j p_{ij}(s, t) = u^2 \frac{\partial}{\partial u}Q_i(u; s, t),$$

(notice that $p_{i,-1}(s, t) = 0$);

$$\sum_{j=0}^{\infty} u^j(j + 1)p_{i,j+1}(s, t) = \sum_{j=1}^{\infty} u^{j-1}jp_{i,j}(s, t)$$

$$= \sum_{j=0}^{\infty} \frac{\partial}{\partial u}u^j p_{ij}(s, t) = \frac{\partial}{\partial u}Q_i(u; s, t),$$

and

$$\sum_{j=0}^{\infty} u^j jp_{i,j}(s, t) = u \sum_{j=0}^{\infty} u^{j-1}jp_{i,j}(s, t) = u\frac{\partial}{\partial u}Q_i(u; s, t).$$

This shows that $Q_i(u; s, t)$ satisfies the PDE given by:

$$\frac{\partial}{\partial t}Q_i(u; s, t) = g(u, t)\frac{\partial}{\partial u}Q_i(u; s, t).$$

The initial condition is $Q_i(u; s, s) = u^i$. As shown in [3] and [6, Chap. 2], the solution of the above PDE under the initial condition $Q_i(u; s, s) = u^i$ is (Exercise 4.2)

$$Q_i(x; s, t) = \left(1 + \frac{(x - 1)}{\xi(t) - (x - 1)\zeta(t)}\right)^i \tag{4.5}$$

where $\xi(t) = \exp\{-\int_0^t [b(x) - d(x)]dx\}$ and $\zeta(t) = \int_0^t b(y)\xi(y)dy$.

Hence, if $i = 1$, then

$$p_{1j}(t) = P\{X(t) = j | X(0) = 1\}$$

$$= \begin{cases} 1 - \dfrac{1}{\xi(t) + \zeta(t)}, & \text{if } j = 0\,; \\[2ex] \dfrac{\xi(t)}{(\xi(t) + \zeta(t))^2} \left(\dfrac{\zeta(t)}{\xi(t) + \zeta(t)}\right)^{j-1}, & \text{if } j > 0\,. \end{cases}$$

$$EX(t) = [\xi(t)]^{-1} = \exp\left\{\int_0^t [b(x) - d(x)]dx\right\}$$

and

$$\text{Var}\, X(t) = \{\xi(t) + 2\zeta(t) - 1\}/\xi(t)^2\,.$$

To derive results for $i > 1$, put:

$$a_1(s,t) = \zeta(s,t)/[\xi(s,t) + \zeta(s,t)],\, a_2(s,t) = [1 - a_1(s,t)]/\zeta(s,t)$$

and $a_3(s,t) = 1 - [\zeta(s,t)]^{-1}$. Then,

$$p_{ij}(s,t) = [a_1(s,t)]^j \sum_{k=0}^{i} \binom{i}{k} \binom{j+k-1}{j} [a_2(s,t)]^k [a_3(s,t)]^{i-k}\,,$$

with $\binom{l}{0} = 1$ for all real l and $\binom{k-1}{k} = 0$ for all integer $k > 0$.

Example 4.9.   The number of initiated tumor cells in the two-stage model of carcinogenesis. In Example 1.13, we have described the non-homogeneous MVK two-stage model of carcinogenesis. Denote the birth rate and death rate of each I cell at time t by $b(t)$ and $d(t)$ respectively and let $\alpha_0(t)$ be the mutation rate from $N \to I$ at time t. That is, the probabilities that an I cell at time t will give rise to 2 I cells, 0 I cells and 1 I cell and 1 tumor cell at time $t + \Delta t$ are given respectively by $\{b(t)\Delta t + o(\Delta t), d(t)\Delta t + o(\Delta t)\}$ and $\alpha_0(t)\Delta t + o(\Delta t)$, respectively. Under the assumption that the number $N(0) = N_0$ of normal stem cells at time 0 is very large, then to order of $O(N_0^{-1})$, one may assume the number $N(t)$ of normal stem cells at t $(t > 0)$ as a deterministic function of t; for proof, see ([16, 27, 28]). In these cases, $I(t)$, the number of initiated cells at time t, follows a non-homogeneous birth-death process with immigration. In this process, given i I cells at time t, the birth rate and the death rate are given by $ib(t) + \lambda(t)$, where $\lambda(t) = N_0\alpha_0(t)$ and $id(t)$

respectively. Thus, the infinitesimal parameters are $\alpha_{i,i+1}(t) = ib(t) + \lambda(t)$, $\alpha_{i,i-1}(t) = id(t)$, and $\alpha_{ij}(t) = 0$ if $|i - j| > 2$. Hence the Kolmogorov forward equations become:

$$\frac{d}{dt}p_{ij}(s,t) = p_{i,j-1}(s,t)[(j-1)b(t) + \lambda(t)] + p_{i,j+1}(s,t)(j+1)d(t)$$

$$- p_{ij}(t)[jb(t) + \lambda(t) + jd(t)], \tag{4.6}$$

for $i, j = 0, 1, \ldots, \infty$ with $p_{ij}(s,s) = \delta_{ij}$.

Denote the pfg of $p_{ij}(s,t)$ by:

$$Q_i(u; s, t) = \sum_{k=0}^{\infty} u^k p_{ik}(s,t).$$

Then, by multiplying both sides of Eq. (4.6) by u^j and summing over j from $j = 0$ to $j = \infty$, we obtain the following partial differential equation for $Q_i(u; s, t)$:

$$\frac{\partial}{\partial t}Q_i(u; s, t) = g(u, t)\frac{\partial}{\partial u}Q_i(u; s, t) + (u - 1)\lambda(t)Q_i(u; s, t)$$

where

$$g(u, t) = (u-1)[ub(t) - d(t)] = (u-1)[(u-1)b(t) + \gamma(t)] \text{ and } \gamma(t) = b(t) - d(t).$$

The initial condition is $Q_i(u; s, s) = u^i$. The solution of the above PDE under the initial condition $Q_i(u; s, s) = u^i$ is very difficult in general cases. When $N_0\alpha_0(t)$ is finite, then as shown by Tan and Brown [16, 28], to the order of $O(N_0^{-1})$, $Q_i(u; s, t)$ given $I(s) = i$ is

$$Q_i(u; s, t) = [f(u; s, t)]^i \exp\left\{\int_s^t N(x)\alpha_0(x)[f(u; x, t) - 1]dx\right\}, \tag{4.7}$$

where $N(t)$ is the number of normal stem cells at time t and $f(u; s, t)$ is the pgf of a Arley–Feller stochastic birth death process with birth rate $b(t)$ and death rate $d(t)$.

By results from Example 4.8,

$$f(x; s, t) = 1 + \frac{(x - 1)}{\xi(s, t) - (x - 1)\zeta(s, t)}$$

where $\xi(s, t) = \exp\{-\int_s^t [b(x) - d(x)]dx\}$ and $\zeta(s, t) = \int_s^t b(y)\xi(s, y)dy$.

Assume $i = 1$ and put $p_j(s,t) = p_{1,j}(s,t)$. To derive $p_j(s,t)$ from Eq. (4.7), denote for $j = 0, 1, \ldots, \infty$:

$$q_j(s,t) = \int_s^t N(x)\alpha_0(x)[\xi(x,t) + \zeta(x,t)]^{-(j+1)}$$

$$\times [\delta_{j0} + (1 - \delta_{j0})\xi(x,t)\zeta(x,t)^{j-1}]dx.$$

Then,

$$\exp\left\{ \int_s^t N(x)\alpha_0(x)[f(u;x,t) - 1]dx \right\}$$

$$= \exp\left\{ -q_0(s,t) + \sum_{j=1}^{\infty} u^j q_j(s,t) \right\} = \sum_{j=0}^{\infty} u^j r_{(2,j)}(s,t),$$

where

$$r_{(2,0)}(s,t) = \exp[-q_0(s,t)],$$

and for $j = 1, \ldots, \infty$,

$$r_{(2,j)}(s,t) = \sum_{i=0}^{j-1} \frac{j-i}{j} r_{(2,i)}(s,t)q_{j-i}(s,t).$$

From Example 4.8, we have:

$$r_{(1,j)}(s,t) = \begin{cases} 1 - \dfrac{1}{\xi(s,t) + \zeta(s,t)}, & \text{if } j = 0; \\[2ex] \dfrac{\xi(s,t)}{(\xi(s,t) + \zeta(s,t))^2} \left(\dfrac{\zeta(s,t)}{\xi(s,t) + \zeta(s,t)} \right)^{j-1}, & \text{if } j > 0. \end{cases}$$

Hence,

$$p_j(s,t) = \sum_{k=0}^{j} r_{(1,k)}(s,t)r_{(2,j-k)}(s,t).$$

For deriving results for the general case $i > 1$, put:

$$a_1(s,t) = \zeta(s,t)/[\xi(s,t) + \zeta(s,t)], \quad a_2(s,t) = [1 - a_1(s,t)]/\zeta(s,t)$$

and $a_3(s,t) = 1 - [\zeta(s,t)]^{-1}$. Then,

$$[f(u; s,t)]^i = \sum_{j=0}^{\infty} u^j r_{(1,j)}^{(i)}(s,t) \,,$$

where

$$r_{(1,j)}^{(i)}(s,t) = [a_1(s,t)]^j \sum_{k=0}^{i} \binom{i}{k} \binom{j+k-1}{j} [a_2(s,t)]^k [a_3(s,t)]^{i-k} \,,$$

with $\binom{l}{0} = 1$ for all real l and $\binom{k-1}{k} = 0$ for all integer $k > 0$.

Using this result, we obtain:

$$p_{ij}(s,t) = \sum_{k=0}^{j} r_{(1,k)}^{(i)}(s,t) r_{(2,j-k)}(s,t) \,.$$

(For details, see Tan and Brown [16].)

4.4. Kolmogorov Equations for Finite Markov Chains with Continuous Time

Suppose that the chain is finite with state space $S = \{1, \ldots, k\}$. Then one may express the Kolmogorov equations in matrix notations. To illustrate, let $P(s,t)$ be the $k \times k$ matrix with (i,j)th element given by $p_{i,j}(s,t), i,j = 1, \ldots, k$. Then $P(s,t)$ is the transition matrix for the transition of states at time s to states at time t. Let $A(t) = (a_{ij}(t))$ be the $k \times k$ matrix with (i,j)th element given by $a_{ij}(t) = \delta_{ij}\alpha_{ii} + (\delta_{ij} - 1)\alpha_{ij}$.

Denote by

$$\frac{\partial}{\partial t}P(s,t) = \left(\frac{\partial}{\partial t}p_{ij}(s,t)\right) \quad \text{and} \quad \frac{\partial}{\partial s}P(s,t) = \left(\frac{\partial}{\partial s}p_{ij}(s,t)\right).$$

Then, in matrix notation, the Kolmogorov forward equations and the Kolmogorov backward equations are expressed respectively by the following matrix equations:

$$\frac{\partial}{\partial t}P(s,t) = -P(s,t)A(t) \tag{4.8}$$

and

$$-\frac{\partial}{\partial s}P(s,t) = -A(s)P(s,t)\,. \tag{4.9}$$

The initial condition is $P(s,s) = I_k$, the k-dimensional identity matrix.

If the chain is homogeneous such that $\alpha_{ij}(t) = \alpha_{ij}$, $P(s,t) = P(t-s)$ and $A(t) = A$, then the above Kolmogorov forward and backward equations reduce respectively to:

$$\frac{d}{dt}P(t) = -P(t)A \tag{4.10}$$

and

$$\frac{d}{dt}P(t) = -AP(t)\,. \tag{4.11}$$

The initial condition is $P(0) = I_k$.

Define the matrix exponential function $e^{-At} = \sum_{j=0}^{\infty} \frac{1}{j!}(-t)^j A^j$, then the solution of (4.10) and (4.11) is given by

$$P(t) = e^{-At}\,.$$

If A is diagonable with real distinct eigenvalues $\lambda_1 < \cdots < \lambda_r$ $(r \leq k)$, then as shown in Subsec. 2.11.2

$$A = \sum_{i=1}^{r} \lambda_i E_i$$

where $E_i = \prod_{j\neq i} \frac{1}{\lambda_i - \lambda_j}(A - \lambda_j I_k)$, $i = 1,\ldots,r$. (Note $E_i^2 = E_i$, $E_i E_j = 0$ if $i \neq j$ and $\sum_{i=1}^{r} E_i = I_k$.) Hence

$$P(t) = e^{-At} = \sum_{i=1}^{r} e^{-\lambda_i t} E_i\,.$$

Note that if $r = k$, then $E_i = \underset{\sim}{q}_i \underset{\sim}{r}'_i$, where $\underset{\sim}{q}_i$ and $\underset{\sim}{r}_i$ are the right and the left eigenvectors corresponding to the eigenvalue λ_i respectively.

For homogeneous finite Markov chains with continuous time, one may derive the matrix $P(t)$ of transition probabilities alternatively by noting the results:

$$(1): \quad P(\Delta t) = I_k - A\Delta t + o(\Delta t),$$

$$(2): \quad P(t) \;= P(n\Delta t) = P(\Delta t)^n$$

$$= \{I_k - A\Delta t + o(\Delta t)\}^n$$

$$= \left\{I_k - \frac{1}{n}At + o(t/n)\right\}^n,$$

by putting $t = n\Delta t$ and

$$(3): \; P(t) = \lim_{n \to \infty} \left\{I_k - \frac{1}{n}At + o(t/n)\right\}^n = e^{-At}.$$

Example 4.10.   **Markov chain models of HIV epidemic.** Consider the Markov chain model described in Fig. 1.1 for HIV epidemic with transition rates $\{\gamma_i(s,t) = \gamma_i, \beta_i(s,t) = \beta_i, \omega_i(s,t) = \omega_i, = 1, \ldots, k\}$. Satten and Longini [15] assumed $k = 6$ so that the 7th stage is the AIDS stage and the 6th stage is defined by $I_6, 200/\text{mm}^3 > \text{CD4}$ counts. Assuming that there are no other transitions and that death occurs only after AIDS, then we have a finite Markov chain with 7 states $S = \{I_i, i = 1, \ldots, 7\}$ with $I_7 = A$ being a absorbing state and all other states being transient states. The estimates of these transition rates have been obtained in [15] by using the San Francisco Men's Health Study (SFMHS) data and are given in Table 4.1.

Because of the 1993 AIDS definition by CDC [29], we will merge the I_6 stage with the AIDS stage. Then we have a finite Markov chain with state space $\Omega = \{I_i, i = 1, \ldots, 6\}$ with $I_6 = A$ being the AIDS stage. (The rate for

Table 4.1. The estimated values of transition rates by Satten and Longini [15].

$\gamma_1 = 0.0381$	$\beta_2 = 0.0030$	$\omega_1 = \omega_2 = 0$
$\gamma_2 = 0.0478$	$\beta_3 = 0.0087$	$\omega_3 = 0.0016$
$\gamma_3 = 0.0399$	$\beta_4 = 0.0064$	$\omega_4 = 0.0025$
$\gamma_4 = 0.0417$	$\beta_5 = 0.0167$	$\omega_5 = 0.0038$
$\gamma_5 = 0.0450$	$\beta_6 = 0.0071$	$\omega_6 = 0.0647$

$I_5 \to A$ is now $\gamma_5 + \omega_5$.) The infinitisemal matrix of this chain is:

$$\begin{pmatrix} B & -\underset{\sim}{w} \\ \underset{\sim}{0}' & 0 \end{pmatrix},$$

where $\underset{\sim}{w}' = (\omega_i, i = 1, \ldots, 4, \gamma_5 + \omega_5)$ and

$$B = \begin{bmatrix} \gamma_1 + \omega_1 & -\gamma_1 & 0 & 0 & 0 \\ -\beta_2 & \beta_2 + \gamma_2 + \omega_2 & -\gamma_2 & 0 & 0 \\ 0 & -\beta_3 & \beta_3 + \gamma_3 + \omega_3 & -\gamma_3 & 0 \\ 0 & 0 & -\beta_4 & \beta_4 + \gamma_4 + \omega_4 & -\gamma_4 \\ 0 & 0 & 0 & -\beta_5 & \beta_5 + \gamma_5 + \omega_5 \end{bmatrix}.$$

On substituting the estimates from Table 4.1, we obtain

$$\underset{\sim}{w}' = (0, 0, 0.0016, 0.0025, 0.0488)$$

and

$$B = \begin{bmatrix} 0.0381 & -0.0381 & 0 & 0 & 0 \\ -0.0030 & 0.0508 & -0.0478 & 0 & 0 \\ 0 & -0.0087 & 0.0502 & -0.0399 & 0 \\ 0 & 0 & -0.0064 & 0.0506 & -0.0417 \\ 0 & 0 & 0 & -0.0167 & 0.0655 \end{bmatrix}.$$

The eigenvalues of B are given by $\lambda_1 = 0.0179$, $\lambda_2 = 0.0714$, $\lambda_3 = 0.0445$, $\lambda_4 = 0.0899$, $\lambda_5 = 0.0315$. The left eigenvectors $\underset{\sim}{u}_i$ and the right eigenvectors $\underset{\sim}{v}_i$ of λ_i are given respectively by:

$$\underset{\sim}{u}_1' = (0.0265, -0.1780, 0.5564, 1.4756, 1.2937),$$

$$\underset{\sim}{u}_2' = (-0.0270, 0.2988, -0.5882, -0.2868, 2.0403),$$

$$\underset{\sim}{u}_3' = (0.0773, -0.1637, -0.4577, 0.8119, 1.6089),$$

$$\underset{\sim}{u}_4' = (0.0044, -0.0767, 0.3253, -1.4451, 2.4697),$$

$$\underset{\sim}{u}_5' = (-0.0789, -0.1724, -0.0360, 1.1826, 1.4523),$$

$$\underset{\sim}{v}_1' = (2.0219, 1.0701, 0.6088, 0.2162, 0.0759),$$

$$\underset{\sim}{v}_2' = (-1.6768, 1.4639, -0.5245, -0.0342, 0.0975),$$

$$\underset{\sim}{v}_3' = (5.3891, -0.8991, -0.4576, 0.1087, 0.0862),$$

$$\underset{\sim}{v}_4' = (0.4154, -0.5648, 0.4360, -0.2593, 0.1775),$$

$$\underset{\sim}{v}_5' = (-6.1220, -1.0532, -0.0400, 0.1761, 0.0866),$$

and

$$\{c_i = \underset{\sim}{u}_i' \underset{\sim}{w}, i = 1, \ldots, 5\} = (0.0669, 0.0979, 0.0998, 0.1174, 0.0738).$$

Denote $E_i = \underset{\sim}{v}_i \underset{\sim}{u}_i'$, then B and e^{-Bt} have the following spectral expansions respectively:

$$B = \sum_{i=1}^{5} \lambda_i E_i, \quad \text{and} \quad e^{-Bt} = \sum_{i=1}^{5} e^{-\lambda_i t} E_i.$$

It follows that

$$P(t) = e^{-At} = \begin{pmatrix} e^{-Bt} & (I_5 - e^{-Bt}) B^{-1} \underset{\sim}{w} \\ 0 & 1 \end{pmatrix}$$

$$= \begin{pmatrix} \sum_{j=1}^{5} e^{-\lambda_i t} E_i & \underset{\sim}{a} \\ 0 & 1 \end{pmatrix},$$

where

$$\underset{\sim}{a} = (I_5 - e^{-Bt}) B^{-1} \underset{\sim}{w} = (I_5 - e^{-Bt}) \underset{\sim}{1}_5$$

$$= \sum_{i=1}^{5} (1 - e^{-\lambda_i t}) E_i \underset{\sim}{1}_5 = \sum_{i=1}^{5} (1 - e^{-\lambda_i t}) c_i \underset{\sim}{v}_i, \text{ where } c_i = \underset{\sim}{u}' \underset{\sim}{1}_5.$$

Example 4.11. The nucleotide substitution Markov chain in molecular evolution. In Example 4.3, it is shown that the nucleotide substitution in Eukaryotes can be described by a finite homogeneous Markov chains with continuous time. The state space of this chain is $S = \{A, C, T, G\}$ and the transition events are described by Fig. 4.1 with transition rates $\{\alpha, \beta, \gamma, \delta, \epsilon, \lambda, \kappa, \sigma\}$. In this Markov chain, the matrix B of infinitesimal parameters is given by:

$$
B = \begin{array}{c} \\ A \\ G \\ C \\ T \end{array}
\begin{array}{cccc}
A & G & C & T
\end{array}
\left(\begin{array}{cccc}
\alpha + \gamma + \lambda & -\alpha & -\gamma & -\lambda \\
-\epsilon & \epsilon + \gamma + \lambda & -\gamma & -\lambda \\
-\delta & -\kappa & \beta + \delta + \kappa & -\beta \\
-\delta & -\kappa & -\sigma & \delta + \kappa + \sigma
\end{array}\right).
$$

Hence the matrix $P(t)$ of transition probabilities from states at time 0 to states at time t is $P(t) = e^{-Bt}$. We will show that the matrix B has four distinct real eigenvalues $\{\nu_i, i = 1, 2, 3, 4\}$ so that B is diagonable.

Let $\underset{\sim}{x}_i$ and $\underset{\sim}{y}_i$ denote the right eigenvector and the left eigenvector of B for ν_i respectively, $i = 1, 2, 3, 4$. Denote by $E_i = (\underset{\sim}{x}'_i \underset{\sim}{y}_i)^{-1} \underset{\sim}{x}_i \underset{\sim}{y}'_i, i = 1, 2, 3, 4$. Then $\{E_i^2 = E_i, E_i E_j = 0, i \neq j, \sum_{i=1}^{4} E_i = I_4\}$ and the matrices B and $P(t)$ have the following spectral expansions:

$$
B = \sum_{i=1}^{4} \nu_i E_i \quad \text{and} \quad P(t) = \sum_{i=1}^{4} e^{-\nu_i t} E_i.
$$

To derive ν_i, notice first that all rows of B sum up to 0 so that $\nu_1 = 0$ is one eigenvalue with right eigenvector $\underset{\sim}{x}_1 = (1, 1, 1, 1)' = \underset{\sim}{1}_4$, a 4×1 column of 1's. To derive the other eigenvalues, let M_1 be a 4×4 matrix defined by:

$$
M_1 = \begin{pmatrix}
1 & 0 & 0 & 1 \\
0 & 1 & 0 & 1 \\
0 & 0 & 1 & 1 \\
0 & 0 & 0 & 1
\end{pmatrix}.
$$

Then the inverse M_1^{-1} of M_1 is $M_1^{-1} = \begin{pmatrix} 1 & 0 & 0 & -1 \\ 0 & 1 & 0 & -1 \\ 0 & 0 & 1 & -1 \\ 0 & 0 & 0 & 1 \end{pmatrix}$, and

$$C = M_1^{-1} B M_1 = \begin{pmatrix} \gamma + \lambda + \alpha + \delta & \kappa - \alpha & \sigma - \gamma & 0 \\ \delta - \epsilon & \gamma + \lambda + \epsilon + \kappa & \sigma - \gamma & 0 \\ 0 & 0 & \delta + \kappa + \beta + \sigma & 0 \\ -\delta & -\kappa & -\sigma & 0 \end{pmatrix}$$

Since the characteristic function of B is $\phi(x) = |B - xI_4| = |M_1^{-1} B M_1 - xI_4|$, so

$$\phi(x) = |C - xI_4|$$

$$= (\delta + \kappa + \beta + \sigma - x)$$

$$\times \{(\gamma + \lambda + \alpha + \delta - x)(\gamma + \lambda + \epsilon + \kappa - x) - (\kappa - \alpha)(\delta - \epsilon)\}$$

$$= (\delta + \kappa + \beta + \sigma - x)(\gamma + \lambda + \alpha + \epsilon - x)(\gamma + \lambda + \delta + \kappa - x)$$

$$= 0.$$

It follows that the other three eigenvalues are

$$\{\nu_2 = \gamma + \lambda + \delta + \kappa, \nu_3 = \gamma + \lambda + \alpha + \epsilon, \nu_4 = \delta + \kappa + \beta + \sigma\}.$$

This shows that all eigenvalues of B are real and distinct so that B is diagonable. The right eigenvector $\underset{\sim}{x}_i$ of ν_i for $i = 2, 3, 4$ can readily be derived by solving the system of equations $(B - \nu_i I_4)\underset{\sim}{x}_i = 0$ which is unique up to non-zero constant multiplier. After some straightforward algebra, these eigenvectors are obtained as:

$$\underset{\sim}{x}_2' = (1, 1, -a, -a) \text{ with } a = (\delta + \kappa)/(\gamma + \lambda),$$

$$\underset{\sim}{x}_3' = (a_1, a_2, 1, 1) \text{ with } a_1 = \frac{\alpha(\delta + \kappa - \nu_3) + \kappa(\gamma + \lambda)}{\alpha\delta - \epsilon\kappa},$$

$$\text{and } a_2 = \frac{(-\epsilon)(\delta + \kappa - \nu_3) - \delta(\gamma + \lambda)}{\alpha\delta - \epsilon\kappa},$$

$$\underset{\sim}{x}_4' = (1, 1, a_3, a_4) \text{ with } a_3 = \frac{\beta(\gamma + \lambda - \nu_4) + \lambda(\delta + \kappa)}{\beta\gamma - \lambda\sigma},$$

$$\text{and } a_4 = \frac{(-\sigma)(\gamma + \lambda - \nu_4) - \gamma(\delta + \kappa)}{\beta\gamma - \lambda\sigma}.$$

To derive the left eigenvectors of the ν_i's of B, put $Q = (\underset{\sim}{x}_1, \underset{\sim}{x}_2, \underset{\sim}{x}_3, \underset{\sim}{x}_4)$. Then Q is non-singular and the rows of the inverse Q^{-1} of Q are the left eigenvectors of the ν_i's respectively. To derive Q^{-1}, notice the result that if the square matrix G is partitioned by $G = \begin{pmatrix} G_{11} & G_{12} \\ G_{21} & G_{22} \end{pmatrix}$ and if $\{G, G_{11}, G_{22}\}$ are non-singular, then

$$G^{-1} = \begin{pmatrix} G_{11.2}^{-1} & -G_{11}^{-1}G_{12}G_{22.1}^{-1} \\ -G_{22}^{-1}G_{21}G_{11.2}^{-1} & G_{22.1}^{-1}, \end{pmatrix}$$

where $G_{ii.j} = G_{ii} - G_{ij}G_{jj}^{-1}G_{ji}, i \neq j$.

Using this result, we obtain:

$$Q^{-1} = \begin{pmatrix} \frac{1 + aa_2}{(1 + a)(a_2 - a_1)} & -\frac{1 + aa_1}{(1 + a)(a_2 - a_1)} & \frac{a + a_4}{(1 + a)(a_4 - a_3)} & -\frac{a + a_3}{(1 + a)(a_4 - a_3)} \\ -\frac{1 - a_2}{(1 + a)(a_2 - a_1)} & \frac{1 - a_1}{(1 + a)(a_2 - a_1)} & \frac{1 - a_4}{(1 + a)(a_4 - a_3)} & -\frac{1 - a_3}{(1 + a)(a_4 - a_3)} \\ -\frac{1 + a}{(1 + a)(a_2 - a_1)} & \frac{1}{(a_2 - a_1)} & 0 & 0 \\ 0 & 0 & -\frac{1}{a_4 - a_3} & \frac{1}{a_4 - a_3} \end{pmatrix}.$$

On substituting the $\{a, a_i, i = 1, 2, 3, 4\}$'s, we obtain the four row vectors of Q^{-1} as:

$$\underset{\sim}{y}_1' = \left\{ \frac{1}{(\nu_2\nu_3)}[\delta(\gamma + \lambda) + \epsilon(\delta + \kappa), \kappa(\gamma + \lambda) + \alpha(\delta + \kappa)], \right.$$

$$\left. \frac{1}{(\nu_2\nu_4)}[\sigma(\gamma + \lambda) + \gamma(\delta + \kappa), \beta(\gamma + \lambda) + \lambda(\delta + \kappa)] \right\};$$

$$\underset{\sim}{y}'_2 = \frac{\gamma + \lambda}{\gamma + \lambda + \delta + \kappa} \left\{ \frac{1}{\alpha + \epsilon - \delta - \kappa}(\epsilon - \delta, \alpha - \kappa)\,, \right.$$

$$\left. \frac{1}{\beta + \sigma - \gamma - \lambda}(\gamma - \sigma, \lambda - \beta) \right\};$$

$$\underset{\sim}{y}'_3 = \frac{\alpha\delta - \epsilon\kappa}{(\gamma + \lambda + \alpha + \epsilon)(\alpha + \epsilon - \delta - \kappa)}(-1,1,0,0)\,;$$

$$\underset{\sim}{y}'_4 = \frac{\gamma\beta - \sigma\lambda}{(\beta + \sigma + \delta + \kappa)(\beta + \sigma - \gamma - \lambda)}(0,0,-1,1)\,.$$

Since $\underset{\sim}{x}'_i \underset{\sim}{y}_i = 1, i = 1,2,3,4$, so we have:

$$P(t) = \underset{\sim}{1}_4 \underset{\sim}{y}'_1 + \sum_{i=2}^{4} e^{-\nu_i t} \underset{\sim}{x}_i \underset{\sim}{y}'_i\,. \tag{4.12}$$

It follows that $\lim_{t \to \infty} P(t) = \underset{\sim}{1}_4 \underset{\sim}{y}'_1$.

Example 4.12.   The Moran genetic model. In Example 4.2, we have shown that the Moran's model is a finite homogeneous birth-death process with birth rate $b_j = (M - j)\lambda_2\{\frac{j}{M}(1 - \alpha_1) + (1 - \frac{j}{M})\alpha_2\}$ and $d_j = j\lambda_1\{(1 - \alpha_2) - \frac{j}{M}(1 - \alpha_1 - \alpha_2)\}$. Hence $P(t) = (P_{jk}(t)) = e^{-At}$, where A is given by:

$$\text{where } A = \begin{pmatrix} b_0 & -b_0 & & & & & & \\ -d_1 & (b_1 + d_1) & -b_1 & & & \underset{\sim}{0} & & \\ & -d_2 & (b_2 + d_2) & -b_2 & & & & \\ & & \ddots & & \ddots & \ddots & & \\ & & & \ddots & \ddots & \ddots & & \\ & & & & \ddots & \ddots & & \ddots \\ & \underset{\sim}{0} & & & & \ddots & & \ddots & & \ddots \\ & & & & & & -d_{M-1} & (b_{M-1} + d_{M-1}) & -b_{M-1} \\ & & & & & & & -d_M & d_M \end{pmatrix}$$

$$\tag{4.13}$$

with b_j and d_j as given above.

When $\lambda_1 = \lambda_2 = \lambda$ (no selection), the eigenvalues and eigenvectors of A have been obtained by Karlin and McGregor [30] using Hahn polynomials. The Hahn polynomial $Q_n(x) = Q_n(x; \alpha, \beta, N)$ with parameters (α, β, N) is defined as:

$$Q_n(x) = Q_n(x; \alpha, \beta, N) = F_{(3,2)}(-n, -x, n + \alpha + \beta + 1; \alpha + 1, -N + 1; 1)$$

$$= \sum_{k=0}^{n} \frac{(-n)_k(-x)_k(n + \alpha + \beta + 1)_k}{(\alpha + 1)_k(-N + 1)_k(k!)} ,$$

where $\alpha > -1, \beta > -1, N > n, (c)_0 = 1$ and $(c)_k = c(c + 1) \cdots (c + k - 1)$ for $k \geq 1$.

The following two properties of Hahn polynomials have been proved in [31, 32] and will be used to derive the eigenvalues and eigenvectors of A:

(1) Orthogonality relationships:

$$\sum_{x=0}^{N-1} Q_n(x)Q_m(x)\rho(x) = \left\{ \begin{array}{ll} 0, & \text{if } n \neq m \\ \dfrac{1}{\pi_m}, & \text{if } n = m \end{array} \right\} \tag{4.14}$$

and

$$\sum_{n=0}^{N-1} Q_n(x)Q_n(y)\pi_n = \left\{ \begin{array}{ll} 0, & \text{if } x \neq y \\ \dfrac{1}{\rho(x)}, & \text{if } x = y \end{array} \right\} . \tag{4.15}$$

Where

$$\rho(x) = \rho(x; \alpha, \beta, N) = \frac{\binom{\alpha+x}{x}\binom{\beta+N-1-x}{N-1-x}}{\binom{N+\alpha+\beta}{N-1}}, \qquad x = 0, 1, 2, \ldots, N - 1,$$

and

$$\pi_0 = 1, \ \text{for } n = 1, 2, \ldots, N - 1,$$

$$\pi_n = \pi_n(\alpha, \beta, N)$$

$$= \frac{\binom{N-1}{n}}{\binom{N+\alpha+\beta+n}{n}} \times \frac{\Gamma(\beta + 1)}{\Gamma(\alpha + 1)\Gamma(x + \beta + 1)}$$

$$\times \frac{\Gamma(n + \alpha + 1)\Gamma(n + \alpha + \beta + 1)}{\Gamma(n + \beta + 1)\Gamma(n + 1)} \times \frac{(2n + \alpha + \beta + 1)}{(\alpha + \beta + 1)} .$$

(2) $Q_n(x)$ satisfies the following difference equation

$$\lambda_n Q_n(x) = D(x)Q_n(x-1) - (B(x)+D(x))Q_n(x) + B(x)Q_n(x+1), \qquad (4.16)$$

where

$$B(x) = (N-1-x)(\alpha+1+x), D(x) = x(N+\beta-x), \text{ and}$$

$$\lambda_n = n(n+\alpha+\beta+1).$$

Case 1: If $\alpha_1 > 0, \alpha_2 > 0$ but $1 > \alpha_1 + \alpha_2$, then the birth rate b_j and death rate d_j of Moran's model can be rewritten as:

$$b_j = \frac{\lambda}{M}(1-\alpha_1-\alpha_2)[(M+1)-1-j]\left[\left(\frac{M\alpha_2}{1-\alpha_1-\alpha_2}-1\right)+1+j\right]$$

$$= \frac{\lambda}{M}(1-\alpha_1-\alpha_2)(N-1-j)(\alpha+1+j),$$

and

$$d_j = \frac{\lambda}{M}(1-\alpha_1-\alpha_2)j\left[(M+1)+\left(\frac{M\alpha_1}{1-\alpha-\alpha_2}-1\right)-j\right]$$

$$= \frac{\lambda}{M}(1-\alpha_1-\alpha_2)j[N+\beta-j],$$

where

$$N = M+1, \alpha = \frac{M\alpha_2}{1-\alpha_1-\alpha_2}-1 \text{ and } \beta = \frac{M\alpha_1}{1-\alpha_1-\alpha_2}-1.$$

Putting $\mu_n = \frac{\lambda}{M}(1-\alpha_1-\alpha_2)n(n+\alpha+\beta+1)$, then we have for $j = 0, 1, 2, \ldots, M = N-1$:

$$\mu_n Q_n(j) = -b_j Q_n(j+1) + (b_j+d_j)Q_n(j) - d_j Q_n(j-1).$$

Noting (4.16), it is immediately seen that $\mu_n = \frac{\lambda}{M}(1-\alpha_1-\alpha_2)n(n+\alpha+\beta+1), n = 0, 1, 2, \ldots, M = N-1$ are the eigenvalues of A with corresponding right eigenvectors

$$\underset{\sim}{r}'_n = (Q_n(0), Q_n(1), \ldots, Q_n(M)).$$

Let $\underset{\sim}{\zeta}_n$ be the left eigenvector of A corresponding to the eigenvalue μ_n. Then $\underset{\sim}{r}'_n \underset{\sim}{\zeta}_m = \delta_{nm} C_n$ for some constant C_n. Using the orthogonality relationship

(4.14), the left eigenvector $\underset{\sim}{\zeta}_n$ of μ_n of A is proportional to

$$\underset{\sim}{\zeta}_n \propto \underset{\sim}{s}_n = \{\rho(0)Q_n(0), \rho(1)Q_n(1), \ldots, \rho(M)Q_n(M)\}',$$

and $\underset{\sim}{r}'_n \underset{\sim}{s}_n = \sum_{x=0}^{M} \rho(x)Q_n^2(x) = \pi_n^{-1}, n = 0, 1, 2, \ldots, M$. Since all the eigenvalues μ_n are distinct, so

$$P(t) = \sum_{n=0}^{M} e^{-\mu_n t} \frac{1}{(\underset{\sim}{r}'_n \underset{\sim}{s}_n)} \underset{\sim}{r}_n \underset{\sim}{s}'_n ;$$

or,

$$P_{jk}(t) = \rho(k) \sum_{n=0}^{M} \exp\left\{ -\frac{\lambda}{M}(1 - \alpha_1 - \alpha_2)n(n + \alpha + \beta + 1) \right\} \pi_n Q_n(j)Q_n(k),$$

$$(4.17)$$

where $j, k = 0, 1, 2, \ldots, M$.

Formula (4.17) was first derived by Karlin and McGregor [30].

Case 2: $\alpha_1 = \alpha_2 = 0$ (no mutation).

In this case, $b_0 = d_0 = b_M = d_M = 0$ so that 0 and M are absorbing states. Furthermore, $p_j = \frac{j}{M}$ and $q_j = 1 - \frac{j}{M}, j = 0, 1, 2, \ldots, M$. Hence, matrix A in (4.13) becomes $A = \frac{\lambda}{M} \begin{pmatrix} 0 & \underset{\sim}{0}' & 0 \\ -\underset{\sim}{w}_1 & B & -\underset{\sim}{w}_2 \\ 0 & \underset{\sim}{0}' & 0 \end{pmatrix}$, where the $\underset{\sim}{w}_j$ are $(M-1) \times 1$ columns given by $\underset{\sim}{w}_1 = (M-1, 0, \ldots, 0)'$, $\underset{\sim}{w}_2 = (0, \ldots, 0, M-1)'$ and where

$$B = \begin{pmatrix} 2(M-1), & -(M-1), & 0 & & & & \\ -2(M-2), & 2\cdot 2(M-2), & -2(M-2) & & & & \\ & -3(M-3), & 2\cdot 3(M-3), & -3(M-3) & & & \underset{\sim}{0} \\ & & \ddots & \ddots & \ddots & & \\ & & & \ddots & \ddots & \ddots & \\ & & & & \ddots & \ddots & \ddots \\ & \underset{\sim}{0} & & & -2(M-2), & 2\cdot 2(M-2), & -2(M-2) \\ & & & & & -(M-1), & 2(M-1) \end{pmatrix}.$$

Obviously, 0 is an eigenvalue of A with algebraic multiplicity 2 while the other eigenvalues of A are given by $\mu_n = \frac{\lambda}{M}\sigma_n, n = 1, 2, \ldots, M - 1,$ where the σ_n are the eigenvalues of B. Furthermore, the two independent left eigenvector of 0 of A are $\underset{\sim}{x}_1' = (1, 0, 0, \ldots, 0)$ and $\underset{\sim}{x}_2' = (0, 0, 0, \ldots, 1)$; the two right eigenvector of 0 of A are obviously $\underset{\sim}{y}_1' = \{1, (B^{-1}\underset{\sim}{w}_1)', 0\}$ and $\underset{\sim}{y}_2' = \{0, (B^{-1}\underset{\sim}{w}_2)', 1\}$. Notice that $\underset{\sim}{x}_i'\underset{\sim}{y}_j = \delta_{ij}$ $(i, j = 1, 2)$ as it should be.

Now $\sigma_n \neq 0$ as $\det(B) \neq 0$. Let $\underset{\sim}{u}_n$ and $\underset{\sim}{v}_n$ denote the left eigenvector and the right eigenvector of B corresponding to the eigenvalue σ_n respectively. Then, it is easily seen that the left eigenvector and the right eigenvector of A corresponding to the eigenvalue μ_n are given respectively by:
$\underset{\sim}{y}_n' = (-\frac{1}{\sigma_n}\underset{\sim}{u}_n'\underset{\sim}{w}_1, \underset{\sim}{u}_n', -\frac{1}{\sigma_n}\underset{\sim}{u}_n'\underset{\sim}{w}_2)$ and $\underset{\sim}{x}_n' = (0, \underset{\sim}{v}_n', 0)$.

To obtain the eigenvalues and the eigenvectors of B, notice that the first and the $(M - 1)$th columns of B are given respectively by

$$\underset{\sim}{c}_1 = \{2(M - 1), -2(M - 2), 0, \ldots, 0\}', \text{ and}$$

$$\underset{\sim}{c}_{M-1} = \{0, 0, \ldots, 0, -2(M - 2), 2(M - 1)\}';$$

and, for $j = 2, 3, \ldots, M - 2$, the jth column is

$$\underset{\sim}{c}_j = \{\underset{\sim}{0}_{(j-2)}', -(j-1)(M-j+1), 2j(M-j), -(j+1)(M-j-1), \underset{\sim}{0}_{(M-j-2)}\}',$$

where $\underset{\sim}{0}_k'$ is a row of k $0'$ s.

Hence, for $j = 1, \ldots, M - 1$, the $(j-1)$th element, the $(j+1)$th element and the jth element of $\underset{\sim}{c}_j$ are given respectively by $c_j(j - 1) = -\delta_{j-1}, c_j(j + 1) = -\beta_{j-1}, c_j(j) = \delta_{j-1} + \beta_{j-1} + 2 = 2j(M - j)$, where

$$\delta_{j-1} = (j - 1)(M - j + 1) = (j - 1)[(M - 1) - (j - 1) + 1],$$

$$\beta_{j-1} = (M - 1 - j)(1 + j) = [(M - 1) - 1 - (j - 1)][1 + 1 + (j - 1)].$$

(For $j = 1$, the first and second elements of $\underset{\sim}{c}_j$ are $c_1(1) = \delta_{j-1} + \beta_{j-1} + 2 = 2(M - 1)$ and $c_1(2) = -\beta_{j-1} = -2(M - 2)$ respectively.)

Let $N = M - 1, \alpha = \beta = 1$ and put:

$$\xi_{n-1} = (n - 1)(n - 1 + 1 + 1 + 1), R_{n-1}(j - 1) = Q_{n-1}(j - 1; 1, 1, M - 1),$$

for $n = 1, 2, \ldots, M - 1$ and $j = 1, 2, \ldots, M - 1$.

Then, we have:

$$-\xi_{n-1}R_{n-1}(j-1) = \delta_{j-1}R_{n-1}(j-2) - (\delta_{j-1} + \beta_{j-1})R_{n-1}(j-1)$$
$$+ \beta_{j-1}R_{n-1}(j)\,,$$
$$n = 1,\ldots,N(= M-1);\ j = 1,\ldots,N(= M-1)\,.$$

Or, since $\xi_{n-1} + 2 = n(n+1)$,

$$n(n+1)R_{n-1}(j-1) = (\xi_{n-1} + 2)R_{n-1}(j-1)$$
$$= -(j-1)(M-j+1)R_{n-1}(j-2)$$
$$+ 2j(M-j)R_{n-1}(j-1) - (j+1)(M-j-1)R_{n-1}(j)$$
$$= c_j(j-1)R_{n-1}(j-2) + c_j(j)R_{n-1}(j-1)$$
$$+ c_j(j+1)R_{n-1}(j)\,, \tag{4.18}$$

where $j = 1,\ldots,N = M-1$ and $n = 1,\ldots,N = M-1$.

Formula (4.18) implies that $\sigma_n = n(n+1), n = 1,\ldots,N = M-1$ are the eigenvalues of B and the left eigenvector of B corresponding to the eigenvalue σ_n is

$$\underset{\sim}{u}'_n = \{R_{n-1}(0), R_{n-1}(1), \ldots, R_{n-1}(M-2)\}\,,$$

where $R_{n-1}(j) = Q_{n-1}(j;1,1,M-1)$ is valid for $j = 0,1,\ldots,M-2$ and $n = 1,2,\ldots,N = M-1$.

For deriving $\underset{\sim}{v}_n$, define:

$$\rho_0(x) = \rho(x;1,1,M-1) = \frac{\binom{1+2}{x}\binom{1+M-1-1-x}{(M-1)-1-x}}{\binom{M-1+1+1}{M-1-1}} = \frac{6(1+x)(M-x-1)}{(M+1)M(M-1)}\,,$$

$$x = 0,1,2,\ldots,M-2;\ \pi_0^{(*)} = 1 \text{ and for } n = 1,2,\ldots,M-2,$$

$$\pi_n^{(*)} = \pi_n(1,1,M-1) = \frac{\binom{M-2}{n}}{\binom{M-1+1+1+n}{n}} \times \frac{\Gamma(2)}{\Gamma(2)\Gamma(3)} \times \frac{\Gamma(n+2)\Gamma(n+3)}{\Gamma(n+2)\Gamma(n+1)}$$

$$\times \frac{(2n+1+1+1)}{(1+1+1)} = \frac{\binom{M-2}{n}}{\binom{M+n+1}{n}} \times \frac{1}{6}(n+2)(n+1)(2n+3)\,.$$

Then, by (4.14),

$$\sum_{x=0}^{N-1} R_n(x)R_m(x)\rho_0(x) = \left\{ \begin{array}{ll} 0, & \text{if } n \neq m \\ \dfrac{1}{\pi_n^{(*)}}, & \text{if } n = m \end{array} \right\}, \quad n = 0, 1, \ldots, M-2(= N-1).$$

This result implies that the right eigenvector $\underset{\sim}{v}_n$ of σ_n of B is proportional to

$$\underset{\sim}{q}_n = \{\rho_0(0)R_{n-1}(0), \rho_0(1)R_{n-1}(1), \ldots, \rho_0(M-2)R_{n-1}(M-2)\}'$$

$$= \frac{6}{M(M^2-1)}\{(M-1)R_{n-1}(0), 2(M-2)R_{n-1}(1),$$

$$\ldots, r(M-r)R_{n-1}(r-1), \ldots, (M-1)R_{n-1}(M-2)\}',$$

Now, $\underset{\sim}{u}'_n \underset{\sim}{v}_n = 1$ and $\underset{\sim}{u}'_n \underset{\sim}{q}_n = \sum_{x=0}^{M-2} \rho_0(x)[R_{n-1}(x)]^2 = (\pi_{n-1}^*)^{-1}$ by (4.14), where $\pi_{n-1}^{(*)} = \frac{\binom{M-2}{n-1}}{\binom{M+n}{n-1}}\frac{1}{6}n(n+1)(2n+1), n = 1, 2, \ldots, M-1$. Hence $\pi_{n-1}^{(*)} \underset{\sim}{u}'_n \underset{\sim}{q}_n = 1$ so that we have:

$$\underset{\sim}{v}'_n = \pi_{n-1}^{(*)} \underset{\sim}{q}'_n = c_n[(M-1)R_{n-1}(0), 2(M-2)Q_{n-1}(1), \ldots, r(M-r)$$

$$Q_{n-1}(r-1), \ldots, (M-1)Q_{n-1}(M-2)],$$

where

$$c_n = \frac{\binom{M-2}{n-1}n(n+1)(2n+1)}{\binom{M+n}{n-1}(M+1)M(M^2-1)}.$$

Notice that $\sigma_n \neq 0$ and are all distinct so that B is diagonable and A has M distinct eigenvalues $\{\mu_1 = 0, \mu_{j+1} = \sigma_j, j = 1, \ldots, M-1\}$. For $j = 2, \ldots, M$, the left $\underset{\sim}{x}_j$ and right eigenvectors $\underset{\sim}{y}_j$ are given respectively by

$$\underset{\sim}{x}'_j = \left(-\frac{1}{\sigma_j}\underset{\sim}{u}'_j \underset{\sim}{w}_1, \underset{\sim}{u}'_j, -\frac{1}{\sigma_j}\underset{\sim}{u}'_j \underset{\sim}{w}_2\right) \quad \text{and} \quad \underset{\sim}{y}'_j = (0, \underset{\sim}{v}'_j, 0).$$

Define the matrix E_0 by:

$$E_0 = \begin{pmatrix} 1 & 0 \\ B^{-1}\underset{\sim}{w}_1 & B^{-1}\underset{\sim}{w}_2 \\ 0 & 1 \end{pmatrix} \begin{pmatrix} 1,0,\ldots,0 \\ 0,0,\ldots,1 \end{pmatrix}$$

$$= \begin{pmatrix} 1 & \underset{\sim}{0}'_{M-1} & 0 \\ B^{-1}\underset{\sim}{w}_1 & 0_{(M-1,M-1)} & B^{-1}\underset{\sim}{w}_2 \\ 0 & \underset{\sim}{0}'_{M-1} & 1 \end{pmatrix}$$

$$= \begin{pmatrix} 1 & \underset{\sim}{0}'_{M-1} & 0 \\ \displaystyle\sum_{j=1}^{M-1}\frac{1}{\sigma_j}(\underset{\sim}{u}'_j\underset{\sim}{w}_1)\underset{\sim}{v}_j & 0_{M-1,M-1} & \displaystyle\sum_{j=1}^{M-1}\frac{1}{\sigma_j}(\underset{\sim}{u}'_j\underset{\sim}{w}_2)\underset{\sim}{v}_j \\ 0 & \underset{\sim}{0}'_{M-1} & 1 \end{pmatrix},$$

where $0_{(p,q)}$ is a $p \times q$ matrix of 0's; and for $j = 1,\ldots,M-1$, define the matrices E_j by:

$$E_j = \underset{\sim}{y}_{j+1}\underset{\sim}{x}'_{j+1} = \begin{pmatrix} 0 \\ \underset{\sim}{v}_j \\ 0 \end{pmatrix} \left(-\frac{1}{\sigma_j}\underset{\sim}{u}'_j\underset{\sim}{w}_1, \ \underset{\sim}{u}'_j, \ -\frac{1}{\sigma_j}\underset{\sim}{u}'_j\underset{\sim}{w}_2 \right)$$

$$= \begin{pmatrix} 0 & \underset{\sim}{0}'_{M-1} & 0 \\ -\frac{1}{\sigma_j}(\underset{\sim}{u}'_j\underset{\sim}{w}_1)\underset{\sim}{v}_j & \underset{\sim}{v}_j\underset{\sim}{u}'_j & -\frac{1}{\sigma_j}(\underset{\sim}{u}'_j\underset{\sim}{w}_2)\underset{\sim}{v}_j \\ 0 & \underset{\sim}{0}'_{M-1} & 0 \end{pmatrix}.$$

Then we have:

$$P(t) = e^{-At} = E_0 + \sum_{j=1}^{M-1} e^{-\frac{\lambda}{M}\sigma_j t}E_j = \begin{pmatrix} 1 & \underset{\sim}{0}'_{M-1} & 0 \\ \underset{\sim}{a}_1 & \displaystyle\sum_{j=1}^{M-1} e^{-\frac{\lambda}{M}\sigma_j t}\underset{\sim}{v}_j\underset{\sim}{u}'_j, & \underset{\sim}{a}_2 \\ 0 & \underset{\sim}{0}'_{M-1} & 1 \end{pmatrix},$$

where $\underset{\sim}{a}_j = \sum_{j=1}^{M-1}(1 - e^{-\frac{\lambda}{M}\sigma_j t})\frac{1}{\sigma_j}\underset{\sim}{v}_j(\underset{\sim}{u}'_j\underset{\sim}{w}_j), j = 1,2.$

4.5. Complements and Exercises

Exercise 4.1.   Prove Kolmogorov forward equation given by Eq. 4.2.

Exercise 4.2.   Prove Eq. (4.5) for the pgf of $P\{X(t) = j|X(0) = i\}$ of the Feller–Arley birth-death process.

Exercise 4.3.   Derive the Kolmogorov forward equation for the stochastic logistic birth-death process with birth rate $b_i(t) = ib(1 - i/M)$ and death rate $d_i(t) = id(1 - i/M)$. Derive the pgf of $P\{X(t) = j|X(0) = i\}$ of this process.

Exercise 4.4.   Consider a continuous-time Galton–Watson branching process $\{X(t), t \geq 0\}$ with progeny distribution $\{p_j, j = 0, 1, \ldots, \}$ and with survival parameter λ ($\lambda > 0$) as described in Example 4.7. Let $g(x)$ denote the PGF of the progeny distribution and $\phi(x, t) = E\{x^{X(t)}|X(0) = 1\}$ the PGF of $X(t)$ given $X(0) = 1$.

(a) By using the Chapman–Kolmogorov equation, show that

$$\phi(x, t + \tau) = \phi\{\phi(x, \tau), t\}\,.$$

(b) Noting the results $\phi(x, \Delta t) = x + u(x)\Delta t + o(\Delta t)$, where $u(x) = \lambda[g(x) - x]$, then

$$\phi(x, t + \Delta t) = \phi\{\phi(x, \Delta t), t\}$$

$$= \phi\{x + u(x)\Delta t + o(\Delta t), t\}\,.$$

Hence show that $\phi(x, t)$ satisfies the following partial differential equation:

$$\frac{\partial}{\partial t}\phi(x, t) = u(x)\frac{\partial}{\partial x}\phi(x, t), \quad \phi(x, 0) = x\,.$$

Exercise 4.5.   In a large population, assume that the probability density of the survival time T of each individual is given by:

$$f(t) = \lambda\, e^{-\lambda t}, \quad t \geq 0, \quad \lambda > 0\,.$$

Suppose that when each individual dies at time t, immediately it either leaves beyond two individuals, or die (no progenies) with probabilities $p(t)$ and $q(t) = 1 - p(t)$ respectively. Let $X(t)$ be the number of individuals at time t. Show that $\{X(t), t \geq 0\}$ is a Feller–Arley birth-death process with birth rate $b_i(t) = i\lambda\, p(t)$ and death rate $d_i(t) = i\lambda\, q(t)$.

Let $\phi(z,t) = E[z^{X(t)}|X(0) = 1]$. If $p(t) = p$ so that $q(t) = 1 - p = q$, show that $\phi(z,t)$ satisfies the following integral equation:

$$\phi(z,t) = ze^{-\lambda t} + \int_0^t \lambda \, e^{-\lambda(t-x)}\{p[\phi(z,x)]^2 + q\}dx\,.$$

Hence show that $\phi(z,t)$ satisfies the following Ricatti equation:

$$\frac{\partial}{\partial t}\phi(z,t) = b[\phi(z,t)]^2 - \lambda\,\phi(z,t) + d\,,$$

where $\{b = p\lambda, d = q\lambda\}$ and $\phi(z,0) = z$.

Show that the solution of $\phi(z,t)$ is given by Eq. (4.5) with $i = 1$.

Exercise 4.6.   Two-types population growth model. Consider a large population consisting of two types of individuals, say normal type (or Type-1) and mutant type (or Type-2). Suppose that the following conditions hold:

(i) The probability density of the survival time T_i ($i = 1,2$) of the Type-i individual is given by:

$$f_i(t) = \lambda_i e^{-\lambda_i t}, t \ge 0, \lambda_i > 0\,.$$

(ii) When a normal individual dies at time t, immediately it either leaves beyond two normal individuals, or 1 normal individual and 1 mutant individual, or die (no progenies) with probabilities $\{p_1(t), r(t)\}$ and $q_1(t) = 1 - p_1(t) - r(r)$, respectively, where $0 \le p_1(t) + r(t) \le 1$.

(iii) When a mutant individual dies at time t, immediately it either leaves beyond two mutant individuals, or die (no progenies) with probabilities $p_2(t)$ and $q_2(t) = 1 - p_2(t)$ respectively.

(iv) All individuals in the population produce progenies or die by following the above probability laws independently of one another.

Let $X_i(t)$ denote the number of Type-i individuals at time t and put $\underset{\sim}{X}(t) = [X_1(t), X_2(t)]'$.

(a) Show that the above process $\{\underset{\sim}{X}(t), t \ge 0\}$ is equivalent to the cancer tumor drug resistant model described in Example 4.5 with

$$\{b_1(t) = \lambda_1 p_1(t), d_1(t) = \lambda_1 q_1(t), \alpha(t) = \lambda_1 r(t), b_2(t) = \lambda_2 p_2(t), d_2(t) = \lambda_2 q_2(t)\}\,.$$

(b) Let $\psi_i(z,t) = E[z^{X_i(t)}|X_1(0) = 1, X_2(0) = 0], i = 1,2$. Show that $\psi_1(z,t)$ is the pgf of a Feller–Arley birth-death process with birth rate $b_1(t)$ and death rate $d_1(t)$ as given in Example 4.8.

(c) Suppose that the parameters $\{p_i(t) = p_i, q_i(t) = q_i, i = 1,2\}$ are independent of time t. Then $r(t) = r = 1 - p_1 - q_1$ and $\alpha(t) = \alpha = \lambda_1 r$ are independent of t. Show that $\psi_2(z,t) = \phi(z,t)$ satisfies the following integral equation:

$$\phi(z,t) = e^{-\lambda_1 t} + \int_0^t \lambda_1 e^{-\lambda_1(t-x)}\{p_1[\phi(z,x)]^2 + r\psi_1(z,x)\phi(z,x)\}dx\,.$$

Hence, show that ϕ satisfies the following equation:

$$\frac{\partial}{\partial t}\phi(z,t) = b_1[\phi(z,t)]^2 + (\alpha\psi_1(z,t) - \lambda_1)\phi(z,t)\,,$$

with $\phi(z,0) = 1$.
With $\psi_1(z,t)$ available from (b), solve the above equation to derive $\psi_2(z,t) = \phi(z,t)$.

Exercise 4.7.   Continuous-time multiple branching processes. Consider a large population with k different types. Let $X_i(t)$ $(i = 1,\ldots,k)$ denote the number of the ith type at time t. Suppose that the following conditions hold:

(1) The probability density of the survival time of Type-i individuals is

$$h_i(t) = \lambda_i e^{-\lambda_i t}, t \geq 0, \lambda_i > 0, i = 1,\ldots,k\,.$$

(2) When a Type-i individual dies, with probability $q_i(\underset{\sim}{j}) = q_i(j_1,\ldots,j_k)$ it immediately leaves beyond j_r progenies of Type-r, $r = 1,\ldots,k$, where the $j_r(r = 1,\ldots,k)$'s are non-negative integers.

(3) All individuals in the population follow the above probability laws for proliferation independently of one another.

Then $\{\underset{\sim}{X}(t) = [X_1(t),\ldots,X_k(t)]', t \geq 0\}$ is a k-dimensional continuous-time multiple branching process with state space $S = \{\underset{\sim}{i} = (i_1,\ldots,i_k)', i_j = 0,1,\ldots,\infty, j = 1,\ldots,k\}$, with progeny distributions $\{q_i(\underset{\sim}{j}) = q_i(j_1,\ldots,j_k), \underset{\sim}{j} \in S, i = 1,\ldots,k\}$ and with survival parameters $\{\lambda_i, i = 1,\ldots,k\}$. One may assume $P\{X_i(0) = 1\} = 1$ for some $i = 1,\ldots,k$. Let M be the matrix of expected number of progeny distributions per generation and $\underset{\sim}{g}(\underset{\sim}{x})$ the vector of pgf's of progeny distributions. Then, as in discrete-time multiple branching processes, one defines $\underset{\sim}{X}(t)$ as singular if $\underset{\sim}{g}(\underset{\sim}{x}) = A\underset{\sim}{x}$ for some matrix A and define $\underset{\sim}{X}(t)$ as positive regular if there exists an integer r such that all elements of M^r are positive.

For illustration, assume $k = 2$ and define the generating functions:

$$\phi_{10}(x_1, x_2; t) = \phi_{10}(\underset{\sim}{x}; t) = E\left\{\left[\prod_{i=1}^{2} x_i^{X_i(t)}\right] |X_1(0) = 1, X_2(0) = 0\right\},$$

$$\phi_{01}(x_1, x_2; t) = \phi_{01}(\underset{\sim}{x}; t) = E\left\{\left[\prod_{i=1}^{2} x_i^{X_i(t)}\right] |X_1(0) = 0, X_2(0) = 1\right\},$$

$$\phi_{11}(x_1, x_2; t) = \phi_{11}(\underset{\sim}{x}; t) = E\left\{\left[\prod_{i=1}^{2} x_i^{X_i(t)}\right] |X_1(0) = 1, X_2(0) = 1\right\}.$$

Let $g_i(x_1, x_2) = g_i(\underset{\sim}{x})(i = 1, 2)$ denote the pfg of the progeny distribution of Type-i individuals and put $u_i(x_1, x_2) = u_i(\underset{\sim}{x}) = \lambda_i[g_i(\underset{\sim}{x}) - x_i]$.

(a) Prove the following results:

$$\phi_{11}(\underset{\sim}{x}; t) = \phi_{10}(\underset{\sim}{x}; t)\phi_{01}(\underset{\sim}{x}; t),$$

$$\phi_{10}(\underset{\sim}{x}; \Delta t) = x_1 + u_1(x_1, x_2)\Delta t + o(\Delta t),$$

$$\phi_{01}(\underset{\sim}{x}; \Delta t) = x_2 + u_2(x_1, x_2)\Delta t + o(\Delta t).$$

(b) By using results of (a), show that the infinitesimal parameters from $(i, j) \to (u, v)$ for $\{\underset{\sim}{X}(t), t \geq 0\}$ are given by:

$$\alpha(i, j; u, v) = \begin{cases} 0, & \text{if } u < i - 1, \text{ or} \\ & \text{if } v < j - 1, \text{ or} \\ & \text{if } u = i - 1, v < j, \text{ or} \\ & \text{if } u < i = i - 1, v = j-1; \\ \lambda_1 q_1(0, v - j), & \text{if } u = i - 1, v \geq j; \\ \lambda_2 q_2(u - i, 0), & \text{if } u \geq i, v = j - 1; \\ i\lambda_1[1 - q_1(1, 0)] + j\lambda_2[1 - q_2(0, 1)], & \text{if } u = i, v = j; \\ i\lambda_1 q_1(1, v - j)] + j\lambda_2 q_2(0, v - j + 1)], & \text{if } u = i, v > j; \\ i\lambda_1 q_1(u - i + 1, 0)] + j\lambda_2 q_2(u - i, 1)], & \text{if } u > i, v = j; \\ i\lambda_1 q_1(u-i+1, v-j)] + j\lambda_2 q_2(u-i, v-j+1)], & \text{if } u > i, v > j. \end{cases}$$

(c) By using results of (a), show that for $(i,j) = (1,0)$ or $(i,j) = (0,1)$, the $\phi_{ij}(\underset{\sim}{x};t)$'s satisfy the following equations:

$$\phi_{ij}(x, t + \tau) = \phi_{ij}\{\phi_{10}(x, \tau), \phi_{01}(x, \tau); t\} \text{ for } t > 0, \tau > 0,$$

$$\frac{\partial}{\partial t}\phi_{ij}(x, t) = u_1(\underset{\sim}{x})\frac{\partial}{\partial x_1}\phi_{ij}(x, t) + u_2(\underset{\sim}{x})\frac{\partial}{\partial x_2}\phi_{ij}(x, t),$$

where $\{\phi_{10}(\underset{\sim}{x}, 0) = x_1, \phi_{01}(\underset{\sim}{x}, 0) = x_2\}$.

Exercise 4.8. Multivariate birth-death processes. Let $\{\underset{\sim}{X}(t) = [X_1(t), \ldots, X_k(t)]', t \geq 0\}$ be a k-dimensional Markov process with state space $S = \{\underset{\sim}{i} = (i_1, \ldots, i_k)', i_j = 0, 1, \ldots, \infty, j = 1, \ldots, k\}$, $\underset{\sim}{e}_i$ the $k \times 1$ column vector with 1 at the ith position and with 0 at other positions.

Definition 4.1. $\{\underset{\sim}{X}(t) = [X_1(t), \ldots, X_k(t)]', t \geq 0\}$ is called a k-*dimensional birth-death process with birth rates* $\{b_i(j,t), i = 1, \ldots, k, j = 0, 1, \ldots, \infty; b_i(j,t) \geq 0\}$, *death rates* $\{d_i(j,t), i = 1, \ldots, k, j = 0, 1, \ldots, \infty; d_i(j,t) \geq 0\}$ *and cross-transition rates* $\{\alpha_{i,j}(r,t), i, j = 1, \ldots, k \ (i \neq j), r = 0, 1, \ldots, \infty; \alpha_{i,j}(r,t) \geq 0\}$ *iff the following conditions hold:*

$$P\{\underset{\sim}{X}(t + \Delta t) = \underset{\sim}{j} \mid \underset{\sim}{X}(t) = \underset{\sim}{i}\}$$

$$= \begin{cases} \left[b_r(i_r, t) + \sum_{u \neq r} \alpha_{u,r}(i_u, t) \right] \Delta t + o(\Delta t), & \text{if } \underset{\sim}{j} = \underset{\sim}{i} + \underset{\sim}{e}_r; \\ d_r(i_r, t)\Delta t + o(\Delta t), & \text{if } \underset{\sim}{j} = \underset{\sim}{i} - \underset{\sim}{e}_r; \\ o(\Delta t), & \text{if } |\underset{\sim}{1}'_k(\underset{\sim}{j} - \underset{\sim}{i})| \geq 2. \end{cases}$$

Using Definition 4.1, then the drug-resistant cancer tumor model as described in Example 4.5 is a 2-dimensional birth-death process with birth rates $\{b_i(j,t) = jb_i(t), i = 1, 2\}$, death rates $\{d_i(j,t) = jd_i(t), i = 1, 2\}$ and cross transition rates $\{\alpha_{1,2}(j,t) = j\alpha(t), \alpha_{u,v}(j,t) = 0 \text{ if } (u,v) \neq (1,2)\}$. Similarly, for the k-dimensional multi-event model of carcinogenesis as described in Example 4.6, $\underset{\sim}{X}(t) = \{I_0(t) = N(t), I_i(t), i = 1, \ldots, k-1\}$ is a k-dimensional birth-death process with birth rates $\{b_i(j,t) = jb_i(t), i = 0, 1, \ldots, k-1\}$, death rates $\{d_i(j,t) = jd_i(t), i = 0, 1, \ldots, k-1\}$ and cross transition rates

$\{\alpha_{i,i+1}(j,t) = j\alpha_i(t), i = 0,1,\ldots,k-1, \alpha_{u,v}(j,t) = 0 \text{ if } (u,v) \neq (i,i+1) \text{ for } i = 0,1,\ldots,k-1\}.$

(a) Show that the Kolmogorov forward equation for the probabilities $P\{I_j(t) = i_j, j = 0,1,\ldots,k-1|I_0(0) = N_0\} = P(i_j, j = 0,1,\ldots,k-1;t)$ in the k-dimensional multi-event model is given by:

$$\frac{d}{dt}P(i_j, j = 0,1,\ldots,k-1;t)$$

$$= P(i_0 - 1, i_j, j = 1,\ldots,k-1;t)(i_0 - 1)b_0(t)$$

$$+ \sum_{j=1}^{k-1} P(i_0, i_1,\ldots,i_{j-1}, i_j - 1, i_{j+1},\ldots,i_{k-1};t)(i_j - 1)b_j(t)$$

$$+ \sum_{j=0}^{k-2} P(i_0, i_1,\ldots,i_j, i_{j+1} - 1, i_{j+2},\ldots,i_{k-1};t)i_j\alpha_j(t)$$

$$+ \sum_{j=0}^{k-1} P(i_0, i_1,\ldots,i_{j-1}, i_j + 1, i_{j+1},\ldots,i_{k-1};t)(i_j + 1)d_j(t)$$

$$- P(i_0, i_1,\ldots,i_{k-1};t)\left\{\sum_{j=0}^{k-1} i_j[b_j(t) + d_j(t)] + \sum_{j=0}^{k-2} i_j\alpha_j(t)\right\},$$

for $i_j = 0,1,\ldots,\infty, j = 0,1,\ldots,k-1$.

(b) Denote by $\phi(x_0, x_1,\ldots,x_{k-1};t) = \phi(\underset{\sim}{x})$ the pgf of the probabilities $P\{I_j(t) = i_j, j = 0,1,\ldots,k-1|I_0(0) = N_0\} = P(i_j, j = 0,1,\ldots,k-1;t)$ in the k-dimensional multi-event model as described in Example 4.6. Show that $\phi(\underset{\sim}{x})$ satisfies the following partial differential equation with initial condition $\phi(\underset{\sim}{x};0) = x_0^{N_0}$:

$$\frac{\partial}{\partial t}\phi(\underset{\sim}{x};t) = \sum_{i=0}^{k-2}\left\{[(x_i - 1)(x_ib_i(t) - d_i(t)) + x_i(x_{i+1} - 1)\alpha_i]\frac{\partial}{\partial x_i}\phi(\underset{\sim}{x};t)\right\}$$

$$+ \left\{(x_{k-1} - 1)[x_{k-1}b_{k-1}(t) - d_{k-1}(t)]\frac{\partial}{\partial x_{k-1}}\phi(\underset{\sim}{x};t)\right\}.$$

References

[1] G. M. Grimmett and D. R. Stirzaker, *Probability and Random Processes, Second Edition*, Clarendon Press, Oxford (1992).

[2] S. Karlin, *A First Course in Stochastic Processes*, Academic Press, New York (1968).

[3] W. Y. Tan, *A Stochastic Gompertz birth-death process*, Statist. and Prob. Lett. **4** (1986) 25–28.

[4] W. Y. Tan, *Stochastic logistic growth and applications*, in Logistic Distributions, ed. B. N. Balakrishnan, Marcel Dekker, Inc., New York (1991) 397–426.

[5] W. Y. Tan and S. Piatadosi, *On stochastic growth process with application to stochastic logistic growth*, Statistica Sinica **1** (1991) 527–540.

[6] W. Y. Tan, *Stochastic Models of Carcinogenesis*, Marcel Dekker, New York (1991).

[7] W. Y. Tan, *Stochastic Modeling of AIDS Epidemiology and HIV Pathogenesis*, World Scientific, Singapore (2000).

[8] W. Y. Tan and H. Wu, *Stochastic modeling of the dynamics of CD4+ T cell infection by HIV and some Monte Carlo studies*, Math. Biosciences **147** (1998) 173–205.

[9] P. A. P. Moran, *Random processes in genetics*, Proc. Camb. Phil. Soc. **54** (1958) 60–72.

[10] M. Kimura, *A simple method for estimating evolutionary rate of base substitution through comparative studies of nucleotide sequencs*, J. Mol. Evol. **16** (1980) 111–120.

[11] I. M. Longini, W. S. Clark, R. H. Byers, J. W. Ward, W. W. Darrow, G. H. Lemp and H. W. Hethcote, *Statistical analysis of the stages of HIV infection using a Markov model*, Statistics in Medicine **8** (1989) 831–843.

[12] I. M. Longini, W. S. Clark, L. I. Gardner and J. F. Brundage, *The dynamics of CD4+ T–lymphocyte decline in HIV-infected individuals: A Markov modeling approach*, J. AIDS **4** (1991) 1141–1147.

[13] I. M. Longini, W. S. Clark, G. A. Satten, R. H. Byers and J. M. Karon, *Staged Markov models based on CD4+ T-lymphocytes for the natural history of HIV infection*, in: *Models for Infectious Human Diseases: Their Structure and Relation to Data*, eds. V. Isham and G. Medley, Cambridge University Press, Cambridge (1996) 439–459.

[14] G. Satten and I. M. Longini, *Estimation of incidence of HIV infection using cross-sectional marker survey*, Biometrics **50** (1994) 675–688.

[15] G. Satten and I. M. Longini, *Markov chain with measurement error: Estimating the 'true' course of marker of the progression of human immunodeficiency virus disease*, Appl. Statist. **45** (1996) 275–309.

[16] W. Y. Tan and C. C. Brown, *A stochastic model for drug resistance and immunization, I. One drug case*, Math. Biosciences **97** (1989) 145–160.

[17] S. H. Moolgavkar, *A population perspective on multistage carcinogenesis*, in: *Multistage Carcinogenesis*, eds. C. C. Harris, S. Hirohashi, N. Ito, H. C. Pitot, T. Sugimura, M. Terada and J. Yokota., CRC Press, Boca Raton, Florida (1992).

[18] K. M. Kinzler and B. Vogelstein, *Colorectal tumors*, in *The Genetic Basis of Human Cancer*, eds. B. Vogelstein and K. M. Kinzler, McGraw-Hill, New York (1998) 565–587.

[19] M. P. Little, *Are two mutations sufficient to cause cancer? Some generalizations of the two-mutation model of carcinogenesis of Moolgavkar, Venson and Knudson, and of the multistage model of Armitage and Doll*, Biometrics **51** (1995) 1278–1291.

[20] M. P. Little, *Generalizations of the two-mutation and classical multi-stage models of carcinogenesis fitted to the Japanese atomic bomb survivor data*, J. Radiol. Prot. **16** (1996) 7–24.

[21] M. P. Little, C. R. Muirhead, J. D. Boice Jr. and R. A. Kleinerman, *Using multistage models to describe radiation-induced leukaemia* J. Radiol. Prot. **15** (1995) 315–334.

[22] M. P. Little, C. R. Muirhead and C. A. Stiller, *Modelling lymphocytic leukaemia incidence in England and Wales using generalizations of the two-mutation model of carcinogenesis of Moolgavkar, Venzon and Knudson*, Statistics in Medicine **15** (1996) 1003–1022.

[23] K. C. Chu, *Multi-event model for carcinogenesis: A model for cancer causation and prevention*, in: *Carcinogenesis: A Comprehensive Survey Volume 8: Cancer of the Respiratory Tract-Predisposing Factors*, M. J. Mass, D. G. Ksufman, J. M. Siegfied, V. E. Steel and S. Nesnow (eds.), Raven Press, New York (1985) 411–421.

[24] K. C. Chu, C. C. Brown, R. E. Tarone and W. Y. Tan, *Differentiating between proposed mechanisms for tumor promotion in mouse skin using the multi-vent model for cancer*, Jour. Nat. Cancer Inst. **79** (1987) 789–796.

[25] G. L. Yang and C. W. Chen, *A stochastic two-stage carcinogenesis model: A new approach to computing the probability of observing tumor in animal bioassays*, Math. Biosci. **104** (1991) 247–258.

[26] W. Y. Tan and W. C. Chen, *Stochastic models of carcinogenesis. Some new insight*, Math Comput. Modeling **28** (1998) 49–71.

[27] W. Y. Tan and C. C. Brown, *A nonhomogenous two stages model of carcinogenesis*, Math. Modeling **9** (1987) 631–642.

[28] W. Y. Tan *On the distribution of number of mutants in cell population with both forward and backward mutations*, SIAM J. App. Math. **49** (1989) 186–196.

[29] CDC. *1993 Revised Classification System for HIV Infection and Expanded Surveillance Case Definition for AIDS Among Adolescents and Adults*, MMWR **41** (1992), No. RR17.

[30] S. Karlin and J. L. McGregor, *On a genetic model of Moran*, Proc. Camb. Phil. Soc. **58** (2) (1962) 299–311.

[31] A. Erdelyi (editor.), *Higher Transcendental Functions, Vol. 2.*, McGraw-Hill, New York (1953).

[32] S. Karlin and J. L. McGregor, *The Hahn polynomials, formulas and an application*, Scripta Math. **26** (1961) 33–45.

Chapter 5

Absorption Probabilities and Stationary Distributions in Continuous-Time Markov Chain Models

In Chap. 4, we have discussed some general results of Markov chains with continuous time. For the applications of these chains, we present in this chapter some results on absorption probabilities, first absorption times, stationary distributions as well as some other topics of importance. We will illustrate these results by examples from genetics, cancer and AIDS.

5.1. Absorption Probabilities and Moments of First Absorption Times of Transient States

Consider a homogeneous Markov chain $\{X(t), t \in T = [0, \infty)\}$ with continuous time and with state space $S = \{0, 1, \ldots, \infty\}$. Assume that the chain contains both transient states and persistent states and that the persistent states are grouped into k closed sets $\{C_j, j = 1, \ldots, k\}$. In this section we will illustrate how to derive the absorption probabilities of transient states into persistent states.

Let α_{ij} be the transition rates (infinitesimal parameters) of the chain. (Note that $\alpha_{ij}(t) = \alpha_{ij}$ are independent of t as the chain is homogeneous.) Let C_T be the set of transient states. To find the absorption probability of $i \in C_T$ into a persistent state, denote by $\omega_i(t) = \Pr\{X(t) \in C_T | X(0) = i\}$. Since persistent states go only to persistent states, then $\omega_i(t - \Delta t) - \omega_i(t) \cong g_i(t)\Delta t$, where $g_i(t) = -\frac{d\omega_i(t)}{dt}$, is the probability that starting with $X(0) = i \in C_T$, the chain is absorbed into a persistent state during $[t, t + \Delta t)$ for the first time. If

$\int_0^\infty g_i(t)dt = 1$, then $g_i(t)$ is the pdf of the first absorption time T_i of $i \in C_T$ into a persistent state. Given that with probability one, $i \in C_T$ will eventually be absorbed into a persistent state, then one may evaluate the mean μ_i and the variance V_i of T_i. Now, we have:

$$\int_0^\infty tg_i(t)dt = \lim_{N\to\infty} \lim_{\Delta t\to 0} \sum_{n=1}^{N}(n\Delta t)\{\omega_i[(n-1)\Delta t] - \omega_i(n\Delta t)\}$$

$$= \lim_{N\to\infty} \lim_{\Delta t\to 0} \left\{ \sum_{n=0}^{N-1} \omega_i(n\Delta t)\Delta t - (N\Delta t)\omega_i(N\Delta t) \right\}$$

$$= \int_0^\infty \omega_i(t)dt$$

and

$$\int_0^\infty t^2 g_i(t)dt = \lim_{N\to\infty} \lim_{\Delta t\to 0} \sum_{n=1}^{N}(n\Delta t)^2[\omega_i((n-1)\Delta t) - \omega_i(n\Delta t)]$$

$$= \lim_{N\to\infty} \lim_{\Delta t\to 0} \left\{ \sum_{n=1}^{N}[(n-1)\Delta t]^2\omega_i[(n-1)\Delta t] \right.$$

$$- \sum_{n=1}^{N}(n\Delta t)^2\omega_i(n\Delta t) + 2\sum_{n=1}^{N}[(n-1)\Delta t]\omega_i[(n-1)\Delta t]\Delta t$$

$$\left. + \sum_{n=1}^{N}\omega_i[(n-1)\Delta t](\Delta t)^2 \right\}$$

$$= 2\int_0^\infty t\omega_i(t)dt\,.$$

Hence, the mean μ_i and the variance V_i of T_i are given respectively by the following formulas:

$$\mu_i = \int_0^\infty tg_i(t)dt = \int_0^\infty \omega_i(t)dt \qquad (5.1)$$

and

$$V_i = \int_0^\infty t^2 g_i(t)dt - \mu_i^2 = 2\int_0^\infty t\omega_i(t)dt - \mu_i^2\,. \qquad (5.2)$$

To find the absorption probability into C_j of $i \in C_T$, let $g_{i,j}(t)\Delta t$ be the probability of absorption into C_j of $i \in C_T$ during $[t, t + \Delta t)$ for the first time and $g_i(t) = \sum_{j=1}^{k} g_{i,j}(t)$. Then $g_i(t)\Delta t$ is the probability of first absorption time into a persistent state of $i \in C_T$ during $[t, t + \Delta t)$. Let $\nu_i(j) = \sum_{l \in C_j} \alpha_{il}$ and $\nu_i = \sum_{j=1}^{k} \nu_i(j)$. Then, to order of $o(\Delta t)$, $\nu_i(j)\Delta t$ is the probability that the state $i \in C_T$ at time t will be absorbed into C_j during $[t, t + \Delta t)$; and, to order of $o(\Delta t)$, $\nu_i\Delta t$ is the probability that the state $i \in C_T$ at time t will be absorbed into a persistent state during $[t, t + \Delta t)$.

Hence,

$$g_{i,j}(t) = \sum_{l \in C_T} p_{il}(t)\nu_l(j)$$

and

$$g_i(t) = \sum_{j=1}^{k} g_{i,j}(t) = \sum_{l \in C_T} p_{il}(t)\nu_l .$$

(Notice also that $g_i(t)\Delta t \cong \omega_i(t - \Delta t) - \omega_i(t)$ for small Δt.)

The ultimate absorption probability into C_j of $i \in C_T$ is therefore:

$$\rho_i(j) = \int_0^{\infty} g_{i,j}(t)dt .$$

5.1.1.   *The case when C_T is finite*

Assume that C_T is finite with r elements. With no lose of generality, assume that the first r states of S are transient states. Then the absorption probabilities and moments of first absorption times can be expressed in matrix notation. To this end, denote by $Q(t)$ the $r \times r$ matrix of transition probabilities $P\{X(t) = j | X(0) = i\}$ of transient states (i.e. $i, j = 1, \ldots, r$) and by B be the $r \times r$ matrix with the (i, j)th element given by $b_{ij} = \delta_{ij}\alpha_{ii} + (\delta_{ij} - 1)\alpha_{ij}$, $i, j = 1, \ldots, r$. Then, $Q(n\Delta t) = Q(\Delta t)^n$ by the Chapman–Kolmogorov equation; and to order of $o(\Delta t)$ $(\Delta t > 0)$,

$$Q(\Delta t) = I_r - B\Delta t + o(\Delta t) .$$

It follows that with $t = n\Delta t$, we have:

$$Q(t) = \lim_{\Delta t \to 0} \{I_r - B\Delta t + o(\Delta t)\}^n = \lim_{n \to \infty} \left\{ I_r - \frac{Bt}{n} + o(\Delta t) \right\}^n = e^{-Bt},$$

where e^{-Bt} is the matrix exponential function defined by $e^{-Bt} = \sum_{j=0}^{\infty} \frac{1}{j!}(-t)^j B^j$.

Furthermore, since $\alpha_{ii} = \sum_{l \neq i} \alpha_{il} = \sum_{l \neq i; l \in C_T} \alpha_{il} + \sum_{j=1}^{k} \sum_{l \in C_j} \alpha_{il} = \sum_{l \neq i; l \in C_T} \alpha_{il} + \nu_i$, we have:

$$\nu_i = \alpha_{ii} - \sum_{l \neq i; l \in C_T} \alpha_{il} = \sum_{l=1}^{r} b_{il},$$

and $\underset{\sim}{\nu} = \{\nu_1, \ldots, \nu_r\}' = B \underset{\sim}{1}_r$, where $\underset{\sim}{1}_r$ is the $r \times 1$ column of 1's.

To express the results in matrix notation, denote by:

$$\underset{\sim}{\omega}(t) = \{\omega_1(t), \ldots, \omega_r(t)\}', \qquad \underset{\sim}{q}'_i = \text{the } i\text{th row of } Q \,;$$

$$\underset{\sim}{\nu}(j) = \{\nu_1(j), \ldots, \nu_r(j)\}', \qquad \underset{\sim}{\nu} = \{\nu_1, \ldots, \nu_r\}' \,;$$

$$\underset{\sim}{g}_j(t) = \{g_{1,j}(t), \ldots, g_{r,j}(t)\}', \qquad \underset{\sim}{g}(t) = \{g_1(t), \ldots, g_r(t)\}' \,;$$

and

$$\underset{\sim}{\rho}(j) = \{\rho_1(j), \ldots, \rho_r(j)\}', \qquad \underset{\sim}{\rho} = \{\rho_1, \ldots, \rho_r\}' \,.$$

Then, since persistent states go only to persistent states, we have:

$$\omega_i(t) = \sum_{l \in C_T} p_{il}(t) = \underset{\sim}{q}'_i(t)\underset{\sim}{1}_r, \qquad\qquad \underset{\sim}{\omega}(t) = Q(t)\underset{\sim}{1}_r \,;$$

$$g_{i,j}(t) = \sum_{l \in C_T} p_{il}(t)\nu_l(j) = \underset{\sim}{q}'_i(t)\underset{\sim}{\nu}(j), \qquad\qquad \underset{\sim}{g}_j(t) = Q(t)\underset{\sim}{\nu}(j) \,;$$

$$g_i(t) = \sum_{l \in C_T} p_{il}(t)\nu_l = \underset{\sim}{q}'_i(t)\underset{\sim}{\nu} = \underset{\sim}{q}'_i(t)B\underset{\sim}{1}_r, \qquad \underset{\sim}{g}(t) = Q(t)\underset{\sim}{\nu} = Q(t)B\underset{\sim}{1}_r \,.$$

On substituting the result $Q(t) = e^{-Bt}$, we obtain:

$$\underset{\sim}{\omega}(t) = e^{-Bt}\underset{\sim}{1}_r \,, \tag{5.3}$$

$$\underset{\sim}{g}_j(t) = e^{-Bt}\underset{\sim}{\nu}(j) \,, \tag{5.4}$$

and

$$\underset{\sim}{g}(t) = Q(t)\underset{\sim}{\nu} = e^{-Bt}B\underset{\sim}{1}_r \,. \tag{5.5}$$

Equation (5.5) is the matrix exponential distribution first derived by Tan [1].

Denote by $\int_0^\infty \underset{\sim}{f}(t)dt = \{\int_0^\infty f_1(t)dt, \ldots, \int_0^\infty f_r(t)dt\}'$ for any vector $\underset{\sim}{f}(t) = \{f_1(t), \ldots, f_r(t)\}'$ of integrable functions $f_i(t), i = 1, \ldots, r$. Then, from Eqs. (5.4) and (5.5), we obtain:

$$\underset{\sim}{\rho}(j) = \int_0^\infty \underset{\sim}{g}_j(t)dt = \left\{ \int_0^\infty e^{-Bt}B\,dt \right\} B^{-1}\underset{\sim}{\nu}(j) = B^{-1}\underset{\sim}{\nu}(j), \qquad (5.6)$$

$$\underset{\sim}{\rho} = \int_0^\infty \underset{\sim}{g}(t)dt = \int_0^\infty e^{-Bt}B\underset{\sim}{1}_r\,dt = -\int_0^\infty de^{-Bt}\underset{\sim}{1}_r = \underset{\sim}{1}_r. \qquad (5.7)$$

Equation (5.7) shows that for each $i \in C_T$, $\rho_i = 1$. That is, starting with any $i \in C_T$, with probability one the chain will eventually be absorbed into a persistent state as time progresses. Let $\underset{\sim}{e}'_i$ denote the $1 \times r$ row with 1 in the ith position $(i = 1, \ldots, r)$ and 0 at other positions. Then by Eq. (5.6), the probability of ultimate absorption into C_j of $i \in C_T$ is $\underset{\sim}{e}'_i B^{-1}\underset{\sim}{\nu}(j)$.

Using Eqs. (5.1) and (5.3), we obtain the vector $\underset{\sim}{U}$ of the means of the first absorption times of transient states as:

$$\underset{\sim}{U} = \int_0^\infty \underset{\sim}{\omega}(t)dt = \int_0^\infty e^{-Bt}\underset{\sim}{1}_r\,dt = B^{-1}\underset{\sim}{1}_r = (\nu_1, \ldots, \nu_r)'. \qquad (5.8)$$

Denote by $\frac{1}{2}\underset{\sim}{\eta} = \{\int_0^\infty t\omega_1(t)dt, \ldots, \int_0^\infty t\omega_r(t)dt\}' = \int_0^\infty t\underset{\sim}{\omega}(t)dt$. Then, by Eq. (5.2), the vector of variances of first passage times of transient states is $\underset{\sim}{V} = \underset{\sim}{\eta} - \underset{\sim}{U}_{sq}$, where $\underset{\sim}{U}_{sq} = (\nu_1^2, \ldots, \nu_r^2)'$.

Now, by Eq. (5.3), we have:

$$\frac{1}{2}\underset{\sim}{\eta} = \int_0^\infty te^{-Bt}\underset{\sim}{1}_r\,dt = -\int_0^\infty td(e^{-Bt})B^{-1}\underset{\sim}{1}_r$$

$$= \int_0^\infty e^{-Bt}B^{-1}\underset{\sim}{1}_r\,dt = B^{-2}\underset{\sim}{1}_r = B^{-1}\underset{\sim}{U}.$$

It follows that

$$\underset{\sim}{V} = 2B^{-1}\underset{\sim}{U} - \underset{\sim}{U}_{sq}. \qquad (5.9)$$

Example 5.1. Moran's model of genetics with no selection and no mutation. Consider the Moran's model of genetics as described in

Examples 4.2 and 4.12. Assume that there are no selection and no mutation. Then, the states 0 and M are absorbing states while all other states are transient states. In this case, with $\{u_j, v_j, w_i\}$ being given in Example 4.12, the transition matrix $P(t)$ is given by:

$$P(t) = \begin{pmatrix} 1 & 0'_{M-1} & 0 \\ F_1(t) & \sum_{j=1}^{M-1} e^{-\frac{\lambda}{M}j(j+1)t} v_j u'_j, & F_2(t) \\ 0 & 0'_{M-1} & 1 \end{pmatrix},$$

where

$$F_i(t) = \sum_{j=1}^{M-1} (1 - e^{-\frac{\lambda}{M}j(j+1)t}) \frac{1}{j(j+1)} v_j(u'_j w_i), \quad i = 1, 2.$$

Since $\sum_{k=1}^{2} w_k = B 1_{M-1} = \sum_{j=1}^{M-1} j(j+1) v_j u'_j 1_{M-1}$, we have:

$$F(t) = \sum_{i=1}^{2} F_i(t) = \sum_{j=1}^{M-1} (1 - e^{-\frac{\lambda}{M}j(j+1)t}) v_j u'_j 1_{M-1}.$$

Using these results, we have:

(1) The vectors of absorptions of transient states into 0, M and persistent states (0 or M) at or before time t are given by $F_1(t)$, $F_2(t)$ and $F(t)$ respectively.

(2) The vectors of first time absorptions into 0, M and persistent states (0 or M) of transient states at time t are given respectively by:

$$g_0(t) = \frac{d}{dt} F_1(t) = \frac{\lambda}{M} \sum_{j=1}^{M-1} e^{-\frac{\lambda}{M}j(j+1)t} v_j u'_j w_1,$$

$$g_M(t) = \frac{d}{dt} F_2(t) = \frac{\lambda}{M} \sum_{j=1}^{M-1} e^{-\frac{\lambda}{M}j(j+1)t} v_j u'_j w_2,$$

$$g(t) = \frac{d}{dt} F(t) = \frac{\lambda}{M} \sum_{j=1}^{M-1} j(j+1) e^{-\frac{\lambda}{M}j(j+1)t} v_j u'_j 1_{M-1}.$$

(3) The vectors of ultimate absorptions probabilities into 0, M and persistent states (0 or M) of transient states are given respectively by:

$$\underset{\sim}{\rho}(0) = \lim_{t \to \infty} \underset{\sim}{F}_1(t) = \sum_{j=1}^{M-1} \{j(j+1)\}^{-1} \underset{\sim}{v}_j \underset{\sim}{u}'_j \underset{\sim}{w}_1,$$

$$\underset{\sim}{\rho}(M) = \lim_{t \to \infty} \underset{\sim}{F}_2(t) = \sum_{j=1}^{M-1} \{j(j+1)\}^{-1} \underset{\sim}{v}_j \underset{\sim}{u}'_j \underset{\sim}{w}_2,$$

$$\underset{\sim}{\rho} = \lim_{t \to \infty} \underset{\sim}{F}(t) = \sum_{j=1}^{M-1} \underset{\sim}{v}_j \underset{\sim}{u}'_j \underset{\sim}{1}_{M-1} = \underset{\sim}{1}_{M-1}.$$

(4) The vectors of mean values and variances of first absorption times of transient states into persistent states (0 or M) are given respectively by:

$$\underset{\sim}{U} = \int_0^\infty t\,\underset{\sim}{g}(t)dt = \sum_{j=1}^{M-1} \left\{\frac{\lambda}{M}j(j+1)\right\}^{-1} \underset{\sim}{v}_j \underset{\sim}{u}'_j \underset{\sim}{1}_{M-1} = \{\mu_1, \ldots, \mu_{M-1}\}',$$

$$\underset{\sim}{V} = \underset{\sim}{\eta} - \{\mu_1^2, \ldots, \mu_{M-1}^2\}', \text{ where}$$

$$\underset{\sim}{\eta} = \int_0^\infty t^2\,\underset{\sim}{g}(t)dt = 2 \sum_{j=1}^{M-1} \left\{\frac{\lambda}{M}j(j+1)\right\}^{-2} \underset{\sim}{v}_j \underset{\sim}{u}'_j \underset{\sim}{1}_{M-1}.$$

Example 5.2. The HIV epidemic and the HIV incubation distribution. In Examples 4.4 and 4.10, we have described a Markov chain model with continuous time for the HIV epidemic. The transition rates of this chain have been estimated by Satten and Longini [2] by using data of the SF Men's Health study. With $\{c_i, \lambda_i, \underset{\sim}{u}_i, \underset{\sim}{v}_i\}$ being given in Example 4.10, the transition matrix $P(t)$ is given by:

$$P(t) = \begin{pmatrix} \sum_{i=1}^5 e^{-\lambda_i t} E_i & \underset{\sim}{F}(t) \\ 0 & 1 \end{pmatrix},$$

where $E_i = \underset{\sim}{v}_i \underset{\sim}{u}_i'$ and

$$F(t) = (I_4 - e^{-Bt})\underset{\sim}{1}_5 = \sum_{i=1}^{5}(1 - e^{-\lambda_i t})E_i\underset{\sim}{1}_5$$

$$= \sum_{i=1}^{5} d_i(1 - e^{-\lambda_i t})\underset{\sim}{v}_i, \quad \text{where } d_i = \underset{\sim}{u}_i'\underset{\sim}{1}_5, \ i = 1, \ldots, 5.$$

Hence, we have:

(1) $F(t)$ is the vector of cdf's of the first absorption times of transient states.

(2) The vector $g(t)$ of pdf's of first absorption times of transient states is,

$$g(t) = e^{-Bt}B\underset{\sim}{1}_5 = \sum_{i=1}^{5} d_i\lambda_i e^{-\lambda_i t}\underset{\sim}{v}_i.$$

(3) The vector $\underset{\sim}{U}$ of mean absorption times and the vector $\underset{\sim}{V}$ of variances of mean absorption times are given respectively by:

$$\underset{\sim}{U} = \sum_{i=1}^{5} \lambda_i d_i \left\{ \int_0^{\infty} t e^{-\lambda_i t} dt \right\} \underset{\sim}{v}_i$$

$$= \sum_{i=1}^{5} \{d_i/\lambda_i\}\underset{\sim}{v}_i = \{\mu_1, \ldots, \mu_5\}',$$

$$\underset{\sim}{V} = \underset{\sim}{\eta} - \{\mu_1^2, \ldots, \mu_5^2\}',$$

where

$$\underset{\sim}{\eta} = \sum_{i=1}^{5} \lambda_i d_i \left\{ \int_0^{\infty} t^2 e^{-\lambda_i t} dt \right\} \underset{\sim}{v}_i = \sum_{i=1}^{5} \{2d_i/\lambda_i^2\}\underset{\sim}{v}_i.$$

In HIV epidemic, the random time period from HIV infection to AIDS onset is defined as the HIV incubation period, to be denoted by T_{inc}. In the above formulation, T_{inc} is then the first absorption time of I_1 to A. The conditional probability density function $f_{\text{inc}}(t)$ of T_{inc} given that the individual dies after AIDS is called the HIV incubation distribution. In the above formulation, T_{inc} is then the first absorption time of I_1 to A. Thus, $f_{\text{inc}}(t) = g_1(t) = \underset{\sim}{e}_1' g(t)$, where $\underset{\sim}{e}_i$ is a 5×1 column with 1 at the ith position and with 0 at other positions. Hence,

$$f_{\text{inc}}(t) = \underset{\sim}{e}_1' e^{-B\,t} B \underset{\sim}{1}_5 = \sum_{i=1}^{5} \lambda_i e^{-\lambda_i t} (\underset{\sim}{e}_1' \underset{\sim}{v}_i) d_i$$

$$= \sum_{i=1}^{5} r_i \lambda_i e^{-\lambda_i t} .$$

where $r_i = (\underset{\sim}{e}_1' \underset{\sim}{v}_i) d_i, i = 1, \ldots, 5$.

Example 5.3. The continuous time Galton–Watson branching process. Consider the continuous time branching process described in Example 4.7 for mutants. To evaluate the extinction probability of mutants in this chain, we consider an embedded branching process $\{Y(n), n \in T = (0, 1, \ldots, \infty)\}$ with progeny distribution $\{q_i, i = 0, 1, \ldots, \infty\}$. Let $f_n(s)$ be the pgf of $Y(n)$ given $Y(0) = 1$. Then, $f_1(s) = h(s) = s + u(s)\Delta t$ and $f_n(s) = f[f_{n-1}(s)]$ for $n \geq 2$. Using this embedded Markov chain, then by Theorem 2.10, the probability x_0 that the mutant at 0 is lost eventually is the smallest non-negative root of the function equation $f_1(x) = x$. But $f_1(x) = x$ iff $u(x) = 0$ iff $g(x) = x$ and this hold for all $\Delta t > 0$. Thus, the extinction probability x_0 is the smallest non-negative root of the functional equation $g(x) = x$. Further, by Theorem 2.11, $x_0 = 1$ iff $(\frac{f_1(s)}{ds})_{s=1} = f_1'(1) \leq 1$ or iff $(\frac{u(s)}{ds})_{s=1} = u'(1) \leq 0$ or iff $(\frac{g(s)}{ds})_{s=1} = g'(1) \leq 1$; and this is true for all $\Delta t > 0$. Notice that $\mu = g'(1)$ is the expected value of the progeny distribution $\{p_j, j = 0, 1, \ldots, \infty\}$. Hence with probability one the mutant gene at $t = 0$ will eventually be lost iff the expected value μ of the progeny distribution $\{p_j, j = 0, 1, \ldots, \infty\}$ is ≤ 1. It follows that if $\mu \leq 1$, then starting with any state i at time 0, with probability one the state will eventually be absorbed into the state 0 as time progresses. On the other hand, if $\mu > 1$, with positive probability, the chain will stay forever in transient states.

To find the extinction probability x_0 when $\mu > 1$, one need to solve the functional equation $g(x) = x$. If $p_j = e^{-\lambda} \frac{\lambda^j}{j!}, j = 0, 1, \ldots, \infty$, then the functional equation $g(x) = x$ becomes $x = e^{-\lambda(1-x)}$ and $\mu = \lambda$. Put $\lambda = 1 + s$ $(s > 0)$. If $s > 0$ is very small, then x_0 is close to 1 so that $\delta = 1 - x_0$ is very small. It is shown in Example 2.16 that if $\delta = 1 - x_0$ is very small, then to order of $o(\delta^3)$, x_0 is approximated by $x_0 \cong \frac{2s}{1+s}$.

If $\mu \leq 1$, then one may define the pdf of first absorption time of the mutant at time 0 and also find the mean μ of this first absorption time. Let $t = n\Delta t$ and let $x_0(n)$ denote the probability that starting with $X(0) = 1$, the chain is absorbed into 0 during $[t, t + \Delta t)$ for the first time. Let $\omega(t)$ denote the probability that starting with $X(0) = 1$, the chain is in a transient state by time $t + \Delta t$. Then, $\omega(t) = \omega(n\Delta t) = 1 - x_0(n)$ and $\omega(t - \Delta t) = \omega[(n-1)\Delta t] = 1 - x_0(n-1)$.

Now, by using the imbedded chain $Y(n)$, $x_0(n) = f_n(0) = f_1[f_{n-1}(0)] = h[x_0(n-1)] = x_0(n-1) + \{g[x_0(n-1)] - x_0(n-1)\}\lambda\Delta t$. Hence,

$$x_0(n) - x_0(n-1) = \omega(t - \Delta t) - \omega(t)$$

$$= \{g[x_0(n-1)] - x_0(n-1)\}\lambda\Delta t$$

$$= \{g[1 - \omega(t - \Delta t)] - 1 + \omega(t - \Delta t)\}\lambda\Delta t.$$

Hence $f(t) = \{g[x_0(t)] - x_0(t)\}\lambda = \{g[1 - \omega(t)] - 1 + \omega(t)\}\lambda$ is the pdf of the first absorption time of the mutant at 0. Notice that since $x_0(0) = 0$ and $\lim_{N\to\infty} x_0(N) = x_0 = 1$ as $\mu \leq 1$,

$$\int_0^\infty f(t)dt = \lim_{\Delta t\to 0}\lim_{N\to\infty}\sum_{n=1}^N \{g[x_0(n-1)] - x_0(n-1)\}\lambda\Delta t$$

$$= \lim_{\Delta t\to 0}\lim_{N\to\infty}\sum_{n=1}^N \{x_0(n) - x_0(n-1)\} = \lim_{\Delta t\to 0}\lim_{N\to\infty} x_0(N) = 1.$$

Denote by $\omega(n\Delta t) = \omega_0(n)$. Then the mean μ of first absorption time is

$$\int_0^\infty tf(t)dt = \lim_{\Delta t\to 0}\lim_{N\to\infty}\sum_{n=1}^N n\{g[x_0(n-1)] - x_0(n-1)\}\lambda\Delta t$$

$$= \lim_{\Delta t\to 0}\lim_{N\to\infty}\sum_{n=1}^N n\{\omega_0(n-1) - \omega_0(n)\}$$

$$= \lim_{\Delta t\to 0}\lim_{N\to\infty}\sum_{n=1}^N \{(n-1)\omega_0(n-1) - n\omega_0(n) + \omega_0(n-1)\}$$

$$= \lim_{\Delta t\to 0}\sum_{n=0}^\infty \omega_0(n).$$

Notice that for each fixed Δt, $x_0(n) = e^{\lambda(x_0(n-1)-1)}$ and $\omega_0(n) = 1 - x_0(t)$; and it is shown in Example 2.16 that $\sum_{n=0}^{\infty} \omega_0(n) = \infty$. This holds for all Δt. Hence $\mu = \infty$ if $p_j = e^{-\lambda}\frac{\lambda^j}{j!} j = 0, 1, \ldots, \infty$.

5.2. The Stationary Distributions and Examples

Let $\{X(t), t \in T = [0, \infty)\}$ be a Markov chain with continuous time T and with state space $S = \{0, 1, \ldots, \infty\}$. If $X(t)$ is homogeneous, then the transition rates (i.e. the infinitesimal parameters) $\{\alpha_{ij}, i \in S, j \in S\}$ are independent of t. In these cases, as in Markov chain with discrete time, one may define stationary distributions for $X(t)$ (Note: Stationary distribution can be defined only if the chain is homogeneous. However, Example 5.4 shows that stationary distributions may not exist although the chain is homogeneous.)

Definition 5.1. A probability density $\{\pi_i, i \in S\}$ over S is called *the stationary distribution* of $X(t)$ iff for every $t > 0$, the following condition holds:

$$\pi_j = \sum_{i=0}^{\infty} \pi_i p_{ij}(t), \quad j = 0, 1, \ldots, \infty. \tag{5.10}$$

By using embedded Markov chains and results of Markov chain with discrete time, one may readily find the stationary distributions if they exist. The following theorem provides a method for finding the stationary distribution when it exists.

Theorem 5.1. *The probability density* $\{\pi_i, i \in S\}$ *over* S *is a stationary distribution of* $X(t)$ *iff the following condition holds:*

$$\pi_j \alpha_{jj} = \sum_{i \neq j} \pi_i \alpha_{ij}, \quad j = 0, 1, \ldots, \infty. \tag{5.11}$$

Notice that if the chain is finite with n elements, then Eq. (5.11) is equivalent to the matrix equation

$$\underset{\sim}{\pi}'A = \underset{\sim}{0},$$

where A is the $n \times n$ matrix with (i, j)th element given by $\delta_{ij}\alpha_{jj} - (1 - \delta_{ij})\alpha_{ij}, i, j = 1, \ldots, n$.

Proof.   Suppose that $\{\pi_i, i \in S\}$ over S is a stationary distribution of $X(t)$. Then, for any $\Delta t > 0$,

$$\pi_j = \sum_{i=0}^{\infty} \pi_i p_{ij}(\Delta t), \quad j = 0, 1, \ldots, \infty. \tag{5.12}$$

But, to order of $o(\Delta t), p_{ij}(\Delta t) \cong \alpha_{ij}\Delta t$ if $i \neq j$, and $p_{jj}(\Delta t) \cong 1 - \alpha_{jj}\Delta t$. Hence, on substituting these results, Eq. (5.12) becomes

$$\pi_j = \sum_{i=0}^{\infty} \pi_i p_{ij}(\Delta t)$$

$$= \sum_{i \neq j} \pi_i \alpha_{ij}\Delta t + \pi_j[1 - \alpha_{jj}\Delta t] + o(\Delta t), \quad j = 0, 1, \ldots, \infty.$$

It follows that

$$\pi_j \alpha_{jj} = \sum_{i \neq j} \pi_i \alpha_{ij} + o(\Delta t)/\Delta t, \quad j = 0, 1, \ldots, \infty.$$

Letting $\Delta t \to 0$ gives Eq. (5.11).

Conversely, suppose that Eq. (5.11) holds. Then, for any Δt, one obtains easily:

$$\pi_j = \sum_{i=0}^{\infty} \pi_i p_{ij}(\Delta t) + o(\Delta t), \quad j = 0, 1, \ldots, \infty.$$

Choose $\Delta t > 0$ such that $t = n\Delta t$. Through the embedded Markov chain, one has:

$$\pi_j = \sum_{i=0}^{\infty} \pi_i p_{ij}(t) + o(\Delta t), \quad j = 0, 1, \ldots, \infty.$$

Letting $\Delta t \to 0$ then gives Eq (5.10) so that $\{\pi_i, i \in S\}$ over S is a stationary distribution of $X(t)$. $\qquad\square$

As in Markov chain with discrete time, the stationary distribution may or may not exist. If the stationary distributions exist, it may not be unique. One may summarize the results as follows. These results are easily proved by using embedded chains and by noting the result that all persistent states in the embedded chain are aperiodic.

(1) If the chain is homogeneous, irreducible and persistent, then the limiting distribution and hence the stationary distribution exist iff the persistent states are positive. Furthermore, the stationary distribution is unique and is given by $\lim_{t\to\infty} p_{ij}(t) = \pi_j, j \in S$. Notice that an irreducible homogeneous chain may not contain persistent states in which case the stationary distribution does not exist. (If the irreducible chain does not contain persistent states, then the chain must have infinitely many states.)

(2) If the chain contains more than one closed sets and if stationary distributions exist, then there are infinitely many stationary distributions.

(3) For finite Markov chains, the stationary distributions always exist although it may not be unique. Example 5.4 shows that if the chain is infinite, then stationary distributions may not exist.

Example 5.4. Stationary distributions in some birth-death processes. Consider a homogeneous birth-death process $\{X(t), t \in T = [0, \infty)\}$ with birth rate b_i $(b_i > 0, i = 0, 1, \ldots,)$ and death rate d_i $(d_0 = 0, d_i > 0$ for $i = 1, \ldots)$. In this case, $\alpha_{ij} = b_i$ if $j = i + 1; = d_i$ if $j = i - 1; = 0$ if $|i - j| \geq 2$; and $\alpha_{ii} = b_i + d_i$. Thus, Eq. (5.11) become:

$$\pi_j(b_j + d_j) = \pi_{j-1}b_{j-1} + \pi_{j+1}d_{j+1}, j = 0, 1, \ldots, \text{ with } \pi_{-1} = 0.$$

Putting $j = 0$, then $\pi_1 = \frac{b_0}{d_1}\pi_0$; putting $j = 1$ yields $\pi_2 = \frac{b_0 b_1}{d_1 d_2}\pi_0$. By mathematical induction, it can easily be shown that $\pi_j = c_j\pi_0, j = 0, 1, \ldots, \infty$, where the c_j is defined by:

$$c_0 = 1, c_j = \prod_{i=0}^{j-1} \frac{b_i}{d_{i+1}} \text{ for } j = 1, 2, \ldots, \infty.$$

The condition $\sum_{i=0}^{\infty} \pi_i = 1$ then leads to:

$$\pi_0 = \frac{1}{1 + \sum_{j=1}^{\infty} c_j}.$$

Notice that π_0 is finite iff $\sum_{j=0}^{\infty} c_j < \infty$. Thus the stationary distribution $\{\pi_j, j = 0, 1, \ldots, \infty\}$ exists iff $\sum_{j=0}^{\infty} c_j < \infty$. As an example, assume $b_j = jb + \lambda$ with $b > 0$ and $\lambda > 0$ and $d_j = jd, d > 0$. (This is the so-called Feller–Arley birth-death process with immigration.) Then we have, for $j = 1, 2, \ldots,$:

$$c_j = (b/d)^j \prod_{i=0}^{j-1} \frac{i + (\lambda/b)}{i + 1}.$$

Thus, if $0 < r = b/d < 1$ and if $\lambda/b \leq 1$,

$$\sum_{j=0}^{\infty} c_j < \sum_{j=0}^{\infty} r^j < \infty \, ;$$

in these cases, the stationary distribution exists and is unique as given above. On the other hand, if $r = (b/d) > 1$ and if $\lambda/b \geq 1$, then the series $\sum_{j=0}^{\infty} c_j$ diverges so that the stationary distribution does not exist.

Example 5.5. Nucleotide substitution model in molecular evolution. In Examples 4.3 and 4.11, we have described the nucleotide substitution model in Eukaryotes. This is a finite homogeneous Markov chains with continuous time. Since the transition rates are positive, the chain is irreducible. Further, with $\{\nu_i, \underset{\sim}{x}_i, \underset{\sim}{y}_i\}$ being given in Example 4.11, the transition matrix $P(t)$ is given by:

$$P(t) = \underset{\sim}{1}_4 \underset{\sim}{y}'_1 + \sum_{i=2}^{4} e^{-\nu_i t} \underset{\sim}{x}_i \underset{\sim}{y}'_i \, .$$

It follows that $\lim_{t \to \infty} P(t) = \underset{\sim}{1}_4 \underset{\sim}{y}'_1$. Thus, $\underset{\sim}{y}_1$ is the density of the stationary distribution.

Example 5.6. Moran's genetic model with no selection. Consider the Moran's genetic model with $\lambda_i = \lambda$ so that there are no selection between the two types. This is a finite homogeneous Markov chain with continuous time. The transition matrix of this chain has been derived in Example 4.12 in terms of Hahn polynomials. In this example, we derive the stationary distributions of this chain.

Case 1: If $\alpha_1 > 0, \alpha_2 > 0$ but $1 > \alpha_1 + \alpha_2$, then the chain is irreducible so that there is an unique stationary distribution for the states. Also, all states are persistent states and are positive. With $\{\rho(k), \pi_n, Q_n(j)\}$ being given in Example 4.12, the transition probabilities of this chain are given by:

$$P_{jk}(t) = \rho(k) \sum_{n=0}^{M} \exp\left\{-\frac{\lambda}{M}(1 - \alpha_1 - \alpha_2)n(n + \alpha + \beta + 1)t\right\} \pi_n Q_n(j) Q_n(k)$$

$$= \rho(k) \left\{1 + \sum_{n=1}^{M} \exp\left\{-\frac{\lambda}{M}(1 - \alpha_1 - \alpha_2)n(n + \alpha + \beta + 1)t\right\} \pi_n Q_n(j) Q_n(k)\right\} \, .$$

Let $t \to \infty$, then $\lim_{t\to\infty} P_{jk}(t) = \rho(k), k = 0, \ldots, M$. Hence, $\{\rho(j), j = 0, \ldots, M\}$ is the unique stationary distribution.

Case 2: $\alpha_1 = \alpha_2 = 0$ (no mutation). In this case, 0 and M are absorbing states while all other states are transient states. Because the two independent left eigenvector of 0 of A are $\underset{\sim}{x}'_1 = (1, 0, 0, \ldots, 0)$ and $\underset{\sim}{x}'_2 = (0, 0, 0, \ldots, 1)$, the stationary distributions of this chain are

$$\underset{\sim}{g}' = \alpha(1, 0, \ldots, 0) + (1 - \alpha)(0, \ldots, 0, 1),$$

where $0 \le \alpha \le 1$ is a real number. That is, there are uncountable many stationary distributions.

Notice that, as shown in Example 4.12, the transition matrix in this case is given by:

$$P(t) = e^{-At} = E_0 + \sum_{j=1}^{M-1} e^{-\frac{\lambda}{M}j(j+1)t} E_j.$$

Thus, $\lim_{t\to\infty} P(t) = E_0$; further, $\underset{\sim}{x}_i E_0 = \underset{\sim}{x}_i, i = 1, 2$.

5.3. Finite Markov Chains and the HIV Incubation Distribution

Consider a homogeneous Markov chains $\{X(t), t \in T = [0, \infty)\}$ with state space S. If the state space contains only n elements $(1 \le n < \infty)$ so that the chain is finite, then the chain must contain persistent states. These persistent states can be grouped into k disjoint closed sets $(C_j, j = 1, \ldots, k; k \ge 1)$. The chain may or may not contain transient states; for describing the general structure, however, we assume that the chain contains transient states and let C_T denote the set of transient states. Suppose that there are $n_j (n_j \ge 1)$ states in $C_j, j = 1, \ldots, k$ and that there are $r(r > 0)$ transient states. (Note $\sum_{j=1}^{k} n_j + r = n$.) Also, with no loss of generality, we assume that the first r states are transient states. Let $\alpha_{ij} i \in S, j \in S$ denote transition rates and let A denote the $n \times n$ matrix whose (u, v)th element is $\delta_{uv} \alpha_{uu} - (1 - \delta_{uv})\alpha_{uv}(u, v = 1, \ldots, n)$. Then, $P(\Delta t) = I_n - A\Delta t + o(\Delta t)$. As in Markov chain with discrete time, we also have the following proposition which shows that A can be put in the following canonical form:

5.3.1. *Some general results in finite Markov chains with continuous time*

In this subsection we give some general results for finite Markov chains with continuous time. We first prove the following canonical form of the transition matrices.

Proposition 5.1. (Canonical Form of A). *Let A be the matrix of infinitesimal parameters for a finite Markov chain with continuous time and suppose that the chain contains transient states. Then A can be expressed in the following canonical form.*

$$
A = \begin{array}{c} \\ C_T \\ C_1 \\ C_2 \\ \\ \\ C_k \end{array}
\begin{array}{cccccc}
C_T & C_1 & C_2 & \ldots & C_k \\
\left(\begin{array}{ccccc}
B & -D_1 & -D_2 & \ldots & -D_k \\
0 & A_1 & 0 & \ldots & 0 \\
0 & 0 & A_2 & \ldots & 0 \\
\vdots & \ddots & & \vdots & \vdots \\
0 & & 0 & 0 & A_k
\end{array} \right)
\end{array},
$$

where A_i is the matrix of transition rates for states in C_i, D_i the matrix of transition rates from states in C_T to states in C_i, and B the matrix of transient rates for states in C_T. We notice that $D = (D_1, D_2, \ldots, D_k) \neq 0$ as $\sum_{i=1}^{k} D_i \underset{\sim}{1}_{n_i} = B \underset{\sim}{1}_r$ and B is not 0. Also, if $n_j = 1$, then $A_j = 0$.

Notice that the above canonical form of A is similar to the the canonical form of the one-step transition matrix of probabilities of finite homogeneous Markov chains with discrete time given in Sec. 2.8. Indeed, this canonical form of A can be proved by that of the one-step transition matrix through the embedded Markov chain.

To prove the above canonical form, notice that through the embedded Markov chain, we have, by results from Sec. 2.8:

$$
P(\Delta t) = \begin{pmatrix}
Q(\Delta t) & H_1(\Delta t) & H_2(\Delta t) & \ldots & H_k(\Delta t) \\
 & P_1(\Delta t) & & & 0 \\
 & & P_2(\Delta t) & & \\
 & & & \ddots & \\
 & 0 & & & P_k(\Delta t)
\end{pmatrix}.
$$

By definition, $P(\Delta t) = I_n - A\Delta t + o(\Delta t)$ and the C_j are closed sets. It follows that to order of $o(\Delta t)$, $P_j(\Delta t) = I_{n_j} - A_j\Delta t\ j = 1,\ldots,k; Q(\Delta t) = I_r - B\Delta t$ and $H_j(\Delta t) = D_j\Delta t$, $j = 1,\ldots,k$. This shows that the matrix A has the above canonical form.

From the above canonical form, it appears that the A matrix in finite homogeneous Markov chains with continuous time plays the role of the matrix P of the one-step transition probabilities in finite homogeneous Markov chains with discrete time. The major difference between the A matrix and the P matrix is that in P the elements of each row sum up to 1 whereas in A the elements of each row sum up to 0. That is,

$$P_i\underset{\sim}{1}_{n_i} = \underset{\sim}{1}_{n_i}, \ i = 1,\ldots,k; \quad \sum_{j=1}^{k} R_j\underset{\sim}{1}_{n_j} + Q\underset{\sim}{1}_r = \underset{\sim}{1}_r;$$

but

$$A_i\underset{\sim}{1}_{n_i} = \underset{\sim}{0}_{n_i}, \ i = 1,\ldots,k; \quad B\underset{\sim}{1}_r - \sum_{j=1}^{k} D_j\underset{\sim}{1}_{n_j} = \underset{\sim}{0}_r,$$

where $\underset{\sim}{0}_m$ is a $m \times 1$ column of 0.

As in finite homogeneous Markov chain with discrete time, through the canonical form of A, one may also derive formulas for absorption probabilities and moments of first absorption times of transient states as given in Sec. 5.1. To this end, notice that for $m = 1,\ldots,$

$$A^m = \begin{pmatrix} B^m & -E_1(m) & -E_2(m) & \ldots & -E_k(m) \\ & A_1^m & 0 & \ldots & 0 \\ 0 & 0 & A_2^m & \ldots & 0 \\ \vdots & \ddots & & \vdots & \vdots \\ 0 & & \ldots & 0 & A_k^m \end{pmatrix},$$

$$\text{where} \quad \begin{cases} E_i(1) = D_i, & i = 1, 2,\ldots,k; \\ E_i(m) = \sum_{s=0}^{m-1} B^s D_i A_i^{m-1-s}, & i = 1, 2,\ldots,k; \quad m = 2, 3,\ldots. \end{cases}$$

Let $t = m\Delta t, \Delta t > 0$. Then, by Chapman–Kolmogorov equation, $P(t)$ can be expressed as:

$$P(t) = \lim_{\Delta t \to 0} [P(\Delta t)]^m = \lim_{\Delta t \to 0} [I_n - A\Delta t + o(\Delta t)]^m$$

$$= \lim_{m \to \infty} [I_n - At/m + o(t/m)]^m = e^{-At}$$

$$= I_n + \sum_{m=1}^{\infty} \frac{1}{m!}(-t)^m A^m .$$

On substituting the above canonical form for A^m, we obtain:

$$P(t) = I_n + \sum_{m=1}^{\infty} \frac{1}{m!}(-t)^m A^m = \begin{pmatrix} e^{-Bt} & -G_1(t) & -G_2(t) & \ldots & -G_k(t) \\ & e^{-A_1 t} & 0 & \ldots & 0 \\ 0 & 0 & e^{-A_2 t} & \ldots & 0 \\ \vdots & \ddots & & \vdots & \vdots \\ 0 & & \ldots & 0 & e^{-A_k t} \end{pmatrix},$$

where

$$G_i(t) = \sum_{m=1}^{\infty} \frac{1}{m!}(-t)^m E_i(m), i = 1, \ldots, k .$$

Since $\underset{\sim}{F}_j(t) = -G_j(t)\underset{\sim}{1}_{n_j}$ is the vector of absorption into C_j of transient states at or before time t, so $\underset{\sim}{F}(t) = \sum_{j=1}^{k} \underset{\sim}{F}_j(t) = -\sum_{j=1}^{k} G_j(t)\underset{\sim}{1}_{n_j}$ is the vector of cdf of the first absorption times $(T_i, i \in C_T)$ into persistent states of transient states at time t.

Since $A_j \underset{\sim}{1}_{n_j} = \underset{\sim}{0}_{n_j}$, so

$$\underset{\sim}{F}_j(t) = -G_j(t)\underset{\sim}{1}_{n_j} = -\sum_{m=1}^{\infty} \frac{1}{m!}(-t)^m E_j(m)\underset{\sim}{1}_{n_j}$$

$$= -\sum_{m=1}^{\infty} \frac{1}{m!}(-t)^m B^{m-1} D_j \underset{\sim}{1}_{n_j}$$

$$= \{I_n - e^{-Bt}\}B^{-1}D_j\underset{\sim}{1}_{n_j}, \qquad j = 1, \ldots, k . \tag{5.13}$$

Since $\sum_{j=1}^{k} D_j \underset{\sim}{1}_{n_j} = B \underset{\sim}{1}_r$ as the elements of each row of A sum up to 0, so

$$\underset{\sim}{F}(t) = \sum_{j=1}^{k} \underset{\sim}{\omega}_j(t) = \{I_n - e^{-Bt}\} B^{-1} \sum_{j=1}^{k} D_j \underset{\sim}{1}_{n_j}$$

$$= \{I_n - e^{-Bt}\} \underset{\sim}{1}_r . \tag{5.14}$$

It follows that the vector of pdf's of first time absorption of the transient states is

$$\underset{\sim}{g}(t) = \frac{d}{dt} \underset{\sim}{F}(t) = e^{-Bt} B \underset{\sim}{1}_r . \tag{5.15}$$

Notice that formulae (5.15) is also derived alternatively in Sec. 5.1.

Using Eq. (5.13), the vector of ultimate absorption probabilities of the transient states into C_j is

$$\underset{\sim}{\rho}(j) = \lim_{t \to \infty} \underset{\sim}{F}_j(t) = B^{-1} D_j \underset{\sim}{1}_{n_j}, \ j = 1, \ldots, k .$$

Notice again that

$$\underset{\sim}{\rho} = \sum_{j=1}^{k} \underset{\sim}{\rho}(j) = \lim_{t \to \infty} \underset{\sim}{F}(t) = \underset{\sim}{1}_r .$$

Using Eq. (5.15), one may obtain:

$$\underset{\sim}{U} = \int_0^\infty t e^{-Bt} B \underset{\sim}{1}_r dt = - \int_0^\infty t d(e^{-Bt}) \underset{\sim}{1}_r$$

$$= \int_0^\infty e^{-Bt} \underset{\sim}{1}_r dt = B^{-1} \underset{\sim}{1}_r ,$$

$$\underset{\sim}{\eta} = \int_0^\infty t^2 e^{-Bt} B \underset{\sim}{1}_r dt = - \int_0^\infty t^2 d(e^{-Bt}) \underset{\sim}{1}_r$$

$$= 2 \int_0^\infty t e^{-Bt} \underset{\sim}{1}_r dt = 2B^{-2} \underset{\sim}{1}_r = 2B^{-1} \underset{\sim}{U} ,$$

and $\underset{\sim}{V} = \underset{\sim}{\eta} - \underset{\sim}{U}_{sq}.$

5.3.2. *Non-homogeneous finite chain with continuous time*

The above results extend readily to piece-wise non-homogeneous cases. In these cases, the time interval $[0, \infty)$ is partitioned into m non-overlapping sub-intervals $L_i = [t_{i-1}, t_i), i = 1, \ldots, m$ with $(t_0 = 0, t_m = \infty)$. The transition rates are given by:

$$\alpha_{ij}(t) = \alpha_{ij}(u), \text{ if } t \in L_u.$$

Let $P(0, t) = P(t)$ and for $u = 1, \ldots, m$, put

$$A(u) = \begin{pmatrix} B_u & -D_1(u) & -D_2(u) & \ldots & -D_k(u) \\ & A_1(u) & & & 0 \\ & & A_2(u) & & \\ & & & \ddots & \\ & 0 & & & A_k(u) \end{pmatrix}.$$

Then, for $t \in L_u, u = 1, \ldots, m, P(t)$ is given by:

$$P(t) = \left\{ \prod_{j=1}^{u-1} e^{-A(j)\tau_j} \right\} e^{-A(u)(t - t_{u-1})}, \tag{5.16}$$

where $\tau_j = t_j - t_{j-1}$ and where $\prod_{j=1}^{0} R(j)$ is defined as I_n.
Using the above canonical form, we have for $j = 1, \ldots, m$:

$$e^{-A(j)t} = \begin{pmatrix} e^{-B_j t} & -G_1(j;t) & -G_2(j;t) & \ldots & -G_k(j;t) \\ & e^{-A_1(j)t} & 0 & \ldots & 0 \\ 0 & 0 & e^{-A_2(j)t} & \ldots & 0 \\ \vdots & \ddots & & \vdots & \vdots \\ 0 & & & 0 & e^{-A_k(j)t} \end{pmatrix},$$

where

$$G_i(u;t) = \sum_{j=1}^{\infty} \frac{1}{j!}(-t)^j E_i(u,j), \quad i = 1, \ldots, k,$$

where

$$E_i(u;1) = D_i(u), \ i = 1, 2, \ldots, k, \ u = 1, \ldots, m,$$

$$E_i(u;j) = \sum_{s=0}^{j-1} B_u^s D_i(u) A_i^{j-1-s}(u),$$

$$i = 1, 2, \ldots, k; \quad u = 1, \ldots, m, \ j = 2, 3, \ldots.$$

To derive $P(t)$ for $t \in L(u), u = 1, \ldots, m$, denote by:

$$Q(t) = \left\{ \prod_{j=1}^{u-1} e^{-B_j \tau_j} \right\} e^{-B_u(t-t_{u-1})}, R_i(t) = \left\{ \prod_{j=1}^{u-1} e^{-A_i(j)\tau_j} \right\} e^{-A_i(u)(t-t_{u-1})},$$

and

$$F_i(t) = -\sum_{v=1}^{u-1} \left\{ \prod_{j=1}^{v-1} e^{-B_j \tau_j} \right\} G_i(v;t)$$

$$- \left\{ \prod_{j=1}^{u-1} e^{-B_j \tau_j} \right\} G_i(u; t - t_{u-1}), \quad i = 1, \ldots, k.$$

On multiplying out the matrices in the Eq. (5.16) and simplifying, we obtain for $t \in L_u$:

$$P(t) = \begin{pmatrix} Q(t) & F_1(t) & F_2(t) & \ldots & F_k(t) \\ & R_1(t) & 0 & \ldots & 0 \\ 0 & 0 & R_2(t) & \ldots & 0 \\ \vdots & \ddots & & \vdots & \vdots \\ 0 & & \ldots & 0 & R_k(t) \end{pmatrix}.$$

Since $A_i \underset{\sim}{1}_{n_i} = 0_{n_i}$, we have $E_i(u;j)\underset{\sim}{1}_{n_i} = B_u^{j-1} \underset{\sim}{D}_i(u)$, where $\underset{\sim}{D}_i(u) = D_i(u)\underset{\sim}{1}_{n_i}$, and

$$-G_i(u,t)\underset{\sim}{1}_{n_i} = -\sum_{j=1}^{\infty} \frac{1}{j!}(-t)^j B_u^{j-1}\underset{\sim}{D}_i(u)$$

$$= (I_r - e^{-B_u t})B_u^{-1}\underset{\sim}{D}_i(u), \qquad i = 1,\ldots,k.$$

Using these results and noting $\sum_{i=1}^{k} D_i(u) = B_u \underset{\sim}{1}_r, u = 1,\ldots,m$, we obtain:

(1) For $t \in L_u$ $(u = 1,\ldots,m)$, the vectors of absorption probabilities into the jth closed set and into persistent states of transient states at or before time t are given respectively by:

$$\underset{\sim}{F}_i(t) = F_i(t)\underset{\sim}{1}_{n_i} = \sum_{v=1}^{u-1} \left\{ \prod_{j=1}^{v-1} e^{-B_j \tau_j} \right\} (I_r - e^{-B_v \tau_v})B_v^{-1}\underset{\sim}{D}_i(v)$$

$$+ \left\{ \prod_{j=1}^{u-1} e^{-B_j \tau_j} \right\} [I_r - e^{-B_u(t-t_{u-1})}]B_u^{-1}\underset{\sim}{D}_i(u), \qquad i = 1,\ldots,k.$$

$$\underset{\sim}{F}(t) = \sum_{i=1}^{k} \underset{\sim}{F}_i(t) = \sum_{v=1}^{u-1} \left\{ \prod_{j=1}^{v-1} e^{-B_j \tau_j} \right\} (I_r - e^{-B_v \tau_v})\underset{\sim}{1}_r$$

$$+ \left\{ \prod_{j=1}^{u-1} e^{-B_j \tau_j} \right\} [I_r - e^{-B_u(t-t_{u-1})}]\underset{\sim}{1}_r.$$

(2) The vectors of first time absorption probabilities into the ith closed set and into persistent states of transient states at time t for $t \in L_u$ $(u = 1,\ldots,m)$ are given respectively by:

$$\underset{\sim}{g}_i(t) = \frac{d}{dt}\underset{\sim}{F}_i(t) = \left\{ \prod_{j=1}^{u-1} e^{-B_j \tau_j} \right\} e^{-B_u(t-t_{u-1})}\underset{\sim}{D}_i(u), \quad i = 1,\ldots,k,$$

$$\underset{\sim}{g}(t) = \frac{d}{dt}\underset{\sim}{F}(t) = \sum_{j=1}^{k} \underset{\sim}{g}_j(t) = \left\{ \prod_{j=1}^{u-1} e^{-B_j \tau_j} \right\} e^{-B_u(t-t_{u-1})}B_u\underset{\sim}{1}_r.$$

(3) The vectors of mean absorption times and variances of first time of transient states are given respectively by:

$$\underset{\sim}{U} = \int_0^\infty t\underset{\sim}{g}(t)dt = \sum_{u=1}^{m-1} \left\{ \prod_{j=1}^{u-1} e^{-B_j\tau_j} \right\} \int_{t_{u-1}}^{t_u} xe^{-B_u(x-t_{u-1})} B_u \underset{\sim}{1}_r dx$$

$$+ \left\{ \prod_{j=1}^{m-1} e^{-B_j\tau_j} \right\} \int_{t_{m-1}}^\infty xe^{-B_m(x-t_{m-1})} B_m \underset{\sim}{1}_r dx$$

$$= \sum_{u=1}^{m-1} \left\{ \prod_{j=1}^{u-1} e^{-B_j\tau_j} \right\} \{ B_u^{-1}(I_r - e^{-B_u\tau_u}) + (t_{u-1}I_r - t_u e^{-B_u\tau_u}) \} \underset{\sim}{1}_r$$

$$+ \left\{ \prod_{j=1}^{m-1} e^{-B_j\tau_j} \right\} (t_{m-1}I_r + B_m^{-1}) \underset{\sim}{1}_r = (\mu_1, \ldots, \mu_r)',$$

$$\underset{\sim}{V} = \underset{\sim}{\eta} - (\mu_1^2, \ldots, \mu_r^2)', \text{ where},$$

$$\underset{\sim}{\eta} = \int_0^\infty t^2 \underset{\sim}{g}(t)dt = \sum_{u=1}^{m-1} \left\{ \prod_{j=1}^{u-1} e^{-B_j\tau_j} \right\} \int_{t_{u-1}}^{t_u} x^2 e^{-B_u(x-t_{u-1})} B_u \underset{\sim}{1}_r dx$$

$$+ \left\{ \prod_{j=1}^{m-1} e^{-B_j\tau_j} \right\} \int_{t_{m-1}}^\infty x^2 e^{-B_m(x-t_{m-1})} B_m \underset{\sim}{1}_r dx$$

$$= \sum_{u=1}^{m-1} \left\{ \prod_{j=1}^{u-1} e^{-B_j\tau_j} \right\} \{ 2B_u^{-2}(I_r - e^{-B_u\tau_u}) + t_{u-1}(t_{u-1}I_r + 2B_u^{-1})$$

$$- \tau_u(\tau_u I_r + 2B_u^{-1})e^{-B_u\tau_u} \} \underset{\sim}{1}_r$$

$$+ \left\{ \prod_{j=1}^{m-1} e^{-B_j\tau_j} \right\} [2B_m^{-2} + t_{m-1}(t_{m-1}I_r + 2B_m^{-1})] \underset{\sim}{1}_r.$$

Example 5.7. The HIV incubation distribution under treatment. In this example, we will apply the above theory to derive the HIV incubation distribution under AZT treatment. Thus, as in Examples 4.4 and 4.10, we consider a large population consisting of k HIV infected stages ($I_i, i = 1, \ldots, k$) and AIDS cases ($A = I_{k+1}$). We assume that there are no backward transition and the other transition rates are given by:

Transition	$I_i \to I_{i+1}$	$I_i \to A$	Treatment status
First Interval	$\gamma_i(1)$	$\omega_i(1)$	No treatment
Second Interval	$\gamma_i(2) = \theta\gamma_i(1)$	$\omega_i(2) = \theta\omega_i(1)$	Treated by AZT

The above is a piece-wise non-homogeneous Markov chain with continuous time with $m = 2$. This chain has been studied by Longini *et al.* [3, 4] and by Tan [5]. For this Markov chain, the $A(i)$ matrix is given by:

$$A(i) = \begin{pmatrix} B_i & -\underset{\sim}{w}_i \\ \underset{\sim}{0}'_k & 0 \end{pmatrix},$$

where

$$B_i = \begin{pmatrix} \gamma_1(i) + \omega_1(i) & -\gamma_1(i) & 0 & \cdots & 0 \\ 0 & \gamma_2(i) + \omega_2(i) & -\gamma_2(i) & \cdots & 0 \\ 0 & 0 & \gamma_3(i) + \omega_3(i) & -\gamma_3(i) & \cdots \\ & & & \ddots & \\ & 0 & & & \gamma_k(i) + \omega_k(i) \end{pmatrix},$$

and $\underset{\sim}{w}'_i = \{\omega_1(i), \ldots, \omega_k(i)\}$.

Thus, for an individual who has contracted HIV to become an I_1 person at time 0 and who has been treated by AZT since t_1, the transition matrix $P(t)$ is:

$$P(t)$$

$$= \begin{cases} \begin{pmatrix} e^{-B_1 t} & (I_k - e^{B_1 t})\underset{\sim}{1}_k \\ \underset{\sim}{0}'_k & 1 \end{pmatrix}, & \text{if } t \leq t_1 \\[2em] \begin{pmatrix} e^{-B_1 t_1} \times e^{-B_2(t-t_1)}, & \{(I_k - e^{B_1 t_1}) + e^{-B_1 t_1}(I_k - e^{B_2(t-t_1)})\}\underset{\sim}{1}_k \\ \underset{\sim}{0}'_k, & 1 \end{pmatrix}, & \text{if } t > t_1. \end{cases}$$

For this individual, the vector of pdf of first absorption time of $I_i(i = 1, \ldots, k)$ into A is:

$$
\underset{\sim}{g}(t) = \begin{cases} e^{-B_1 t} B_1 \underset{\sim}{1}_k = e^{-B_1 t} \underset{\sim}{w}_1, & \text{if } t \leq t_1 ; \\[2ex] e^{B_1 t_1} \times e^{-B_2(t-t_1)} B_2 \underset{\sim}{1}_k = e^{B_1 t_1} \times e^{-B_2(t-t_1)} \underset{\sim}{w}_2, & \text{if } t > t_1 . \end{cases}
$$

Hence, the pdf of the HIV incubation distribution is:

$$
f_{\text{inc}}(t) = \begin{cases} \underset{\sim}{e}_1' e^{-B_1 t} \underset{\sim}{w}_1, & \text{if } t \leq t_1 , \\[2ex] \underset{\sim}{e}_1' e^{B_1 t_1} \times e^{-B_2(t-t_1)} \underset{\sim}{w}_2, & \text{if } t > t_1 , \end{cases}
$$

where $\underset{\sim}{e}_1' = (1, 0, \ldots, 0)$.

Now, it is obvious that the eigenvalues of B_i are $\{\lambda_j(i) = \gamma_j(i) + \omega_j(i), j = 1, \ldots, k\}$. If $\lambda_j(i) \neq \lambda_u(i)$ for all $j \neq u$, then as shown in Lemma 5.1, a left eigenvector $\underset{\sim}{q}_j(i)$ and a right eigenvector $\underset{\sim}{p}_j(i)$ of B_i for $\lambda_j(i)$ are given respectively by:

$$
\underset{\sim}{p}_j(i) = [B_{j1}(i), B_{j2}(i), \ldots, B_{jk}(i)]'
$$

and

$$
\underset{\sim}{q}_j(i) = [C_{j1}(i), C_{j2}(i), \ldots, C_{jk}(i)]' ,
$$

where for $j, r = 1, \ldots, k$,

$$
B_{jr}(i) = \begin{cases} \displaystyle\prod_{l=r}^{j-1} \{\gamma_l(i)/[\lambda_l(i) - \lambda_j(i)]\}, & \text{if } r \leq j , \\[3ex] 0, & \text{if } r > j , \end{cases}
$$

and

$$
C_{rj}(i) = \begin{cases} \displaystyle\prod_{l=r+1}^{j} \{\gamma_{l-1}(i)/[\lambda_l(i) - \lambda_r(i)]\}, & \text{if } r \leq j , \\[3ex] 0, & \text{if } r > j , \end{cases}
$$

with $\prod_{l=l+1}^{i}$ being defined as 1.

Put $E_j(i) = \underset{\sim}{p}_j(i)\underset{\sim}{q}'_j(i)$. Then $E_j^2(i) = E_j(i), E_j(i)E_u(i) = 0$ if $j \neq u$ and $\sum_{j=1}^{k} E_j(i) = I_k$. It follows that

$$B_i = \sum_{j=1}^{k} \lambda_j(i)E_j(i), \; e^{-B_i t} = \sum_{j=1}^{k} e^{-\lambda_j(i)t}E_j(i)\,.$$

Denote $A_{uv}^{(i)}(j) = B_{ju}(i)C_{jv}(i)$. Then, by the above definition,

$$A_{uv}^{(i)}(j) = \begin{cases} 0, & \text{if } j > v \text{ or } u > j\,; \\[2mm] \left\{\prod_{l=u}^{v-1} \lambda_l(i)\right\}\left\{\prod_{\substack{l=u \\ l \neq j}}^{v} [\lambda_l(i) - \lambda_j(i)]^{-1}\right\}, & \text{if } u \leq j \leq v\,. \end{cases}$$

Let $h_{uv}^{(i)}(t) = \sum_{j=u}^{v} e^{-\lambda_j(i)t} A_{uv}^{(i)}(j)$ for $1 \leq u \leq v \leq k, i = 1, 2$, then we have:

(1) For $0 \leq t \leq t_1$,

$$f_{\text{inc}}(t) = \underset{\sim}{e}'_1 e^{-B_1 t}\underset{\sim}{w}_1 = \sum_{j=1}^{k} e^{-\lambda_j(1)t}(\underset{\sim}{e}'_1\underset{\sim}{p}_j(1))(\underset{\sim}{q}'_j(1)\underset{\sim}{w}_1)$$

$$= \sum_{j=1}^{k} e^{-\lambda_j(1)t} B_{j1}(1)\left\{\sum_{i=1}^{k} \omega_i(1)C_{ji}(1)\right\}$$

$$= \sum_{j=1}^{k} e^{-\lambda_j(1)t} B_{j1}(1)\left\{\sum_{i=j}^{k} \omega_i(1)C_{ji}(1)\right\}$$

$$= \sum_{i=1}^{k} \omega_i(1) \sum_{j=1}^{i} e^{-\lambda_j(1)t} B_{j1}(1)C_{ji}(1)$$

$$= \sum_{i=1}^{k} \omega_i(1) \sum_{j=1}^{i} e^{-\lambda_j(1)t} A_{1i}^{(1)}(j)$$

$$= \sum_{i=1}^{k} \omega_i(1) h_{1i}^{(1)}(t)\,.$$

(2) For $t > t_1$,

$$f_{\text{inc}}(t) = \underset{\sim}{e}'_1 e^{-B_1 t_1} e^{-B_2(t-t_1)} \underset{\sim}{w}_2 = \sum_{i=1}^{k} e^{-\lambda_i(1)t_1}(\underset{\sim}{e}'_1 \, 1 p_i(1))$$

$$\times \left\{ \sum_{j=1}^{k} e^{-\lambda_j(2)(t-t_1)}(\underset{\sim}{q}'_i(1)\underset{\sim}{p}_j(2))(\underset{\sim}{q}'_j(2)\underset{\sim}{w}_2) \right\}$$

$$= \sum_{i=1}^{k} e^{-\lambda_i(1)t_1} B_{i1}(1) \left\{ \sum_{j=1}^{k} e^{-\lambda_j(2)(t-t_1)} \left[\sum_{u=i}^{k} C_{iu}(1)B_{ju}(2) \right] \right.$$

$$\left. \times \left[\sum_{v=j}^{k} \omega_v(2)C_{jv}(2) \right] \right\} = \sum_{i=1}^{k}\sum_{u=i}^{k} e^{-\lambda_i(1)t_1} B_{i1}(1)C_{iu}(1)$$

$$\times \left\{ \sum_{j=1}^{k}\sum_{v=j}^{k} e^{-\lambda_j(2)(t-t_1)}\omega_v(2)B_{ju}(2)C_{jv}(2) \right\}$$

$$= \sum_{u=1}^{k}\sum_{i=1}^{u} e^{-\lambda_i(1)t_1} A_{1u}^{(1)}(i) \left\{ \sum_{v=u}^{k}\sum_{j=1}^{v} e^{-\lambda_j(2)(t-t_1)}\omega_v(2)A_{uv}^{(2)}(j) \right\}$$

$$= \sum_{v=1}^{k}\omega_v(2) \left\{ \sum_{u=1}^{v}\sum_{i=1}^{u} e^{-\lambda_i(1)t_1} A_{1u}^{(1)}(i) \right\} \left\{ \sum_{j=u}^{v} e^{-\lambda_j(2)(t-t_1)} A_{uv}^{(2)}(j) \right\}$$

$$= \sum_{v=1}^{k}\omega_v(2) \left\{ \sum_{u=1}^{v} h_{1u}^{(1)}(t_1)h_{uv}^{(2)}(t-t_1) \right\} .$$

Lemma 5.1. *Let B be a $k \times k$ upper bi-diagonal matrix with distinct diagonal elements $\lambda_1,\ldots,\lambda_k$ and upper off-diagonal elements $-\gamma_1,\ldots,-\gamma_{k-1}$. Then, $\lambda_1,\ldots,\lambda_k$ are the eigenvalues and a right eigenvector $\underset{\sim}{p}_j$ and a left eigenvector $\underset{\sim}{q}_j$ corresponding to the eigenvalue λ_j are given respectively by*

$$\underset{\sim}{p}_j = [B_{j1}, B_{j2}, \ldots, B_{jk}]'$$

and

$$\underset{\sim}{q}_j = [C_{j1}, C_{j2}, \ldots, C_{jk}]',$$

where for $j, r = 1, \ldots, k$,

$$B_{jr} = \begin{cases} \displaystyle\prod_{l=r}^{j-1} [\gamma_l/(\lambda_l - \lambda_j)], & \text{if } r \leq j, \\ \\ 0, & \text{if } r > j, \end{cases}$$

and

$$C_{rj} = \begin{cases} \displaystyle\prod_{l=r+1}^{j} [\gamma_{l-1}/(\lambda_l - \lambda_r)], & \text{if } r \leq j, \\ \\ 0, & \text{if } r > j, \end{cases}$$

with $\prod_{l=l+1}^{i}$ being defined as 1.

Proof.   We prove only the right eigenvectors since the proof of the left eigenvector is quite similar.

To prove the right eigenvectors, consider the matrix equations:

$$\begin{bmatrix} \lambda_1 & -\gamma_1 & & & \\ & \lambda_2 & -\gamma_2 & & O \\ & & \ddots & \ddots & \\ & O & & \lambda_{k-1} & -\gamma_{k-1} \\ & & & & \lambda_k \end{bmatrix} \begin{bmatrix} x_1 \\ \vdots \\ x_{k-1} \\ x_k \end{bmatrix} = \begin{bmatrix} \lambda_1 x_1 - \gamma_1 x_2 \\ \lambda_2 x_2 - \gamma_2 x_3 \\ \vdots \\ \lambda_{k-1} x_{k-1} - \gamma_{k-1} x_k \\ \lambda_k x_k \end{bmatrix} = \lambda_i \begin{bmatrix} x_1 \\ \vdots \\ x_{k-1} \\ x_k \end{bmatrix},$$

$i = 1, \ldots, k.$

Since $\lambda_i \neq \lambda_j$ for all $i \neq j$, if $i < k$, then $x_{i+1} = \cdots = x_k = 0$. It follows that a right eigenvector for λ_1 is

$$\underset{\sim}{x}' = x_1 [1, 0, \ldots, 0] = x_1 [B_{11}, 0, \ldots, 0],$$

$x_1 \neq 0$; or $\underset{\sim}{p}_1$ is a right eigenvector for λ_1.

Let $1 < i \leq k$. Then the above equations give $\lambda_{i-1} x_{i-1} - \gamma_{i-1} x_i = \lambda_i x_{i-1}$ which yields $x_{i-1} = [\gamma_{i-1}/(\lambda_{i-1} - \lambda_i)] x_i = B_{i,i-1} x_i$; similarly, $\lambda_{i-2} x_{i-2} - \gamma_{i-2} x_{i-1} = \lambda_i x_{i-2}$ yields $x_{i-2} = [\gamma_{i-2}/(\lambda_{i-2} - \lambda_i)] x_{i-1} = [\gamma_{i-2}/(\lambda_{i-2} - \lambda_i)] B_{i,i-1} x_i = B_{i,i-2} x_i$. Suppose now $x_j = B_{ij} x_i$ for $1 < j \leq i - 1$, then $\lambda_{j-1} x_{j-1} - \gamma_{j-1} x_j = \lambda_i x_{j-1}$ so that $x_{j-1} = [\gamma_{j-1}/(\lambda_{j-1} - \lambda_i)] x_j =$

$[\gamma_{j-1}/(\lambda_{j-1} - \lambda_i)]\, B_{ij} x_i = B_{i,j-1} x_i$. By mathematical induction, we have that $x_j = B_{ij} x_i$ for all $j = 1, \ldots, i-1$. Hence, for $1 < i \leq k$, a right eigenvector for λ_i is $\underset{\sim}{x}' = x_i\, [B_{i0}, B_{i1}, \ldots, B_{ik}]$. Thus, $\underset{\sim}{p}_i$ is a right eigenvector of B corresponding to the eigenvalue λ_i, $i = 1, \ldots, k$.

For deriving the HIV incubation distributions under AZT treatment under general conditions, let t_i be the time to start the ith round of treatment and with t_1 being the first time for AZT treatment, $t_0 = 0 < t_1 < \cdots$. (Longini *et al.* [3] assumed t_1=March 1987 since that was the time AZT was made available in the United States.) Let u_j be the probability that each infected person receives AZT treatment at t_j with $q_j = 1 - u_j$. Assume that whenever a person is treated by AZT at t_j, then he/she is always treated by AZT at t for $t \geq t_j$. Let p_j be the probability that each infected person is actually treated at t_j. Then $p_0 = 0, p_1 = u_1, p_2 = p_1 + q_1 u_2, \ldots, p_j = p_{j-1} + \bar{p}_j$, for $j = 1, 2, \ldots, n$, where $\bar{p}_j = u_j \prod_{i=1}^{j-1} q_i$. That is, $\bar{p}_j = p_j - p_{j-1}$ $(j = 1, \ldots,)$ is the probability that the person starts treatment at t_j. Now $\sum_{j=1}^{n} \bar{p}_j = p_n$ and for all integer n, $\sum_{j=1}^{n} \bar{p}_j + (\prod_{i=1}^{n} q_i) = \sum_{j=1}^{n-1} \bar{p}_j + (\prod_{i=1}^{n-1} q_i) = \cdots = \bar{p}_1 + q_1 = 1$; so $(\prod_{i=1}^{n} q_i) = 1 - p_n$.

Using these results, for HIV infected people who contracted HIV at time $t_0 = 0$, the probability density of the HIV incubation distribution is:

(1) If $t < t_1$, then

$$f_{\text{inc}}(t) = \sum_{i=1}^{k} \omega_i(1) h_{1i}^{(1)}(t)\,.$$

(2) If $t_n \leq t < t_{n+1}$ for $n = 1, \ldots,$

$$f_{\text{inc}}(t) = (1 - p_n) \left\{ \sum_{i=1}^{k} \omega_i(1) h_{1i}^{(1)}(t) \right\}$$
$$+ \sum_{i=1}^{n} (p_i - p_{i-1}) \sum_{v=1}^{k} \omega_v(2) \left\{ \sum_{u=1}^{v} h_{1u}^{(1)}(t_i) h_{uv}^{(2)}(t - t_i) \right\}\,.$$

The above results are first given by Longini *et al.* [3].

5.4. Stochastic Differential Equations for Markov Chains with Continuous Time

As in Markov chains with discrete time, in many cases one may use stochastic differential equations to represent and characterize Markov chains with continuous time. As shown in Chap. 8, this will then provide an avenue for making inferences about unknown parameters in Markov chains with continuous time and to validate the model. We now illustrate the basic theories by using some examples.

5.4.1. *The Feller–Arley stochastic birth-death processes*

Consider a Feller–Arley birth-death processes $\{X(t), t \in T = [0, \infty)\}$ with state space $S = (0, 1, \ldots, \infty)$ and with birth rate $b_j(t) = jb(t)$ and death rate $d_j(t) = jd(t)$. To develop a stochastic differential equation representation, let $B(t)$ and $D(t)$ denote the numbers of birth and death during $[t, t + \Delta t)$ respectively. We will show that the above stochastic birth-death process is equivalent to assuming that the conditional probability distribution of $\{B(t), D(t)\}$ given $X(t)$ is a two-dimensional multinomial distribution with parameters $\{X(t), b(t)\Delta t, d(t)\Delta t\}$. That is,

$$\{B(t), D(t)\}|X(t) \sim ML\{X(t); b(t)\Delta t, d(t)\Delta t\}.$$

These distribution results are equivalent to the following proposition.

Proposition 5.2. *X(t) satisfies the following stochastic equation:*

$$X(t + \Delta t) = X(t) + B(t) - D(t). \tag{5.17}$$

To prove this proposition, let $\phi(u, t)$ denote the pgf of $X(t)$. Then, by Eq. (5.17), we obtain:

$$\phi(u, t + \Delta t) = E\{u^{X(t+\Delta t)}\}$$

$$= E\{u^{X(t)}E[u^{B(t)-D(t)}|X(t)]\}$$

$$= E\{u^{X(t)}[1 + (u - 1)b(t)\Delta t + (u^{-1} - 1)d(t)\Delta t]^{X(t)}\}$$

$$= E\{u^{X(t)}[1 + X(t)u^{-1}g(u, t)\Delta t + o(\Delta t)]\}$$

$$= \phi(u, t) + g(u, t)\frac{\partial}{\partial u}\phi(u, t)\Delta t + o(\Delta t),$$

where $g(u,t) = (u-1)ub(t) + (1-u)d(t) = (u-1)[(u-1)b(t) + \gamma(t)]$ with $\gamma(t) = b(t) - d(t)$.

Subtracting $\phi(u,t)$ from both sides of the above equation, dividing by Δt and letting $\Delta t \to 0$, one obtains the following equation for the pgf of $X(t)$:

$$\frac{\partial}{\partial t}\phi(u,t) = g(u,t)\frac{\partial}{\partial u}\phi(u,t).$$

Since the above equation is precisely the same equation for the pgf of $X(t)$ derived by using the Kolmogorov equation as given in Sec. 4.3, the proposition is proved. $\qquad\square$

Let $\epsilon(t)\Delta t = [B(t) - X(t)b(t)\Delta t] - [D(t) - X(t)d(t)\Delta t]$. Then Eq. (5.17) gives:

$$dX(t) = X(t + \Delta t) - X(t) = B(t) - D(t)$$

$$= X(t)\gamma(t)\Delta t + \epsilon(t)\Delta t, \tag{5.18}$$

where $\gamma(t) = b(t) - d(t)$.

Equation (5.18) is the stochastic differential equation for $X(t)$. Let Y_j be the observed number on $X(t)$ at times $t_j, j = 1, \ldots, n$. Then,

$$Y_j = X(t_j) + e_j, j = 1, \ldots, n. \tag{5.19}$$

Combining Eqs. (5.18) and (5.19), we have a state space model for the birth-death process $X(t)$. In this state space model, the stochastic system model is given by Eq. (5.18) whereas the observation model is given by Eq. (5.19).

5.4.2. *The number of initiated cancer tumor cells in the two-stage model of carcinogenesis*

In Subsec. 5.4.1, we have developed a stochastic differential equation representation for Feller–Arley birth-death processes. We have shown that during small time intervals, the stochastic birth and death are equivalent to multinomial distribution. In this section, we will prove that the stochastic birth, death and mutation (immigration) process is also equivalent to a multinomial distribution. Since as shown in Example 4.9, the number of initiated cancer tumor cells $I(t)$ in the two-stage model of carcinogenesis is closely approximated by a stochastic birth-death process with immigration with birth rate $b_j(t) = jb(t) + \lambda(t)$ and with death rate $d_j(t) = jd(t)$, we can therefore develop a stochastic differential equation representation for the number of I cells. To this end, let $B(t)$ and $D(t)$ denote the numbers of birth and death of I cells during $[t, t+\Delta t)$ respectively. Let $M_N(t)$ denote the number of mutations from $N \to I$ during $[t, t + \Delta t)$. Then, $M_N(t)$ is distributed as Poisson with mean $\lambda(t)\Delta t$ and conditional on $X(t)$, the probability distribution of $\{B(t), D(t)\}$ is two-dimensional multinomial with parameters $\{X(t), b(t)\Delta t, d(t)\Delta t\}$ independently of $M_N(t)$.

Using this setup, we now prove that the multinomial distribution results for the birth-death-mutation process is equivalent to the following proposition.

Proposition 5.3. *$X(t)$ satisfies the following stochastic equation:*

$$X(t + \Delta t) = X(t) + M_N(t) + B(t) - D(t)\,. \tag{5.20}$$

The proof of this proposition is almost exactly the same as that of Proposition 5.2 and hence is left as an exercise.

Let $\epsilon(t)\Delta t = [M_N(t) - \lambda(t)\Delta t] + [B(t) - X(t)b(t)\Delta t] - [D(t) - X(t)d(t)\Delta t]$. Then Eq. (5.20) is equivalent to the following stochastic differential equation:

$$dX(t) = X(t + \Delta t) - X(t) = M_N(t) + B(t) - D(t)$$

$$= [\lambda(t) + X(t)\gamma(t)]\Delta t + \epsilon(t)\Delta t\,, \tag{5.21}$$

where $\gamma(t) = b(t) - d(t)$ and $\lambda(t) = N(t)\alpha_0(t)$.

5.4.3. *The number of sensitive and resistant cancer tumor cells under chemotherapy*

In Example 4.5, we have considered the numbers of sensitive (T_1) and resistant (T_2) cancer tumor cells under chemotherapy. Let $T_i(t)(i = 1, 2)$ denote the number of T_i cancer tumor cells at time t. Then $\{T_i(t), i = 1, 2, t \in T = [0, \infty)\}$ is a two-dimensional Markov chain with state space $S = \{(i, j), i, j = 0, 1, \ldots, \infty\}$. By using results from Subsec. 5.4.2, in this section we will develop a stochastic differential equation representation for this process. To this end, assume that the $T_i(i = 1, 2)$ cells follow a Gompertz stochastic birth-death process with birth rate $jb_i(t)$ with $b_i(t) = \eta_i(t)e^{-\delta_i t}(\eta_i(t) \geq 0; \delta_i \geq 0)$ and death rate $jd_i(t)$ with $d_i(t) = \zeta_i e^{-\delta_i t}(\zeta_i \geq 0)$. Assume that the mutation rate from T_i to T_j $(i \neq j; i, j = 1, 2)$ is $\alpha_i(t)$; that is, during $[t, t + \Delta t)$ the probability that one T_i cell will yield one T_i cell and one $T_j(j \neq i)$ cell is $\alpha_i(t)\Delta + o(\Delta t)$. Let $B_i(t)$ and $D_i(t)$ denote the numbers of birth and death of $T_i(i = 1, 2)$

tumor cells during $[t, t + \Delta t)$ respectively. Let $M_i(t)$ denote the number of mutations from $T_i \to T_j (i \neq j)$ during $[t, t + \Delta t)$. Then, the conditional probability distribution of $\{B_i(t), D_i(t), M_i(t)\}$ given $T_i(t)$ is a three-dimensional multinomial distribution with parameters $\{T_i(t), b_i(t)\Delta t, d_i(t)\Delta t, \alpha_i(t)\Delta t\}$ independently of $\{B_j(t), D_j(t), M_j(t)\}$ for $i \neq j$.

Using the above distribution results, we have the following proposition.

Proposition 5.4. *The* $\{T_i(t), \ i = 1, 2\}$ *satisfy the following stochastic equations*:

$$T_1(t + \Delta t) = T_1(t) + M_2(t) + B_1(t) - D_1(t), \tag{5.22}$$

$$T_2(t + \Delta t) = T_2(t) + M_1(t) + B_2(t) - D_2(t). \tag{5.23}$$

The proof of this proposition is almost exactly the same as that of Propositions 5.2 and 5.3 and hence is left as an exercise.

For $i \neq j$ and $i, j = 1, 2$, let

$$\epsilon_i(t)\Delta t = [M_j(t) - T_j(t)\alpha_j(t)\Delta t] + [B_i(t) - T_i(t)b_i(t)\Delta t] - [D_i(t) - T_i(t)d_i(t)\Delta t].$$

Then Eqs. (5.22)–(5.23) are equivalent to the following stochastic differential equations:

$$dT_1(t) = T_1(t + \Delta t) - T_1(t) = M_2(t) + B_1(t) - D_1(t)$$

$$= [T_2(t)\alpha_2(t) + T_1(t)\gamma_1(t)]\Delta t + \epsilon_1(t)\Delta t, \tag{5.24}$$

$$dT_2(t) = T_2(t + \Delta t) - T_2(t) = M_1(t) + B_2(t) - D_2(t)$$

$$= [T_1(t)\alpha_1(t) + T_2(t)\gamma_2(t)]\Delta t + \epsilon_2(t)\Delta t, \tag{5.25}$$

where $\gamma_i(t) = b_i(t) - d_i(t), i = 1, 2$.

5.4.4. *Finite Markov chains with continuous time*

In Example 4.2, we have shown that the Moran's genetic model is a finite stochastic birth and death process which is a special case of Markov chains with continuous time. In Example 4.3, we have shown that the nucleotide substitution model is a finite Markov chain with continuous time. In this section we will develop a stochastic differential equation representation for general finite Markov chains with continuous time. To this end, let $\{X(t), t \in T = [0, \infty)\}$ be a finite Markov chain with state space $S = (1, \ldots, k+1)$. Let the transition rates of this chain be given by $\alpha_{ij}(t)$ for $i \neq j$ and $\alpha_{ii}(t) = \sum_{j \neq i} \alpha_{ij}(t)$. Let $X_i(t)(i = 1, \ldots, k+1)$ be the number of the state i at time t and $Z_{ij}(t)$ $(i, j = 1, \ldots, N)$ the number of state j at time $t + \Delta t$ arising from the state i during $[t, t + \Delta t)$. Then, the conditional probability distribution of $\{Z_{ij}(t), j = 1, \ldots, k+1, j \neq i\}$ given $X_i(t)$ is k-dimensional multinomial with parameters $\{X_i(t), \alpha_{ij}(t)\Delta t, j = 1, \ldots, k+1, j \neq i\}$ independently of $\{Z_{rj}(t), j = 1, \ldots, k+1, j \neq r\}$ for all $i \neq r$. Further, $Z_{ii}(t) = X_i(t) - \sum_{j \neq i} Z_{ij}(t)$ and hence the conditional distribution of $Z_{ii}(t)$ given $X_i(t)$ is binomial with parameters $\{X_i(t), 1 - \alpha_{ii}(t)\Delta t\}$.

Using the above distribution results, we have the following stochastic representation:

$$X_j(t + \Delta t) = \sum_{i=1}^{k+1} Z_{ij}(t) \text{ for } j = 1, \ldots, k+1. \tag{5.26}$$

Let $\epsilon_j(t)\Delta t = \sum_{i \neq j}[Z_{ij}(t) - X_i(t)\alpha_{ij}\Delta t] + \{Z_{jj}(t) - X_j(t)[1 - \alpha_{jj}(t)\Delta t]\}$, $j = 1, \ldots, k+1$.

Then Eq. (5.26) are equivalent to the following set of stochastic differential equations:

$$dX_j(t) = X_j(t + \Delta t) - X_j(t)$$

$$= \left\{ -X_j(t)\alpha_{jj}(t) + \sum_{i \neq j} X_i(t)\alpha_{ij}(t) \right\}\Delta t + \epsilon_j(t)\Delta t,$$

$$= -\left\{ \sum_{i=1}^{k+1} X_i(t)a_{ij}(t) \right\}\Delta t + \epsilon_j(t)\Delta t, \qquad \text{for } j = 1, \ldots, k+1, \tag{5.27}$$

where $a_{ij}(t) = \delta_{ij}\alpha_{ii}(t) + (\delta_{ij} - 1)\alpha_{ij}(t)$ for all $i, j = 1, \ldots, k+1$.

Put $\underset{\sim}{X}(t) = \{X_1(t), \ldots, X_{k+1}(t)\}'$ and $\underset{\sim}{\epsilon}(t) = \{\epsilon_1(t), \ldots, \epsilon_{k+1}(t)\}'$ and let $A'(t) = (a_{ij}(t))$ be the $(k+1) \times (k+1)$ matrix with (i, j)-element being given by $a_{ij}(t)$. Then, in matrix notation, Eq. (5.27) become:

$$d\underset{\sim}{X}(t) = -A(t)\underset{\sim}{X}(t)\Delta t + \underset{\sim}{\epsilon}(t)\Delta t. \tag{5.28}$$

If some data are available from the system, then one may develop a state space model for the system with the above equation as the equation for the stochastic system model. This is the hidden Markov model as defined in Chap. 1.

5.5. Complements and Exercises

Exercise 5.1. Absorption in birth-death processes. Consider a homogeneous birth-death process $\{X(t), t \geq 0\}$ with state space $S = (0, 1, \ldots, \infty)$. Suppose that the birth rates and death rates are given by $\{b_i(t) = \theta(t)b_i > 0, d_i = 0, i = 0, 1, \ldots, \}$, where $b_i > 0$ for all $i = 0, 1, \ldots$, and $\theta(t)$ is a continuous function of t.

(a) Show that given $X(0) = i$ for $i < k$, with probability one the process will eventually reach k as time progresses.

(b) Let $f_i(t)$ be the density that the process will reach $k(i < k)$ during $[t, t + \Delta t)$ given $X(0) = i$. (That is, the probability that given $X(0) = i$, the process will reach k for the first time during $[t, t + \Delta t)$ is $f_i(t)\Delta t + o(\Delta t)$.) Let A be the $k \times k$ matrix given by:

$$
A = \begin{pmatrix}
b_0 & -b_0 & 0 & \cdots & & 0 \\
0 & b_1 & -b_1 & \cdots & & 0 \\
0 & & & & & 0 \\
0 & 0 & \ddots & b_{k-1} & -b_{k-1} \\
0 & 0 & \cdots & 0 & b_k
\end{pmatrix}.
$$

Put $\underset{\sim}{f}(t) = \{f_0(t), f_1(t), \ldots, f_{k-1}(t)\}'$. Show that $\underset{\sim}{f}(t)$ is given by:

$$
\underset{\sim}{f}(t) = \theta(t)e^{-Ag(t)} A \underset{\sim}{1}_k ,
$$

where $g(t) = \int_0^t \theta(x)dx$.

This is the special case of absorption probabilities for birth-death process derived by Tan [4].

Exercise 5.2. Consider a stochastic logistic birth-death process with state space $S = (0, 1, \ldots, M)$ and with bith rate $b_i(t) = ib(1 - i/M)(b > 0)$ and death rate $d_i(t) = id(1 - i/M)(d > 0)$.

(a) Derive the probabilities of ultimate absorption into the states 0 and M respectively.

(b) Derive the vectors of the means and the variances of first absorption times of transient states.

Exercise 5.3. Consider a birth-death process $\{X(t), t \geq 0\}$ with state space $S = (0, 1, \ldots, \infty)$ and with birth rates $b_i(t) = b_i$ and death rates $d_i(t) = d_i$. If $\{b_0 = 0, b_i > 0, d_{i-1} > 0, i = 1, 2, \ldots, \}$, then the state 0 is an absorbing state.

(a) Given $X(0) = i(i > 0)$, let u_i denote the probability of ultimate absorption into 0. Show that the u_i's satisfy the following difference equation with initial condition $u_0 = 1$:

$$u_i = \frac{b_i}{b_i + d_i} u_{i+1} + \frac{d_i}{b_i + d_i} u_{i-1}, \quad \text{or}$$

$$u_{i+1} - u_i = \frac{d_i}{b_i}(u_i - u_{i-1}), \qquad i = 1, 2, \ldots .$$

Hence, show that the $\{u_n, n = 1, \ldots, \}$ are given by

$$u_{m+1} = u_1 + (u_1 - 1) \sum_{i=1}^{m} \left(\prod_{j=1}^{i} \frac{d_j}{b_j} \right)$$

$$= \frac{\sum_{i=m+1}^{\infty} \left(\prod_{j=1}^{i} [d_j/b_j] \right)}{1 + \sum_{i=1}^{\infty} \left(\prod_{j=1}^{i} [d_j/b_j] \right)}, \qquad m \geq 1;$$

$$u_1 = \frac{\sum_{i=1}^{\infty} \left(\prod_{j=1}^{i} [d_j/b_j] \right)}{1 + \sum_{i=1}^{\infty} \left(\prod_{j=1}^{i} [d_j/b_j] \right)}.$$

(b) Assume $\sum_{i=1}^{\infty} \left(\prod_{j=1}^{i} [d_j/b_j] \right) = \infty$ so that with probability one the process will eventually be absored into tyhe state 0. Given $X(0) = i(i > 0)$, let w_i denote the mean time untill absorption. Show that the w_i's satisfy the following difference equation with initial condition $w_0 = 0$:

$$w_i = \frac{1}{b_i + d_i} + \frac{b_i}{b_i + d_i} w_{i+1} + \frac{d_i}{b_i + d_i} w_{i-1}, \qquad \text{or}$$

$$w_i - w_{i+1} = \frac{1}{b_i} + \frac{d_i}{b_i}(w_{i-1} - w_i), \qquad i = 1, \ldots .$$

Denote by

$$\rho_i = \frac{\prod_{j=1}^{i-1} b_j}{\prod_{j=1}^{i} d_j}, \qquad \text{for } i = 1, 2, \ldots .$$

By using results from (b), show that the $w_m(m \geq 1)$'s are given by:

$$w_m = \begin{cases} \infty, & \text{if } \sum_{i=1}^{\infty} \rho_i = \infty\,; \\[3mm] \sum_{i=1}^{\infty} \rho_i + \sum_{r=1}^{m-1} \left(\prod_{k=1}^{r} \frac{d_k}{b_k} \right) \sum_{j=r+1}^{\infty} \rho_j, & \text{if } \sum_{i=1}^{\infty} \rho_i < \infty\,. \end{cases}$$

Exercise 5.4. In Example 5.5, the stationary distribution of the nucleotide substitution was derived by using the spectrual expansion. Prove the result by using Theorem 5.1.

Exercise 5.5. In Example 5.6, the stationary distribution of the Moran's genetic model was derived by using the spectrual expansion. Prove the result by using Theorem 5.1 for the case with mutation but no selection.

Exercise 5.6. Let $\{\underset{\sim}{X}(t) = [X_1(t),\ldots,X_k(t)]', t \geq 0\}$ be a k-dimensional Markov chain with continusous time and with state space $S = \{\underset{\sim}{i} = (i_1,\ldots,i_k)', i_j = 0,1,\ldots,\infty, j = 1,\ldots,k\}$. Let the transition rates from $\underset{\sim}{i}$ to $\underset{\sim}{j} = (j_1,\ldots,j_k)'$ be given by $\alpha(\underset{\sim}{i},\underset{\sim}{j})$ for all $\{\underset{\sim}{i} \in S, \underset{\sim}{j} \in S, \underset{\sim}{i} \neq \underset{\sim}{j}\}$. Assume that $\underset{\sim}{X}(t)$ is irreducible. Then $g(i_1,\ldots,i_k) = g(\underset{\sim}{i})$ is the stationary distribution iff

$$g(\underset{\sim}{j})\alpha(\underset{\sim}{j},\underset{\sim}{j}) = \underset{\underset{\sim}{i} \neq \underset{\sim}{j}}{\sum\sum} g(\underset{\sim}{i})\alpha(\underset{\sim}{i},\underset{\sim}{j})\,.$$

Prove the result for the case $k = 2$.

Exercise 5.7. Consider a two-dimensional continuous-time multiple branching process $\{\underset{\sim}{X}(t), t \geq 0\}$ with progeny distributions $\{q_i(j_1,j_2), j_1,j_2 = 0,1,\ldots,\infty, i = 1,2\}$ and with survival parameters $\{\lambda_i = \lambda, i = 1,2\}$. Then the transition rates $\alpha(i,j;u,v)$ are given in Exercise 4.7. For fixed Δt, let one time unit correspond to Δt. Then, for a Type-1 individual at time t, the probability that this individual will give rise to u Type-1 individual and v Type-2 individual at time $t + \Delta t$ is:

$$\xi_1(u,v) = \begin{cases} \lambda\, q_1(u,v)\Delta t, & \text{if } (u,v) \neq (1,0)\,; \\[2mm] 1 - \lambda[1 - q_1(1,0)]\Delta t, & \text{if } (u,v) = (1,0)\,. \end{cases}$$

Similarly, for a Type-2 individual at time t, the probability that this individual will give rise to u Type-1 individual and v Type-2 individual at time $t + \Delta t$ is:

$$\xi_2(u, v) = \begin{cases} \lambda\, q_2(u, v)\Delta t, & \text{if } (u, v) \neq (0, 1)\,; \\ 1 - \lambda[1 - q_2(0, 1)]\Delta t, & \text{if } (u, v) = (0, 1)\,. \end{cases}$$

Then one may define an embedded two-dimensional multiple branching process $\{\underset{\sim}{Y}(t) = [Y_1(t), Y_2(t)]', t \geq 0\}$. This branching process has discrete time and has progeny distributions $\{\xi_i(u, v), u, v = 0, 1, \ldots, \infty, i = 1, 2\}$.

(a) Show that the pgf of the progeny distribution $\xi_i(u, v)(i = 1, 2)$ of $\underset{\sim}{Y}(t)$ is $h_i(x_1, x_2) = h_i(\underset{\sim}{x}) = x_i + u_i(\underset{\sim}{x})\Delta t$, where $u_i(\underset{\sim}{x}) = \lambda[g_i(\underset{\sim}{x}) - x_i](i = 1, 2)$ and $g_i(\underset{\sim}{x})$ is the pfg of the original progeny distribution $q_i(\underset{\sim}{x})$. Hence, show that the matrix M_ξ of mean progenies of $\underset{\sim}{Y}(t)$ is

$$M_\xi = (1 - \lambda\Delta t)I_2 + \lambda\Delta t\, M\,,$$

where M is the matrix of mean numbers of the original progeny distribution $q_i(\underset{\sim}{x})$.

(b) Show that if $\underset{\sim}{X}(t)$ is non-singular and positive regular, so is the process $\underset{\sim}{Y}(t)$.

(c) Assume that $\underset{\sim}{X}(t)$ is non-singular and positive regular and let γ_0 be the largest positive eigenvalue of M. Let μ_i be the probability of ultimate absorption into 0 of $X_i(t)$ and put $\underset{\sim}{\mu} = (\mu_1, \mu_2)'$. Using Theorem 2.12 in Exercise 2.12 and the above embbeded branching process, show that $\underset{\sim}{\mu} = \underset{\sim}{1}_2$ iff $\gamma_0 \leq 1$.

Exercise 5.8. Consider a large population consisting of two niches. Suppose that each niche contains two types of individuals, say A_1 and A_2 and that the following conditions hold:

(1) The probability density of the survival time of Type-i individuals in each niche is

$$h(t) = \lambda\, e^{-\lambda\, t}, \quad t \geq 0, \lambda > 0\,.$$

(2) When an individual in the ith niche dies, with probability $p_i(j) = e^{-\gamma_i}\gamma_i^{\,j}/j!\ (j = 0, 1, \ldots,)$ it leaves beyond j progenies immediately.

(3) When an individual in the ith niche yields n progenies, the probability that there are n_i ($i = 1, 2, n_1 + n_2 = n$) Type-i progenies is

$$\eta(n_1, n_2) = \binom{n}{n_1} p_{i_1}^{n_1} (1 - p_{i_1})^{n_2}. \tag{5.29}$$

(4) All individuals in the population follow the above probability laws for proliferation independently of one another.

Let $X_i(t)$ ($i = 1, 2$) denote the number of the ith type at time t. Then $\{ \underset{\sim}{X}(t) = [X_1(t), X_2(t)]', t \geq 0 \}$ is a two-dimentsional continuous-time multiple branching process with state space $S = \{(i, j), i, j = 0, 1, \ldots, \infty\}$.

(a) Derive the pgf of the progeny distributions.

(b) By using Theorem 2.12 given in Exercise 2.12 (or Part (c) of Exercise 5.7), derive the necessary and sufficient condition that the population will eventually be extinct as time progresses.

(Note: The model in this exercise is the continuous analog of the model proposed by Pollak [5].)

Exercise 5.9. Consider a homogeneous Markov chain $\{X(t), t \geq 0\}$ with state space $S = (e_i, i = 1, \ldots, k+1)$. Let the transition rate of this chain from $e_i \to e_j$ be given by

$$\alpha(i, j) = \begin{cases} \gamma_i, & \text{if } j = i + 1, \\ 0, & \text{if } j \neq i + 1. \end{cases}$$

Then, starting with $X(0) = e_i (i < k + 1)$, with probability one the chain will eventually be absorbed into the state e_{k+1}. Let $f(t)$ be the density of absorption into e_{k+1} at t given $X(0) = e_1$.

(a) If $\gamma_i = \gamma$, show that $f(t)$ is given by:

$$f(t) = \frac{1}{\Gamma(k)} \gamma^k t^{k-1} e^{-\gamma t}, t \geq 0.$$

(b) If $\gamma_i \neq \gamma_j$ for all $i \neq j$, show that $f(t)$ is given by:

$$f(t) = \left(\prod_{i=1}^{k} \gamma_i \right) \sum_{i=1}^{k} A(i) e^{-\gamma_i t}, \ t \geq 0,$$

where $A(i) = \prod_{j \neq i} (\gamma_j - \gamma_i)^{-1}$.

(c) Using results (a) to derive the HIV incubation distribution in Example 5.7 for the special case $\{\gamma_i(1) = \gamma(1), \omega_i(1) = \omega(1), i = 1, \ldots, k\}$.

Exercise 5.10. Prove Proposition 5.3.

Exercise 5.11. Prove Proposition 5.4.

References

[1] W. Y. Tan, *On the absorption probability and absorption time of finite homogeneous birth-death processes*, Biometrics **32** (1976) 745–752.

[2] G. Satten and Ira M. Jr. Longini, *Markov Chain With Measurement Error: Estimating the 'True' Course of Marker of the Progression of Human Immunodeficiency Virus Disease*, Appl. Statist. **45** (1996) 275–309.

[3] I. M. Longini, R. H. Byers, N. A. Hessol and W. Y. Tan. *Estimation of the stage-specific numbers of HIV infections via a Markov model and backcalculation*, Statistics in Medicine **11** (1992) 831–843.

[4] W. Y. Tan, *On first passage probability distributions in continuous time Markov processes*, Utilitas Mathematica **26** (1984) 89–102.

[5] E. Pollak, *Some effects of two types of migration on the survival of a gene*, Biometrics **28** (1972) 385–400.

Chapter 6

Diffusion Models in Genetics, Cancer and AIDS

In many biomedical systems, the population size is usually very large so that the state variables are closely approximated by continuous variables. If time is also continuous, then one is entertaining a stochastic process with continuous state space and continuous parameter space. As we shall see, many problems in cancer, genetics and AIDS can be considered as stochastic processes which have both continuous state space and continuous parameter space. This is especially true for evolution processes describing changes of frequencies of certain types or genes in large populations. The latter is justified by the observation that evolution is a slow process which takes place over millions of years and that the population size is usually very large. In this chapter we will thus consider a class of stochastic processes with continuous state space and continuous parameter space involving only the first two moments of the changes. In particular we will consider Markov processes with continuous state space and continuous parameter space which have been referred to as diffusion processes. As we shall see, many processes in biomedical systems can be approximated by diffusion processes; this includes stochastic processes describing population growth, stochastic processes of carcinogenesis, some stochastic process in infectious diseases as well as evolutionary processes involving changes of gene frequencies in natural populations.

6.1.   The Transition Probabilities

Let $\{X(t), t \geq 0\}$ be a diffusion process with state space $S = [a, b]$ and with coefficients $\{m(x, t), v(x, t)\}$ as defined in Definition 1.7. For $0 \leq s < t$, let $f(x, y; s, t)$ be the conditional probability density function (pdf) of $X(t)$ at y given $X(s) = x, x \in S$ and put:

$$F(x, y; s, t) = \int_a^y f(x, z; s, t)dz, \quad \text{for } y \geq a.$$

Then the initial conditions become:

$$f(x, y; s, s) = \delta(y - x), \quad \text{the Dirac's delta function},$$

where the Dirac's delta function $\delta(x)$ is defined by $\int_{-\infty}^{\infty} \delta(x)g(x)dx = g(0)$ for any integrable function $g(x)$ and

$$F(x, y; s, s) = H(y - x),$$

where $H(x) = 1$ if $x > 0$ and $H(x) = 0$ if $x \leq 0$.

The Chapman–Kolmogorov equations become: For every $x \in S, y \in S$ and for every $\infty > t \geq r \geq s \geq 0$,

$$f(x, y; s, t) = \int_a^b f(x, z; s, r)f(z, y; r, t)dz, \tag{6.1a}$$

and

$$F(x, y; s, t) = \int_a^b f(x, z; s, r)F(z, y; r, t)dz$$

$$= \int_a^b F(z, y; r, t)d_z F(x, z; s, r), \tag{6.1b}$$

where $d_z F(x, z; s, r) = f(x, z; s, r)dz$.

As in Markov chains, the diffusion process $\{X(t), t \geq 0\}$ is called a homogeneous diffusion process iff $f(x, y; s, t) = f(x, y; t - s)$ (or $F(x, y; s, t) = F(x, y; t - s)$). That is, $f(x, y; s, t)$ (or $F(x, y; s, t)$) depends on the times s and t only through the difference $t - s$ of times. For homogeneous diffusion process, therefore, one may always start with the original time 0. Notice also that for the diffusion process $\{X(t), t \geq 0\}$ to be homogeneous, it is necessary that the coefficients $\{m(x, t) = m(x), v(x, t) = v(x)\}$ must be independent of time t. For homogeneous diffusion processes, the initial conditions then become $f(x, y; 0) = \delta(y - x)$ and $F(x, y; 0) = H(y - x)$. The Chapman–Kolmogorov equations become:

For every $x \in S, y \in S$ and for every $\infty > t \geq r \geq 0$,

$$f(x, y; t) = \int_a^b f(x, z; r) f(z, y; t - r) dz , \qquad (6.2a)$$

and

$$F(x, y; t) = \int_a^b f(x, z; r) F(z, y; t - r) dz$$

$$= \int_a^b F(z, y; t - r) d_z F(x, z; r) . \qquad (6.2b)$$

6.2.   The Kolmogorov Forward Equation

Using (6.1a) or (6.1b) (or (6.2a) or (6.2b) for homogeneous diffusion processes), as in Markov chains with continuous time, one may derive Kolmogorov forward and backward equations for evaluating $f(x, y; s, t)$. These equations are useful also to derive stationary distributions and the absorption probabilities and the mean time to absorptions; see Chap. 7 for detail.

The following theorem shows that the conditional pdf $f(x, y; s, t)$ (or $f(x, y; t - s)$) satisfies the Kolmogorov forward equation (6.3a) (or (6.3b)). This equation is called the *forward* equation because the derivatives are taken with respect to the forward time t.

Theorem 6.1.   *Let* $\{X(t), t \geq 0\}$ *be a diffusion process with state space* $S = [a, b]$ *and with diffusion coefficients* $\{m(x, t), v(x, t)\}$. *Suppose that for all* $y \in [a, b]$ *and for all* t, *both* $\frac{\partial}{\partial y}\{m(y, t) f(x, y; s, t)\}$ *and* $\frac{\partial^2}{\partial y^2}\{v(y, t) f(x, y; s, t)\}$ *exist. Then,* $\frac{\partial}{\partial t} f(x, y; s, t)$ *exists and* $f(x, y; s, t)$ *satisfies the following partial differential equation with initial condition* $f(x, y; s, s) = \delta(y - x)$:

$$\frac{\partial}{\partial t} f(x, y; s, t) = -\frac{\partial}{\partial y}\{m(y, t) f(x, y; s, t)\}$$

$$+ \frac{1}{2}\frac{\partial^2}{\partial y^2}\{v(y, t) f(x, y; s, t)\} , \qquad (6.3a)$$

with initial condition $f(x, y; s, s) = \delta(y - x)$, *the Dirac's delta function.*

If the process is homogeneous, then the above Kolmogorov forward equation reduces to:

$$\frac{\partial}{\partial t} f(x, y; t) = -\frac{\partial}{\partial y}\{m(y)f(x, y; t)\} + \frac{1}{2}\frac{\partial^2}{\partial y^2}\{v(y)f(x, y; t)\}, \qquad (6.3b)$$

with initial condition $f(x, y; 0) = \delta(y - x)$.

(The above equation has also been referred to as *Fokker–Plank* equation.)

We prove the theorem only for the case in which both $m(x, t)$ and $v(x, t)$ are polynomials in x. A proof for general cases is given in Subsec. 6.8.1.

Proof.   To prove the above theorem for the case that both $m(x, t)$ and $v(x, t)$ are polynomials in x, write simply $f(x, y; s, t) = f(y, t)$ when there is no confusion. Let $\phi(u, t)$ be the conditional characteristic function (cf) of $X(t)$ given $X(s) = x$. Then, with $i = \sqrt{-1}$, we have,

$$\phi(u, t) = \int_a^b e^{iuy} f(y, t) dy,$$

and by the Fourier inversion formulae,

$$f(y, t) = \frac{1}{2\pi} \int_{-\infty}^{\infty} e^{-iuy} \phi(u, t) du.$$

(For the proof of the Fourier inversion formulae and general theories, see [1, Chap. 4].)

To prove the above forward equation, we will first show that if $m(y, t)$ and $v(y, t)$ are polynomials in y, then $\phi(u, t)$ satisfies the following partial differential equation:

$$\frac{\partial}{\partial t}\phi(u, t) = (iu)m\left(\frac{\partial}{\partial(iu)}, t\right)\phi(u, t) + \frac{(iu)^2}{2}v\left(\frac{\partial}{\partial(iu)}, t\right)\phi(u, t). \qquad (6.4)$$

In the above equation, $m(\frac{\partial}{\partial(iu)}, t)$ and $v(\frac{\partial}{\partial(iu)}, t)$ are operators operating on $\phi(u, t)$. As an example, if $m(y, t) = b_0 + b_1 y + b_2 y^2$, then

$$m\left(\frac{\partial}{\partial(iu)}, t\right) = b_0 + b_1\frac{\partial}{\partial(iu)} + b_2\frac{\partial^2}{\partial(iu)^2}.$$

To prove Eq. (6.4), notice that by the Chapman–Kolmogorov equation,

$$\phi(u, t+dt) = \int_a^b e^{iuy} f(x, y; s, t+dt) dy$$

$$= \int_a^b e^{iuy} \left\{ \int_a^b f(x, z; s, t) f(z, y; t, t+dt) dz \right\} dy$$

$$= \int_a^b e^{iuz} \left\{ \int_a^b e^{iu(y-z)} f(z, y; t, t+dt) dy \right\} f(x, z; s, t) dz \,.$$

Expanding $e^{iu(y-z)}$ in Taylor series to give

$$e^{iu(y-z)} = 1 + (iu)(y-z) + \frac{(iu)^2}{2}(y-z)^2 + \frac{(iu)^3}{3!}(y-z)^3 + \cdots,$$

we obtain:

$$\int_a^b e^{iu(y-z)} f(z, y; t, t+dt) dy = 1 + (iu)m(z, t)dt + \frac{(iu)^2}{2} v(z, t)dt + o(dt) \,.$$

It follows that

$$\phi(u, t+dt) = \int_a^b e^{iuz} \left\{ \int_a^b e^{iu(y-z)} f(z, y; t, t+dt) dy \right\} f(x, z; s, t) dz$$

$$= \int_a^b e^{iuz} \left\{ 1 + (iu)m(z, t)dt + \frac{(iu)^2}{2} v(z, t)dt + o(dt) \right\}$$

$$\times f(x, z; s, t) dz$$

$$= \phi(u, t) + \left\{ (iu)m \left(\frac{\partial}{\partial(iu)}, t \right) + \frac{(iu)^2}{2} v \left(\frac{\partial}{\partial(iu)}, t \right) \right\}$$

$$\times \phi(u, t)dt + o(dt) \,.$$

From the above equation, on both sides, subtracting $\phi(u, t)$, dividing by dt and then letting $dt \to 0$, we obtain:

$$\lim_{\Delta t \to 0} \frac{1}{\Delta t} \{ \phi(u, t+\Delta t) - \phi(u, t) \}$$

$$= \left\{ (iu)m \left(\frac{\partial}{\partial(iu)}, t \right) \phi(u, t) + \frac{(iu)^2}{2} v \left(\frac{\partial}{\partial(iu)}, t \right) \right\} \phi(u, t) \,.$$

Similarly, by following exactly the same procedure, we obtain:

$$\lim_{\Delta t \to 0} \frac{1}{\Delta t} \{\phi(u,t) - \phi(u, t - \Delta t)\}$$

$$= \left\{ (iu)m\left(\frac{\partial}{\partial(iu)}, t\right)\phi(u,t) + \frac{(iu)^2}{2} v\left(\frac{\partial}{\partial(iu)}, t\right)\right\}\phi(u,t).$$

This shows that $\frac{\partial}{\partial t}\phi(u,t)$ exists and $\phi(u,t)$ satisfies Eq. (6.4) for $\phi(u,t)$. By the Fourier inversion formulae, this also shows that $\frac{\partial}{\partial t}f(x,y;s,t)$ exists.

To prove Eq. (6.3a), on both sides of Eq. (6.4), multiply by $\frac{1}{2\pi}e^{-iuy}$ and then integrate u from $-\infty$ to ∞. Then, by the Fourier inversion formulae, on the left side, we obtain

$$\frac{1}{2\pi}\int_{-\infty}^{\infty} e^{-iuy}\left\{\frac{\partial}{\partial t}\phi(u,t)\right\}du = \frac{\partial}{\partial t}\left\{\frac{1}{2\pi}\int_{-\infty}^{\infty} e^{-iuy}\phi(u,t)du\right\}$$

$$= \frac{\partial}{\partial t}f(x,y;s,t).$$

On the right side, the first term is, by applying results from Lemma 6.1 given below:

$$\frac{1}{2\pi}\int_{a}^{b} e^{-iuy}\left\{(iu)m\left(\frac{\partial}{\partial(iu)}, t\right)\phi(u,t)\right\}du$$

$$= -\frac{\partial}{\partial y}\frac{1}{2\pi}\int_{-\infty}^{\infty} e^{-iuy}\left\{m\left(\frac{\partial}{\partial(iu)}, t\right)\phi(u,t)\right\}du$$

$$= -\frac{\partial}{\partial y}\{m(y,t)f(x,y;s,t)\}.$$

On the right side, the second term is, by applying results from Lemma 6.1 given below:

$$\frac{1}{2\pi}\int_{-\infty}^{\infty} e^{-iuy}\left\{\frac{(iu)^2}{2} v\left(\frac{\partial}{\partial(iu)}, t\right)\phi(u,t)\right\}du$$

$$= \frac{1}{2}\frac{\partial^2}{\partial y^2}\frac{1}{2\pi}\int_{-\infty}^{\infty} e^{-iuy}\left\{v\left(\frac{\partial}{\partial(iu)}, t\right)\phi(u,t)\right\}du$$

$$= \frac{1}{2}\frac{\partial^2}{\partial y^2}\{v(y,t)f(x,y;s,t)\}.$$

This shows that Eq. (6.3a) holds. $\qquad\square$

Lemma 6.1. *Let $P(x)$ be a polynomial in x and let $\phi(u)$ be the cf of the pdf $f(x)$. Then,*

$$\frac{1}{2\pi}\int_{-\infty}^{\infty} e^{-iux} P\left\{\frac{\partial}{\partial(iu)}\right\}\phi(u)du = P(x)f(x)\,.$$

We will prove the result of the above lemma by mathematical induction. Thus, first we assume $P(x) = ax + b$, where a and b are constants. Then, by using integration by part,

$$\frac{1}{2\pi}\int_{-\infty}^{\infty} e^{-iux}\left\{a\frac{\partial}{\partial(iu)} + b\right\}\phi(u)du = bf(x) + a\frac{1}{2\pi i}\int_{-\infty}^{\infty} e^{-iux}d\phi(u)$$

$$= bf(x) + ax\frac{1}{2\pi}\int_{-\infty}^{\infty} e^{-iux}\phi(u)du$$

$$= (b + ax)f(x)\,.$$

Thus, the result holds if $P(x)$ is any polynomial of degree 1. Suppose that the results hold for any polynomial $P_n(x)$ of degree n. We need to show that the results also hold for any polynomial $P(x)$ of degree $n + 1$. To prove this, write $P(x)$ as $P(x) = a_{n+1}x^{n+1} + P_n(x), a_{n+1} \neq 0$, where $P_n(x)$ is a polynomial of degree n. Then,

$$\frac{1}{2\pi}\int_{-\infty}^{\infty} e^{-iux} P\left\{\frac{\partial}{\partial(iu)}\right\}\phi(u)du = P_n(x)f(x) + a_{n+1}\frac{1}{2\pi}$$

$$\times \int_{-\infty}^{\infty} e^{-iux}\frac{\partial^{n+1}}{\partial(iu)^{n+1}}\phi(u)du\,.$$

By integration by part, the second term on the right side of the above equation becomes:

$$a_{n+1}\frac{1}{2\pi}\int_{-\infty}^{\infty} e^{-iux}\frac{\partial^{n+1}}{\partial(iu)^{n+1}}\phi(u)du = a_{n+1}x\frac{1}{2\pi}\int_{-\infty}^{\infty} e^{-iux}\frac{\partial^n}{\partial(iu)^n}\phi(u)du$$

$$= a_{n+1}x^{n+1}f(x)\,.$$

It follows from the above results that if $P(x) = a_{n+1}x^{n+1} + P_n(x)$, then,

$$\frac{1}{2\pi}\int_{-\infty}^{\infty} e^{-iux} P\left\{\frac{\partial}{\partial(iu)}\right\}\phi(u)du = P_n(x)f(x) + a_{n+1}x^{n+1}f(x)$$

$$= P(x)f(x)\,.$$

Thus, the result of Lemma 6.1 holds.

Using Theorem 6.1, one may derive the conditional pdf $f(x, y; s, t)$ by solving the Kolmogorov equation (6.3). This is possible only in some cases whereas it is extremely difficult in many other cases. Given below are some examples in which explicit solution of Eq. (6.3) is possible.

Example 6.1.   The Browning motion process. In 1827, R. Brown, a botanist, noticed that when pollen is dispersed in water, the individual particles were in uninterrupted irregular motion. The stochastic process describing this motion is called the Browning motion process. Let 0 denote the position of the particle initially (i.e. at time 0). Let $\{X(t), t \geq 0\}$ denote the distance between the particle and the original position 0 at time t. Suppose that the following postulates hold:

(a) When the time periods are not overlapping, the displacement of the particle during these periods are independent. This is equivalent to stating that the process $\{X(t), t \geq 0\}$ has independent increment.

(b) The displacement during the period $[s, t)$ depends on time only through the time period $t - s$. That is, $X(t)$ has homogeneous increment.

(c) $E[X(t)|X(0) = 0] = 0$ for all $t \geq 0$.

(d) $\lim_{\Delta t \to 0} \frac{1}{\Delta t} E\{[X(t + \Delta t) - X(t)]^2 | X(0) = 0\} = b(t) > 0$.

(e) $\{X(t), t \geq 0\}$ is a continuous process; see Definition 1.7.

Notice that $P(X(0) = 0) = 1$ and condition (a) implies that $X(t)$ is Markov. Conditions (c)–(e) then imply that $\{X(t), t \geq 0\}$ is a diffusion process with state space $(-\infty, \infty)$ and with coefficients $\{m(x, t) = 0, v(x, t) = b(t)\}$. Let $f(x, t)$ denote the conditional pdf given $X(0) = 0$. Then $f(x, t)$ satisfies the following Kolmogorov forward equation (Fokker–Plank equation):

$$\frac{\partial}{\partial t} f(x, t) = \frac{1}{2} b(t) \frac{\partial^2}{\partial x^2} f(x, t). \tag{6.5}$$

The initial condition is $f(x, 0) = \delta(x)$, the Dirac's delta function.

Suppose that the total probability mass is confined to finite intervals in the real line so that $\lim_{x \to \pm\infty} f(x, t) = 0$ and $\lim_{x \to \pm\infty} \frac{\partial}{\partial x} f(x, t) = 0$. Suppose further that $b(t) = \sigma^2$ is independent of time t so that the process is time homogeneous. Then explicit solution $f(x, t)$ of (6.5) can be derived.

To obtain this solution, let $\phi(u,t)$ denote the cf of $f(x,t)$. That is,

$$\phi(u,t) = \int_{-\infty}^{\infty} e^{iux} f(x,t)dx\,,$$

where $i = \sqrt{-1}$.

Multiplying both sides of (6.5) by e^{iux} and integrating over x from $-\infty$ to ∞, we obtain on the left side:

$$\int_{-\infty}^{\infty} e^{iux} \frac{\partial}{\partial t} f(x,t) = \frac{\partial}{\partial t}\phi(u,t)\,.$$

Now, by using integration by parts, we have:

$$\int_{-\infty}^{\infty} e^{iux} \frac{\partial^2}{\partial x^2} f(x,t)dx = \left[e^{iux}\frac{\partial}{\partial x}f(x,t) \right]_{-\infty}^{\infty} - (iu)\int_{-\infty}^{\infty} e^{iux}\frac{\partial}{\partial x}f(x,t)dx$$

$$= -(iu)\left\{ [e^{iux}f(x,t)]_{-\infty}^{\infty} - (iu)\int_{-\infty}^{\infty} e^{iux}f(x,t)dx \right\}$$

$$= (-u^2)\phi(u,t)\,.$$

Hence, $\phi(u,t)$ satisfies the equation

$$\frac{\partial}{\partial t}\phi(u,t) = -u^2\frac{\sigma^2}{2}\phi(u,t)\,. \tag{6.6}$$

The initial condition is $\phi(u,0) = 1$.

The solution of Eq. (6.6) under the condition $\phi(u,0) = 1$ is

$$\phi(u,t) = \exp\left\{ -\frac{1}{2}u^2\sigma^2 t \right\}\,.$$

The above is the cf of a normal density with mean 0 and variance $\sigma^2 t$. Hence,

$$f(x,t) = \frac{1}{\sqrt{2\pi\sigma^2 t}}e^{-\frac{1}{2t\sigma^2}x^2}\,.$$

Example 6.2. The probability distribution of gene frequency in natural populations under genetic drift. As illustrated in Sec. 6.5, the frequency of genes in natural populations can be closely approximated by diffusion processes with state space $[0, 1]$. In the event that there are no mutations, no selection and no immigration and no migration, the stochastic change of gene frequency is then caused by random chances other than major genetic pressures. This random change has been referred to as *Genetic Drift* by Wright [2]. In this case, the frequency $\{X(t), t \geq 0\}$ is a homogeneous diffusion process with state space $[0, 1]$ and with coefficients $\{m(x, t) = 0, v(x, t) = x(1 - x)\}$; for detail see Sec. 6.5. In this case, the Kolmogorov forward equation is

$$\frac{\partial}{\partial t} f(p, x; t) = \frac{1}{2} \frac{\partial^2}{\partial x^2} \{x(1 - x) f(p, x; t)\}, \tag{6.7}$$

where $f(p, x; t)$ is the conditional pdf of $X(t)$ given $X(0) = p$. The initial condition is $f(p, x; 0) = \delta(x - p)$.

To solve the above partial differential equation, for simplicity we write simply $f(p, x; t) = f(x, t)$ by suppressing p and put $f(x, t) = h(t)g(x)$. On substituting this into Eq. (6.7), we obtain:

$$\frac{h'(t)}{h(t)} = \frac{1}{2g(x)} \left\{ \frac{d^2}{dx^2} [x(1 - x)g(x)] \right\} = -\lambda,$$

for some constant λ.

The above results lead to $h(t) \propto e^{-\lambda t}$ and $g(x)$ satisfies the equation:

$$x(1 - x)g''(t) + (2 - 4x)g'(t) + (2\lambda - 2)g(x) = 0. \tag{6.8}$$

In Eq. (6.8), λ and the associated solution $g(x)$ have been referred to as the eigenvalue and eigenfunction of the operator

$$S = x(1 - x)\frac{d^2}{dx^2} + (2 - 4x)\frac{d}{dx} - 2,$$

respectively. (Some general theories of eigenvalues and eigenfunctions in differential equations are given in Subsec. 6.8.3.)

To find the eigenvalues of Eq. (6.8), consider the series solution $g(x) = \sum_{i=0}^{\infty} a_i x^i$. On substituting this series solution into Eq. (6.8), we obtain:

$$a_{k+1} = \frac{1}{(k + 1)(k + 2)}[(k + 1)(k + 2) - 2\lambda]a_k,$$

for $k = 0, 1, \ldots, \infty$.

Thus, in order that $g(x)$ is finite at $x = 1$, we must have that $\lambda = \frac{1}{2}(k + 1)(k + 2) = \lambda_k$, for some non-negative integer k. Thus, Eq. (6.8) has infinitely many eigenvalues given by $\lambda_k = \frac{1}{2}(k + 1)(k + 2), k = 0, 1, \ldots, \infty$.

The eigenfunction $g_k(x)$ corresponding to the eigenvalue λ_k are solutions of the equation:

$$x(1-x)g_k''(t) + (2-4x)g_k'(t) + k(k+3)g_k(x) = 0. \tag{6.9}$$

Notice that the above equation is a special case of the Jacobi equation with $a = b = 2$; see Subsec. 6.8.2 for some general results of Jacobi polynomials. Hence, aside from a constant multiple, $g_k(x)$ is given by $g_k(x) = J_k(x; 2, 2)$.

It follows that the general solution of Eq. (6.7) is

$$f(x,t) = \sum_{k=0}^{\infty} C_k J_k(x; 2, 2) e^{-\frac{1}{2}(k+1)(k+2)t},$$

where the C_j are constants to be determined by the initial condition $f(x, 0) = \delta(x - p)$ if given $X(0) = p$.

To determine C_j, put $t = 0$, multiply both sides of the general solution by $x(1-x)J_j(x; 2, 2)$ and then integrate both sides with respect to x from $x = 0$ to $x = 1$. Then, on the left side, we have:

$$\int_0^1 x(1-x)J_j(x; 2, 2)\delta(x - p)dx$$

$$= \int_{-p}^{1-p} (z + p)(1 - z - p)J_j(z + p; 2, 2)\delta(z)dz$$

$$= p(1-p)J_j(p; a, b).$$

On the right side, because of the orthogonality of Jacobi polynomials, we have:

$$C_j \int_0^1 x(1-x)J_j^2(x; 2, 2)dx = C_j \frac{(j+1)}{(j+2)(2j+3)}.$$

Hence $C_j = \frac{1}{(j+1)}(j+2)(2j+3)p(1-p)J_j(p; 2, 2)$ and the solution of Eq. (6.7) under the initial condition $f(x, y; 0) = \delta(y - x)$ is

$$f(p, x; t) = p(1-p) \sum_{k=0}^{\infty} \frac{1}{(k+1)}(k+2)(2k+3)$$

$$\times J_k(p; 2, 2)J_k(x; 2, 2)e^{-\frac{1}{2}(k+1)(k+2)t}$$

$$= p(1-p) \sum_{j=1}^{\infty} \frac{1}{j(j+1)}(2j+1)T_j(p)T_j(x)e^{-\frac{1}{2}j(j+1)t}, \tag{6.10}$$

where $T_j(x) = T_j(x; 2, 2) = (j + 1)J_{j-1}(x; 2, 2)$ $(j = 1, \ldots, \infty)$ is the Gagenbuaer polynomial.

The above solution was first derived by Kimura [3].

6.3. The Kolmogorov Backward Equation

As in Markov chain with continuous time, the conditional pdf $f(x, y; s, t)$ of the diffusion process $\{X(t), t \geq 0\}$ also satisfies the Kolmogorov backward equation. In fact, we can prove a more general result of backward equation for the transformation $u(x; s, t)$ of $f(x, y; s, t)$. By using this general form, one can show that not only the conditional pdf $f(x, y; s, t)$ but also the absorption probabilities and the moments satisfy the Kolmogorov backward equation.

To illustrate, let $\{X(t), t \geq 0\}$ be a diffusion process with state space $[a, b]$ and with coefficients $\{m(x, t), v(x, t)\}$. Define $u(x; s, t)$ by:

$$u(x; s, t) = \int_a^b f(x, z; s, t)g(z)dz \qquad (6.11)$$

where $g(y)$ is an integrable function defined over $[a, b]$.

Then, by properly choosing $g(x)$, we can attach different meaning to $u(x; s, t)$. For example, if we chose $g(x) = \delta(x - y)$, then $u(x; s, t) = f(x, y; s, t)$ is the conditional density of $X(t)$ at y given $X(s) = x$; if we chose $g(x) = x^k$, then $u(x; s, t)$ is the kth conditional moment of $X(t)$ given $X(s) = x$; if we chose $g(x) = 1$ if $a \leq c \leq x \leq d \leq b$ and $= 0$ for otherwise, then $u(x; s, t)$ is the probability that $c \leq X(t) \leq d$ given $X(s) = x$; if the boundary points a and b are absorbing states while all states x with $a < x < b$ are transient states and if we put $g(x) = \delta(x - a) + \delta(b - x)$, then $u(x; s, t)$ is the probability that $X(t)$ absorbs into a or b at or before time t given $X(s) = x$.

Theorem 6.2.   (The Kolmogorov backward equation for $u(x; s, t)$).
The $u(x; s, t)$ as defined above satisfies the following backward equation:

$$-\frac{\partial}{\partial s}u(x; s, t) = m(x, s)\frac{\partial}{\partial x}u(x; s, t) + \frac{1}{2}v(x, s)\frac{\partial^2}{\partial x^2}u(x; s, t), \qquad (6.12a)$$

with initial condition $u(x; s, s) = g(x)$. (Note that $f(x, y; s, s) = \delta(y - x)$.)

If the diffusion process $\{X(t), t \geq 0\}$ is time homogeneous, then $u(x; s, t) = u(x; t - s)$ and the above Kolmogorov backward equation reduces to:

$$\frac{\partial}{\partial t} u(x; t) = m(x) \frac{\partial}{\partial x} u(x; t) + \frac{1}{2} v(x) \frac{\partial^2}{\partial x^2} u(x; t) \,, \tag{6.12b}$$

with initial condition $u(x; 0) = g(x)$.

Proof. By the Chapman–Kolmogorov equation, we have

$$f(x, y; s - \Delta s, t) = \int_a^b f(x, z; s - \Delta s, s) f(z, y; s, t) dz \,.$$

Hence

$$u(x; s - \Delta s, t) = \int_a^b \left\{ \int_a^b f(x, z; s - \Delta s, s) f(z, y; s, t) dz \right\} g(y) dy$$

$$= \int_a^b f(x, z; s - \Delta s, s) \left\{ \int_a^b f(z, y; s, t) g(y) dy \right\} dz$$

$$= \int_a^b f(x, z; s - \Delta s, s) u(z; s, t) dz \,.$$

Expanding $u(z; s, t) = \int_a^b f(z, y; s, t) g(y) dy$ in Taylor series around x to give

$$u(z; s, t) = u(x; s, t) + (z - x) \frac{\partial}{\partial x} u(x; s, t) + \frac{1}{2}(z - x)^2 \frac{\partial^2}{\partial x^2} u(x; s, t) + \cdots \,,$$

we obtain:

$$u(x; s - \Delta s, t) = \int_a^b f(x, z; s - \Delta s, s) u(z; s, t) dz = u(x; s, t)$$

$$+ \left\{ \int_a^b (z - x) f(x, z; s - \Delta s, s) dz \right\} \frac{\partial}{\partial x} u(x; s, t)$$

$$+ \left\{ \frac{1}{2} \int_a^b (z - x)^2 f(x, z; s - \Delta s, s) dz \right\} \frac{\partial^2}{\partial x^2} u(x; s, t)$$

$$+ o(\Delta s) = u(x; s, t) + m(x, s) \Delta s \frac{\partial}{\partial x} u(x; s, t)$$

$$+ \frac{1}{2} v(x, s) \Delta s \frac{\partial^2}{\partial x^2} u(x; s, t) + o(\Delta s) \,. \tag{6.13}$$

From both sides of Eq. (6.13), subtracting $u(x; s, t)$, dividing by Δs and letting $\Delta s \to 0$, we obtain:

$$- \lim_{\Delta s \to 0} \frac{1}{\Delta s} \{ u(x; s, t) - u(x; s - \Delta s, t) \}$$

$$= m(x, s) \frac{\partial}{\partial x} u(x; s, t) + \frac{1}{2} v(x, s) \frac{\partial^2}{\partial x^2} u(x; s, t). \tag{6.14}$$

Similarly, by using exactly the same procedure, we obtain:

$$- \lim_{\Delta s \to 0} \frac{1}{\Delta s} \{ u(x; s + \Delta s, t) - u(x; s, t) \}$$

$$= m(x, s) \frac{\partial}{\partial x} u(x; s, t) + \frac{1}{2} v(x, s) \frac{\partial^2}{\partial x^2} u(x; s, t). \tag{6.15}$$

Combining Eqs. (6.14) and (6.15), we prove that $\frac{\partial}{\partial s} u(x; s, t)$ exists and that $u(x; s, t)$ satisfies the backward equation as given by Eq. (6.12a). Because $f(x, y; s, s) = \delta(y - x)$, the initial condition is $u(x; s, s) = g(x)$. $\qquad \square$

In diffusion processes, the backward equation as given in Theorem 6.2 is useful in many ways: First, theoretically it can be used to derive the probability density functions. (Explicit solution is possible only in some cases; in many cases, however, the explicit solution is extremely difficult to derive even if it is not impossible.) Second, it can be used to derive the probabilities of fixation and absorptions into absorbing states; see Chap. 7. Third, it can be used to derive the mean of the first passage times and the mean so-join times; see Chap. 7. In this section we give some examples for which explicit solution is possible.

Example 6.3. The solution of backward equation for the probability distribution of gene frequency under genetic drift in natural populations. In Example 6.2, we have considered the problem of genetic drift under the Wright model. In this case, the stochastic process $\{X(t), t \geq 0\}$ is a homogeneous diffusion process with state space $[0, 1]$ and with coefficients $\{m(y, t) = 0, v(y, t) = y(1 - y)\}$. Hence, the Kolmogorov backward equation for the conditional pdf $f(x, y; t)$ is

$$\frac{\partial}{\partial t} f(x, y; t) = x(1 - x) \frac{1}{2} \frac{\partial^2}{\partial x^2} f(x, y; t). \tag{6.16}$$

The initial condition is $f(x, y; 0) = \delta(y - x)$.

To solve the above equation, for simplicity we write $f(x, y; t) = f(x, t)$ by suppressing y and let $f(x; t) = g(x)e^{-\lambda t}$, where λ is a constant. Then $g(x)$ satisfies the equation:

$$x(1 - x)\frac{d^2}{dx^2}g(x) + 2\lambda\, g(x) = 0 \,. \tag{6.17}$$

In the above equation, the λ's (say λ_j) satisfying the above equation are the eigenvalues of the operator $S = \frac{1}{2}x(1-x)\frac{d^2}{dx^2}$ and the solutions $g_j(x)$ associated with λ_j are the eigenfunctions corresponding to the eigenvalue λ_j. The general solution of the above equation is then given by:

$$f(x, t) = \sum_j C_j g_j(x)e^{-\lambda_j t} \,. \tag{6.18}$$

To find these eigenvalues and the associated eigenfunctions, consider the series solution $g(x) = \sum_{i=0}^{\infty} a_i x^i$. On substituting this series solution into Eq. (6.17) and equating to zero the coefficient of x^k for $k = 0, 1, \ldots$, we obtain $a_0 = 0$ and for $k = 0, 1, \ldots, \infty$,

$$a_{k+1} = \frac{1}{(k + 1)k}[k(k - 1) - 2\lambda]a_k \,.$$

It follows that $a_1 = 0$ and we can assume $a_2 = 1$. Further, in order that the solution is finite at $x = 1$, we require $2\lambda - k(k - 1) = 0$, or $\lambda_k = \frac{1}{2}k(k - 1)$, $k = 2, \ldots, \infty$. We also require that for all $j, k = 2, \ldots, \infty$,

$$\int_0^1 [x(1 - x)]^{-1}g_j(x)g_k(x)dx$$

is finite; see Subsec. 6.8.3. Under these restrictions, the eigenvalues of the operator S are given by:

$$\lambda_k = \frac{1}{2}k(k - 1) \quad \text{for } k = 2, 3, \ldots \,.$$

Given $\lambda_k = \frac{1}{2}k(k - 1)$ $(k \geq 2)$, write a_j as $a_j^{(k)}$. Then, $a_2^{(k)} = 1$ and $a_j^{(k)} = 0$ for all $j = k + 1, \ldots, \infty$; moreover, for $j \leq k, k = 3, \ldots, \infty$, the $a_j^{(k)}$'s are derived from the iterative equation:

$$a_{j+1}^{(k)} = \frac{1}{(j + 1)j}[j(j - 1) - k(k - 1)]a_j^{(k)} \,, \tag{6.19}$$

for $j = 2, \ldots, k - 1$ with $a_2^{(k)} = 1$.

An eigenfunction corresponding to λ_k is $g_k(x) = \sum_{i=1}^{k} a_i^{(k)} x^i$. Using Eq. (6.19) iteratively with $a_2^{(k)} = 1, k = 2, \ldots, \infty$, we obtain:

$$g_2(x) = x(1-x) = x(1-x)J_0(x; 2, 2),$$

$$g_3(x) = 2x(1-x)(1-2x) = x(1-x)J_1(x; 2, 2),$$

$$g_4(x) = x(1-x)(1-5x+5x^2) = \frac{1}{3}x(1-x)J_2(x; 2, 2), \ldots .$$

The general solution of $f(x; t) = f(x, y; t)$ is

$$f(x, y; t) = \sum_{i=2}^{\infty} C_i g_i(x) e^{-\frac{1}{2}i(i-1)t}.$$

Now the operator S defined by $S = \frac{1}{2}x(1-x)\frac{d^2}{dx^2}$ satisfies the Green formulae given by:

$$[Sg(x)]f(x) - [Sf(x)]g(x) = \frac{1}{2}x(1-x)\frac{d}{dx}\{g'(x)f(x) - f'(x)g(x)\}.$$

(See Subsec. 6.8.3 for the Green Formulae and its applications.)

It follows that for all solutions $g(x)$ which vanish at $x = 0$ or $x = 1$, S is self adjoint. In these cases, as shown in Subsec. 6.8.3, for all $\lambda_j \neq \lambda_k, j, k = 2, \ldots, \infty$,

$$\int_0^1 [x(1-x)]^{-1} g_j(x)g_k(x)dx = 0.$$

Since $g_2(x) = x(1-x)J_0(x; 2, 2)$ and $g_3(x) = x(1-x)J_1(x; 2, 2)$, by the orthogonality of Jacobi polynomials $J(x; 2, 2)$, we have for $j = 2, \ldots,$

$$\frac{g_j(x)}{x(1-x)} = b_j J_{j-2}(x; 2, 2), \quad b_j \text{ being a constant}.$$

On substituting these results into Eq. (6.18), the solution $f(x, y; t)$ then becomes:

$$f(x, y; t) = \sum_{i=2}^{\infty} D_{i-2}x(1-x)J_{i-2}(x; 2, 2)e^{-\frac{1}{2}i(i-1)t}$$

$$= \sum_{i=0}^{\infty} D_i x(1-x)J_i(x; 2, 2)e^{-\frac{1}{2}(i+1)(i+2)t}, \qquad (6.20)$$

where the D_i's are constants to be determined by the initial condition $f(x, y; 0) = \delta(y - x)$.

Multiplying both sides of Eq. (6.20) by $J_k(x; 2, 2)$, putting $t = 0$ and integrating the function from 0 to 1, we obtain:

$$J_k(y; 2, 2) = D_k \int_0^1 x(1 - x) J_k^2(x; 2, 2) dx$$

$$= D_k \binom{2k + 2}{k} B(k + 2, k + 2)$$

$$= D_k \frac{k + 1}{(k + 2)(2k + 3)} .$$

Hence $D_k = (k + 2)(2k + 3) J_k(y; 2, 2)/(k + 1)$ so that

$$f(x, y; t) = x(1 - x) \sum_{i=0}^{\infty} \frac{(i + 2)(2i + 3)}{i + 1} J_i(x; 2, 2) J_i(y; 2, 2) e^{-\frac{1}{2}(i+1)(i+2)t}$$

$$= x(1 - x) \sum_{i=1}^{\infty} \frac{(2i + 1)}{i(i + 1)} T_i(x) T_i(y) e^{-\frac{1}{2}i(i+1)t} , \tag{6.21}$$

where $T_k(x) = T_k(x; 2, 2) = (k + 1) J_{k-1}(x; 2, 2)$ is the Gagenbauer polynomial.

Notice that the solution in (6.21) is exactly the same as that given by (6.10) for the Kolmogorov forward equation as it should be; see Remark 6.1.

Remark 6.1. It can be shown that under the conditions given in Theorems 6.1 and 6.2, the solution of the forward equation is the same as the solution of the backward equation and is unique. The proof is quite complicated and hence will not be given here.

Example 6.4. The solution of backward equation for the distribution of gene frequency under mutation and genetic drift in natural populations. Consider a large diploid population with one locus and two alleles A and a. Let $Y(t)$ be the frequency of gene A at time t. Assuming that there are gene mutations from $A \to a$ and from $a \to A$ with rate $\beta_1 dt$ and $\beta_2 dt$ during $[t, t + dt)$ respectively. Assume that the β_i's are independent of time. Then, under the Wright model, $\{Y(t), t \geq 0\}$ is a homogeneous diffusion process with state space $S = [0, 1]$ and with coefficients $\{m(y, t) = \beta_2 - (\beta_1 + \beta_2)y, v(y, t) = y(1 - y)\}$; for detail see Example 6.9 in Sec. 6.5. It follows that the conditional probability density $f(x, y; s, t) = f(x, y; t - s)$ of $Y(t)$ given $Y(s) = x$ satisfies the following backward equation: (We suppress y by writing $f(x, y; t)$ as $f(x, y; t) = f(x; t)$ when there is no confusion.)

$$\frac{\partial}{\partial t} f(x; t) = [\beta_2 - (\beta_1 + \beta_2)x] \frac{\partial}{\partial x} f(x; t) + \frac{1}{2} x(1 - x) \frac{\partial^2}{\partial x^2} f(x; t), \qquad (6.22)$$

with initial condition $f(x; 0) = \delta(y - x)$. (The Dirac's δ function.)

To solve the above equation, put $f(x; t) = h(t)g(x)$. Then

$$\frac{h'(t)}{h(t)} = \frac{1}{g(x)} [\beta_2 - (\beta_1 + \beta_2)x] \frac{d}{dx} g(x) + \frac{1}{2} x(1 - x) \frac{d^2}{dx^2} g(x) = -\lambda$$

for some constant λ.

This leads to $h(t) \propto e^{-\lambda t}$. Further, $g(x)$ satisfies the equation:

$$x(1 - x) \frac{d^2}{dx^2} g(x) + 2[\beta_2 - (\beta_1 + \beta_2)x] \frac{d}{dx} g(x) + 2\lambda\, g(x) = 0. \qquad (6.23)$$

Let $a = 2\beta_2$ and $b = 2\beta_1$ and consider the series solution of the above equation $g(x) = \sum_{i=0}^{\infty} a_i x^i$. On substituting this series solution into Eq. (6.23), we obtain:

$$a_{k+1} = \frac{1}{(k + 1)(k + a)} \{k(k + a + b - 1) - 2\lambda\} a_k,$$

for $k = 0, 1, \ldots$.

Set $a_0 = 1$. Under the condition that the solutions of Eq. (6.23) are finite at $x = 1$ and $x = 0$, the eigenvalues are then given by $\lambda_k = \frac{1}{2} k(k + a + b - 1)$, $k = 0, 1, \ldots$. The eigenfunction $g_k(x)$ corresponding to λ_k satisfies the equation:

$$x(1 - x) \frac{d^2}{dx^2} g_k(x) + [a - (a + b)x] \frac{d}{dx} g_k(x)$$

$$+ k(k + a + b - 1) g_k(x) = 0. \qquad (6.24)$$

Notice that the above equation is the equation for the Jacobi polynomial $J_k(x; a, b)$. Hence, aside from a constant multiple, $g_k(x) = J_k(x; a, b)$. The general solution of Eq. (6.22) is then given by:

$$f(x, y; t) = \sum_{i=0}^{\infty} C_i J_i(x; a, b) e^{-\frac{1}{2} i(i + a + b - 1)t}, \qquad (6.25)$$

where the C_j's are constants to be determined by the initial condition $f(x, y; 0) = \delta(y - x)$.

In the above equation, put $t = 0$, multiply both sides by $x^{a-1}(1 - x)^{b-1} J_k(x; a, b)$ and then integrate out x from 0 to 1. This gives

$$y^{a-1}(1 - y)^{b-1} J_k(y; a, b) = C_k \binom{2k + a + b - 2}{k} B(k + a, k + b).$$

It follows that the solution of Eq. (6.22) is

$$f(x, y; t) = y^{a-1}(1 - y)^{b-1} \sum_{k=0}^{\infty} \frac{1}{\binom{2k+a+b-2}{k} B(k + a, k + b)}$$

$$\times J_k(x; a, b) J_k(y; a, b) e^{-\frac{1}{2}k(k+a+b-1)t}. \tag{6.26}$$

In the above solution, notice that the first term is the Beta density given by $f_0(y) = \frac{1}{B(a,b)} y^{a-1}(1 - y)^{b-1}$ which is the limit of $f(x, y; t)$ as $t \to \infty$.

6.4. Diffusion Approximation of Models from Genetics, Cancers and AIDS

In this section we will show that most of the systems in cancer and AIDS can be approximated by diffusion processes. Similarly, most of the evolutionary processes can be approximated by diffusion processes.

Example 6.5. Approximation of stochastic growth by diffusion process. Consider the growth of a biological population such as the population of bacteria. Let $X(t)$ be the number of individuals in the population at time t and let M be the maximum population size. Then, realistically, $\{X(t), t \geq 0\}$ is a stochastic process with state space $S = \{0, 1, 2, \ldots, M\}$. This follows from the fact that because of the changing environment which involve many random variables, the population growth involves many variables which are subjected to random variations. To find the probability law for $X(t)$, observe that M is usually very large so that $Y(t) = \frac{X(t)}{M}$ changes continuously in $[0, 1]$ approximately. This suggests that one may approximate $Y(t)$ by a continuous process with $Y(t)$ changing continuously in the interval $[0, 1]$.

To illustrate how to approximate $\{Y(t), t > 0\}$ by a diffusion process, suppose that $Y(t)$ can be described by the following stochastic differential equation:

$$\frac{d}{dy}Y(t) = \epsilon Y(t)g[Y(t)] + u(t)\,, \tag{6.27a}$$

where $\epsilon > 0, g(x)$ is an ordinary function of x and $u(t)$ a random variable so that $\{Y(t), t \geq 0\}$ is a stochastic process. Or, to cover cases in which $\frac{dY(t)}{dt}$ does not exist, we consider alternatively the stochastic equation:

$$dY(t) = \epsilon Y(t)g[Y(t)]dt + Z(t)\sqrt{dt}\,, \tag{6.27b}$$

where $Z(t) = u(t)\sqrt{dt}$.

The above Eqs. (6.27a) and (6.27b) describe the mechanism by means of which the population grows. In Eqs. (6.27a) and (6.27b), the first term for given $Y(t)$ is the systematic change which is often used by scientists to describe the population growth deterministically while the second term $u(t)$ and $Z(t)\sqrt{dt}$ are the random noises which are assumed to have zero expectation; the random noises are introduced to take into account the effects of random disturbances for the increment during $[t, t + dt)$. For example, if $g(x) = 1 - x$ and if $u(t) = 0$, then Eq. (6.27b) gives a two-parameter logistic growth function as defined in [4, 5].

To find the probability law for $Y(t)$, assume that $E[Z(t)|Y(t) = y] = 0$, $\text{Var}[Z(t)|Y(t) = y] = \sigma^2(y, t) > 0$ and $E\{|Z(t)|^r|Y(t) = y\} = o(dt, y)$ for $r \geq 3$, where $\lim_{dt \to 0} \frac{o(dt,y)}{dt} = 0$ uniformly for $0 < y < 1$ and for $t \geq 0$; also we assume that the conditional pdf $f(y, t)$ of $Y(t)$ given $Y(t_0) = y_0$ satisfies $\lim_{y \to 0} f(y, t) = \lim_{y \to 1} f(y, t) = 0$ for all $t \geq 0$.

Given these conditions, we proceed to show that the stochastic process $\{Y(t), t \geq 0\}$ is in fact a diffusion process with state space $[0, 1]$ and coefficients $m(y, t) = \epsilon yg(y)$ and $v(y, t) = \sigma^2(y, t)$.

To prove this, notice that with $dY(t) = Y(t + \Delta t) - Y(t)$, we have:

$$\psi(y, t; \Delta t) = E\{\exp[-\theta dY(t)]|Y(t) = y\}$$

$$= 1 - \theta\epsilon yg(y)\Delta t + \frac{1}{2}\theta^2\sigma^2(y, t)\Delta t + o(\Delta t, y)\,,$$

where $\lim_{\Delta t \to 0} \frac{o(\Delta t, y)}{\Delta t} = 0$ uniformly for $0 \leq y \leq 1$ and for $t \geq 0$. For real $\theta > 0$, put:

$$\phi(\theta, t) = E\{\exp[-\theta Y(t)]|Y(t_0) = y_0\} = \int_0^1 e^{-\theta y} f(y, t) dy \,.$$

Then,

$$\frac{\partial}{\partial t} \phi(\theta, t) = \int_0^1 e^{-\theta y} \left[\frac{\partial f(y, t)}{\partial t}\right] dy = \lim_{\Delta t \to 0} \frac{1}{\Delta t} \{\phi(\theta, t + \Delta t) - \phi(\theta, t)\}$$

$$= \lim_{\Delta t \to 0} \frac{1}{\Delta t} E\{[\exp(-\theta Y(t + \Delta t)) - \exp(-\theta Y(t))]|Y(t_0) = y_0\}$$

$$= \lim_{\Delta t \to 0} \frac{1}{\Delta t} \int_0^1 e^{-\theta y} \{\psi(y, t; \Delta t) - 1\} f(y, t) dy$$

$$= \lim_{\Delta t \to 0} \frac{1}{\Delta t} \int_0^1 e^{-\theta y} \left\{-\theta \epsilon y g(y) \Delta t + \frac{1}{2} \theta^2 \sigma^2(y, t) \Delta t + o(\Delta t)\right\} f(y, t) dy$$

$$= \int_0^1 e^{-\theta y} \left\{-\theta \epsilon y g(y) + \frac{1}{2} \theta^2 \sigma^2(y, t)\right\} f(y, t) dy \,.$$

Now integration by parts gives:

$$\int_0^1 e^{-\theta y} [-\theta y g(y)] f(y, t) dy = \int_0^1 y g(y) f(y, t) de^{-\theta y}$$

$$= -\int_0^1 e^{-\theta y} \left\{\frac{\partial}{\partial y} [y g(y) f(y, t)]\right\} dy \,;$$

$$\int_0^1 e^{-\theta y} \frac{1}{2} \theta^2 \sigma^2(y, t) f(y, t) dy = -\frac{1}{2} \int_0^1 \theta \sigma^2(y, t) f(y, t) de^{-\theta y}$$

$$= \frac{1}{2} \int_0^1 e^{-\theta y} \theta \left\{\frac{\partial}{\partial y} [\sigma^2(y, t) f(y, t)]\right\} dy$$

$$= -\frac{1}{2} \int_0^1 \left\{\frac{\partial}{\partial y} [\sigma^2(y, t) f(y, t)]\right\} de^{-\theta y}$$

$$= \frac{1}{2} \int_0^1 e^{-\theta y} \left\{\frac{\partial^2}{\partial y^2} [\sigma^2(y, t) f(y, t)]\right\} dy \,.$$

It follows that

$$\int_0^1 e^{-\theta y}\left\{\frac{\partial}{\partial t}f(y,t)\right\}dy$$

$$= \int_0^1 e^{-\theta y}\left\{-\frac{\partial}{\partial y}[\epsilon y g(y)f(y,t)] + \frac{1}{2}\frac{\partial^2}{\partial y^2}[\sigma^2(y,t)f(y,t)]\right\}dy$$

or

$$\int_0^1 e^{-\theta y}\left\{\frac{\partial}{\partial t}f(y,t) + \frac{\partial}{\partial y}[\epsilon y g(y)f(y,t)] - \frac{1}{2}\frac{\partial^2}{\partial y^2}[\sigma^2(y,t)f(y,t)]\right\}dy = 0.$$

This holds for all real $\theta > 0$. Hence,

$$\frac{\partial}{\partial t}f(y,t) = -\frac{\partial}{\partial y}[\epsilon y g(y)f(y,t)] + \frac{1}{2}\frac{\partial^2}{\partial y^2}[\sigma^2(y,t)f(y,t)]. \tag{6.28}$$

Obviously, the initial condition is $f(y,t_0) = \delta(y-y_0)$, the Dirac's δ-function. This shows that $\{Y(t), t > 0\}$ is a diffusion process with state space $[0, 1]$ and coefficients $\epsilon y g(y)$ and $\sigma^2(y,t)$. If $\sigma^2(y,t) = \sigma^2(t)y(1-y)$ and if $g(y) = 1-y$ so that the above is a two-parameter stochastic logistic growth process, the above approximation was first obtained in [4, 5] by using an alternative approach.

Example 6.6. Diffusion approximation of branching process. Let $\{X(t), t \in T = (0,1,\ldots,\infty)\}$ be a simple branching process with progeny distribution $\{p_j, j = 0,\ldots,\infty\}$. (To avoid trivial cases, we will assume that $0 < p_0, p_1 < 1$ because if $p_0 = 0$, then the mutant gene is certainly to be lost in the next generation; if $p_0 = 1$, then the mutant gene will never be lost from the population.)

Let $\Delta t = 1/N$ correspond to one generation and put $Y(t) = X(t)/N = X(t)\epsilon, \epsilon = 1/N$. If N is very large and if the mean of the progeny distribution is $1 + \frac{1}{N}\alpha + O(N^{-2})$, then we will show in the following that to the order of $O(N^{-2})$, $Y(t)$ is approximated by a diffusion process with parameter space $T = \{t \geq 0\}$ and state space $S = \{y \geq 0\}$. The coefficients of this diffusion process are $\{m(y,t) = \alpha y, v(y,t) = \sigma^2 y\}$, where σ^2 is the variance of the progeny distribution.

To prove that $Y(t)$ is a diffusion process as above for large N, let $f(s)$ be the pgf of the progeny distribution. Then $f'(1) = 1 + \frac{1}{N}\alpha + O(N^{-2})$ and

$\sigma^2 = f''(1) + f'(1)[1 - f'(1)]$. Hence, with $i = \sqrt{(-1)}$,

$$f(e^{ize}) = f(1) + ize f'(1) + \frac{1}{2}(iz)^2[f''(1) + f'(1)]\epsilon^2 + O(\epsilon^3)$$

$$= 1 + ize(1 + \alpha\epsilon) + \frac{1}{2}(iz)^2[\sigma^2 + (1 + \alpha\epsilon)^2]\epsilon^2 + O(\epsilon^3).$$

This gives:

$$\log f(e^{ize}) = \log\left\{1 + ize(1 + \alpha\epsilon) + \frac{1}{2}(iz)^2[\sigma^2 + (1 + \alpha\epsilon)^2]\epsilon^2 + O(\epsilon^3)\right\}$$

$$= ize(1 + \alpha\epsilon) + \frac{1}{2}(iz)^2[\sigma^2 + (1 + \alpha\epsilon)^2]\epsilon^2$$

$$- \frac{1}{2}(iz)^2(1 + \alpha\epsilon)^2\epsilon^2 + O(\epsilon^3)$$

$$= ize\left(1 + \alpha\epsilon + \frac{1}{2}ize\sigma^2\right) + O(\epsilon^3).$$

Let $f_n(s)$ be the pfg of $X(n)$ given $X(0) = 1$. Then it has been shown in Example 2.10 that $f_n(s) = f_{n-1}(f(s)) = f(f_{n-1}(s))$ (see Sec. 2.2). Let $\phi(t, z)$ be the characteristic function of $Y(t)$ given $X(0) = 1$. Then,

$$\phi(t + \epsilon, z) = f_{n+1}(e^{ize}) = f_n[f(e^{ize})]$$

$$= f_n(e^{\log f(e^{ize})}) = \phi(t, z + ze(\alpha + iz\sigma^2/2) + O(\epsilon^2))$$

$$= \phi(t, z) + ze(\alpha + iz\sigma^2/2)\frac{\partial}{\partial z}\phi(t, z) + O(\epsilon^2).$$

Hence,

$$\frac{1}{\epsilon}\{\phi(t + \epsilon, z) - \phi(t, z)\} = z(\alpha + iz\sigma^2/2)\frac{\partial}{\partial z}\phi(t, z) + O(\epsilon^2)/\epsilon.$$

Letting $\epsilon \to 0$, then

$$\frac{\partial}{\partial t}\phi(t, z) = z(\alpha + iz\sigma^2/2)\frac{\partial}{\partial z}\phi(t, z). \tag{6.29}$$

The above equation shows that to the order of $O(N^{-2})$, the c.f. $\phi(t, z)$ of $Y(t)$ satisfies the above Eq. (6.29). Let $f(y, t)$ be the pdf of $Y(t)$ given

$X(0) = 1$. Then by the inversion formulae [1, Chap. 4],

$$f(y, t) = \frac{1}{2\pi} \int_{-\infty}^{\infty} e^{-izy} \phi(t, z) dz \, .$$

Multiplying both sides by $\frac{1}{2\pi} e^{-izy}$ and integrating over z from $-\infty$ to ∞, we have on the left side:

$$\frac{1}{2\pi} \int_{-\infty}^{\infty} e^{-izy} \left\{ \frac{\partial}{\partial t} \phi(t, z) \right\} dz = \frac{\partial}{\partial t} \frac{1}{2\pi} \int_{-\infty}^{\infty} e^{-izy} \phi(t, z) dz$$

$$= \frac{\partial}{\partial t} f(y, t) \, ;$$

on the right side, we have:

$$\frac{1}{2\pi} \int_{-\infty}^{\infty} e^{-izy} \left[iz\alpha + \frac{1}{2}(iz)^2 \sigma^2 \right] \left\{ \frac{\partial}{\partial(iz)} \phi(t, z) \right\} dz$$

$$= -\alpha \frac{\partial}{\partial y} \frac{1}{2\pi} \int_{-\infty}^{\infty} e^{-izy} \frac{\partial}{\partial(iz)} \phi(t, z) dz$$

$$+ \frac{\sigma^2}{2} \frac{\partial^2}{\partial y^2} \frac{1}{2\pi} \int_{-\infty}^{\infty} e^{-izy} \frac{\partial}{\partial(iz)} \phi(t, z) dz \, .$$

Now, by integration by parts, we have, under the restriction $\lim_{z \to \pm\infty} \phi(t, z) = 0$:

$$\frac{1}{2\pi} \int_{-\infty}^{\infty} e^{-izy} \frac{\partial}{\partial(iz)} \phi(t, z) dz = \left(-\frac{i}{2\pi} \phi(t, z) e^{-izy} \right)_{z=-\infty}^{z=\infty}$$

$$+ y \frac{1}{2\pi} \int_{-\infty}^{\infty} e^{-izy} \phi(t, z) dz$$

$$= y f(y, t) \, .$$

Hence the right side becomes

$$-\alpha \frac{\partial}{\partial y} \{y f(y, t)\} + \frac{\sigma^2}{2} \frac{\partial^2}{\partial y^2} \{y f(y, t)\} \, .$$

It follows that

$$\frac{\partial}{\partial t} f(y, t) = -\alpha \frac{\partial}{\partial y} \{y f(y, t)\} + \frac{\sigma^2}{2} \frac{\partial^2}{\partial y^2} \{y f(y, t)\} \, . \qquad (6.30)$$

The initial condition is $f(y, 0) = \delta(y - \frac{1}{N})$.

The above approximation was first proved by Feller [6].

Example 6.7.   Diffusion approximation of the initiated cells in the two-stage model of carcinogenesis. In Example 1.13, we have considered the two-stage model of carcinogenesis. In this process, $N(t)$ and $I(t)$ denote the numbers of normal stem cells and the initiated cells at time t respectively. Since the number $N(0) = N_0$ of normal stem cells at the time of birth (i.e. $t = 0$) is usually very large, in Example 4.9 it has been shown that $I(t)$ will follow a birth-death process with immigration with birth rate $jb(t) + \lambda(t)$ and death rate $jd(t)$, where $b(t) > 0, d(t) \geq 0$ and $\lambda(t) = N(t)\alpha_N(t)$ with $\alpha_N(t)$ being the mutation rate of $N \to I$ at time t; see Example 4.9. We will show that to order of $O(N_0^{-2})$, $\{Y(t) = I(t)/N_0, t \in [0, \infty)\}$ is approximated by a diffusion process with state space $[0, \infty)$ and with coefficients $\{m(x, t) = \alpha_N(t) + x\gamma(t), v(x, t) = \frac{1}{N_0}x\omega(t)\}$, where $\gamma(t) = b(t) - d(t)$ and $\omega(t) = b(t) + d(t)$.

To prove this, let $M_N(t)$ denote the number of mutations from $N \to I$ during $[t, t + \Delta t)$; let $B_I(t)$ and $D_I(t)$ denote the numbers of birth and death of I cells during $[t, t + \Delta t)$. Then, as shown in Subsec. 5.4.2

$$I(t + \Delta t) = I(t) + M_N(t) + B_I(t) - D_I(t) \,, \tag{6.31}$$

where

$$\{B_I(t), D_I(t)\} | X(t) \sim ML\{X(t); b(t)\Delta t, d(t)\Delta t\}$$

and

$$M_N(t) \sim \text{Poisson}\{\lambda(t)\Delta t\} \,,$$

independently of $\{B_I(t), D_I(t)\}$.

Let $\phi(u, t)$ denote the characteristic function (cf) of $Y(t)$. That is,

$$\phi(u, t) = Ee^{iuY(t)} \,, \quad \text{where } i = \sqrt{-1} \,.$$

Then, we have:

$$\phi(u, t + \Delta t) = E[e^{iuY(t+\Delta t)}]$$

$$= E[e^{(iu)\frac{M_N(t)}{N_0}}]E\{e^{iuY(t)}\eta[X(t), u, t]\} \,, \tag{6.32}$$

where $\eta(X(t), u, t) = E\{e^{\frac{1}{N_0}iu[B_I(t) - D_I(t)]} | X(t)\}$.

Now, by using the distribution results of $\{B_I(t), D_I(t)\}$ and $M_N(t)$, we have:

$$\eta(X(t), u, t) = \{1 + [e^{\frac{iu}{N_0}} - 1]b(t)\Delta t + [e^{-\frac{iu}{N_0}} - 1]d(t)\Delta t\}^{X(t)}$$

$$= 1 + X(t)\{[e^{\frac{iu}{N_0}} - 1]b(t) + [e^{-\frac{iu}{N_0}} - 1]d(t)\}\Delta t + o(\Delta t)$$

$$= 1 + Y(t)\left\{(iu)\gamma(t) + \frac{(iu)^2}{2N_0}\omega(t)\right\}\Delta t + o(\Delta t) + O(N_0^{-2})\Delta t,$$

and

$$E\{e^{(iu)\frac{M_N(t)}{N_0}}\} = \exp\{\lambda(t)\Delta t(e^{\frac{iu}{N_0}} - 1)\} = 1 + \lambda(t)\Delta t\{e^{\frac{iu}{N_0}} - 1\} + o(\Delta t)$$

$$= 1 + (iu)\frac{\lambda(t)}{N_0}\Delta t + o(\Delta t) + O(N_0^{-2})\Delta t.$$

On substituting these results into Eq. (6.32) and simplifying, we obtain:

$$\phi(u, t + \Delta t) = Ee^{iuY(t)}\left\{1 + (iu)\frac{\lambda(t)}{N_0}\Delta t + Y(t)\left[(iu)\gamma(t) + \frac{(iu)^2}{2N_0}\omega(t)\right]\Delta t\right\}$$

$$+ o(\Delta t) + O(N_0^{-2})\Delta t$$

$$= \phi(u, t)\left\{1 + (iu)\frac{\lambda(t)}{N_0}\Delta t\right\} + \left[(iu)\gamma(t) + \frac{(iu)^2}{2N_0}\omega(t)\right]$$

$$\times \Delta t\frac{\partial}{\partial(iu)}\phi(u, t) + o(\Delta t) + O(N_0^{-2})\Delta t.$$

On both sides of the above equation, subtracting $\phi(u, t)$, dividing by Δt and letting $\Delta t \to 0$, we obtain:

$$\frac{\partial}{\partial t}\phi(u, t) = (iu)\frac{\lambda(t)}{N_0}\phi(u, t) + \left[(iu)\gamma(t) + \frac{(iu)^2}{2N_0}\omega(t)\right]\frac{\partial}{\partial(iu)}\phi(u, t) + O(N_0^{-2})$$

$$= (iu)m\left(\frac{\partial}{\partial(iu)}, t\right)\phi(u, t) + \frac{(iu)^2}{2}v\left(\frac{\partial}{\partial(iu)}, t\right)\phi(u, t) + O(N_0^{-2})$$

$$\tag{6.33}$$

where

$$m\left(\frac{\partial}{\partial(iu)}, t\right) = \frac{\lambda(t)}{N_0} + \gamma(t)\frac{\partial}{\partial(iu)}$$

and

$$v\left(\frac{\partial}{\partial(iu)}, t\right) = \frac{\omega(t)}{N_0}\frac{\partial}{\partial(iu)}.$$

Let $f(x, y : s, t)$ denote the pdf of $Y(t)$ at y given $Y(s) = x$. Then, $f(x, y; s, s) = \delta(y - x)$, the Dirac's delta function and by the Fourier inversion formulae [1, Chap. 4], $f(x, y; s, t)$ is given by:

$$f(x, y; s, t) = \frac{1}{2\pi}\int_{-\infty}^{\infty} e^{-iuy}\phi(u, t)du.$$

On multiplying both sides of Eq. (6.33) by $\frac{1}{2\pi}e^{-iuy}$ and integrating over u from $-\infty$ to ∞, on the left side we obtain:

$$\frac{\partial}{\partial t}f(x, y; s, t).$$

On the right side, since $m(y, t)$ is a polynomial in y, the first term is, by using results from Lemma 6.1:

$$\frac{1}{2\pi}\int_{-\infty}^{\infty} e^{-iuy}(iu)m\left[\frac{\partial}{\partial(iu)}, t\right]\phi(u, t)du$$

$$= -\frac{1}{2\pi}\frac{\partial}{\partial y}\int_{-\infty}^{\infty} e^{-iuy}m\left[\frac{\partial}{\partial(iu)}, t\right]\phi(u, t)du$$

$$= -\frac{\partial}{\partial y}\{m(y, t)f(x, y; s, t)\}.$$

Since $v(y, t)$ is a polynomial in y, the second term on the right side is, by using results from Lemma 6.1:

$$\frac{1}{2\pi}\int_{-\infty}^{\infty} e^{-iuy}\frac{(iu)^2}{2}v\left[\frac{\partial}{\partial(iu)}, t\right]\phi(u, t)du$$

$$= \frac{1}{2}\frac{\partial^2}{\partial y^2}\left\{\frac{1}{2\pi}\int_{-\infty}^{\infty} e^{-iuy}v\left[\frac{\partial}{\partial(iu)}, t\right]\phi(u, t)du\right\}$$

$$= \frac{1}{2}\frac{\partial^2}{\partial y^2}\{v(y, t)f(x, y; s, t)\}.$$

Thus, we obtain:

$$\frac{\partial}{\partial t} f(x, y; s, t) = -\frac{\partial}{\partial y}\{m(y, t)f(x, y; s, t)\} + \frac{1}{2}\frac{\partial^2}{\partial y^2}\{v(y, t)f(x, y; s, t)\}$$

$$+ O(N_0^{-2}). \tag{6.34}$$

Notice that the above equation is the Kolmogorov forward equation of a diffusion process with state space $[0, \infty)$ and with coefficients $\{m(x, t) = \frac{\lambda(t)}{N_0} + x\gamma(t), v(x, t) = x\frac{\omega(t)}{N_0}\}$, where $\gamma(t) = b(t) - d(t)$ and $\omega(t) = b(t) + d(t)$. Hence the result is proved.

Example 6.8. Diffusion approximation of fraction of infective people in the SIR model. In Example 1.6, we have considered a large population of homosexual men who are at risk for AIDS. Let $S(t), I(t)$ and $A(t)$ denote the numbers of S people, I people and AIDS cases at time t in this population. Let $Y(t) = I(t)/N(t)$, where $N(t) = S(t) + I(t)$. In this section, we will show that for large $N(t)$, $Y(t)$ is closely approximated by a diffusion process if the following assumptions hold:

(i) We assume that there are no sexual contact or IV drug contact with AIDS cases.

(ii) There is only one sexual activity level so that each person in the population has the same number of different sexual partners per unit time. Thus, the number of different sexual partners of each S person during the time interval $[t, t + \Delta t)$ can be expressed by $c(t)\Delta t$, where $c(t)$ is a non-negative real number.

(iii) We assume that AIDS spread mainly through sexual contact between S people and I people, ignoring other transmission avenues. (This is approximately true in the city of San Francisco since sexual contact between homosexual men accounts for over 90% of AIDS cases in the homosexual population in that city; see [8].)

(vi) We assume that the immigration and recruitment rates $(\nu_S(t), \nu_I(t))$ are equal to the death and emigration rates $(\mu_S(t), \mu_I(t))$ for both S people and I people, respectively. It follows that as a close approximation, one may assume $N(t) \cong N$. That is, approximately, $N(t)$ is independent of t.

(v) The infection duration of an I person is defined as the time elapsed since this I person acquired the HIV. We assume that the infection duration of the I people have no significant impacts on the HIV transmission and progression. It follows that one may let $\gamma(t)\Delta t$ be the probability of $I \to A$ during $[t, t + \Delta t)$.

(vi) There are no people who are immune to the disease and there are no reverse transitions from A to I and from I to S.

(vii) We assume that people pick up their sexual partners randomly from the population (random mixing or proportional mixing). Let $p_S(t)\Delta t$ be the probability of $S \to I$ during $[t, t+\Delta t)$. Since $N(t) \cong N$ is usually very large,

$$p_S(t)\Delta t = c(t)q(t)\frac{I(t)}{N(t)}\Delta t = Y(t)\alpha(t)\Delta t,$$

where $q(t)$ is the per partner probability of HIV transmission from the I person to the S person given sexual contacts between these two people during $[t, t+\Delta t)$ and $\alpha(t) = c(t)q(t)$.

To approximate $Y(t)$, denote by:

(1) $R_S(t)(R_I(t))$ = Number of Immigration and Recruitment of S People (I People) During $[t, t+\Delta t)$.

(2) $D_S(t)(D_I(t))$ = Number of Death and Emigration of S People (I People) During $[t, t+\Delta t)$.

(3) $F_S(t)(R_I(t))$ = Number of $S \to I$ ($I \to A$) Transitions During $[t, t+\Delta t)$.

Then, we have the following stochastic equations for $S(t)$ and $I(t)$:

$$S(t + \Delta t) = S(t) + R_S(t) - F_S(t) - D_S(t), \tag{6.35}$$

$$I(t + \Delta t) = I(t) + R_I(t) + F_S(t) - F_I(t) - D_I(t). \tag{6.36}$$

In the above equations all variables are random variables. Given $\{S(t), I(t)\}$, the conditional probability distributions of these random variables are given by:

$$\{F_S(t), D_S(t)\}|[S(t), I(t)] \sim ML\{S(t), Y(t)\alpha(t)\Delta t, \mu_S(t)\Delta t\};$$

$$\{F_I(t), D_I(t)\}|I(t) \sim ML\{I(t), \gamma(t)\Delta t, \mu_I(t)\Delta t\};$$

$$R_S(t)|S(t) \sim B\{S(t), \nu_S(t)\Delta t\};$$

and

$$R_I(t)|I(t) \sim B\{I(t), \nu_I(t)\Delta t\}.$$

Further, given $\{S(t), I(t)\}$, $\{[F_S(t), D_S(t)], [F_I(t), D_I(t)], R_S(t), R_I(t)\}$ are independently distributed of one another.

Theorem 6.3. **(Diffusion approximation of $Y(t)$).** *Given the above assumptions, to order of $O(N^{-2})$, the conditional pdf $f(x, y; s, t)$ of $Y(t)$ given $Y(s) = x = \frac{i}{N}$ satisfies the following partial differential equations:*

$$\frac{\partial}{\partial t} f(x, y; s.t) = -\frac{\partial}{\partial y}\{m(y, t) f(x, y; s, t)\} + \frac{1}{2N}\frac{\partial^2}{\partial y^2}\{v(y, t) f(x, y; s, t)\},$$

$$(6.37)$$

where $m(y, t) = \alpha(t)y(1 - y) + y\nu_I(t) - y[\gamma(t) + \mu_I(t)]$ and $v(y, t) = \alpha(t)y(1 - y) + y\nu_I(t) + y[\gamma(t) + \mu_I(t)]$.

That is, to order $O(N^{-2})$, $Y(t)$ is approximated by a diffusion process with state space $S = [0, 1]$ and with coefficients $\{m(y, t), \frac{1}{N}v(y, t)\}$.

Proof. To prove the above theorem, notice first that $S(t) = N - I(t) = N[1 - Y(t)]$, $F_S(t)|[S(t), I(t)] \sim B\{S(t); Y(t)\alpha(t)\Delta t\}$ and $F_I(t) + D_I(t)|I(t) \sim B\{I(t); [\gamma(t) + \mu_I(t)]\Delta t\}$. Let $\phi(u, t)$ denote the cf of $Y(t)$. That is, $\phi(u, t) = E[e^{iuY(t)}], i = \sqrt{(-1)}$.

Then,

$$\phi(u, t + \Delta t) = E[e^{iuY(t+\Delta t)}]$$

$$= E\left(\exp\left\{\frac{iu}{N}[I(t) + R_I(t) + F_S(t) - F_I(t) - D_I(t)]\right\}\right)$$

$$= E\{e^{iuY(t)}[\eta_S(Y(t))\zeta_I(Y(t))\eta_I(Y(t))]\} \qquad (6.38)$$

where

$$\eta_S[Y(t)] = E\{e^{\frac{1}{N}iu\, F_S(t)}|[S(t), I(t)]\},$$

$$\zeta_I[Y(t)] = E\{e^{\frac{1}{N}iu\, R_I(t)}|I(t)\},$$

and

$$\eta_I[Y(t)] = E\{e^{-\frac{1}{N}iu[F_I(t)+D_I(t)]}|I(t)\}.$$

Since $F_S(t)|[S(t), I(t)] \sim B\{S(t), Y(t)\alpha(t)\Delta t\}$,

$$\eta_S[Y(t)] = \{1 + Y(t)\alpha(t)\Delta t[e^{\frac{iu}{N}} - 1]\}^{S(t)}$$

$$= 1 + S(t)Y(t)\alpha(t)\Delta t[e^{\frac{iu}{N}} - 1] + o(\Delta t)$$

$$= 1 + Y(t)[1 - Y(t)]\alpha(t)\left\{(iu) + \frac{(iu)^2}{2N}\right\}\Delta t$$

$$+ o(\Delta t) + O(N^{-2})\Delta t\,, \tag{6.39}$$

where $O(N^{-2})$ is of the same order as N^{-2} (i.e. $\lim_{N\to\infty}\{N^2 O(N^{-2})\} =$ Constant.).

Since $R_I(t)|I(t) \sim B\{I(t); \nu_I(t)\Delta t\}$,

$$\zeta_I[Y(t)] = \{1 + \nu_I(t)\Delta t[e^{\frac{iu}{N}} - 1]\}^{I(t)}$$

$$= 1 + NY(t)\nu_I(t)\Delta t[e^{\frac{iu}{N}} - 1] + o(\Delta t)$$

$$= 1 + Y(t)\nu_I(t)\left\{(iu) + \frac{(iu)^2}{2N}\right\}\Delta t$$

$$+ o(\Delta t) + O(N^{-2})\Delta t\,. \tag{6.40}$$

Since $F_I(t) + D_I(t)|I(t) \sim B\{I(t); [\gamma(t) + \mu_I(t)]\Delta t\}$,

$$\eta_I[Y(t)] = \{1 + [\gamma(t) + \mu_I(t)][e^{-\frac{iu}{N}} - 1]\Delta t\}^{I(t)}$$

$$= 1 + NY(t)[\gamma(t) + \mu_I(t)]\Delta t[e^{-\frac{iu}{N}} - 1] + o(\Delta t)$$

$$= 1 + Y(t)[\gamma(t) + \mu_I(t)]\left\{-(iu) + \frac{(iu)^2}{2N}\right\}\Delta t$$

$$+ o(\Delta t) + O(N^{-2})\Delta t\,. \tag{6.41}$$

On substituting the results of Eqs. (6.39)–(6.41) into Eq. (6.38), we obtain:

$$\eta_S[Y(t)]\zeta_I[Y(t)]\eta_I[Y(t)] = 1 + Y(t)[1 - Y(t)]\alpha(t)\left\{(iu) + \frac{(iu)^2}{2N}\right\}\Delta t$$

$$+ Y(t)\nu_I(t)\left\{(iu) + \frac{(iu)^2}{2N}\right\}\Delta t + Y(t)[\gamma(t) + \mu_I(t)]$$

$$\times \left\{-(iu) + \frac{(iu)^2}{2N}\right\}\Delta t + o(\Delta t) + O(N^{-2})\Delta t$$

$$= 1 + (iu)m[Y(t), t]\Delta t + \frac{(iu)^2}{2N}v[Y(t), t]\Delta t$$

$$+ o(\Delta t) + O(N^{-2})\Delta t\,.$$

It follows that we have:

$$\phi(u, t + \Delta t) = E e^{iuY(t)} \left\{ 1 + (iu)m[Y(t), t]\Delta t + \frac{(iu)^2}{2N} v[Y(t), t]\Delta t \right.$$

$$\left. + o(\Delta t) + O(N^{-2})\Delta t \right\}$$

$$= \phi(u, t) + (iu)m\left[\frac{\partial}{\partial(iu)}, t\right]\phi(u, t)\Delta t + \frac{(iu)^2}{2N}v$$

$$\times \left[\frac{\partial}{\partial(iu)}, t\right]\phi(u, t)\Delta t + o(\Delta t) + O(N^{-2})\Delta t. \qquad (6.42)$$

From both sides of Eq. (6.42), subtracting $\phi(u, t)$ and dividing by Δt, we obtain:

$$\frac{1}{\Delta t}\{\phi(u, t + \Delta t) - \phi(u, t)\} = (iu)m\left[\frac{\partial}{\partial(iu)}, t\right]\phi(u, t) + \frac{(iu)^2}{2N}v$$

$$\times \left[\frac{\partial}{\partial(iu)}, t\right]\phi(u, t) + \frac{o(\Delta t)}{\Delta t} + O(N^{-2})\Delta t.$$

In the above equation, by letting $\Delta t \to 0$ we obtain:

$$\frac{\partial}{\partial t}\phi(u, t) = (iu)m\left[\frac{\partial}{\partial(iu)}, t\right]\phi(u, t)$$

$$+ \frac{(iu)^2}{2N}v\left[\frac{\partial}{\partial(iu)}, t\right]\phi(u, t) + O(N^{-2}). \qquad (6.43)$$

To prove Eq. (6.37), notice that by the Fourier inversion formulae [1, Chap. 4],

$$f(x, y; s, t) = \frac{1}{2\pi}\int_{-\infty}^{\infty} e^{-iuy}\phi(u, t)du.$$

On both sides of Eq. (6.43), multiplying by $\frac{1}{2\pi}e^{-iuy}$ and integrating over u from $-\infty$ to ∞, on the left side we obtain:

$$\frac{\partial}{\partial t}f(x, y; s, t).$$

On the right side, since $m(y,t)$ is a polynomial in y, the first term is, by using results from Lemma 6.1:

$$\frac{1}{2\pi}\int_{-\infty}^{\infty} e^{-iuy}(iu)m\left[\frac{\partial}{\partial(iu)},t\right]\phi(u,t)du$$

$$= -\frac{1}{2\pi}\frac{\partial}{\partial y}\int_{-\infty}^{\infty} e^{-iuy}m\left[\frac{\partial}{\partial(iu)},t\right]\phi(u,t)du$$

$$= -\frac{\partial}{\partial y}\{m(y,t)f(x,y;s,t)\}.$$

Since $v(y,t)$ is a polynomial in y, the second term on the right side is, by using results from Lemma 6.1:

$$\frac{1}{2\pi}\int_{-\infty}^{\infty} e^{-iuy}\frac{(iu)^2}{2N}v\left[\frac{\partial}{\partial(iu)},t\right]\phi(u,t)du$$

$$= \frac{1}{2N}\frac{\partial^2}{\partial y^2}\left\{\frac{1}{2\pi}\int_{-\infty}^{\infty} e^{-iuy}v\left[\frac{\partial}{\partial(iu)},t\right]\phi(u,t)du\right\}$$

$$= \frac{1}{2N}\frac{\partial^2}{\partial y^2}\{v(y,t)f(x,y;s,t)\}. \tag{6.44}$$

Thus, we obtain:

$$\frac{\partial}{\partial t}f(x,y;s,t) = -\frac{\partial}{\partial y}\{m(y,t)f(x,y;s,t)\} + \frac{1}{2N}\frac{\partial^2}{\partial y^2}\{v(y,t)f(x,y;s,t)\}$$

$$+ O(N^{-2}). \tag{6.45}$$

The initial condition is $f(x,y;s,s) = \delta(y-x)$. $\qquad\square$

6.5.   Diffusion Approximation of Evolutionary Processes

It is universally recognized that in most of the cases, the evolution process is a Markov process. This follows from the observation that the frequencies of the types at any generation very often depend only on those of the previous generation and are independent of past history. To develop theories of evolution, the following facts have also been observed:

(a) The population size is usually very large.

(b) The evolution process is a slow process, taking place over millions of years.

(c) Evolution does go on so that the variance of the process is always positive.

From these observations, by properly changing the time scale and the scale for the state variables, the evolution process can be closely approximated by diffusion processes. According to the calculation by Ewens [8], in some cases such approximations are extremely good, so good that one is surprised to see how the mathematical principle works for nature.

Applications of the diffusion processes to genetics and evolution theories was initiated by Sir R.A. Fisher in 1922 who likened the evolution principle to the diffusion of gases in quantum mechanics [9]. To see the connection, suppose that there are two types, say, type 1 and type 2, in the natural population. Then, the evolution theory is the random process to change the relative frequencies of these two types. The basic factors which cause these changes are the mutation, selection, immigration and migration, finite population size and other random factors. This can be likened to two connected containers of gases and notice that if we apply external or internal disturbing factors such as heat and pressure on one container, the gases of this container will flow into the other container. In evolution theories, these disturbing factors are called genetic pressures, which are mutations (internal disturbing factor), selection (external disturbing factor), immigration and migration, genetic drift as well as other random factors.

To illustrate how diffusion processes can be used to approximate the evolutionary processes, consider a large diploid population with N individuals. Suppose that there are two alleles A and a at the A-locus and let $X(t)$ denote the number of A allele at generation t. Then $\{X(t), t \in T\}$ is a Markov chain with discrete parameter space $T = (0, 1, \ldots, \infty)$ and with state space $S = \{0, 1, \ldots, 2N\}$. The one-step transition probability is

$$p_{ij}(t, t+1) = \binom{2N}{j} [p(i,t)]^j [1 - p(i,t)]^{2N-j},$$

where $p(i,t)$ is the probability of A allele for generating the genotypes at generation $t + 1$. (Notice that the $p(i,t)$ are in general functions of $X(t) = i$ and t.) This is the so-called Wright model as described in Example 1.11.

In the above Wright model, put $Y(t) = X(t)/(2N)$. When $N \to \infty$, then $Y(t)$ changes continuously from 0 to 1. Let one generation correspond to $\Delta t = (2N)^{-1}$. Then, for large N, one may expect that $Y(t)$ is a continuous process with parameter space $T = [0, \infty)$ and with state space $S = [0, 1]$. The following theorem shows that to the order $O(N^{-2})$, $Y(t)$ is closely approximated by a diffusion process.

Theorem 6.4. *Let $\{X(t), t = 0, 1, 2, \ldots\}$ be a Markov chain with state space $S = \{0, 1, 2, \ldots, M = 2N\}$. Put $\Pr\{X(n) = j | X(m) = k\} = p_{kj}(m, n)$. Denote by $s = \frac{m}{M}, t = \frac{n}{M}, x = \frac{k}{M}$ and $y = \frac{j}{M}$ and let $f(x, y; s, t)$ be the conditional pdf of $Y(t) = \frac{X(n)}{M}$. Suppose that the following conditions hold:*

$$\text{(i)} \quad \sum_{j=0}^{M} \frac{1}{M^r}(j - k)^r p_{kj}(n, n+1) = \begin{cases} \dfrac{1}{M}m(x, t) + o(M^{-1}), & \text{if } r = 1, \\[2mm] \dfrac{1}{M}v(x, t) + o(M^{-1}), & \text{if } r = 2, \\[2mm] o(M^{-1}), & \text{if } r \geq 3, \end{cases}$$

where $m(x, t)$ and $v(x, t)$ are polynomials in x and are continuous functions of $t = \frac{n}{M}$,

(ii) For large M (i.e. $M \to \infty$), $\frac{\partial}{\partial t} f(x, y; s, t)$, $\frac{\partial}{\partial y}[m(y, t)f(x, y; s, t)]$ and $\frac{\partial^2}{\partial y^2}[v(y, t)f(x, y; s, t)]$ exist and are continuous functions of y in $[0, 1]$ for every $t > 0$.

Then to order of $O(M^{-2})$, $f(x, y; s, t)$ satisfies the following Kolomogorov forward equation (Fokker–Planck equation):

$$\frac{\partial}{\partial t} f(x, y; s, t) = -\frac{\partial}{\partial y}[m(y, t)f(x, y; s, t)]$$

$$+ \frac{1}{2}\frac{\partial^2}{\partial y^2}[v(y, t)f(x, y; s, t)], \tag{6.46}$$

with $f(x, y; s, s) = \delta(y - x)$, the Dirac's δ-function. (See Remark 6.2.)

Remark 6.2. In Theorems 6.4 and 6.5, it is assumed that the population size at generation t is N independent of t. These theorems also hold if the size is N_t provided that N_t is very large for all t and that $\bar{N}/N_t = 1 + O(\bar{N}^{-1})$, where $\bar{N}$ is the harmonic mean of the N_t's. In these cases, one simply replaces N by $\bar{N}$ in these theorems. In population genetics, $\bar{N}$ has been referred to as the effective population size; see [2].

Proof of Theorem 6.4. To prove Theorem 6.4, define for real $\theta > 0$:

$$\phi_{m,n}(k,\theta) = \sum_{j=0}^{M} e^{-\theta \frac{j}{M}} p_{kj}(m,n)$$

where $p_{kj}(m,n) = \Pr\{X(n) = j | X(m) = k\}$.

In the limits as $M \to \infty$, $Y(t) = \frac{X(n)}{M}$ can be assumed to change continuously so that for large M, we write $p_{kj}(m,n)$ as $p_{kj}(m,n) \cong f(x,y;s,t)\Delta y$, where $\Delta y = M^{-1}, s = \frac{m}{M}, t = \frac{n}{M}, x = \frac{k}{M}$ and $y = \frac{j}{M}$. This implies that as $M \to \infty$ so that $\Delta y \to 0$,

$$\lim_{M \to \infty} \phi_{m,n}(k,\theta) = \phi(x,\theta;s,t) = \int_0^1 e^{-\theta y} f(x,y;s,t)dy.$$

Notice that if $f(x,y;s,t)$ has continuous first partial derivative with respect to t, so does $\phi(x,\theta;s,t)$. Now, by the Chapman–Kolmogorov equation,

$$\phi_{m,n+1}(k,\theta) = \sum_{j=0}^{M} e^{-\theta \frac{j}{M}} p_{kj}(m,n+1) = \sum_{j=0}^{M} e^{-\theta \frac{j}{M}} \sum_{r=0}^{M} p_{kr}(m,n)p_{rj}(n,n+1)$$

$$= \sum_{r=0}^{M} e^{-\theta \frac{r}{M}} p_{kr}(m,n) \sum_{j=0}^{M} e^{-\frac{\theta}{M}(j-r)} p_{rj}(n,n+1)$$

$$= \sum_{r=0}^{M} e^{-\theta \frac{r}{M}} p_{kr}(m,n) \sum_{j=0}^{M} \left\{ 1 - \frac{\theta}{M}(j-r) + \frac{1}{2}\left(\frac{\theta}{M}\right)^2 (j-r)^2 \right.$$

$$\left. + 0\left[\frac{1}{M^3}(j-r)^3\right] \right\} p_{rj}(n,n+1) = \phi_{m,n}(k,\theta)$$

$$+ \sum_{r=0}^{M} e^{-\theta \frac{r}{M}} p_{kr}(m,n) \sum_{j=0}^{M} \left\{ -\frac{\theta}{M}(j-r) + \frac{1}{2}\left(\frac{\theta}{M}\right)^2 (j-r)^2 \right\}$$

$$\times p_{rj}(n,n+1) + o(M^{-1})$$

$$= \phi_{m,n}(k,\theta) - \frac{\theta}{M} \sum_{r=0}^{M} e^{-\theta \frac{r}{M}} m\left(\frac{r}{M}, \frac{n}{M}\right) p_{kr}(m,n)$$

$$+ \frac{\theta^2}{2M} \sum_{r=0}^{M} e^{-\theta \frac{r}{M}} v\left(\frac{r}{M}, \frac{n}{M}\right) p_{kr}(m,n) + o(M^{-1}). \tag{6.47}$$

Put $\Delta t = \Delta z = M^{-1}$ and for large M, write $\phi_{m,n}(k,\theta) \cong \phi(s,x;t,\theta)$ and $p_{kr}(m,n) \cong f(x,z;s,t)\Delta z$, where $z = \frac{r}{M}$. Then, as $M \to \infty$, the second and third terms on the right side of the above equation converge, respectively, to the following integrals:

$$\lim_{M\to\infty} \sum_{r=0}^{M} e^{-\theta\frac{r}{M}} m\left(\frac{r}{M},\frac{n}{M}\right) p_{kr}(m,n) = \int_0^1 e^{-\theta z} m(z,t) f(x,z;s,t) dz$$

and

$$\lim_{M\to\infty} \sum_{r=0}^{M} e^{-\theta\frac{r}{M}} v\left(\frac{r}{M},\frac{n}{M}\right) p_{kr}(m,n) = \int_0^1 e^{-\theta z} v(z,t) f(x,z;s,t) dz \,.$$

In Eq. (6.47), subtract both sides by $\phi_{m,n}(k,\theta) \cong \phi(x,\theta;s,t)$, dividing by $\Delta t = M^{-1}$ and noting $\lim_{\Delta t\to 0} \frac{o(\Delta t)}{\Delta t} = 0$, we obtain:

$$\lim_{\Delta t\to 0} \frac{1}{\Delta t}\{\phi(x,\theta;s,t+\Delta t) - \phi(x,\theta;s,t)\}$$

$$= \frac{\partial}{\partial t}\phi(x,\theta;s,t)$$

$$= -\theta \int_0^1 e^{-\theta z} m(z,t) f(x,z;s,t) dz + \frac{1}{2}\theta^2 \int_0^1 e^{-\theta z} v(z,t) f(x,z;s,t) dz \,.$$

Now,

$$\frac{\partial}{\partial t}\phi(x,\theta;s,t) = \frac{\partial}{\partial t} \int_0^1 e^{-\theta z} f(x,z;s,t) dz$$

$$= \int_0^1 e^{-\theta z} \left\{ \frac{\partial}{\partial t} f(x,z;s,t) \right\} dz \,;$$

further, integration by part gives:

$$-\theta \int_0^1 e^{-\theta z} m(z,t) f(x,z;s,t) dz = \int_0^1 m(z,t) f(x,z;s,t) de^{-\theta z}$$

$$= -\int_0^1 e^{-\theta z} \frac{\partial}{\partial z}[m(z,t) f(x,z;s,t)] dz \,,$$

and

$$\frac{1}{2}\theta^2 \int_0^1 e^{-\theta z} v(z,t) f(x,z;s,t) dz = -\frac{1}{2}\theta \int_0^1 v(z,t) f(x,z;s,t) de^{-\theta z}$$

$$= \frac{1}{2}\theta \int_0^1 e^{-\theta z} \left\{ \frac{\partial}{\partial z}[v(z,t) f(x,z;s,t)] \right\} dz$$

$$= -\frac{1}{2} \int_0^1 \left\{ \frac{\partial}{\partial z}[v(z,t) f(x,z;s,t)] \right\} de^{-\theta z}$$

$$= \frac{1}{2} \int_0^1 e^{-\theta z} \left\{ \frac{\partial^2}{\partial z^2}[v(z,t) f(x,z;s,t)] \right\} dz \, .$$

It follows that

$$\int_0^1 e^{-\theta z} \left\{ \frac{\partial}{\partial t} f(x,z;s,t)] \right\} dz$$

$$= \int_0^1 e^{-\theta z} \left\{ -\frac{\partial}{\partial z}[m(z,t) f(x,z;s,t)] + \frac{1}{2}\frac{\partial^2}{\partial z^2}[v(z,t) f(x,z;s,t)] \right\} dz \, .$$

This holds for all real $\theta > 0$ so that

$$\frac{\partial}{\partial t} f(x,y;s,t) = -\frac{\partial}{\partial y}[m(y,t) f(x,y;s,t)] + \frac{1}{2}\frac{\partial^2}{\partial y^2}[v(y,t) f(x,y;s,t)] \, .$$

Since $f(x,y;s,t)$ is a probability density function, one must have $f(x,y;s,s) = \delta(y-x)$, the Dirac's δ-function. $\qquad\qquad\square$

Theorem 6.5.   *Let $\{X(t), t = 0,1,2,\ldots\}$ be a Markov chain with state space $S = \{0,1,2,\ldots,M\}$ as given in Theorem 6.3. Suppose that condition* (i) *given in Theorem 6.3 holds and that in the limit as $M \to \infty$, $f(x,y;s,t)$ is continuous in s and $f(x,y;s,t)$ has continuous first and second partial derivatives with respect to x. Then $\frac{\partial}{\partial s} f(x,y;s,t)$ exists and $f(x,y;s,t)$ also satisfies the following Kolomogorov backward equation:*

$$-\frac{\partial}{\partial s} f(x,y;s,t) = m(x,s)\frac{\partial}{\partial x} f(x,y;s,t)$$

$$+ \frac{1}{2}v(x,s)\frac{\partial^2}{\partial x^2} f(x,y;s,t) \, , \qquad\qquad (6.48)$$

with $f(x,y;s,s) = \delta(y-x)$, the Dirac's δ-function. (See Remark 6.2).

Proof.   To show that $f(x, y; s, t)$ satisfies the backward equation (6.32), observe that

$$\phi_{m-1,n}(k, \theta) = \sum_{j=0}^{M} e^{-\theta \frac{j}{M}} p_{kj}(m-1, n)$$

$$= \sum_{j=0}^{M} e^{-\theta \frac{j}{M}} \sum_{j=0}^{M} p_{kr}(m-1, m) p_{rj}(m, n)$$

$$= \sum_{r=0}^{M} p_{kr}(m-1, m) \left\{ \sum_{j=0}^{M} e^{-\theta \frac{j}{M}} p_{rj}(m, n) \right\}$$

$$= \sum_{r=0}^{M} \phi_{m,n}(r, \theta) p_{kr}(m-1, m).$$

Now, in the limit as $M \to \infty$, we may write $\phi_{m-1,n}(k, \theta)$ and $\phi_{m,n}(r, \theta)$ as $\phi_{m-1,n}(k, \theta) \cong \phi(x, \theta; s - \Delta s, t)$ and $\phi_{m,n}(r, \theta) \cong \phi(z, \theta; s, t)$, where $\Delta s = M^{-1}, s = \frac{m}{M}, t = \frac{n}{M}, x = \frac{k}{M}$ and $z = \frac{r}{M}$. Under the assumption that $f(z, y; s, t)$ has continuous first and second partial derivatives with respect to z, $\phi(z, \theta; s, t)$ has continuous first and second derivatives with respect to z. Thus, one may expand $\phi(z, \theta; s, t)$ in Taylor series with respect to z around $x = \frac{k}{M}$ to give:

$$\phi(z, \theta; s, t) = \phi(x, \theta; s, t) + \frac{1}{M}(r - k) \frac{\partial}{\partial x} \phi(x, \theta; s, t)$$

$$+ \frac{1}{2} \frac{1}{M^2}(r - k)^2 \frac{\partial^2}{\partial x^2} \phi(x, \theta; s, t) + O \left\{ \frac{1}{M^3}(r - k)^3 \right\}.$$

Thus, in the limit as $M \to \infty$,

$$\phi(x, \theta; s - \Delta s, t) = \sum_{r=0}^{M} \left\{ \phi(x, \theta; s, t) + \frac{1}{M}(r - k) \frac{\partial}{\partial x} \phi(x, \theta; s, t) + \frac{1}{2} \frac{1}{M^2}(r - k)^2 \right.$$

$$\left. \times \frac{\partial^2}{\partial x^2} \phi(x, \theta; s, t) + O \left[\frac{1}{M^3}(r - k)^3 \right] \right\} p_{kr}(m-1, m)$$

$$= \phi(x, \theta; s, t) + \frac{1}{M} m(x, s) \frac{\partial}{\partial x} \phi(x, \theta; s, t)$$

$$+ \frac{1}{2M} v(x, s) \frac{\partial^2}{\partial x^2} \phi(x, \theta; s, t) + o(M^{-1}).$$

Letting $M \to \infty$ so that $\Delta s = M^{-1} \to 0$, we obtain:

$$- \lim_{\Delta s \to 0} \frac{1}{\Delta s} \{\phi(x, \theta; s, t) - \phi(x, \theta; s - \Delta s, t)\}$$

$$= m(x, s) \frac{\partial}{\partial x} \phi(x, \theta; s, t) + \frac{1}{2} v(x, s) \frac{\partial^2}{\partial x^2} \phi(x, \theta; s, t) \,. \qquad (6.49)$$

Similarly, for $n > m$, we have:

$$Q_{m,n}(k, \theta) = \sum_{j=0}^{M} e^{-\theta \frac{j}{M}} p_{kj}(m, n)$$

$$= \sum_{j=0}^{M} e^{-\theta \frac{j}{M}} \sum_{r=0}^{M} p_{kr}(m, m + 1) p_{rj}(m + 1, n)$$

$$= \sum_{r=0}^{M} p_{kr}(m, m + 1) \sum_{j=0}^{M} e^{-\theta \frac{j}{M}} p_{rj}(m + 1, n)$$

$$= \sum_{r=0}^{M} p_{kr}(m, m + 1) \phi_{m+1,n}(r, \theta) \,.$$

In the limit as $M \to \infty$, write $\phi_{m+1,n}(r, \theta)$ as $\phi_{m+1,n}(r, \theta) \cong \phi(z, \theta; s + \Delta s, t)$, where $z = \frac{r}{M}$, and expand $\phi(z, \theta; s + \Delta s, t)$ in Taylor series with respect to z around $x = \frac{k}{M}$ to give:

$$\phi(z, \theta; s + \Delta s, t) = \phi(x, \theta; s + \Delta s, t) + \frac{1}{M}(r - k) \frac{\partial}{\partial x} \phi(x, \theta; s + \Delta s, t)$$

$$+ \frac{1}{2} \frac{1}{M^2}(r - k)^2 \frac{\partial^2}{\partial x^2} \phi(x, \theta; s + \Delta s, t) + O\left\{\frac{1}{M^3}(r - k)^3\right\} \,.$$

This yields

$$\phi(s, x; t, \theta) = \sum_{r=0}^{M} p_{kr}(m, m + 1) \left\{ \phi(x, \theta; s + \Delta s, t) \right.$$

$$+ \frac{1}{M}(r - k) \frac{\partial}{\partial x} \phi(x, \theta; s + \Delta s, t) + \frac{1}{2} \frac{1}{M^2}(r - k)^2$$

$$\left. \times \frac{\partial^2}{\partial x^2} \phi(x, \theta; s + \Delta s, t) + O\left[\frac{1}{M^3}(r - k)^3\right] \right\}$$

$$= \phi(x, \theta; s + \Delta s, t) + \frac{1}{M} m(x, s) \frac{\partial}{\partial x} \phi(x, \theta; s + \Delta s, t)$$

$$+ \frac{1}{2M} v(x, s) \frac{\partial^2}{\partial x^2} \phi(x, \theta; s + \Delta s, t) + o(M^{-1}).$$

Since $\phi(x, \theta; s, t)$ is continuous in s, so, as $M \to \infty$ so that $\Delta s = M^{-1} \to 0$,

$$- \lim_{\Delta s \to 0} \frac{1}{\Delta s} \{\phi(x, \theta; s + \Delta s, t) - \phi(x, \theta; s, t)\}$$

$$= m(x, s) \frac{\partial}{\partial x} \phi(x, \theta; s, t) + \frac{1}{2} v(x, s) \frac{\partial^2}{\partial x^2} \phi(x, \theta; s, t). \qquad (6.50)$$

Combining Eqs. (6.49) and (6.50), we prove that $\frac{\partial}{\partial s} \phi(x, \theta; s, t)$ exists and $\phi(x, \theta; s, t)$ satisfies,

$$- \frac{\partial}{\partial s} \phi(x, \theta; s, t) = m(x, s) \frac{\partial}{\partial x} \phi(x, \theta; s, t) + \frac{1}{2} v(x, s) \frac{\partial^2}{\partial x^2} \phi(x, \theta; s, t).$$

Since $\phi(x, \theta; s, t) = \int_0^1 e^{-\theta y} f(x, y; s, t) dy$,

$$- \frac{\partial}{\partial s} \phi(x, \theta; s, t) = - \int_0^1 e^{-\theta y} \left\{ \frac{\partial}{\partial s} f(x, y; s, t) \right\} dy$$

$$= \int_0^1 e^{-\theta y} m(x, s) \left[\frac{\partial}{\partial x} f(x, y; s, t) \right] dy$$

$$+ \frac{1}{2} \int_0^1 e^{-\theta y} v(x, s) \left[\frac{\partial^2}{\partial x^2} f(x, y; s, t) \right] dy.$$

This holds for all real $\theta > 0$ so that $\frac{\partial}{\partial s} f(x, y; s, t)$ exists and $f(x, y; s, t)$ satisfies the equation

$$- \frac{\partial}{\partial s} f(x, y; s, t) = m(x, s) \frac{\partial}{\partial x} f(x, y; s, t) + \frac{1}{2} v(x, s) \frac{\partial^2}{\partial x^2} f(x, y; s, t). \qquad (6.51)$$

Since $f(x, y; s, t)$ is a probability density function, obviously $f(x, y; s, s) = \delta(y - x)$, Dirac's δ-function. $\qquad \square$

Example 6.9. The Wright model in population genetics. In Example 1.11, we have considered the Wright model in population genetics. In this model, there are two alleles A and a and $\{X(t), t \in T\}$ is the number of A

allele in a large diploid population of size N, where $T = \{0, 1, \ldots, \infty\}$. This is a Markov model with one-step transition probabilities given by:

$$\Pr\{X_1(t+1) = j | X(t) = i\} = \binom{2N}{j} p_{t+1}^j q_{t+1}^{2N-j},$$

where p_{t+1} is the frequency of A allele at generation $t+1$ and $q_{t+1} = 1 - p_{t+1}$. It follows that with $x = \frac{i}{2N}$,

$$E\left\{\left(\frac{j}{2N} - \frac{i}{2N}\right)^r \bigg| X(t) = i\right\} = O(N^{-2}), \quad \text{for } r \geq 3\,;$$

$$E\left\{\left(\frac{j}{2N} - \frac{i}{2N}\right) \bigg| X(t) = i\right\} = p_{t+1} - x = \frac{1}{2N}m(x,t) + O(N^{-2}),$$

where

$$m(x,t) = (2N)(p_{t+1} - x) + O(N^{-1})\,;$$

and

$$E\left\{\left(\frac{j}{2N} - \frac{i}{2N}\right)^2 \bigg| X(t) = i\right\} = (p_{t+1} - x)^2 + E\left\{\left(\frac{j}{2N} - p_{t+1}\right)^2 \bigg| X(t) = i\right\}$$

$$= (p_{t+1} - x)^2 + \frac{1}{2N}p_{t+1}(1 - p_{t+1})$$

$$= \frac{1}{2N}v(x,t) + O(N^{-2}),$$

where

$$v(x,t) = (2N)(x - p_{t+1})^2 + p_{t+1}(1 - p_{t+1}) + O(N^{-1}).$$

If $m(x,t)$ is a bounded functions of x and t for all $i \geq 0$ and for all $t \geq 0$, then, by Theorems 6.4 and 6.5, to order of $O(N^{-2})$, $\{Y(t) = \frac{X(t)}{2N}, t \geq 0\}$ is a diffusion process with state space $S = [0, 1]$ and with diffusion coefficients $\{m(x,t), v(x,t)\}$ (cf. Remark 6.3).

Case 1: Genetic drift. In this case, we assume that there are no mutations, no immigration and migration and no selection. Then, given $X(t) = i$, $p_{t+1} =$

$\frac{i}{2N} = x$ so that $m(x,t) = 0$ and $v(x,t) = x(1-x)$. In this case, the diffusion process is a homogeneous process.

Case 2: With mutation only. In this case we assume that there are no immigration, no migration and no selection but there are mutations from A to a with rate $u(t)$ and from a to A with rate $v(t)$. Because the mutation processes are rare events, one may assume that $u(t) = \frac{\beta_1(t)}{2N} + o(N^{-1})$ and $v(t) = \frac{\beta_2(t)}{2N} + o(N^{-1})$, where $\beta_i(t)(i = 1,2)$ are bounded functions of t for all $t \geq 0$. Then, with $x = \frac{i}{2N}$:

$$p_{t+1} = (1 - u(t))\frac{i}{2N} + \left\{1 - \frac{i}{2N}\right\}v(t)$$

$$= x + \{(1-x)\beta_2(t) - x\beta_1(t)\}\frac{1}{2N} + O(N^{-2}).$$

It follows that

$$p_{t+1} - x = \{(1-x)\beta_2(t) - x\beta_1(t)\}\frac{1}{2N} + O(N^{-2}).$$

This leads to $m(x,t) = \{(1-x)\beta_2(t) - x\beta_1(t)\}$.

Since $(2N)(p_{t+1} - x)^2 = \frac{1}{2N}m^2(x,t) = O(n^{-1})$ and since

$$p_{t+1}(1 - p_{t+1}) = \left[x + \frac{1}{2N}m(x,t)\right]\left[1 - x - \frac{1}{2N}m(x,t)\right] + O(N^{-1})$$

$$= x(1-x) + O(N^{-1}),$$

so, we have $v(x,t) = x(1-x)$.

Case 3: With immigration and migration only. In this case we assume that there are no mutations and no selection but there are immigration and migration. To model this, we follow Wright [2] to assume that the population exchanges the A allele with outside at the rate of $m(t)$ per generation at time t. If $x_I(t)$ is the frequency of the A allele at time t among the immigrants, then the frequency of the A allele at generation $t + 1$ is

$$p_{t+1} = x + m(t)[x_I(t) - x] \quad \text{if given} \quad X(t) = i \text{ so that } Y(t) = \frac{i}{2N} = x.$$

Since the proportion of immigration and migration is usually very small, one may assume $m(t) = \frac{1}{2N}\omega(t)$, where $\omega(t)$ is a bounded function of t; then,

$$m(x,t) = \omega(t)[x_I(t) - x] = -\omega(t)[1 - x_I(t)]x + \omega(t)x_I(t)(1 - x).$$

Similarly, as in Case 2, one can easily show that

$$v(x,t) = x(1 - x).$$

From above, it is obvious that Case 3 is exactly the same as Case 2 if one writes $\omega(t)x_I(t) \sim \beta_2(t)$ and $\omega(t)[1 - x_I(t)] \sim \beta_1(t)$.

Case 4: With selection only. In this case, we assume that there are no mutations, no immigration and no migration but there are selection among different genotypes. To model this case, we assume that the selective values of the three genotypes at the tth generation are given by:

$$
\begin{array}{ccc}
AA & Aa & aa \\
1 + s_1(t) & 1 + s_2(t) & 1
\end{array}
$$

where $s_i(t) = \frac{1}{2N}\alpha_i(t) + O(N^{-2})$ with $\alpha_i(t)$ being a bounded function of time t, $i = 1, 2$.

Then we have:

$$
\begin{aligned}
p_{t+1} &= \frac{1}{1 + s_1(t)x^2 + 2s_2(t)x(1 - x)} \\
&\quad \times \{[1 + s_1(t)]x^2 + [1 + s_2(t)]x(1 - x)\} + O(N^{-2}) \\
&= x + \frac{1}{1 + s_1(t)x^2 + 2s_2(t)x(1 - x)}\{x + s_1(t)x^2 + s_2(t)x(1 - x)\} \\
&\quad - x + O(N^{-2}) = x + x(1 - x)\{s_2(t)(1 - 2x) + s_1(t)x\} + O(N^{-2}) \\
&= x + \{x(1 - x)[\alpha_1(t)x + \alpha_2(t)(1 - 2x)]\}\frac{1}{2N} + O(N^{-2}) \\
&= x + m(x,t)(2N)^{-1} + O(N^{-2}),
\end{aligned}
$$

where $m(x,t) = x(1 - x)[\alpha_1(t)x + \alpha_2(t)(1 - 2x)]$.

Using this result, as in Cases 2 and 3, we obtain:

$$v(x,t) = x(1 - x) + O(N^{-1}).$$

Case 5: The case with mutations and selection. In this case, we assume that there are mutations as given in Case 2 and there are selections between the genotypes as given in Case 4. If mutations take place after selection, the frequency p_{t+1} of A at the next generation given $X(t) = i$ is then:

$$p_{t+1} = [1 - u(t)]\frac{[1 + s_1(t)]x^2 + [1 + s_2^2(t)]x(1 - x)}{1 + s_1(t)x^2 + 2s_2(t)x(1 - x)}$$

$$+ v(t)\frac{[1 + s_2(t)]x(1 - x) + (1 - x)^2}{1 + s_1(t)x^2 + 2s_2(t)x(1 - x)}$$

where $x = i/(2N)$.

Hence, on substituting $s_i(t) = \alpha_i(t)/(2N)$ and $\{u(t) = \beta_1(t)/(2N), v(t) = \beta_2(t)/(2N)\}$, we obtain:

$$p_{t+1} - x = \frac{1}{1 + s_1(t)x^2 + 2s_2(t)x(1 - x)}\{[1 + s_1(x)]x^2 + [1 + s_2(t)]x(1 - x)\}$$

$$- x - \beta_1(t)x/(2N) + \beta_2(t)(1 - x)/(2N) + O(N^{-2})$$

$$= \{x(1 - x)[\alpha_2(t)(1 - 2x) + \alpha_1(t)x] - \beta_1(t)x$$

$$+ \beta_2(t)(1 - x)\}\frac{1}{2N} + O(N^{-2})$$

$$= m(x, t)\frac{1}{2N} + O(N^{-2}),$$

where

$$m(x, t) = x(1 - x)[\alpha_2(t)(1 - 2x) + \alpha_1(t)x] - \beta_1(t)x + \beta_2(t)(1 - x).$$

Hence, as in Cases 2–4, we obtain

$$v(x, t) = x(1 - x) + O(N^{-1}).$$

(By assuming mutations occurring first and then selection, one may obtain exactly the same $m(x, t)$ and $v(x, t)$ as above; see Exercise 6.4.)

Case 6: The general case. In the general case, we assume that there are mutations as given in Case 2, immigration and migration as in Case 3 and there are selections between the genotypes as given in Case 4. Then by the same approach, we have:

$$p_{t+1} = x + \{x(1-x)[\alpha_2(t)(1-2x) + \alpha_1(t)x] - \beta_1(t)x + \beta_2(t)(1-x)$$

$$- \omega(t)[1 - x_I(t)]x + \omega(t)x_I(t)(1-x)\}\frac{1}{2N} + O(N^{-2})$$

$$= x + m(x,t)\frac{1}{2N} + O(N^{-2}),$$

where $m(x,t) = x(1-x)[\alpha_2(t)(1-2x) + \alpha_1(t)x] - \gamma_1(t)x + \gamma_2(t)(1-x)$, $\gamma_1(t) = \beta_1(t) + \omega(t)[1 - x_I(t)]$, and $\gamma_2(t) = \beta_2(t) + \omega(t)x_I(t)$, and

$$v(x,t) = x(1-x) + O(N^{-1}).$$

Remark 6.3. In the above cases, the variance is $v(x) = x(1-x)$. This variance has also been used in [10, 11]. Notice that the variance of the frequency of the gene under binomial distribution is $\frac{1}{2N}x(1-x) = x(1-x)\Delta t$, where $\Delta t = \frac{1}{2N}$. Hence the above is the correct variance. In Crow and Kimura [12], instead of using $v(x) = x(1-x)$, they have used $v(x) = \frac{1}{2N}x(1-x)$.

6.6. Diffusion Approximation of Finite Birth-Death Processes

By using the Kolmogorov forward equations and by using similar procedures, it can readily be shown that finite birth-death processes are closely approximated by diffusion processes. To this end, let $\{X(t), t \in T = [0,\infty)\}$ be a finite birth-death process with birth rate $b_j(t)$, death rate $d_j(t)$ and with state space $S = \{0, 1, 2, \ldots, M\}$. Let $Y(t) = \frac{1}{M}X(t)$ and suppose that $b_k(t) = M\sum_{j=0}^{n_1}\beta_j(t)(\frac{k}{M})^j$ and $d_k(t) = M\sum_{j=0}^{n_2}\delta_j(t)(\frac{k}{M})^j$, where $\beta_j(t) \geq 0, \delta_j(t) \geq 0$ are independent of k. Then, the following theorem shows that to the order of $O(M^{-2})$, $\{Y(t), t \geq 0\}$ follows a diffusion process with state space $S = [0,1]$ and with coefficients $\{m(y,t), v(y,t)\}$, where

$$m(y,t) = \sum_{j=0}^{n_1}\beta_j(t)y^j - \sum_{j=0}^{n_2}\delta_j(t)y^j,$$

and

$$v(y,t) = \sum_{j=0}^{n_1}\beta_j(t)y^j + \sum_{j=0}^{n_2}\delta_j(t)y^j.$$

Theorem 6.6. *Let $f(p, x; s, t)$ be the conditional pdf of $Y(t)$ given $Y(s) = p$. Then, to the order of $O(M^{-2})$, $f(p, y; s, t)$ satisfies the following partial differential equation:*

$$\frac{\partial}{\partial t} f(p, y; s, t) = -\frac{\partial}{\partial y}\{m(y, t)f(p, y; s, t)\} + \frac{1}{2M}\frac{\partial^2}{\partial y^2}\{v(y, t)f(p, y; s, t)\},$$

$$(6.52)$$

where $0 \leq y \leq 1$, with $f(p, y; s, s) = \delta(y - p)$, the Dirac's δ-function.

Proof. By using the Kolmogorov forward equation, the proof is very similar to that of Theorem 6.4. Hence we leave it as an exercise; see Exercise 6.5. □

Theorem 6.7. *Let $Y(t) = \frac{1}{M}X(t)$ and let $b_k(t)$ and $d_k(t)$ be as given in Theorem 6.6. If, to the order of $O(M^{-2})$, the conditional probability density $f(p, y; s, t)$ of $Y(t)$ given $\frac{1}{M}X(s) = p$ is an analytic function of p for $0 \leq p \leq 1$, then, to the order of $O(M^{-2})$, $f(p, y; s, t)$ also satisfies*

$$-\frac{\partial}{\partial s} f(p, y; s, t) = m(p, s)\frac{\partial}{\partial p} f(p, y; s, t) + \frac{1}{2M}v(p, s)\frac{\partial^2}{\partial p^2} f(p, y; s, t),$$

$$(6.53)$$

where $f(p, y; s, s) = \delta(y - p)$.

Proof. By using the Kolmogorov backward equation, the proof of Theorem 6.7 is very similar to that of Theorem 6.5. Hence we leave it as an exercise; see Exercise 6.6. □

Example 6.10. Applications to the analysis of Moran's model of genetics. In the Moran's model of genetics as described in Example 4.2 in Sec. 4.2, we have

$$b_j = (M - j)\lambda_2\left\{\alpha_2 + \frac{1}{M}j(1 - \alpha_1 - \alpha_2)\right\}$$

$$= M\left\{\alpha_2\lambda_2 + \lambda_2\left(1 - \alpha_1 - 2\alpha_2\right)\frac{j}{M} - \lambda_2(1 - \alpha_1 - \alpha_2)\left(\frac{j}{M}\right)^2\right\}$$

and

$$d_j = j\lambda_1 \left\{ (1 - \alpha_2) - \frac{1}{M} j (1 - \alpha_1 - \alpha_2) \right\}$$

$$= M \left\{ \lambda_1 (1 - \alpha_2) \frac{j}{M} - \lambda_1 (1 - \alpha_1 - \alpha_2) \left(\frac{j}{M} \right)^2 \right\}.$$

By Theorems 6.6 and 6.7, $Y(t) = \frac{1}{M} X(t)$ can be approximated by a diffusion process (valid to the order of $O(M^{-2})$) with

$$m(x) = \alpha_2 \lambda_2 + \lambda_2 (1 - \alpha_1 - 2\alpha_2)x - \lambda_2 (1 - \alpha_1 - \alpha_2)x^2$$

$$- \lambda_1 (1 - \alpha_2)x + \lambda_1 (1 - \alpha_1 - \alpha_2)x^2$$

$$= \lambda_2 (1 - x)[x(1 - \alpha_1) + (1 - x)\alpha_2] - \lambda_1 x[(1 - \alpha_2)(1 - x) + \alpha_1 x]$$

and

$$\frac{1}{M} v(x) = \frac{1}{M} \{ \lambda_2 (1 - x)[x(1 - \alpha_1) + (1 - x)\alpha_2]$$

$$+ \lambda_1 x[(1 - \alpha_2)(1 - x) + \alpha_1 x] \}.$$

6.7.   Complements and Exercises

Exercise 6.1.   Consider a two-stage model of carcinogenesis under constant chemotherapy and immuno-stimulation [13]. Then one may assume that $\xi(t) = d(t) - b(t) > 0$, where $b(t)$ and $d(t)$ are the birth rate and death rate of initiated cells (I cells). Let $N(0) = N_0$ and let $\alpha_N(t)$ denote the mutation rate from $N \to I$. If N_0 is very large, then it is shown in Example 6.7 that to the order of $O(N_0^{-2})$, $\{Y(t) = I(t)/N_0, t \geq 0\}$ is a diffusion process with state space $S = [0, \infty)$ and with diffusion coefficient

$$\left\{ m(y,t) = \alpha_N(t) - y\xi(t) = \frac{1}{N_0}\lambda(t) - y\xi(t)\,, v(y,t) = \frac{1}{N_0}y\omega(t) \right\},$$

where $\omega(t) = d(t) + b(t)$. Assume that $\{b(t) = b, d(t) = d, \alpha_N(t) = \alpha_N\}$ are independent of time t. Denote by $\{\gamma_1 = 2\lambda/\omega, \gamma_2 = 2N_0\xi/\omega\}$. Show that the solution of the Kolmogorov backward equation gives the conditional pdf

$f(x, y; t)$ of $Y(t)$ given $Y(0) = x$ as:

$$f(x, y; t) = g(y) \left[\sum_{k=0}^{\infty} e^{-k\xi t} L_k^{(\gamma_1)}(\gamma_2 x) L_k^{(\gamma_1)}(\gamma_2 y) \begin{pmatrix} k + \gamma_1 - 1 \\ k \end{pmatrix}^{-1} \right],$$

where

$$g(y) = \frac{\gamma_2^{\gamma_1}}{\Gamma(\gamma_1)} y^{\gamma_1 - 1} e^{-\gamma_2 y}, \quad 0 \le y \le \infty,$$

and where $L_k^{(\gamma)}(y) = \frac{1}{k!} \sum_{j=0}^{k} (-1)^j \binom{k}{j} y^j \frac{\Gamma(k+\gamma)}{\Gamma(j+\gamma)}$ is the Laguerre polynomial with degree k and with parameter γ [14, Chap. 1].

Laguerre polynomials $L_k^{(\gamma)}(y)$ are orthogonal polynomial in y orthogonal with respect to the weight function

$$h(y) = \frac{1}{\Gamma(\gamma)} y^{\gamma - 1} e^{-\gamma y}, \quad 0 \le y \le \infty.$$

(For basic properties of Laguerre polynomials, see [14, Chap. 1].

Exercise 6.2. Let $\{Y(t), t \ge 0\}$ be a continuous Markov process with state space $S = [a, b]$. Assume that $\Delta Y(t) = Y(t + \Delta t) - Y(t) = Z(t)g[Y(t), t] + \epsilon(t)$, where $g(x, t)$ is a deterministic continuous function of x and t and where $\{Z(t), \epsilon(t)\}$ are independently distributed random variables satisfying the conditions:

$$E[Z(t)] = a(t)\Delta t + o(\Delta t), E[\epsilon(t)] = 0;$$

$$\text{Var}[Z(t)] = \sigma_Z^2(t)\Delta t + o(\Delta t),$$

$$\text{Var}[\epsilon(t)] = \sigma^2 \Delta t + o(\Delta t),$$

$$E\{[Z(t)]^k\} = o(\Delta t) \quad \text{and}$$

$$E\{[\epsilon(t)]^k\} = o(\Delta t), \quad \text{for } k = 3, 4, \dots.$$

Show that $\{Y(t), t \ge 0\}$ is a diffusion process with state space $S = [a, b]$ and with diffusion coefficients

$$\{m(x, t) = a(t)g(x, t), \ v(x, t) = \sigma_Z^2(t)g^2(x, t) + \sigma^2\}.$$

Exercise 6.3. Consider the Wright model for one locus with two alleles $A : a$ as defined in Example 6.9. Assume that there are no selection but there are mutations and immigration and migration as given by Cases 2 and 3 in Example 6.9. Then, as shown in Example 6.9, to the order of $O(N^{-2})$, the frequency $Y(t)$ of the A allele is a diffusion process with state space $[0,1]$ and with diffusion coefficients

$$\{m(y) = \gamma_2(1-y) - \gamma_1 y, \ v(y) = y(1-y)\}\,.$$

Let $\mu_k(x,t) = E\{[Y(t)]^k | Y(0) = x\}$ be the kth moment around 0 of $Y(t)$ given $Y(0) = x$. Then, as shown in Sec. 6.3, $\mu_k(x,t)$ satisfy the Kolmogorov backward equation with initial condition $\mu_k(x,0) = \delta(1-x)$. As shown by Example 6.4, this equation can be solved in terms of Jacobi polynomials. Derive these moments.

Exercise 6.4. Derive the diffusion coefficients in Case 5 of the Wright model in Example 6.9.

Exercise 6.5. Prove Theorem 6.6.

Exercise 6.6. Prove Theorem 6.7.

Exercise 6.7. Let $\{X(t), t \geq 0\}$ be a diffusion process with state space $S = [0,1]$ and with diffusion coefficients $\{m(x) = 0, v(x) = V_s x^2 (1-x)^2\}$. This is the model for random fluctuation of selection intensity introduced by Kimura [15] in population genetics. The Kolmogorov forward equation of this process is given by

$$\frac{\partial}{\partial t} f(p,x;t) = \frac{1}{2} V_s \frac{\partial^2}{\partial x^2} \{x^2(1-x)^2 f(p,x;t)\}, \quad 0 < x < 1\,.$$

(a) Let $\xi = \xi(x)$ be a function of x and make the transformation $u(p,\xi;t) = e^{\lambda t} g(x) f(p,x;t) = e^{\lambda t} g^{(*)}(\xi) f(p,x;t)$. Show that if $\{\xi = \xi(x) = \log x/(1-x), \lambda = V_s/8\}$ and if $g(x) = \frac{1}{2}[x(1-x)]^{3/2} = \frac{1}{2}e^{\frac{3}{2}\xi}$, then $u(p,x;t)$ satisfies the following heat equation:

$$\frac{\partial}{\partial t} u(p,\xi;t) = \frac{1}{2} V_s \frac{\partial^2}{\partial \xi^2} u(p,\xi;t)\,, \quad -\infty < \xi < \infty\,, \tag{6.54}$$

with initial condition $u(p,\xi;0) = g(x)\delta(x-p) = g^{(*)}(\xi)\delta(x-p)$.

If one does not know $\{g(x), \xi(x)\}$, then the following hint leads to the solution:

(Hint: With the help of Eq. (6.54), work out $\frac{\partial}{\partial t} u(p, \xi; t)$ and $\frac{1}{2} V_s \frac{\partial^2}{\partial \xi^2} u(p, \xi; t)$ as a linear combinations of $f(p, x; t), f'(p, x; t)$ and $f''(p, x; t)$ and equal coefficients of $f(p, x; t), f'(p, x; t)$ and $f''(p, x; t)$ from $\frac{\partial}{\partial t} u(p, \xi; t)$ to those of $\frac{1}{2} V_s \frac{\partial^2}{\partial \xi^2} u(p, \xi; t)$ respectively.)

(b) Using results of (a), show that the solution $f(p, x; t)$ is

$$
f(p, x; t) = \frac{1}{\sqrt{2\pi V_s t}} \int_{-\infty}^{\infty} e^{-\frac{1}{2V_s t}(\xi - x)^2} u(p, \xi; 0) d\xi
$$

$$
= \frac{1}{\sqrt{2\pi V_s t}} \frac{\exp(-V_s t/8)}{\{x(1-x)\}^{2/3}}
$$

$$
\times \int_0^1 \exp\left\{ -\frac{1}{2V_s t} \left(\log \frac{x(1-z)}{(1-x)z} \right)^2 \right\} [z(1-z)]^{1/2} f(p, z; 0) dz
$$

$$
= \frac{1}{\sqrt{2\pi V_s t}} \frac{[p(1-p)]^{1/2}}{\{x(1-x)\}^{2/3}}
$$

$$
\times \exp\left\{ -V_s t/8 - \frac{1}{2V_s t} \left(\log \frac{x(1-p)}{(1-x)p} \right)^2 \right\}.
$$

6.8.   Appendix

In the appendices, we first provide a general proof of Theorem 6.1. Then, to make the book self-contained, we give in Subsec. 6.8.2 some general results of Jacobi polynomials. In Subsec. 6.8.3, we provide some general results and discussions of eigenvalues and eigenfunctions of differential equations as well as the Green's formulae for differential equations.

6.8.1.   *A general proof of Theorem 6.1*

Let $Q(x)$ be an arbitrary continuous function with continuous first and second derivatives $Q'(x)$ and $Q''(x)$ in $[a, b]$ and with $Q(x) = 0$ for $x \neq [a, b]$. By continuity of $Q(x)$ in $[a, b]$, one must have $Q'(a) = Q'(b) = Q''(a) = Q''(b) = Q(a) = Q(b) = 0$. Note that the class of functions $Q(x)$ satisfying the above conditions is not empty; for example, for $-\infty < a < b < \infty$, one may chose $Q(x) = 0$ for $x \neq [a,b]$ and $Q(x) = (x-a)^3 (b-x)^3 \exp(-\theta x^2)$, with $\theta > 0$ arbitrary.

Consider now the integral

$$I = \int_{-\infty}^{\infty} Q(y) \left[\frac{\partial f(u,y;s,t)}{\partial t} \right] dy = \int_{a}^{b} Q(y) \left[\frac{\partial f(u,y;s,t)}{\partial t} \right] dy \,.$$

Since $Q(y)$ and $[\frac{\partial f(u,y;s,t)}{\partial t}]$ are continuous functions of y, I exists and

$$I = \int_{a}^{b} Q(y) \left[\frac{\partial f(u,y;s,t)}{\partial t} \right] dy = \frac{\partial}{\partial t} \int_{a}^{b} Q(y) f(u,y;s,t) dy$$

$$= \lim_{\Delta t \to 0} \frac{1}{\Delta t} \int_{a}^{b} Q(y)[f(u,y;s,t+\Delta t) - f(u,y;s,t)] dy \,.$$

But, by the Chapman–Kolmogorov equation,

$$f(u,y;s,t+\Delta t) = \int_{-\infty}^{\infty} f(u,z;s,t) f(z,y;t,t+\Delta t) dz \,.$$

Thus,

$$I = \lim_{\Delta t \to 0} \frac{1}{\Delta t} \int_{a}^{b} Q(y) \left\{ \int_{-\infty}^{\infty} f(u,z;s,t) f(z,y;t,t+\Delta t) dz - f(u,y;s,t) \right\} dy \,.$$

Interchanging the order of integration and noting $Q(x) = 0$ for $x \neq [a,b]$, we obtain:

$$I = \lim_{\Delta t \to 0} \frac{1}{\Delta t} \int_{-\infty}^{\infty} f(s,u;t,z) \left\{ \int_{a}^{b} Q(y) f(z,y;t,t+\Delta t) dy - Q(z) \right\} dz \,.$$

Choose $\delta > 0$ in such a way that $y \in [a,b]$ and $|y-z| \leq \delta$ imply $z \in [a,b]$. Also, $Q(x)$ is continuous in $[a,b]$ so that there exists a constant $M > 0$ satisfying $|Q(x)| \leq M$ for all x. For such a $\delta > 0$, we have then:

$$\int_{|y-z|>\delta} Q(y) f(z,y;t,t+\Delta t) dy \leq M \int_{|y-z|>\delta} f(z,y;t,t+\Delta t) dy$$

$$= M \Pr\{|X(t+\Delta t) - X(t)| > \delta | X(t) = z\}$$

$$= o(\Delta t) \,,$$

where $o(\Delta t)$ satisfies $\lim_{\Delta t \to 0} \frac{o(\Delta t)}{\Delta t} = 0$ uniformly for $t \geq 0$ and for $z \in [a,b]$.

It follows that for $\delta > 0$ as chosen above,

$$I = \lim_{\Delta t \to 0} \frac{1}{\Delta t} \int_{-\infty}^{\infty} f(u, z; s, t) \left\{ \int_{|y-z|<\delta} Q(y) f(z, y; t, t + \Delta t) dy - Q(z) \right\} dz \,.$$

Now, $Q''(y)$ is continuous so that for $y \in [a, b]$ and $z \in [a, b]$,

$$Q(y) = Q(z) + (y - z)Q'(z) + \frac{1}{2}(y - z)^2 Q''(z) + O[(y - z)^3] \,.$$

Hence, for $z \in [a, b]$ and $y \in [a, b]$,

$$\lim_{\Delta t \to 0} \frac{1}{\Delta t} \left\{ \int_{|y-z|<\delta} Q(y) f(t, z; t + \Delta t, y) dy - Q(z) \right\}$$

$$= \lim_{\Delta t \to 0} \frac{1}{\Delta t} \left\{ - Q(z) \Pr[|X(t + \Delta t) - X(t)| > \delta | X(t) = z] \right.$$

$$+ Q'(z) \int_{|y-z| \leq \delta} (y - z) f(z, y; t, t + \Delta t) dy$$

$$+ \frac{1}{2} Q''(z) \int_{|y-z| \leq \delta} (y - z)^2 f(z, y; t, t + \Delta t) dy$$

$$+ \left. \int_{|y-z| \leq \delta} O[(y - z)^3] f(z, y; t, t + \Delta t) dy \right\}$$

$$= Q'(z) m(z, t) + \frac{1}{2} Q''(z) v(z, t) \,.$$

It follows that,

$$I = \int_{-\infty}^{\infty} f(s, u; t, z) \left[Q'(z) m(z, t) + \frac{1}{2} Q''(z) v(z, t) \right] dz \,.$$

Integration by parts now gives the results:

$$\int_{-\infty}^{\infty} f(u, z; s, t) m(z, t) Q'(z) dz = \int_{a}^{b} f(u, z; s, t) m(z, t) dQ(z)$$

$$= - \int_{a}^{b} Q(z) \left\{ \frac{\partial}{\partial z} [m(z, t) f(u, z; s, t)] \right\} dz$$

and

$$\int_{-\infty}^{\infty} f(u,z;s,t)\frac{1}{2}Q''(z)v(z,t)dz = \int_a^b \frac{1}{2}v(z,t)f(u,z;s,t)dQ'(z)$$

$$= -\frac{1}{2}\int_a^b Q'(z)\left\{\frac{\partial}{\partial z}[b(z,t)f(u,z;s,t)]\right\}dz$$

$$= -\frac{1}{2}\int_a^b \left\{\frac{\partial}{\partial z}[v(z,t)f(u,z;s,t)]\right\}dQ(z)$$

$$= \frac{1}{2}\int_a^b Q(z)\left\{\frac{\partial^2}{\partial z^2}[v(z,t)f(u,z;s,t)]\right\}dz\,.$$

Thus,

$$I = \int_a^b Q(z)\left[\frac{\partial f(u,z;s,t)}{\partial t}\right]dz$$

$$= \int_a^b Q(z)\left\{-\frac{\partial}{\partial z}[m(z,t)f(u,z;s,t)] + \frac{1}{2}\frac{\partial^2}{\partial z^2}[v(z,t)f(u,z;s,t)]\right\}dz\,;$$

or

$$\int_a^b Q(z)\left\{\frac{\partial}{\partial t}f(u,z;s,t) + \frac{\partial}{\partial z}[m(z,t)f(u,z;s,t)]\right.$$

$$\left. -\frac{1}{2}\frac{\partial^2}{\partial z^2}[v(z,t)f(u,z;s,t)]\right\}dz = 0\,.$$

Since the integrand is a continuous function of z and since $Q(z)$ is arbitrary with continuous first and second derivatives, by the lemma given below,

$$\frac{\partial}{\partial t}f(u,x;s,t) = -\frac{\partial}{\partial x}[m(x,t)f(u,x;s,t)] + \frac{1}{2}\frac{\partial^2}{\partial x^2}[v(x,t)f(u,x;s,t)]\,.$$

The initial condition is obviously $f(u,x;s,s) = \delta(x-u)$.

Lemma 6.2. *Let $f(x)$ be a continuous function defined in $[a,b]$. Assume that $\int_a^b \phi(x)f(x)dx = 0$ whenever $\phi(x)$ satisfies the following two conditions:*

   (i) $\phi(x) = 0$ for $x \neq [a,b]$, and
   (ii) $\phi(x)$ has continuous first and continuous second derivatives in $[a,b]$.

Then, $f(x) \equiv 0$ in $[a,b]$.

Proof of Lemma 6.2. Suppose $f(x) \neq 0$ in $[a, b]$. Then, there exists $x_0 \in [a, b]$ such that $f(x_0) \neq 0$. Assume $f(x_0) > 0$ so that there exists a $\delta > 0$ such that $f(x_0) \geq \delta > 0$. (One may similarly prove the result if $f(x_0) < 0$). Since $f(x)$ is continuous in $[a, b]$, there exists a sub-interval $[\alpha, \beta]$ in $[a, b]$ with $\beta - \alpha > 0$ such that $x_0 \in [\alpha, \beta]$ and $f(x) > 0$ for all $x \in [\alpha, \beta]$. Choose $\epsilon > 0$ such that $\beta > \epsilon$ and $\beta - \alpha - 2\epsilon > 0$ so that $a \leq \alpha < \alpha + \epsilon < \beta - \epsilon < \beta \leq b$. Define $\phi(x) = 0$ if $x \neq [\alpha, \beta]$ and put $\phi(x) = (x - \alpha)^3(\beta - x)^3$ if $\alpha \leq x \leq \beta$. Then, obviously, $\phi(x) = 0$ for $x \neq [a, b]$ and $\phi(x)$ has continuous first derivative and continuous second derivative in $[a, b]$. It follows that $\int_a^b f(x)\phi(x)dx = 0$. But, $\phi(x) \geq 0$ for $\alpha \leq x \leq \beta$ and $\phi(x) > 0$ for $\alpha + \epsilon \leq x \leq \beta - \epsilon$. Hence,

$$0 = \int_a^b f(x)\phi(x)dx = \int_\alpha^\beta f(x)\phi(x)dx \geq \delta \int_{\alpha+\epsilon}^{\beta-\epsilon} \phi(x)dx > 0 \,.$$

This contradicts $\int_a^b f(x)\phi(x)dx = 0$. Thus, one must have $f(x) = 0$ for $x \in [a, b]$. $\qquad\square$

6.8.2. *Jacobi polynomials and some properties*

Jacobi polynomials $J_n(x; a, b)(a > 0, b > 0)$ are orthogonal polynomials in x (n denoting degree) orthogonal with respect to the Beta distribution $f(x) = x^{a-1}(1 - x)^{b-1}/B(a, b), 0 < x < 1$. These polynomials can be derived by using the Gram–Schmidt process described as follows:

(1) Denote by $E[g(x)] = \int_0^1 g(x)f(x)dx$. Then put

$$P_0(x) = 1, \quad P_1(x) = x - E(x) \,.$$

(2) For $k = 2, \ldots, \infty$, put:

$$P_k(x) = x^k - \sum_{j=0}^{k-1} a_{kj}P_j(x)$$

where

$$a_{kj} = E\{x^k P_j(x)\}/E\{P_j^2(x)\}$$

for $j = 0, \ldots, k - 1$.

From the above construction, obviously, one has:

(1) For all $k \neq j$,

$$E\{P_j(x)P_k(x)\} = \int_0^1 P_j(x)P_k(x)f(x)dx = 0.$$

It follows that $J_k(x; a, b)$ is a constant multiple of $P_k(x)$.

(2) For any positive integer k, x^k can be expressed as a linear combination of $P_j(x), j = 0, 1, \ldots, k$ with constant coefficients. It follows that any polynomial in x with degree k can be expressed as a linear combination of $J_i(x; a, b), i = 0, 1, \ldots, k$ with constant coefficients.

6.8.2.1.   *Differential equation for Jacobi polynomials*

The following theorem is useful for explicitly writing down $J_k(x; a, b)$ and for proving some useful results involving Jacobi polynomials.

Theorem 6.8. $J_n(x) = J_n(x; a, b)$ *satisfies the following second order differential equation:*

$$x(1 - x)J_n''(x) + [a - (a + b)x]J_n'(x) + n(n + a + b - 1)J_n(x) = 0. \qquad (6.55)$$

Proof.   To prove the above results, note first that the above equation is equivalent to the following equations:

$$\frac{d}{dx}\{x^a(1 - x)^b J_n'(x)\} = -n(n + a + b - 1)x^{a-1}(1 - x)^{b-1}J_n(x).$$

Next we show that for $J_k(x) = J_k(x; a, b)$ with degree $k < n$, we have:

$$\int_0^1 J_k(x)\frac{d}{dx}\{x^a(1 - x)^b J_n'(x)\}dx = 0.$$

This follows by applying the basic results of integration by parts to give:

$$\int_0^1 J_k(x)\frac{d}{dx}\{x^a(1 - x)^b J_n'(x)\}dx = -\int_0^1 J_k'(x)\left\{x^a(1 - x)^b\frac{d}{dx}J_n(x)\right\}dx$$

$$= \int_0^1 J_n(x)\frac{d}{dx}\{x^a(1 - x)^b J_k'(x)\}dx$$

$$= \int_0^1 \{x^{a-1}(1 - x)^{b-1}J_n(x)\pi_k(x)\}dx = 0$$

as $\pi_k(x)$ is a polynomial in x with degree $k < n$.

Now, obviously,

$$\frac{d}{dx}\{x^a(1-x)^b J'_n(x)\} = x^{a-1}(1-x)^{b-1}\eta_n(x)\,,$$

where $\eta_n(x)$ is a polynomial in x with degree n. The above result then implies that for all $k = 0, 1, \ldots, n-1$,

$$= \int_0^1 \{x^{a-1}(1-x)^{b-1} J_k(x)\eta_n(x)\}dx = 0$$

so that $\eta_n(x) = C J_n(x; a, b)$ for some constant C. It follows that

$$\frac{d}{dx}\{x^a(1-x)^b J'_n(x; a, b)\} = C x^{a-1}(1-x)^{b-1} J_n(x; a, b)\,.$$

Comparing coefficient of $x^{n+a+b-2}$ on both sides gives $C = -n(n+a+b-1)$. This proves the theorem. $\qquad\square$

Notice that from the above theorem, we have the following integral relations between $J_n(x; a, b)$ and $J'_n(x; a, b)$:

$$\frac{d}{dx}\{x^a(1-x)^b J'_n(x)\} = -n(n+a+b-1)x^{a-1}(1-x)^{b-1} J_n(x)\,.$$

6.8.2.2. *An explicit form of Jacobi polynomials*

To obtain an explicit form of $J_n(x; a, b)$, we need to solve the differential equation given by (6.55). Note that with $(c = a, \alpha = -n, \beta = n+a+b-1)$, the above equation is a special case of the Hypergeometric equation given by:

$$x(1-x)f''(x) + [c - (\alpha + \beta + 1)x]f'(t) - \alpha\beta\, f(x) = 0\,. \tag{6.56}$$

To solve Eq. (6.56), consider a series solution

$$f(x) = \sum_{i=0}^{\infty} a_i x^i\,. \tag{6.57}$$

On substituting $f(x)$ in Eq. (6.57) into Eq. (6.56), we obtain:

$$\sum_{i=2}^{\infty} i(i-1)a_i(x^{i-1}-x^i) + \sum_{i=1}^{\infty} ia_i[cx^{i-1}-(\alpha+\beta+1)x^i] - \alpha\beta\sum_{i=0}^{\infty} a_i x^i = 0\,, \tag{6.58}$$

for all $1 \geq x \geq 0$.

Equation (6.58) gives

$$a_{k+1} = \frac{(\alpha + k)(\beta + k)}{(k+1)(c+k)} a_k \quad \text{for } k = 0, 1, \ldots, \infty. \tag{6.59}$$

For any real number a, define $a_{(k)} = 1$ if $k = 0$ and $a_{(k)} = (a+1)\cdots(a+k-1)$ if $k = 1, 2, \ldots, \infty$. If $c \neq 0$, then, from Eq. (6.57), we have, with $a_0 = 1$,

$$a_k = \frac{\alpha_{(k)}\beta_{(k)}}{(k!)c_{(k)}} \quad \text{for } k = 0, 1, \ldots, \infty.$$

It follows that the solution of Eq. (6.56) is given by

$$f(x) = H(\alpha, \beta; c; x) = \sum_{k=0}^{\infty} \frac{\alpha_{(k)}\beta_{(k)}}{(k!)c_{(k)}} x^k.$$

If $\beta = -k$ or $\alpha = -k$ for some positive integer k, then $a_j = 0$ for all $j = k+1, \ldots, \infty$. In this case the solution is a polynomial in x with degree k given by:

$$f(x) = H(\alpha, \beta; c; x) = \sum_{k=0}^{k} \frac{\alpha_{(k)}\beta_{(k)}}{(k!)c_{(k)}} x^k.$$

It follows that $J_n(x; a, b) = CH(-n, n+a+b-1; a; x)$ for some constant C. Now,

$$(-n)_{(k)} = (-n)(-n+1)\cdots(-n+k-1) = (-1)^k(k!)\binom{n}{k},$$

$$\binom{n+a-1}{n} = \frac{1}{n!}(n+a-1)(n+a-2)\cdots(n+a-1-n+1)$$

$$= \frac{1}{n!}a_{(k)}(a+k)_{(n-k)}.$$

If we chose C as $C = \binom{n+a-1}{n}$, then

$$J_n(x; a, b) = \binom{n+a-1}{n}\sum_{k=0}^{n} \frac{(-n)_{(k)}(n+a+b-1)_{(k)}}{(k!)a_{(k)}} x^k$$

$$= \frac{1}{n!}\sum_{k=0}^{n}(-1)^k\binom{n}{k}(n+a+b-1)_{(k)}(a+k)_{(n-k)}x^k. \tag{6.60}$$

(Note: The above choice of C was motivated by the simple form of the Rodrigues's formulae given in Theorem 6.9.)

6.8.2.3. *The Rodrigue's formulae and $E[J_n^2(x; a, b)]$*

For deriving $E[J_n^2(x; a, b)]$, we first give the following Rodrigue's formulae for $J_n(x; a, b) = J_n(x)$.

Theorem 6.9. $J_n(x; a, b)$ *satisfies the following equation:*

$$x^{a-1}(1-x)^{b-1} J_n(x; a, b) = \frac{1}{n!}\frac{d^n}{dx^n}\{x^{n+a-1}(1-x)^{n+b-1}\}. \tag{6.61}$$

The above formulae has been referred to in the literature as the Rodrigue's formulae for Jacobi polynomials.

To prove the above equation, note that by Leibniz' rule,

$$\frac{d^n}{dx^n}\{x^{n+a-1}(1-x)^{n+b-1}\}$$

$$= \sum_{i=0}^{n} \binom{n}{k} \left\{\frac{d^{n-k}}{dx^{n-k}}x^{n+a-1}\right\}\left\{\frac{d^k}{dx^k}(1-x)^{n+b-1}\right\}.$$

Notice that

$$\frac{d^{n-k}}{dx^{n-k}}x^{n+a-1} = (n+a-1)\cdots(n+a-1-(n-k)+1)x^{a-1+k}$$

$$= \{(n-k)!\}\binom{n+a-1}{n-k}x^{a-1+k}$$

and

$$\frac{d^k}{dx^k}(1-x)^{n+b-1} = (-1)^k(n+b-1)\cdots(n+b-1-k+1)(1-x)^{b-1+n-k}$$

$$= (-1)^k(k!)\binom{n+b-1}{k}(1-x)^{b-1+n-k}.$$

Hence we have:

$$\frac{d^n}{dx^n}\{x^{n+a-1}(1-x)^{n+b-1}\} = (n!)x^{a-1}(1-x)^{b-1}\sum_{k=0}^{n}(-1)^k$$

$$\times \binom{n+a-1}{n-k}\binom{n+b-1}{k}x^k(1-x)^{n-k}$$

$$= x^{a-1}(1-x)^{b-1}\xi_n(x)$$

where $\xi_n(x)$ is a polynomial in x of degree n.

Since for any polynomial $\rho_k(x)$ in x of degree k $(k < n)$, $\frac{d^n}{dx^n}\rho_k(x) = 0$, we have, by applying integration by parts n times:

$$\int_0^1 \rho_k(x)\frac{d^n}{dx^n}\{x^{n+a-1}(1-x)^{n+b-1}\}dx$$

$$= (-1)^n \int_0^1 x^{n+a-1}(1-x)^{n+b-1}\frac{d^n}{dx^n}\rho_k(x)dx = 0.$$

It follows that for $k < n$,

$$\int_0^1 x^{a-1}(1-x)^{b-1}\xi_n(x)J_k(x;a,b)dx = 0.$$

Hence, $\xi_n(x) = CJ_n(x;a,b)$ for some constant C and

$$\frac{d^n}{dx^n}\{x^{n+a-1}(1-x)^{n+b-1}\} = Cx^{a-1}(1-x)^{b-1}J_n(x;a,b).$$

Comparing coefficient of $x^{n+a+b-2}$ on both sides of the above equation and noting the explicit form of $J_n(x;a,b)$ given in (6.60), we obtain $C = n!$.

From the above proof, we also derive another explicit form for $J_n(x;a,b)$ as

$$J_n(x;a,b) = \sum_{k=0}^{n}(-1)^k\binom{n+a-1}{n-k}\binom{n+b-1}{k}x^k(1-x)^{n-k}. \qquad (6.62)$$

By using the Rodrigue's formulae, we have that

$$\frac{d^n}{dx^n}J_n(x;a,b) = (-1)^n(n+3)\cdots(2n+a+b-2)$$

$$= (-1)^n(n!)\binom{2n+a+b-2}{n}.$$

Hence, by using integration by parts repeatedly:

$$E\{J_n^2(x; a, b)\} = \frac{1}{B(a, b)} \int_0^1 x^{a-1}(1 - x)^{b-1} J_n^2(x; a, b) dx$$

$$= \frac{1}{B(a, b)(n!)} \int_0^1 J_n(x; a, b) \left\{ \frac{d^n}{dx^n} [x^{n+a-1}(1 - x)^{n+b-1}] \right\} dx$$

$$= \frac{1}{B(a, b)(n!)} (-1)^n \int_0^1 x^{n+a-1}(1 - x)^{n+b-1} \left\{ \frac{d^n}{dx^n} J_n(x; a, b) \right\} dx$$

$$= \frac{1}{B(a, b)} \binom{2n + a + b - 2}{n} \int_0^1 x^{n+a-1}(1 - x)^{n+b-1} dx$$

$$= \frac{B(n + a, n + b)}{B(a, b)} \binom{2n + a + b - 2}{n}. \tag{6.63}$$

If $a = b = 2$, then

$$E\{J_n^2(x; 2, 2)\} = \frac{B(n + 2, n + 2)}{B(2, 2)} \binom{2n + 2}{n} = \frac{6(n + 1)}{(n + 2)(2n + 3)}.$$

6.8.3. *Some eigenvalue and eigenfunction problems in differential equations*

Let S denote the differential operator defined by

$$S = \alpha(x) \frac{d^2}{dx^2} + \beta(x) \frac{d}{dx} + q(x).$$

Consider the equation $S[f(x)] = -\lambda f(x)$. Suppose that this equation is satisfied by some non-zero constants λ_k $(k = 0, 1, \ldots)$ and some functions $f_{k,j}(x)$ $(k = 0, 1, \ldots, j = 1, \ldots, n_k)$ defined over some domain $[a, b]$. That is,

$$S[f_{k,j}(x)] = -\lambda_k f_{k,j}(x).$$

These non-zero constants λ_k are referred to as the eigenvalues of the operator S and the functions $f_{k,j}(x), j = 1, \ldots, n_k$ the eigenfunctions corresponding to the eigenvalue λ_k. In this section, we will give some basic results concerning eigenvalues and eigenfunctions. Specifically, we will prove the following results:

(1) If the operator S is self-adjoint (to be defined below) for functions defined over $[a, b]$, then all eigenvalues are real numbers.

(2) Let $f_{i,j_1}(x)$ and $f_{k,j_2}(x)$ be eigenfunctions corresponding the eigenvalues λ_i and λ_k respectively. If S is self-adjoint and if $\lambda_i \neq \lambda_k$, then

$$\int_a^b \sigma(x) f_{i,j_1}(x) f_{k,j_2}(x) dx = 0,$$

where $\sigma(x) = \{c/\alpha(x)\} \exp\{\int_d^x \beta(y)/\alpha(y) dy\}$ with c and d being some properly chosen constants.

For defining self-adjoint operator and for proving the above properties, we first prove the following result. This result has been referred to as Green's formulae in the literature.

Theorem 6.10. (**The Green's formulae**). *Let $\sigma(x)$ be defined above and let $\kappa(x) = c\exp\{\int_d^x \beta(y)/\alpha(y) dy\}$. Then, for any two twice differentiable functions $f(x)$ and $g(x)$ defined over $[a, b]$, we have:*

$$g(x)S[f(x)] - f(x)S[g(x)] = \frac{1}{\sigma(x)} \frac{d}{dx} \{\kappa(x)[g(x)f'(x) - f(x)g'(x)]\}.$$

Writing $S[f(x)]$ as $S[f(x)] = \frac{1}{\sigma(x)} \{\kappa(x)f'(x)\} + q(x)f(x)$, the above result is straightforward. It is therefore left as an exercise.

Definition 6.1. The operator S is referred to as a *self-adjoint operator* if for any two twice differentiable functions defined over $[a, b]$,

$$\{\kappa(x)[g(x)f'(x) - f(x)g'(x)]\}_a^b = \{\kappa(b)[g(b)f'(b) - f(b)g'(b)]\}$$

$$- \{\kappa(a)[g(a)f'(a) - f(a)g'(a)]\} = 0.$$

Obviously, the above condition holds if $f(a) = f(b) = 0$ and $g(a) = g(b) = 0$; or $f'(a) = c_1 f(a)$, $f'(b) = c_2 f(b)$, $g'(a) = c_1 g(a)$ and $g'(b) = c_2 g(b)$ for some constants c_1 and c_2.

Theorem 6.11. *If the operator S is self-adjoint, then all eigenvalues of S are real.*

Proof. Let $\lambda_k^{(*)}$ be the conjugate of the eigenvalue λ_k and $f^{(*)}(x)$ is the conjugate of $f(x)$. Then,

$$S[f^{(*)}(x)] = -\lambda_k^{(*)} f^{(*)}(x).$$

Hence,

$$\int_a^b \sigma(x)\{f(x)S[f^{(*)}(x)] - f^{(*)}(x)S[f(x)]\}dx$$

$$= (\lambda_k^{(*)} - \lambda_k)\int_a^b \sigma(x)|f(x)|^2 dx\,.$$

On the other hand, since S is self-adjoint,

$$\int_a^b \sigma(x)\{f(x)Sf^{(*)}(x) - f^{(*)}(x)Sf(x)\}dx$$

$$= (\lambda_k^{(*)} - \lambda_k^{(*)})\int_a^b \sigma(x)|f(x)|^2 dx = 0\,.$$

Since $\int_a^b \sigma(x)|f(x)|^2 dx > 0$, so, $\lambda_k^{(*)} = \lambda_k$ for all $k = 1,\ldots$. That is, all eigenvalues are real numbers and all eigenfunctions are real-valued functions. $\qquad\square$

Theorem 6.12. *Let $f_{k,j}(x)$ be eigenfunctions corresponding to the eigenvalue λ_k. If the operator S is self-adjoint, then*

$$\int_a^b \sigma(x)f_{k,j}(x)f_{i,l}(x)dx = 0 \quad for\ all\ k \neq i\,.$$

Proof. Since S is self-adjoint, we have:

$$\int_a^b \sigma(x)\{f_{i,l}(x)S[f_{k,j}(x)] - f_{k,j}(x)S[f_{i,j}(x)]\}dx = 0\,.$$

Thus,

$$\int_a^b \sigma(x)\{f_{i,l}(x)S[f_{k,j}(x)] - f_{k,j}(x)S[f_{i,j}(x)]\}dx$$

$$= (\lambda_i - \lambda_k)\int_a^b \sigma(x)f_{k,j}(x)f_{i,l}(x)dx = 0\,.$$

Since $\lambda_i \neq \lambda_k$, so

$$\int_a^b \sigma(x)f_{k,j}(x)f_{i,l}(x)dx = 0\,. \qquad\square$$

References

[1] M. G. Kendall, A. Stuart and J. K. Ord, *The Advanced Theory of Statistics, Vol. 1* (Fifth Edition), Oxford University Press, New York (1987).

[2] S. Wright, *Evolution And the Genetics of Populations*, Vol. 2, *The Theory of Gene Frequency*, University of Chicago Press, Chicago (1969).

[3] M. Kimura, *Diffusion Models in Population Genetics, Methuen's Monographs on Applied Probability and Statistics*, Vol. 2, Methuen & Co, LTD, London (1964).

[4] W. Y. Tan, *Stochastic logistic growth and applications*, in: *Logistic Distributions*, ed. B. N. Balakrishnan, Marcel Dekker, Inc., New York (1991) 397–426.

[5] W. Y. Tan and S. Piatadosi, *On stochastic growth process with application to stochastic logistic growth*, Statistica Sinica **1** (1991) 527–540.

[6] W. Feller, *Diffusion process in genetics, Proc. 2nd Berkeley Symposium Math. Statist., Probab.*, University of California Press, Berkeley (1951) 227–246.

[7] CDC, *HIV/AIDS: Surveillance Report*, Atlanta, Georgia (1993).

[8] W. J. Ewens, *Numerical results and diffusion approximations in genetic process*, Biometrika **50** (1963) 241–249.

[9] R. A. Fisher, *The Genetical Theory of Natural Selection* (Second Edition), Dover, New York (1958).

[10] W. J. Ewens, *Population Genetics*, Methuen, London (1969).

[11] P. A. P. Moran. *The Statistical Processes of Evolutionary Theory*, Clarendon, London (1962).

[12] J. F. Crow and M. Kimura, *An Introduction to Population genetics Theory*, Harper and Row, New York (1970).

[13] W. Y. Tan and C. C. Brown, *A stochastic model for drug resistance and immunization, I. One drug case*, Math. Biosciences **97** (1989) 145–160.

[14] W. Y. Tan and M. L. Tiku, *Sampling Distributions in Terms of Laguerre Polynomials With Applications*, New Age International Publisher, New Delhi, India (1999).

[15] M. Kimura, *Process leading to quasi-fixation of genes in natural populations due to random fluctuation of selection intensity*, Genetics **39** (1954) 280–295.

Chapter 7

Asymptotic Distributions, Stationary Distributions and Absorption Probabilities in Diffusion Models

In Chap. 6, we have shown that many processes in genetics and biomedical problems can be closely approximated by diffusion processes. Although it is possible to solve the Kolmogorov forward or backward equations to derive the conditional pdf $f(x, y; t)$ in some cases, in most of the cases, the solution is very difficult, if not impossible. If the eigenvalues and eigenfunctions of the equation exist, however, in many cases one may derive asymptotic distributions by approximating these eigenvalues and eigenfunctions. In this chapter, we will thus illustrate how to approximate the eigenvalues and the eigenfunctions whenever exist.

In diffusion processes, in most of the cases the processes will eventually converge to stationary distributions. Hence it is of considerable interest to derive such stationary distributions whenever exists. In this chapter we will illustrate how to derive these stationary distributions and illustrate its applications to some genetic and biomedical models.

In diffusion processes in which there are absorbing states, it is also of considerable interests to compute the absorption probabilities and the moments of first absorption times. In this chapter we will also develop procedures to compute these absorption probabilities and to compute the moments of first absorption times, in particular the mean and the variance of first absorption times.

7.1. Some Approximation Procedures and Asymptotic Distributions in Diffusion Models

In diffusion processes, theoretically one may derive the conditional pdf $f(x, y; s, t)$ by solving the Kolmogorov forward or backward equations. In most practical problems, however, it is often very difficult, if not impossible, to solve these partial differential equations. On the other hand, in many biological systems and in population genetics, in many cases the eigenvalues and eigenfunctions often exist and are real; in these cases one may derive close approximations to the pdf's by approximating the eigenvalues and eigenfunctions. In this section we thus illustrate some basic approaches to derive these approximations. It turns out that for all examples in Chap. 6, the procedures given below are applicable to derive approximate and asymptotic distributions.

To illustrate, suppose that we have a homogeneous diffusion process $\{X(t), t \geq 0\}$ with state space $S = [a, b]$ and with diffusion coefficients $\{m(x), v(x)\}$. Then the backward equation is:

$$\frac{\partial}{\partial t} f(x, y; t) = m(x) \frac{\partial}{\partial x} f(x, y; t) + \frac{1}{2} v(x) \frac{\partial^2}{\partial x^2} f(x, y; t), \qquad (7.1)$$

where $f(x, y; 0) = \delta(y - x)$.

Making the transformation $f(x, y; t) = e^{-\lambda t} \eta(x) h(x)$, where $\eta(x)$ is a given function of x, then $h(x)$ satisfies the following equation:

$$\alpha(x) h''(x) + \beta(x) h'(x) + \left\{ \lambda + \frac{1}{N} q(x) \right\} h(x) = 0, \qquad (7.2)$$

where $\alpha(x) = \frac{1}{2} v(x)$, $\beta(x) = m(x) + v(x) \frac{d}{dx} \log[\eta(x)]$ and $q(x) = N\{m(x) \frac{d}{dx} \log[\eta(x)] + v(x) \frac{\eta''(x)}{2\eta(x)}\}$.

If the eigenvalues $(\lambda_j, j = 1, \dots, \infty)$ of Eq. (7.2) exist and are real, then the general solution of Eq. (7.1) is given by

$$f(x, y; t) = \eta(x) \sum_{j=1}^{\infty} C_j(y) e^{-\lambda_j t} h_j(x),$$

where $h_j(x)$ is an eigenfunction of Eq. (7.2) corresponding to the eigenvalue λ_j and where the $C_j(y)$'s are functions of y and can be determined by the initial condition $f(x, y; 0) = \delta(y - x)$ and the orthogonality of the eigenfunctions $h_j(x)$.

Let A denote the operator $A = \alpha(x)\frac{d^2}{dx^2} + \beta(x)\frac{d}{dx}$. Then Eq. (7.2) is expressed as

$$Ah(x) + \{\lambda + \epsilon q(x)\}h(x) = 0\,,$$

where $\epsilon = \frac{1}{N}$ with large N.

To derive approximations to $\{\lambda_j, h_j(x)\}$, the basic trick is to chose $\eta(x)$ so that the following conditions hold:

(1) The function $q(x)$ is bounded.

(2) The eigenvalues γ_i of the operator A are real and these eigenvalues and its associated eigenfunctions $u_i(x)$ can readily be derived.

Then the eigenvalues λ_j and the eigenfunctions $h_j(x)$ of Eq. (7.2) can readily be approximated by using $\{\gamma_i, u_i(x)\}$ through the following relationships:

$$\lambda_j = \gamma_j + \sum_{i=1}^{\infty} \epsilon^i \gamma_j^{(i)}$$

$$= \gamma_j + \sum_{i=1}^{k} \epsilon^i \gamma_j^{(i)} + O(\epsilon^{k+1})\,; \tag{7.3}$$

$$h_j(x) = u_j(x) + \sum_{i=1}^{\infty} \epsilon^i u_j^{(i)}(x)$$

$$= u_j(x) + \sum_{i=1}^{k} \epsilon^i u_j^{(i)}(x) + O(\epsilon^{k+1})\,, \tag{7.4}$$

for $j = 1, \ldots, \infty$.

Using Eqs. (7.3) and (7.4), by deriving $\gamma_j^{(i)}$ and $u_j^{(i)}(x)$ one may derive $\{\lambda_j, h_j(x)\}$ from the eigenvalues γ_j and eigenfunctions $u_j(x)$. This is called the method of perturbation [1]. Because the first and second eigenvalues dominant for large t as these are the smallest eigenvalues, for large t, we have the following asymptotic distributions:

$$f(x, y; t) \cong \sum_{i=1}^{2} C_i(y)e^{-\lambda_i t}h_i(x) \cong C_1(y)e^{-\lambda_1 t}h_1(x)\,. \tag{7.5}$$

From Eq. (7.5), to derive asymptotic distributions, one need only to approximate the first and second eigenvalues together with their eigenfunctions.

To illustrate how to approximate $\{\lambda_j, h_j(x)\}$ by $\{\gamma_i, u_i(x), i = 1, \ldots, \infty\}$, let $\sigma(x) = \frac{1}{\alpha(x)} \exp\{\int^x \frac{\beta(y)}{\alpha(y)} dy\}$ and denote by $E[r(x)] = \int_a^b r(x)\sigma(x)dx$. Further, we assume that $h_j(x)$ and $u_i(x)$ are normalized eigenfunctions so that $E[h_j^2(x)] = E[u_i^2(x)] = 1$ for all $(i, j = 1, \ldots, \infty)$. Then, we have the following results:

(1) From basic results of eigenfunctions (see Subsec. 6.8.3), $h_j(x)$ and $u_i(x)$ are orthogonal with respect to the weight function $\sigma(x)$. That is, for all $i \neq j$,

$$E[h_i(x)h_j(x)] = \int_a^b \sigma(x)h_i(x)h_j(x)dx = 0$$

and

$$E[u_i(x)u_j(x)] = \int_a^b \sigma(x)u_i(x)u_j(x)dx = 0 \,.$$

(2) The eigenfunctions $\{u_i(x), i = 1, \ldots, \infty\}$ form a basis of all integrable functions in $[a, b]$ so that for all $i, j = 1, \ldots, \infty$,

$$u_i^{(j)}(x) = \sum_k a_{ik}^{(j)} u_k(x) \,,$$

where the $a_{ik}^{(j)}$'s are constants.

(3) Under the assumption $E[u_i^2(x)] = E[h_i^2(x)] = 1$, we have

$$1 = E[h_i^2(x)] = E\left\{ u_i(x) + \sum_{j=1}^{\infty} \epsilon^j u_i^{(j)}(x) \right\}^2$$

$$= E\left\{ u_i^2(x) + \sum_{j=1}^{\infty} \epsilon^j g_{i,j}(x) \right\}$$

$$= 1 + \sum_{j=1}^{\infty} \epsilon^j E[g_{i,j}(x)] \,,$$

where for $j = 0, 1, \ldots, \infty$,

$$g_{i,2j+1}(x) = 2 \sum_{k=0}^{j} u_i^{(k)}(x) u_i^{(2j+1-k)}(x)$$

and

$$g_{i,2j+2}(x) = [u_i^{(j+1)}(x)]^2 + 2\sum_{k=0}^{j} u_i^{(k)}(x)u_i^{(2j+2-k)}(x)$$

with $u_i^{(0)}(x) = u_i(x)$.

It follows that $E[g_{i,j}(x)] = 0$ for all $j = 1,\ldots,\infty$. In particular, with $(j = 1, 2)$:

$$E[u_i(x)u_i^{(1)}(x)] = \frac{1}{2}E[g_{i,1}(x)] = 0\,, \tag{7.6}$$

$$E[g_{i,2}(x)] = E\{[u_i^{(1)}(x)]^2\} + 2E[u_i(x)u_i^{(2)}(x)] = 0\,. \tag{7.7}$$

To derive $\gamma_i^{(j)}$ and the $a_{ik}^{(j)}$, we substitute Eqs. (7.3) and (7.4) for λ_i and $h_i(x)$ respectively into the equation $Ah_i(x) + \{\lambda_i + \epsilon q(x)\}h_i(x) = 0$ to give:

$$A\left\{u_i(x)+\sum_{j=1}^{\infty}\epsilon^j u_i^{(j)}(x)\right\}+\left[\gamma_i+\epsilon q(x)+\sum_{j=1}^{\infty}\epsilon^j\gamma_i^{(j)}\right]\left[u_i(x)+\sum_{j=1}^{\infty}\epsilon^j u_i^{(j)}(x)\right]=0\,.$$

This gives:

$$Au_i(x) + \gamma_i u_i(x) = 0\,,$$

and for $j = 1, 2, \ldots$,

$$Au_i^{(j)}(x) + \gamma_i u_i^{(j)}(x) + [\delta_{1j}\epsilon q(x) + \gamma_i^{(j)}]u_i(x)+\sum_{r=1}^{j-1}[\delta_{1r}q(x) + \gamma_i^{(r)}]u_i^{(j-r)}(x)=0\,,$$

where $\sum_{r=1}^{0}$ is defined as 0.

On substituting $u_i^{(j)}(x) = \sum_k a_{ik}^{(j)} u_k(x)$ into the above equation and noting $Au_k(x) = -\gamma_k u_k(x)$, we obtain for $j = 1,\ldots,\infty$,

$$\sum_k a_{ik}^{(j)}(\gamma_i - \gamma_k)u_k(x) + [\delta_{1j}q(x) + \gamma_i^{(j)}]u_i(x)$$

$$+ \sum_k\left\{\sum_{r=1}^{j-1}[\delta_{1r}q(x) + \gamma_i^{(r)}]a_{ik}^{(j-r)}\right\}u_k(x) = 0\,. \tag{7.8}$$

If $j = 1$, then,

$$\sum_k a_{ik}^{(1)}(\gamma_i - \gamma_k)u_k(x) + [q(x) + \gamma_i^{(1)}]u_i(x) = 0. \tag{7.9}$$

Multiplying both sides of Eq. (7.9) by $\sigma(x)u_i(x)$, integrating from a to b and noting that $E[u_i(x)u_k(x)] = \delta_{ik}$, we obtain

$$\gamma_i^{(1)} = -E[q(x)u_i^2(x)].$$

Multiplying both sides of Eq. (7.9) by $\sigma(x)u_k(x)$, integrating from a to b and noting that $E[u_i(x)u_k(x)] = \delta_{ik}$, we obtain for $k \neq i$,

$$a_{ik}^{(1)} = \frac{1}{\gamma_k - \gamma_i}E[q(x)u_i(x)u_k(x)].$$

To obtain $a_{kk}^{(1)}$, we notice that by (7.6),

$$E[u_i(x)u_i^{(1)}(x)] = 0.$$

Since

$$E[u_i(x)u_i^{(1)}(x)] = E\left\{\sum_k a_{ik}^{(1)}[u_i(x)u_k(x)]\right\} = a_{ii}^{(1)}E[u_i^2(x)] = a_{ii}^{(1)},$$

it follows that $a_{ii}^{(1)} = 0$.

To derive $\gamma_i^{(2)}$ and $a_{ik}^{(2)}$, we put $j = 2$ in Eq. (7.8) to give

$$\sum_k a_{ik}^{(2)}(\gamma_i - \gamma_k)u_k(x) + \gamma_i^{(2)}u_i(x) + [q(x) + \gamma_i^{(1)}]\sum_k a_{ik}^{(1)}u_k(x) = 0. \tag{7.10}$$

Multiplying both sides of Eq. (7.10) by $\sigma(x)u_i(x)$, integrating from a to b and noting that $E[u_i(x)u_k(x)] = \delta_{ik}$, we obtain:

$$\gamma_i^{(2)} = -\sum_k a_{ik}^{(1)}E[q(x)u_i(x)u_k(x)].$$

Multiplying both sides of Eq. (7.10) by $\sigma(x)u_k(x)$, integrating from a to b and noting that $E[u_i(x)u_k(x)] = \delta_{ik}$, we obtain for $k \neq i$,

$$a_{ik}^{(2)} = \frac{1}{\gamma_k - \gamma_i}\left\{\sum_k a_{ir}^{(1)}E[q(x)u_r(x)u_k(x)] + \gamma_i^{(1)}a_{ik}^{(1)}\right\}.$$

To obtain $a_{kk}^{(2)}$, we notice that from Eq. (7.7),

$$0 = E[g_{i,2}(x)] = E\{[u_i^{(1)}(x)]^2\} + 2E[u_i(x)u_i^{(2)}(x)] \,.$$

Since

$$E\{[u_i^{(1)}(x)]^2\} = E\left\{\sum_k a_{ik}^{(1)} u_k(x)\right\}^2 = \sum_k (a_{ik}^{(1)})^2 E[u_k(x)]^2$$

$$= \sum_k (a_{ik}^{(1)})^2 = \sum_{k \neq i} (a_{ik}^{(1)})^2 \,,$$

and

$$E\{u_i(x)u_i^{(2)}(x)]\} = E\left\{\sum_k a_{ik}^{(2)} u_i(x)u_k(x)\right\} = a_{ii}^{(2)} E[u_i(x)]^2 = a_{ii}^{(2)} \,,$$

we obtain

$$a_{ii}^{(2)} = -\frac{1}{2}\sum_{k \neq i} (a_{ik}^{(1)})^2 \,.$$

To summarize, denoting by $E[q(x)u_i(x)u_j(x)] = \langle u_i, qu_j \rangle$, we obtain:

$$\lambda_i = \gamma_i + \sum_{j=1}^{2} \epsilon^i \gamma_i^{(j)} + O(\epsilon^3)$$

$$= \gamma_i - \epsilon\langle u_i, qu_i \rangle - \epsilon^2 \sum_{j \neq i} \frac{1}{\gamma_j - \gamma_i} \langle u_i, qu_j \rangle^2 + O(\epsilon^3)$$

and

$$h_i(x) = u_i(x) + \sum_{j=1}^{2} \epsilon^i u_i^{(j)} + O(\epsilon^3)$$

$$= u_i(x) + \epsilon \sum_{j \neq i} a_{ij}^{(1)} u_j(x) + \epsilon^2 \sum_{j} a_{ij}^{(2)} u_j(x) + O(\epsilon^3)$$

$$= u_i(x) + \epsilon \sum_{j \neq i} \frac{1}{\gamma_j - \gamma_i} \langle u_i, qu_j \rangle u_j(x) + O(\epsilon^2) \,,$$

where

$$a_{ii}^{(1)} = 0, a_{ij}^{(1)} = \frac{1}{\gamma_j - \gamma_i}\langle u_i, qu_j \rangle, \quad \text{for } j \neq i, a_{ii}^{(2)} = -\frac{1}{2}\sum_{j \neq i}(a_{ij}^{(1)})^2,$$

and

$$a_{ij}^{(2)} = \frac{1}{\gamma_j - \gamma_i}\left\{\sum_k a_{ik}^{(1)}\langle u_j, qu_k \rangle + \gamma_i^{(1)}a_{ij}^{(1)}\right\}, \quad \text{for } i \neq j.$$

Better approximation can also be derived. In fact, from Eq. (7.8), by exactly the same approach as above we obtain for $j = 3, 4, \ldots, \infty$,

$$\gamma_i^{(j)} = -\left\{\sum_{r=1}^{j-1}a_{ii}^{(r)}\gamma_i^{(j-r)} + \sum_k a_{ik}^{(j)}E[q(x)u_i(x)u_k(x)]\right\},$$

$$a_{ik}^{(j)} = \frac{1}{\gamma_k - \gamma_i}\left\{\sum_{r=1}^{j-1}\gamma_i^{(r)}a_{ik}^{(j-r)} + \sum_r a_{ir}^{(j-1)}E[q(x)u_r(x)u_k(x)]\right\}, \quad \text{for all } i \neq k.$$

Similarly, by using results $E[g_{i,2j+1}(x)] = 0$ and $E[g_{i,2j}(x)] = 0$ for $j = 1, \ldots, \infty$, we obtain:

$$a_{ii}^{(2j+1)} = -\sum_k\left\{\sum_{r=1}^{j}a_{ik}^{(r)}a_{ik}^{(2j+1-r)}\right\};$$

$$a_{ii}^{(2j)} = -\left\{\frac{1}{2}\sum_k(a_{ik}^{(j)})^2 + \sum_k\sum_{r=1}^{j-1}a_{ik}^{(r)}a_{ik}^{(2j-r)}\right\}.$$

Example 7.1. The Wright model with selection in population genetics. Consider the Wright model for one locus with two alleles $A : a$ in a large population with size N as described in Example 6.9. Assume that there are selections between different genotypes and the selection rates are independent of time t but there are no mutations and no immigration and migration. Then, it is shown in Example 6.9 that to the order of $O(N^{-2})$, the frequency of the A gene is a diffusion process with state space $[0, 1]$ and with diffusion coefficients $\{m(x) = x(1 - x)[\alpha_1 x + \alpha_2(1 - 2x)], v(x) = x(1 - x)\}$, where α_i are the selection intensities. In this case, the Kolmogorov equations are extremely difficult to solve. However, because selection effects are usually very small, it is reasonable to assume the $N\alpha_i$ as finite. Then one may apply the above method to derive approximate and asymptotic solutions.

To proceed, we make the transformation $f(x, y; t) = e^{-\lambda t} e^{-p(x)} h(x)$, where $p(x) = \frac{x^2}{2}\alpha_1 + (x - x^2)\alpha_2$. Then $x(1 - x)\frac{dp(x)}{dx} = m(x)$ and $h(x)$ satisfies the equation:

$$\frac{1}{2}x(1 - x)h''(x) + \{\lambda - \epsilon q(x)\} = 0, \tag{7.11}$$

where

$$\epsilon = \frac{1}{2N}, q(x) = c_0 + \epsilon\{c_1 + c_2 x + c_3 x^2\}, \quad \text{with } \bar{\alpha}_i = N\alpha_i, i = 1, 2,$$

and

$$c_0 = \bar{\alpha}_1 - 2\bar{\alpha}_2, \quad c_1 = 2\bar{\alpha}_2^2, \quad c_2 = 4\bar{\alpha}_2 c_0, \quad c_3 = 2c_0^2.$$

From Example 6.3, we have $\sigma(x) = \frac{1}{x(1-x)}$. Denote the operator $A = \frac{1}{2}x(1 - x)\frac{d^2}{dx^2}$. From Example 6.3, the eigenvalues and the eigenfunctions of $Af(x) + \lambda f(x) = 0$ are given by:

$$\gamma_i = \frac{1}{2}i(i + 1) \quad \text{and} \quad u_i(x) = \{(i + 1)(2i + 1)/i\}^{\frac{1}{2}} x(1 - x)J_{i-1}(x; 2, 2),$$

for $i = 1, 2, \ldots$, where the $J_i(x; p, q)$ are Jacobi polynomials defined in Subsec 6.8.2.

Hence, we have:

$$\langle u_1, qu_1 \rangle = 6 \int_0^1 \sigma(x)q(x)[x(1 - x)]^2 dx$$

$$= 6 \int_0^1 [c_0 + \epsilon(c_1 + c_2 x + c_3 x^2)][x(1 - x)]^2 dx$$

$$= 6\{c_0 B(3, 3) + \epsilon[c_1 B(3, 3) + c_2 B(4, 3) + c_3 B(5, 3)]\}$$

$$= \frac{1}{5}\left[c_0 + \epsilon\left(c_1 + \frac{1}{2}c_2 + \frac{2}{7}c_3\right)\right].$$

Further, for $j = 2, 3, \ldots$, we have to the order of $O(\epsilon)$:

$$\langle u_1, q\, u_j \rangle = c_0\{6(j + 1)(2j + 1)/j\}^{\frac{1}{2}} \int_0^1 x^2(1 - x)^2 J_{j-1}(x; 2, 2)dx$$

$$= \{6(j + 1)(2j + 1)/j\}^{\frac{1}{2}} \frac{c_0}{(j - 1)!} \int_0^1 x(1 - x)$$

$$\times \left[\frac{d^{j-1}}{dx^{j-1}} \{ x^j (1-x)^j \} \right] dx = (-1)^{j-1} \{ 6(j+1)(2j+1)/j \}^{\frac{1}{2}}$$

$$\times \frac{c_0}{(j-1)!} \int_0^1 x^j (1-x)^j \left[\frac{d^{j-1}}{dx^{j-1}} \{ x(1-x) \} \right] dx \,.$$

Thus, to the order of $O(\epsilon)$,

$$\langle u_1, q u_2 \rangle = (-3c_0)\sqrt{5} \int_0^1 x^2 (1-x)^2 (1-2x) dx$$

$$= (-3c_0)\sqrt{5}[B(3,3) - 2B(4,3)] = 0 \,,$$

$$\langle u_1, q u_3 \rangle = (-2c_0)\sqrt{2 \times 7} \int_0^1 x^2 (1-x)^2 dx$$

$$= (-2c_0)\sqrt{14} B(3,3) = -\frac{\sqrt{14}}{15} c_0 \,,$$

$$\langle u_1, q u_j \rangle = 0 \,, \quad \text{for } j = 4, \ldots, \infty \,.$$

From these results, it follows that since $\gamma_3 - \gamma_1 = 5$,

$$\lambda_1 = \gamma_1 + \epsilon \frac{c_0}{5} + \frac{\epsilon^2}{5} \left\{ c_1 + \frac{1}{2} c_2 + \frac{2}{7} c_3 - \frac{14}{225} c_0^2 \right\} + O(\epsilon^3)$$

$$= 1 + \epsilon \frac{c_0}{5} + O(\epsilon^2) \,,$$

and

$$h_1(x) = u_1(x) + \epsilon \sum_{j=2}^{\infty} a_{1j}^{(1)} u_j(x) + O\epsilon^2$$

$$= \sqrt{6} x(1-x) - \epsilon \frac{14\sqrt{6} c_0}{75} x(1-x)(1 - 5x + 5x^2) + O(\epsilon^2) \,.$$

Similarly, we have:

$$\langle u_2, q u_2 \rangle = 30 \int_0^1 \sigma(x) q(x) [x(1-x)(1-2x)]^2 dx$$

$$= 30 \int_0^1 [c_0 + \epsilon(c_1 + c_2 x + c_3 x^2)][x(1-x)(1-2x)]^2 dx$$

$$= 30 \{ [c_0 + \epsilon \, c_1][B(3,3) - 4B(4,3) + 4B(5,3)]$$

$$+ \epsilon c_2[B(4,3) - 4B(5,3) + 4B(6,3)]$$

$$+ \epsilon c_3[B(5,3) - 4B(6,3) + 4B(7,3)]\}$$

$$= \frac{1}{7}\left[c_0 + \epsilon\left(c_1 + \frac{1}{2}c_2 + \frac{1}{3}c_3\right)\right].$$

Further, since $J_1(x; 2, 2) = 2(1 - 2x)$, we have to the order of $O(\epsilon)$:

$$\langle u_2, q\, u_j \rangle = c_0\{30(j+1)(2j+1)/j\}^{\frac{1}{2}} \int_0^1 x^2(1-x)^2(1-2x)J_{j-1}(x; 2, 2)dx$$

$$= \{30(j+1)(2j+1)/j\}^{\frac{1}{2}} \frac{c_0}{(j-1)!} \int_0^1 x(1-x)(1-2x)$$

$$\times \left[\frac{d^{j-1}}{dx^{j-1}}\{x^j(1-x)^j\}\right] dx$$

$$= (-1)^{j-1}\{30(j+1)(2j+1)/j\}^{\frac{1}{2}}$$

$$\times \frac{c_0}{(j-1)!} \int_0^1 x^j(1-x)^j \left[\frac{d^{j-1}}{dx^{j-1}}\{x(1-x)(1-2x)\}\right] dx.$$

Thus, to the order of $O(\epsilon)$, $\langle u_2, qu_1 \rangle = \langle u_1, qu_2 \rangle = 0$, and

$$\langle u_2, qu_3 \rangle = (-3c_0)\sqrt{4 \times 70} \int_0^1 x^3(1-x)^3(1-2x)dx$$

$$= (-6c_0)\sqrt{70}[B(4,4) - 2B(5,4)] = 0,$$

$$\langle u_2, qu_4 \rangle = (-15c_0)\sqrt{6} \int_0^1 x^3(1-x)^3 dx$$

$$= (-15c_0)\sqrt{6}B(4,4) = -\frac{\sqrt{6}}{28}3c_0,$$

$$\langle u_2, qu_j \rangle = 0, \quad \text{for } j = 5, \ldots, \infty.$$

From these results, it follows that since $\gamma_4 - \gamma_2 = 7$,

$$\lambda_2 = \gamma_2 + \epsilon\frac{c_0}{7} + \frac{\epsilon^2}{7}\left\{c_1 + \frac{1}{2}c_2 + \frac{1}{3}c_3 - \frac{27}{392}c_0^2\right\} + O(\epsilon^3)$$

$$= 3 + \epsilon\frac{c_0}{7} + O(\epsilon^2)$$

and

$$h_2(x) = u_2(x) + \epsilon \sum_{j=3}^{\infty} a_{2j}^{(1)} u_j(x) + O(\epsilon^2)$$

$$= \sqrt{30}x(1-x)(1-2x) - \epsilon \frac{9}{56} \sqrt{30}c_0 x(1-x)J_3(x; 2, 2) + O(\epsilon^2).$$

7.2. Stationary Distributions in Diffusion Processes

As in Markov chains, as time progresses, many of the diffusion processes in nature will converge to a steady-state condition. The probability distribution of the process under the steady state condition is the so-called stationary distribution of the process. Notice again that as in Markov chains, for the stationary distribution to exist, the diffusion process must be homogeneous so that the coefficients $\{m(y,t) = m(y), v(y,t) = v(y)\}$ must be independent of time t. This follows from the observation that the stationary distribution is independent of time. However, as shown in Example 7.4, homogeneous diffusion processes may not have stationary distributions.

To proceed, let $\{X(t), t \geq 0\}$ be a homogeneous diffusion process with state space $[a, b]$ and with coefficients $\{m(x), v(x)\}$. Let $f(x, y; t)$ be the conditional pdf of $X(t)$ at y given $X(0) = x$.

Definition 7.1.   The boundary point a is called a *regular boundary point* iff a is accessible from the interior of S and the interior points of S are accessible from a. We define the boundary point a as *accessible from the interior of S* iff for every $\epsilon > 0$ given and for every x_0 satisfying $a < x_0 < b$, there exists a time t such that $\int_a^{a+\epsilon} f(x_0, y; t)dy > 0$; similarly, the interior point $x_0 (a < x_0 < b)$ is *accessible from a* iff for every $\epsilon > 0$, there exists a time t such that $\int_{x_0-\epsilon}^{x_0+\epsilon} f(a, y; t)dy > 0$. The boundary point a is called *an absorbing barrier* iff a is accessible from the interior of S but the interior points of S are not accessible from a. Similarly, b is a regular boundary point iff b is accessible from the interior of S and the interior points of S are accessible from b; b is an absorbing barrier iff b is accessible from the interior of S but the interior points of S are not accessible from b.

Notice that the case in which the boundary points a and b are regular is the analog of irreducible Markov chains in diffusion processes; similarly, the absorbing barrier points are the analog of absorbing states in Markov chains in diffusion process. Hence, once the process reaches the absorbing barrier, it will stay there forever; this is the condition of fixation in diffusion processes.

In the Wright model considered in Example 6.9, if there are mutations from $A \to a$ and from $a \to A$ and/or if there are immigartion and migration, then both 0 and 1 are regular boundary; if there are no mutations and no immigration and migration, then both 0 and 1 are absorbing barriers. On the other hand, if there are no immigration and migration and no mutations from $A \to a$ but there are mutations from $a \to A$, then 0 is a regular boundary but 1 is an absrobing barrier; similarly, if there are no immigration and migration and no mutations from $a \to A$ but there are mutations from $A \to a$, then 0 is an absorbing barrier but 1 is a regular boundary.

Definition 7.2. Let $\{X(t), t \geq 0\}$ be a homogeneous diffusion process with state space $S = [a, b]$ and with coefficients $\{m(x), v(x)\}$. Let $f(x, y; t)$ be the conditional pdf of $X(t)$ given $X(0) = x$. Suppose that the boundary points a and b are regular boundary points. Then the density function $g(y)$ defined in $S = [a, b]$ is defined as a *stationary distribution* iff

$$g(y) = \int_a^b g(x) f(x, y; t) dx, \quad y \in S.$$

For deriving the stationary distribution for homogeneous diffusion processes, we will first prove the following theorem which provides some intuitive insights into the steady state condition and the stationarity of diffusion processes. This theorem was first due to Kimura [2].

Theorem 7.1. *Let $\{X(t), t \geq 0\}$ be a diffusion process with state space $S = [a, b]$ and with coefficients $\{m(x, t), v(x, t)\}$. Let $P(p, x; s, t)dt$ be the probability mass crossing the point x during $[t, t + dt)$ given $X(s) = p$. Then $P(p, x; s, t)$ is given by:*

$$P(p, x; s, t) = m(x, t) f(p, x; s, t) - \frac{1}{2} \frac{\partial}{\partial x} \{v(x, t) f(p, x; s, t)\} \qquad (7.12)$$

where $f(p, x; s, t)$ is the conditional pdf of $X(t)$ given $X(s) = p$.

Proof. To prove this theorem, let $P_{(+)}(p, x; s, t)dt$ be the probability of crossing the point x from the left during $[t, t + dt)$ given $X(s) = p$ and $P_{(-)}(p, x; s, t)dt$ the probability of crossing the point x from the right during $[t, t + dt)$ given $X(s) = p$. Let $g(\Delta\xi|x, \Delta t)$ denote the conditional pdf of the change $\Delta\xi$ during $[t, t + \Delta t)$ given $X(t) = x$. Then, the probability of the event that $X(t) \in (x - \frac{1}{2}dx, x + \frac{1}{2}dx)$ and $\Delta\xi \in (\Delta\xi - \frac{1}{2}d(\Delta\xi), \Delta\xi + \frac{1}{2}d(\Delta\xi))$ given $X(s) = p$ is

$$f(p, x; s; t)g(\Delta\xi|x, \Delta t)dx \, d(\Delta\xi) \,.$$

It follows that,

$$P_{(+)}(p, x; s, t)dt = \int_{x>\xi} \int_{\Delta\xi>x-\xi} f(p, \xi; s, t)g(\Delta\xi|\xi, dt)d(\Delta\xi)d\xi$$

$$= \int_{\Delta\xi>0} \int_{x-\Delta\xi}^{x} f(p, \xi; s, t)g(\Delta\xi|\xi, dt)d\xi \, d(\Delta\xi) \,,$$

and

$$P_{(-)}(p, x; s, t)dt = \int_{x<\xi} \int_{\Delta\xi<x-\xi} f(p, \xi; s, t)g(\Delta\xi|\xi, dt)d(\Delta\xi)d\xi$$

$$= \int_{\Delta\xi<0} \int_{x}^{x-\Delta\xi} f(p, \xi; s, t)g(\Delta\xi|\xi, dt)d\xi \, d(\Delta\xi) \,.$$

Hence, we have:

$$P(p, x; s, t)dt = P_{(+)}(p, x; s, t)dt - P_{(-)}(p, x; s, t)dt$$

$$= \int \int_{x-\Delta\xi}^{x} f(p, \xi; s, t)g(\Delta\xi|\xi, dt)d\xi \, d(\Delta\xi) \,. \qquad (7.13)$$

Expanding $f(p, \xi; s, t)g(\Delta\xi|\xi, dt)$ in Taylor series with respect to ξ around x, we obtain:

$$f(p, \xi; s, t)g(\Delta\xi|\xi, dt) = f(p, x; s, t)g(\Delta\xi|x, dt)$$

$$+ (\xi - x)\frac{\partial}{\partial x}\{f(p, x; s, t)g(\Delta\xi|x, dt)\}$$

$$+ \frac{1}{2}(\xi - x)^2 \frac{\partial^2}{\partial x^2}\{f(p, x; s, t)g(\Delta\xi|x, dt)\}$$

$$+ \frac{1}{3!}(\xi - x)^3 \frac{\partial^3}{\partial x^3}\{f(p, x; s, t)g(\Delta\xi|x, dt)\} + \cdots \,.$$

On substituting this expansion into the inner integral of (7.13), the inner integral of (7.13) becomes:

$$\int_{x-\Delta\xi}^{x} f(p,\xi;s,t)g(\Delta\xi|\xi,dt)d\xi = (\Delta\xi)f(p,x;s,t)g(\Delta\xi|x,dt)$$

$$-\frac{1}{2}(\Delta\xi)^2\frac{\partial}{\partial x}\{f(p,x;t)g(\Delta\xi|x,dt)\}$$

$$+\frac{1}{3!}(\Delta\xi)^3\frac{\partial^2}{\partial x^2}\{f(p,x;t)g(\Delta\xi|x,dt)\}+\cdots.$$

Noting the results

$$\int(\Delta\xi)g(\Delta\xi|x,dt)d(\Delta\xi) = m(x,t)dt + o(dt),$$

$$\int(\Delta\xi)^2 g(\Delta\xi|x,dt)d(\Delta\xi) = v(x,t)dt + o(dt),$$

and

$$\int(\Delta\xi)^r g(\Delta\xi|x,dt)d(\Delta\xi) = o(dt) \quad \text{for } r = 3,4,\ldots,$$

we obtain from Eq. (7.13):

$$P(p,x;s,t)dt = P_{(+)}(p,x;s,t)dt - P_{(-)}(p,x;s,t)dt$$

$$= \int\int_{x-d(\Delta\xi)}^{x} f(p,\xi;s,t)g(\Delta\xi|\xi,dt)d\xi$$

$$= m(x,t)f(p,x;t)dt - \frac{1}{2}\frac{\partial}{\partial x}\left\{f(p,x;t)\int(\Delta\xi)^2 g(\Delta\xi|x,dt)d(\Delta\xi)\right\}$$

$$+\frac{1}{3!}\frac{\partial^2}{\partial x^2}\left\{f(p,x;t)\int(\Delta\xi)^3 g(\Delta\xi|x,dt)d(\Delta\xi)\right\}+\cdots$$

$$= \left\{m(x,t)f(p,x;t) - \frac{1}{2}\frac{\partial}{\partial x}[v(x,t)f(p,x;t)]\right\}dt + o(dt).$$

This shows that

$$P(p,x;s,t) = m(x,t)f(p,x;s,t) - \frac{1}{2}\frac{\partial}{\partial x}[v(x,t)f(p,x;s,t)]. \qquad \square$$

Since $P(p, x; s, t) = 0$ implies that the net flow of probability mass crossing x at time t is 0, if $m(x, t) = m(x)$ and $v(x, t) = v(x)$ are independent of t, then the solution $f(p, x; t) = g(x)$ of the above equation under the constraint $P(p, x; s, t) = 0$ is the stationary distribution of $X(t)$ and is independent of both p and t. In adopting this result, however, we have to be careful in its interpretation. If $x = a$ and $x = b$ are regular boundary points so that starting with $x = a$ or $x = b$, with positive probability the process can go to $a < x < b$ in the process, the solution in this case does provide the stationary distribution as defined in Definition 7.2; but, if $x = a$ and $x = b$ are absorbing barriers, then the stationary distribution may not exist.

Theorem 7.2.   *Let $\{X(t), t \geq 0\}$ be a homogeneous diffusion process with state space $S = [a, b]$ and with coefficients $\{m(x), v(x)\}$. Assume that a and b are regular boundary points.   If the solution $\{g(x), x \in S\}$ of the following equation is unique and if $\lim_{x \to a} m(x)g(x) = \lim_{x \to b} m(x)g(x) = \lim_{x \to a} \frac{d}{dx}[v(x)g(x)] = \lim_{x \to b} \frac{d}{dx}[v(x)g(x)] = 0$, then $\{g(x), x \in S\}$ is the unique stationary distribution of the diffusion process $X(t)$:*

$$-m(x)g(x) + \frac{1}{2}\frac{d}{dx}[v(x)g(x)] = 0, \quad x \in S. \tag{7.14}$$

Proof.   To prove Theorem 7.2, put

$$\int_a^b g(x)f(x, y; t)dx = \eta(y, t). \tag{7.15}$$

We will show that $\eta(y, t) = \eta(y)$ is independent of t and that $\eta(y) = g(y)$ for all $y \in S$.

On both sides of Eq. (7.15), taking partial derivative with respect to t and noting the result of the backward equation, we obtain:

$$\frac{\partial}{\partial t}\eta(y, t) = \frac{\partial}{\partial t}\int_a^b g(x)f(x, y; t)dx = \int_a^b g(x)\frac{\partial}{\partial t}f(x, y; t)dx$$

$$= \int_a^b g(x)\left\{m(x)\frac{\partial}{\partial x}f(x, y; t) + \frac{1}{2}v(x)\frac{\partial^2}{\partial x^2}f(x, y; t)\right\}dx.$$

It follows that by applying integration by parts,

$$\frac{\partial}{\partial t}\eta(y, t) = \int_a^b f(x, y; t)\frac{d}{dx}\left\{-[m(x)g(x)] + \frac{1}{2}\frac{d}{dx}[v(x)g(x)]\right\}dx = 0.$$

This shows that $\eta(y,t) = \eta(y)$ is independent of t. To prove $\eta(y) = g(y)$ for all $y \in S$, denote by $F(x,z;t) = \int_a^z f(x,y;t)dy$. Then, from the forward equation, we have:

$$\frac{\partial}{\partial t}F(x,y;t) = -m(y)f(x,y;t) + \frac{1}{2}\frac{\partial}{\partial y}[v(y)f(x,y;t)].$$

On both sides of Eq. (7.15), multiplying by $-m(y)$, we obtain:

$$\int_a^b g(x)[-m(y)f(x,y;t)]dx = -m(y)\eta(y). \tag{7.16}$$

On both sides of Eq. (7.15), multiplying by $\frac{1}{2}v(y)$ and taking derivative with respect to y, we obtain:

$$\int_a^b g(x)\left\{\frac{1}{2}\frac{\partial}{\partial y}[v(y)f(x,y;t)]\right\}dx = \frac{1}{2}\frac{\partial}{\partial y}[v(y)\eta(y)]. \tag{7.17}$$

Adding Eqs. (7.16) and (7.17) gives

$$-m(y)\eta(y)+\frac{1}{2}\frac{\partial}{\partial y}[v(y)\eta(y)] = \int_a^b g(x)\Big\{-m(y)f(x,y;t)$$

$$+\frac{1}{2}\frac{\partial}{\partial y}[v(y)f(x,y;t)]\Big\}dx$$

$$= \int_a^b g(x)\left\{\frac{\partial}{\partial t}F(x,y;t)\right\}dx$$

$$= \frac{\partial}{\partial t}\int_a^b g(x)\left\{\int_a^y f(x,z;t)dz\right\}dx$$

$$= \int_a^y \left\{\frac{\partial}{\partial t}\int_a^b g(x)f(x,z;t)dx\right\}dz$$

$$= \int_a^y \left\{\frac{\partial}{\partial t}\eta(z)\right\}dz = 0.$$

This shows that $\eta(y) = g(y)$. $\qquad\square$

Now, putting $\zeta(x) = v(x)g(x)$, Eq. (7.14) gives

$$\frac{2m(x)}{v(x)}\zeta(x) = \frac{d}{dx}\zeta(x),\ x \in S.$$

This gives

$$g(x) = \frac{C}{v(x)} \exp\left\{2 \int^x \frac{m(y)}{v(y)} dy\right\}, \tag{7.18}$$

where C is a normalizing constant.

Hence, if the conditions in Theorem 7.2 hold, then the density of the staionary distribution of $X(t)$ is given by Eq. (7.18).

Example 7.2.   The stationary distribution of gene frequency in the Wright model of population genetics. In Examples 1.11 and 6.9, we have considered the Wright model in population genetics for a single locus with two alleles A and a. In this model, $\{X(t), t \in T\}$ is the number of A allele in a large diploid population of size N and the chain is irreducible if there are mutations from $A \to a$ and from $a \to A$ and /or if there are immigration and migration. Let the mutation rates from $A \to a$ and from $a \to A$ be given by $u(t) = \frac{\beta_1}{2N} + O(N^{-2})$ and $v(t) = \frac{\beta_2}{2N} + O(N^{-2})$ respectively. Let $\chi(t) = \frac{1}{2N}\omega + O(N^{-2})$ be the population exchange rate per generation between the A allele of the population and the A allele of the outside population and x_I the frequency of the A allele among the immigrants. Denote the relative fitness of the three genotypes $\{AA, Aa, aa\}$ by $\{1 + \frac{1}{2N}\alpha_1 + O(N^{-2}), 1 + \frac{1}{2N}\alpha_2 + O(N^{-2}), 1\}$ respectively. Then, it is shown in Example 6.9 that to the order of $O(N^{-2})$, $Y(t) = \frac{1}{2N}X(t)$ is a diffusion process with state space $S = [0, 1]$ and with diffusion coefficients given by

$$m(x, t) = x(1 - x)[\alpha_2(1 - 2x) + \alpha_1 x] - \gamma_1 x + \gamma_2(1 - x),$$

where $\gamma_1 = \beta_1(t) + \omega[1 - x_I]$, and $\gamma_2 = \beta_2 + x_I\omega$, and

$$v(x, t) = x(1 - x) + O(N^{-1}).$$

In this process, if there are mutations and/or immigration and migration, then $\gamma_i > 0$ $(i = 1, 2)$ so that 0 and 1 are regular boundaries. Under these conditions, the stationary distribution of $Y(t)$ exists and is given by

$$g(x) = \frac{C_0}{v(x)} \exp\left\{2 \int_0^x \frac{m(y)}{v(y)} dy\right\}, \quad 0 \leq x \leq 1,$$

where C_0 is a normalizing constant such that $\int_0^1 g(x)dx = 1$. Now

$$\int_0^x \frac{m(y)}{v(y)}dy = \int_0^x \left\{ [\alpha_2(1-2y) + \alpha_1 y] - \gamma_1 \frac{1}{1-y} + \gamma_2 \frac{1}{y} \right\} dy$$

$$= \left[\alpha_2(x - x^2) + \frac{1}{2}\alpha_1 x^2 \right] + \gamma_1 \log(1-x) + \gamma_2 \log(x).$$

It follows that

$$g(x) = \frac{C_0}{v(x)} \exp \left\{ 2 \int_0^x \frac{m(y)}{v(y)}dy \right\}$$

$$= \frac{C_0}{v(x)} \exp\{\alpha_1 x^2 + 2\alpha_2 x(1-x) + 2\gamma_1 \log(1-x) + 2\gamma_2 \log(x)\}$$

$$= C_0 x^{2\gamma_2 - 1}(1-x)^{2\gamma_1 - 1} \exp\{\alpha_1 x^2 + 2\alpha_2 x(1-x)\}, \quad 0 \le x \le 1.$$

If there are no selection so that $\alpha_i = 0$ for $i = 1, 2$, then

$$g(x) = \frac{1}{B(2\gamma_2, 2\gamma_1)} x^{2\gamma_2 - 1}(1-x)^{2\gamma_1 - 1}, \quad 0 \le x \le 1.$$

Example 7.3.   The stationary distribution of gene frequency in the Moran model of genetics. In Example 6.10, it is shown that for the Moran's model of genetics, to the order of $O(M^{-2})$), $Y(t) = \frac{1}{M}X(t)$ is a diffusion process with diffusion coefficients:

$$m(x) = (1-x)\lambda_2[x(1-\alpha_1) + (1-x)\alpha_2] - x\lambda_1[(1-x)(1-\alpha_2) + x\alpha_1],$$

$$v(x) = \frac{1}{M}\{(1-x)\lambda_2[x(1-\alpha_1) + (1-x)\alpha_2] + x\lambda_1[(1-x)(1-\alpha_2) + x\alpha_1]\}.$$

In this case, if $\lambda_i \ne 0$ for $i = 1, 2$, then the 0 and 1 are regular boundaries so that one may derive the stationary distribution of $Y(t)$. By Theorem 7.2, this stationary distribution is given by

$$f(x) = c_0[v(x)]^{-1} \exp \left\{ 2 \int^x \frac{m(y)}{v(y)}dy \right\}.$$

Suppose that $\alpha_j = \frac{1}{M}a_j, j = 1, 2, \lambda_1 = \lambda$ and $\lambda_2 = \lambda + \frac{1}{M}\zeta$, where a_j and ζ are independent of M. Then

$$m(x) = \frac{1}{M}[x(1-x)(\zeta + \lambda a_2 - \lambda a_1) + \lambda a_2(1-x)^2 - \lambda a_1 x^2] + O(M^{-2})$$

and

$$v(x) = \frac{1}{M} 2\lambda x(1 - x) + O(M^{-2}).$$

Hence,

$$2\int^x \frac{m(Y)}{v(y)}\,dy = c_1 + \frac{1}{\lambda}\{[\zeta + \lambda(a_2 - a_1)]x + \lambda a_2(\log x - x)$$

$$+ \lambda a_1(x + \log(1 - x))\} + O(M^{-2}).$$

Thus, to the order of $O(M^{-2})$,

$$f(x) = c_1 \frac{1}{x(1 - x)} \exp\left\{\frac{1}{\lambda}\zeta x + a_2 \log x + a_1 \log(1 - x)\right\}$$

$$= c_2 e^{\frac{\zeta}{\lambda}x} x^{a_2 - 1}(1 - x)^{a_1 - 1},$$

where c_1 and c_2 are constants such that $c_2^{-1} = \int_0^1 e^{\frac{1}{\lambda}\zeta x} x^{a_2 - 1}(1 - x)^{a_1 - 1}dx$. If $\zeta = 0$ (no selection), then

$$f(x) = \frac{1}{B(a_2, a_1)} x^{a_2 - 1}(1 - x)^{a_1 - 1}, \quad 0 \le x \le 1.$$

Example 7.4.　The stationary distribution of the number of initiated cells in the two-stage model of carcinogenesis. It is shown in Example 4.9 that for the two-stage model of carcinogenesis, if the number of normal stem cells is very large, then the number of initiated cells $I(t)$ is a birth-death process with immigration with birth rate $jb(t)+\lambda(t)$ and with death rate $jd(t)$. Also, it is shown in Example 6.7 that to the order of $O(N_0^{-2})$, $Y(t) = \frac{X(t)}{N_0}$ is a diffusion process with state space $S = [0, \infty)$ and with diffusion coefficients $\{m(x,t) = \frac{\lambda(t)}{N_0} - \xi(t)x, v(x,t) = \frac{\omega(t)}{N_0}x\}$, where $\{\xi(t) = d(t) - b(t), \omega(t) = b(t) + d(t)\}$. To derive the stationary distribution, we assume that $\{b(t) = b, d(t) = d, \lambda(t) = \lambda\}$ so that $\xi(t) = \xi$ and $\omega(t) = \omega$.

Now it is shown in Example 5.4 that if $b > d$, then the stationary distribution does not exist; it follows that for the diffusion process $Y(t) = \frac{X(t)}{N_0}$, the stationary distribution also does not exist if $b > d$. Intuitively, since $\lambda \ge 0$, $I(t)$ will keep on increasing as time increases if $b > d$, so that it will never reach a steady state condition. To derive the stationary distribution of $Y(t)$, we thus assume $d > b$. This is usually the case when the cancer patients are constantly subjected to chemotherapy or immunotherapy. In these cases, we have $\xi(t) = \xi > 0$ and the stationary distribution exists.

Now, with $\{m(x) = \frac{\lambda}{N_0} - x\xi, v(x) = \frac{\omega}{N_0}x\}$, we have:

$$\int^x \frac{m(y)}{v(y)}dy = C_0 + \frac{1}{2}[\gamma_1 \log(x) - \gamma_2 x],$$

where $\{\gamma_1 = \frac{2\lambda}{\omega}, \gamma_2 = \frac{2N_0\xi}{\omega}\}$ and C_0 is a constant. Hence, the density of the stationary distribution of $Y(t)$ is

$$g(x) = \frac{C_1}{v(x)}e^{2\int^x \frac{m(y)}{v(y)}dy}$$

$$= \frac{\gamma_2^{\gamma_1}}{\Gamma(\gamma_1)}x^{\gamma_1-1}e^{-\gamma_2 x}, \quad 0 \le x \le \infty,$$

where C_1 is a constant such that $g(x)$ is a density.

For the above diffusion process, the Kolmogorov backward equation is

$$\frac{\partial}{\partial t}f(x,y;t) = \left(\frac{\lambda}{N_0} - \delta x\right)\frac{\partial}{\partial x}f(x,y;t) + \frac{1}{2N_0}\omega x \frac{\partial^2}{\partial x^2}f(x,y;t), \qquad (7.19)$$

with initial condition $f(x,y;0) = \delta(y-x)$, the Dirac's δ function.

Using exactly the same approach given in Example 6.2, Eq. (7.19) can readily be solved in terms of Laguerre polynomials; see Exercise 6.1. In fact, making the transformation $f(x,y;t) = e^{-\sigma t}h(x)$, $h(x)$ satisfies the equation:

$$\frac{1}{2N_0}\omega x h''(x) + \left(\frac{\lambda}{N_0} - \xi x\right)h'(x) + \sigma h(x) = 0.$$

The eigenvalues and eigenfunctions of this equation are given by

$$\sigma_k = k\xi, \quad h_k(x) = L_k^{(\gamma_1)}(\gamma_2 x), \quad k = 0, 1, \ldots, \infty,$$

where $L_k^{(\gamma_1)}(y) = \frac{1}{k!}\sum_{j=0}^{k}(-1)^j \binom{k}{j}y^j \frac{\Gamma(k+\gamma_1)}{\Gamma(j+\gamma_1)}$ is the kth degree Laguerre polynomial with parameter γ_1.

Using the same approach as in Example 6.2 and using the orthogonal properties of Laguerre polynomials as given in [3, p. 6], the solution of Eq. (7.19) is,

$$f(x,y;t) = g(y)\left[\sum_{k=0}^{\infty}e^{-k\xi t}L_k^{(\gamma_1)}(\gamma_2 x)L_k^{(\gamma_1)}(\gamma_2 y)\binom{k+\gamma_1-1}{k}^{-1}\right].$$

Noting $L_0^{(\gamma_1)}(x) = 1$, $\lim_{t\to\infty}f(x,y;t) = g(y)$.

7.3. The Absorption Probabilities and Moments of First Absorption Times in Diffusion Processes

Consider a homogeneous diffusion process $\{X(t), t \geq 0\}$ with state space $S = [a, b]$ and wih diffusion coefficients $\{m(x), v(x)\}$. Assume that the two boundary points a and b are absorbing barriers and that starting at any $X(0) = x$ satisfying $a < x < b$, with probability one the process will eventually be absorbed into the boundary points as time progresses. In this section we will illustrate how to derive the absorption probabilities into these boundary points and derive formula to computer the mean and the variance of the first absorption time.

7.3.1. *Absorption probabilities*

Let $u_a(p; t)$ denote the probability that starting with $X(0) = p \in S$, the process is absorbed into the boundary a at or before time t and $u_b(p; t)$ the probability that starting with $X(0) = p \in S$, the process is absorbed into the boundary b at or before time t. Put $\zeta(p; t) = u_a(p; t) + u_b(p; t)$. Then $h(t; p) = \frac{\partial}{\partial t}\zeta(p; t)$ is the pdf of the first absorption time $R(p)$ of the state $X(0) = p \in S$, $U_a(p) = \lim_{t\to\infty} u_a(p; t)$ the ultimate absorption probability of $X(0) = p \in S$ into a and $U_b(p) = \lim_{t\to\infty} u_b(p; t)$ the ultimate absorption probability of $X(0) = p \in S$ into b. In formulae of $u(x; s, t)$ defined by Eq. (6.11), by choosing $g(y) = \delta(y - a), g(y) = \delta(y - b)$ and $g(y) = \delta(y - a) + \delta(y - b)$, the corresponding $u(x; s, t)$ in Sec. 6.3 is $u_a(x; t), u_b(x, t)$ and $\zeta(x; t)$ respectively. Hence, by Theorem 6.2, $u_a(x, t), u_b(x, t)$ and $\zeta(x, t)$ all satisfy the following Kolmogorov backward equations respectively:

$$\frac{\partial}{\partial t}u_a(p; t) = m(p)\frac{\partial}{\partial p}u_a(p; t) + \frac{1}{2}v(p)\frac{\partial^2}{\partial p^2}u_a(p; t), \qquad (7.20)$$

with $u_a(p; 0) = 1$ if $p = a$ and $u_a(p; 0) = 0$ if $p \neq a$;

$$\frac{\partial}{\partial t}u_b(p; t) = m(p)\frac{\partial}{\partial p}u_b(p; t) + \frac{1}{2}v(p)\frac{\partial^2}{\partial p^2}u_b(p; t), \qquad (7.21)$$

with $u_b(p; 0) = 1$ if $p = b$ and $u_b(p; 0) = 0$ if $p \neq b$; and

$$\frac{\partial}{\partial t}\zeta(p; t) = m(p)\frac{\partial}{\partial p}\zeta(p; t) + \frac{1}{2}v(p)\frac{\partial^2}{\partial p^2}\zeta(p; t), \qquad (7.22)$$

with $\zeta(p; 0) = 1$ if $p = a$ or $p = b$ and $\zeta(p; 0) = 0$ if $p \neq a$ and $p \neq b$.

From Eq. (7.20), by letting $t \to \infty$, we have the following equation for the ultimate probability $U_a(p)$ of absorption into $x = a$:

$$m(p)\frac{d}{dp}U_a(p) + \frac{1}{2}v(p)\frac{d^2}{dp^2}U_a(p) = 0, \tag{7.23}$$

with the boundary conditions $U_a(a) = 1$ and $U_a(b) = 0$.

To solve the above equation, notice the result

$$\frac{2m(p)}{v(p)}\frac{d}{dp}U_a(p) + \frac{d^2}{dp^2}U_a(p) = e^{-2\int^p \frac{m(x)}{v(x)}dx}\frac{d}{dp}\left\{e^{2\int^p \frac{m(x)}{v(x)}dx}\frac{d}{dp}U_a(p)\right\} = 0.$$

It follows that $e^{2\int^p \frac{m(x)}{v(x)}dx}\frac{d}{dp}U_a(p) = C_0$, where C_0 is a constant. The solution of (7.23) is then given by

$$U_a(p) = U_a(a) + C_0\int_a^p \psi(x)dx = 1 + C_0\int_a^p \psi(x)dx,$$

where

$$\psi(x) = \exp\left\{-2\int^x \frac{m(y)}{v(y)}dy\right\}.$$

Putting $p = b$ and noting $U_a(b) = 0$, we obtain

$$C_0^{-1} = -\int_a^b \psi(x)dx$$

so that

$$U_a(p) = 1 - \frac{\int_a^p \psi(x)dx}{\int_a^b \psi(x)dx}. \tag{7.24}$$

Similarly, by letting $t \to \infty$ in Eq. (7.21), we obtain:

$$m(p)\frac{d}{dp}U_b(p) + \frac{1}{2}v(p)\frac{d^2}{dp^2}U_b(p) = 0 \tag{7.25}$$

the boundary conditions being $U_b(b) = 1$ and $U_b(a) = 0$.

The solution of Eq. (7.25) under the boundary conditions $U_b(b) = 1$ and $U_b(a) = 0$ is

$$U_b(p) = \frac{\int_a^p \psi(x)dx}{\int_a^b \psi(x)dx}. \tag{7.26}$$

Notice that $U_a(p)+U_b(p) = 1$ so that the ultimate probability of absorption is 1; this is expected as $x = a$ and $x = b$ are absorbing barriers.

The following Theorem provides an avenue to derive the pdf $h(t;p)$ of the time to absorption given $X(0) = p$.

Theorem 7.3.   *The pdf $h(t;p)$ of the first passage time $R(p)$ satisfies the following backward equation:*

$$\frac{\partial}{\partial t}h(t;p) = m(p)\frac{\partial}{\partial p}h(t;p) + \frac{1}{2}v(p)\frac{\partial^2}{\partial p^2}h(t;p)\,, \qquad (7.27)$$

where $h(0;p) = \delta(p - a) + \delta(p - b)$.

Proof.   To prove the theorem, notice that $\zeta(x,t) = \int_a^t h(z;p)dz$ and $\frac{\partial}{\partial t}\int_a^t h(z;p)dz = h(t;p)$. Hence, from Eq. (7.22),

$$h(t;p) = m(p)\frac{\partial}{\partial p}\left[\int_0^t h(z;p)dz\right] + \frac{1}{2}v(p)\frac{\partial^2}{\partial p^2}\left[\int_0^t h(z;p)dz\right]$$

$$= \int_0^t \left\{m(p)\frac{\partial}{\partial p}h(z;p) + \frac{1}{2}v(p)\frac{\partial^2}{\partial p^2}h(z;p)\right\} dz\,.$$

Taking derivative with respect to t on both sides of the above equation, we obtain:

$$\frac{\partial}{\partial t}h(t;p) = m(p)\frac{\partial}{\partial p}h(t;p) + \frac{1}{2}v(p)\frac{\partial^2}{\partial p^2}h(t;p)\,.$$

This proves the theorem. $\square$

7.3.2.   *The first two moments of first passage times in diffusion processes*

Using Theorem 7.3, we can readily derive equations for the mean time $T(p)$ and the variance $\sigma^2(p)$ of $R(p)$.

Theorem 7.4.   *Let $T(p)$ be the expected time for absorption given $X(0) = p$. Then $T(p)$ satisfies the following equation:*

$$m(p)\frac{dT(p)}{dp} + \frac{1}{2}v(p)\frac{d^2T(p)}{dp^2} = -1\,, \qquad (7.28)$$

with boundary conditions $T(a) = T(b) = 0$. (If there is only one boundary a, then $T(a) = 0$.)

Proof. To prove Theorem 7.4, we notice that, if we let $h(t;p)$ be the probability density that absorption to either a or b takes place at time t, then, as shown in Theorem 7.3 above, $h(t;p)$ satisfies the following backward equation:

$$\frac{\partial}{\partial t}h(t;p) = m(p)\frac{\partial}{\partial p}h(t;p) + \frac{1}{2}v(p)\frac{\partial^2}{\partial p^2}h(t;p)\,.$$

Now, by definition,

$$T(p) = \int_0^\infty t\,h(t;p)dt\,.$$

Hence, by interchanging the order of differentiation and integration,

$$m(p)\frac{dT(p)}{dp} + \frac{1}{2}v(p)\frac{d^2T(p)}{dp^2} = \int_0^\infty t\left\{m(p)\frac{\partial}{\partial p}h(t;p) + \frac{1}{2}v(p)\frac{\partial^2}{\partial p^2}h(t;p)\right\}dt$$

$$= \int_0^\infty t\frac{\partial}{\partial t}h(t;p)dt$$

$$= \{th(t;p)\}_0^\infty - \int_0^\infty h(t;p)dt$$

$$= -1\,,$$

as $\lim_{t\to\infty} h(t;p) = 0$ and with the initial conditions being $T(1) = T(0) = 0$. $\square$

To solve (7.28), we notice that

$$m(p)\frac{dT(p)}{dp} + \frac{1}{2}v(p)\frac{d^2T(p)}{dp^2} = 0$$

has two solutions given by

$$U_a(p) = \frac{\int_p^b \psi(x)dx}{\int_a^b \psi(x)dx} \quad \text{and} \quad U_b(p) = \frac{\int_a^p \psi(x)dx}{\int_a^b \psi(x)dx}\,,$$

where $U_a(p) + U_b(p) = 1$. Hence, $U_a'(p) = -U_b'(p)$, and

$$\det\begin{pmatrix} U_a(p) & U_b(p) \\ U_a'(p) & U_b'(p) \end{pmatrix} = \det\begin{pmatrix} U_a(p) & U_b(p) \\ -U_b'(p) & U_b'(p) \end{pmatrix} = U_b'(p) = \frac{\psi(p)}{\int_a^b \psi(x)dx} \neq 0\,.$$

The method of variation of parameters then suggests that

$$T(p) = A(p)U_a(p) + B(p)U_b(p); \quad \text{or} \quad T = AU_a + BU_b,$$

by suppressing p in $A(p)$, $B(p)$, $U_a(p)$ and $U_b(p)$. Differentiation gives

$$T' = AU_a' + A'U_a + BU_b' + B'U_b.$$

We choose A and B such that $A(a) = B(b) = 0$ and $A'U_a + B'U_b = 0$ so that

$$T'' = AU_a'' + A'U_a' + BU_b'' + B'U_b'.$$

Substitution of this into $m(p)T' + \frac{1}{2}v(p)T'' = -1$ gives

$$-1 = \frac{1}{2}v(p)[AU_a'' + A'U_a' + BU_b'' + B'U_b'] + m(p)[AU_a' + BU_b']$$

$$= \left\{\frac{1}{2}v(p)U_a'' + m(p)U_a'\right\}A + \left\{\frac{1}{2}v(p)U_b'' + m(p)U_b'\right\}B$$

$$\quad + \frac{1}{2}v(p)A'U_a' + \frac{1}{2}v(p)B'U_b'$$

$$= \frac{1}{2}v(p)A'U_a' + \frac{1}{2}v(p)B'U_b',$$

or

$$\frac{1}{2}v(p)A'U_a' + \frac{1}{2}v(p)B'U_b' + 1 = 0.$$

Thus, combining with $A'U_a + B'U_b = 0$, we have:

$$\begin{pmatrix} U_a & U_b \\ U_a' & U_b' \end{pmatrix}\begin{pmatrix} A' \\ B' \end{pmatrix} = \begin{pmatrix} 0 \\ -2/v(p) \end{pmatrix},$$

so that

$$\begin{pmatrix} A' \\ B' \end{pmatrix} = \frac{1}{U_a U_b' - U_b U_a'}\begin{pmatrix} 2U_b/v(p) \\ -2U_a/v(p) \end{pmatrix}.$$

Now,

$$U_a' = U_a'(p) = -\frac{\psi(p)}{\int_a^b \psi(x)dx} \quad \text{and} \quad U_b' = \frac{\psi(p)}{\int_a^b \psi(x)dx}.$$

Hence,

$$A' = \frac{2U_b}{v(p)} \cdot \frac{1}{U_a U_b' - U_b U_a'} = \frac{2U_b \int_0^b \psi(x)dx}{v(p)\psi(p)(U_a + U_b)} = \frac{2}{v(p)\psi(p)} \int_a^p \psi(x)dx \,,$$

and

$$A(p) = 2 \int_a^p \frac{1}{v(x)\psi(x)} \int_a^x \psi(y)dydx \,.$$

Similarly,

$$B' = -\frac{2U_a}{v(p)} \cdot \frac{1}{U_a U_b' - U_b U_a'} = \frac{-2}{v(p)\psi(p)} \int_p^b \psi(x)dx \,,$$

so that

$$B(p) = 2 \int_p^b \frac{1}{v(x)\psi(x)} \int_x^b \psi(y)dydx \,.$$

Hence, a particular solution is given by

$$T(p) = A(p)U_a(p) + B(p)U_b(p)$$

$$= 2U_a(p) \int_a^p \frac{1}{v(x)\psi(x)} \int_a^x \psi(y)dydx + 2U_b(p) \int_p^b \frac{1}{v(x)\psi(x)} \int_x^b \psi(y)dydx$$

$$= \int_a^b t(x,p)dx \,, \tag{7.29}$$

where

$$t(x,p) = 2U_a(p)[v(x)\psi(x)]^{-1} \int_a^x \psi(y)dy \,, \quad \text{if } 0 \leq x \leq p \,,$$

and

$$t(x,p) = 2U_b(p)[v(x)\psi(x)]^{-1} \int_x^b \psi(y)dy \,, \quad \text{if } p \leq x \leq 1 \,.$$

Since $U_a(b) = U_b(a) = 0$, we have $T(a) = T(b) = 0$. We next show that the above solution is the only solution satisfying $T(a) = T(b) = 0$. To prove this, let $T_1(p)$ be another solution satisfying $T_1(1) = T_1(0) = 0$; we wish to show that $T_1(p) = T(p)$ for all $a < p < b$. Putting $K = K(p) = T(p) - T_1(p)$, then $K(p)$ satisfies

$$m(p)\frac{d}{dp}K + \frac{1}{2}v(p)\frac{d^2}{dp^2}K = 0\,, \quad \text{with } K(a) = K(b) = 0\,.$$

But this implies that

$$\frac{d}{dp}\left\{ e^{2\int^p \frac{m(x)}{v(x)}\,dx}\frac{d}{dp}K \right\} = 0\,,$$

so that

$$K(p) = c_1 \int_a^p e^{-2\int_a^y \frac{m(x)}{v(x)}\,dx}\,dy + c_2\,.$$

Hence,

$$K(a) = c_2 = 0 \quad \text{and} \quad K(b) = c_1 \int_a^b e^{-2\int_a^y \frac{m(x)}{v(x)}\,dx}\,dy = 0 \Rightarrow c_1 = 0$$

or, $K(p) = 0$ so that $T(p) = T_1(p)$.

Let σ_p^2 be the variance of the first time absorption $R(p)$ given initially $X(0) = p$, and put $\sigma_p^2 = W(p) - (T(p))^2$. Then, we have the following theorem for the computation of $W(p)$:

Theorem 7.5.   *$W(p)$ satisfies the following second order equation:*

$$m(p)\frac{d}{dp}W(p) + \frac{1}{2}v(p)\frac{d^2}{dp^2}W(p) = -2T(p)\,, \tag{7.30}$$

where $W(a) = W(b) = 0$ and $T(p)$ is given by (7.29). (If there is only one boundary a, then $W(a) = 0$.)

The above theorem was first given by Tan [4].

Proof.   Let $h(t;p)$ be the pdf of the first passage time $R(p)$ given $X(0) = p$, then, as shown above, $h(t;p)$ satisfies

$$\frac{\partial}{\partial t}h(t;p) = m(p)\frac{\partial}{\partial p}h(t;p) + \frac{1}{2}v(p)\frac{\partial^2}{\partial p^2}h(t;p)\,.$$

Since

$$W(p) = \int_0^\infty t^2 h(t;p)dt\,,$$

so, by interchanging $\frac{d}{dp}$ and $\int$, integrating by parts and noting that

$$t^2 \zeta(p,t) \to 0 \quad \text{as} \quad t \to \infty\,,$$

$$
\begin{aligned}
m(p)\frac{d}{dp}W(p) + \frac{1}{2}v(p)\frac{d^2}{dp^2}W(p) &= \int_0^\infty t^2 \left\{ m(p)\frac{\partial}{\partial p}h(t;p) + \frac{1}{2}v(p)\frac{\partial^2}{\partial p^2}h(t;p) \right\} dt \\
&= \int_0^\infty t^2 \left(\frac{\partial}{\partial t}h(t;p) \right) dt \\
&= -2 \int_0^\infty t\, h(t;p) dt \\
&= -2T(p)\,. \qquad \qquad \square
\end{aligned}
$$

By using the method of variation of parameters, one can similarly show that the solution of (7.30) is given by

$$W(p) = 4U_a(p) \int_a^p T(x)[\psi(x)v(x)]^{-1} \int_a^x \psi(y)dy\,dx$$

$$+ 4U_b(p) \int_p^b T(x)[\psi(x)v(x)]^{-1} \int_x^b \psi(y)dy\,dx\,. \qquad (7.31)$$

Example 7.5. Absorption probabilities of gene frequency in the Wright model of population genetics. In the Wright model in population genetics for one locus with two alleles A and a, if there are no mutations and no immigration and migration, then 0 and 1 are absorbing barriers. In this case, with probability one the process will eventually be absorbed into the absorbing states. When the population size N is very large, then as shown in Example 6.9, to the order of $O(N^{-2})$ the frequency $Y(t)$ of A gene is a diffusion process with diffusion coefficients given by

$$\{m(x,t) = x(1-x)[x\alpha_1(t) + (1-2x)\alpha_2(t)]\,, \quad v(x) = x(1-x)\}\,,$$

where the $\alpha_i(t)$'s are the selection intensities (See Example 6.9, Case 4). Assume now $\alpha_i(t) = \alpha_i, i = 1, 2$, so that the process is time homogeneous. Then,

$$\phi(x) = \exp\left\{ -2 \int^x \frac{m(y)}{v(y)}dy \right\} = \exp\{-[\alpha_1 x^2 + 2\alpha_2 x(1-x)]\}\,.$$

If there are no selection so that $\alpha_i = 0, i = 1, 2$, then $\phi(x) = 1$. In this case, the ultimate absorption probabilities into 0 and 1 given $Y(0) = p$ are given respectively by:

$$U_0(p) = 1 - \frac{\int_0^p \phi(x)dx}{\int_0^1 \phi(x)dx} = 1 - p,$$

$$U_1(p) = \frac{\int_0^p \phi(x)dx}{\int_0^1 \phi(x)dx} = p.$$

The mean of the first absorption time given $Y(0) = p$ is given by:

$$T(p) = 2U_0(p) \int_0^p [v(x)\phi(x)]^{-1} \left\{ \int_0^x \phi(y)dy \right\} dx$$

$$+ 2U_1(p) \int_p^1 [v(x)\phi(x)]^{-1} \left\{ \int_x^1 \phi(y)dy \right\} dx$$

$$= 2(1-p) \int_0^p (1-x)^{-1}dx + 2p \int_p^1 x^{-1}dx$$

$$= -2(1-p)\log(1-p) - 2p\log(p) = \log\{(1-p)^{2(1-p)}p^{2p}\}^{-1}.$$

The variance of the first absorption time given $Y(0) = p$ is $V(p) = W(p) - T^2(p)$, where $W(p)$ is given by:

$$W(p) = 4U_0(p) \int_0^p T(x)[v(x)\phi(x)]^{-1} \left\{ \int_0^x \phi(y)dy \right\} dx$$

$$+ 4U_1(p) \int_p^1 T(x)[v(x)\phi(x)]^{-1} \left\{ \int_x^1 \phi(y)dy \right\} dx$$

$$= 4(1-p) \int_0^p T(x)(1-x)^{-1}dx + 4p \int_p^1 T(x)x^{-1}dx.$$

Example 7.6. Absorption probabilities of gene frequency in Moran's model of genetics. In the Moran's model of genetics considered in Examples 6.10 and 7.3, if $a_1 = a_2 = 0$ (no mutation), then O and M are absorbing states; we may then apply (7.24), (7.26), (7.29) and (7.31) to compute the ultimate absorption probabilities, the mean absorption times, and the variances of first absorption times, valid to the order of $O(M^{-2})$.

Putting $a_1 = a_2 = 0$ (i.e. $\alpha_1 = \alpha_2 = 0$), $\lambda_1 = \lambda$ and $\lambda_2 = \lambda + \frac{1}{M}\zeta$, then

$$\psi(x) = \exp\left\{-2M \int^x \frac{m(y)}{v(y)} dy\right\} \propto \exp\left\{-\frac{1}{\lambda}\zeta x\right\} .$$

Hence,

$$U_1(p) = \frac{\int_0^p \psi(x)dx}{\int_0^1 \psi(x)dx} = \frac{\int_0^p e^{-\frac{1}{\lambda}\zeta x}dx}{\int_0^1 e^{-\frac{1}{\lambda}\zeta x}dx} = \frac{(1 - e^{-\frac{1}{\lambda}\zeta p})}{(1 - e^{-\frac{1}{\lambda}\zeta})} ,$$

$$U_0(p) = 1 - U_1(p) = \frac{(e^{-\frac{1}{\lambda}\zeta p} - e^{-\frac{1}{\lambda}\zeta})}{(1 - e^{-\frac{1}{\lambda}\zeta})} ,$$

$$T(p) = \frac{M}{\lambda}\left\{ U_0(p) \int_0^p \frac{1}{x(1-x)}(e^{-\frac{1}{\lambda}\zeta x} - 1)dx \right.$$

$$\left. + U_1(p) \int_0^q \frac{1}{x(1-x)}(1 - e^{-\frac{1}{\lambda}\zeta x})dx \right\} ,$$

$$\sigma_p^2 = W(p) - (T(p))^2 ,$$

where

$$W(p) = 2\frac{M}{\lambda}\left\{ U_0(p) \int_0^p \frac{T(x)}{x(1-x)}(e^{\frac{1}{\lambda}\zeta x} - 1)dx \right.$$

$$\left. + U_1(p) \int_0^q \frac{T(x)}{x(1-x)}(1 - e^{-\frac{1}{\lambda}\zeta x})dx \right\} ,$$

and $q = 1 - p$.

Letting $\zeta \to 0$ and applying the L'Hospital rule, we have then:

$$\lim_{\zeta \to 0} U_1(p) = p, \quad \lim_{\zeta \to 0} U_0(p) = 1 - p = q ,$$

$$\lim_{\zeta \to 0} T(p) = \frac{M}{\lambda}\left\{ q \int_0^p \frac{x}{x(1-x)}dx + p \int_0^q \frac{x}{x(1-x)}dx \right\}$$

$$= -\frac{M}{\lambda}\{q \log q + p \log p\} ,$$

and

$$\lim_{\zeta \to 0} W(p) = -2 \left(\frac{M}{\lambda}\right)^2 \left\{ q \int_0^p \frac{x}{x(1-x)} (x \log x + (1-x) \log(1-x)) dx \right.$$

$$\left. + p \int_0^q \frac{x}{x(x+1)} (x \log x + (1-x) \log(1-x)) dx \right\}.$$

These results are comparable to those of the Wright's model with no selection and no mutation as given above.

7.4.　Complements and Exercises

Exercise 7.1.　Consider the Wright model in population genetics for one locus with two alleles $A : a$. Under the assumption of selection, mutation, immigration and migration, the frequency of A gene is a diffusion process with state space $[0, 1]$ and with diffusion coefficients given by (See Example 6.9, Case 6):

$$m(x) = x(1-x)[\alpha_1 x + \alpha_2(1-2x)] - \gamma_1 x + \gamma_2(1-x),$$

and $v(x) = x(1-x)$. Assume that $N\alpha_i$ is finite for $i = 1, 2$, where N is the population size. Using results of Example 6.4 and the theories in Sec. 7.1, derive an approximation to the first two smallest eigenvalues and the associated eigenvectors of the associated differential equation, to the order of $O(N^{-2})$. Hence derive an asymptotic distribution of the conditional density $f(p, x; t)$ to the order of $O(N^{-2})$.

Exercise 7.2.　Consider the Wright model in population genetics for one locus with two alleles $A : a$. Suppose that the genotype aa is lethal and A is dominant over a so that AA and Aa have the same phenotype. Assume that there are mutations from $A \to a$ and from $a \to A$ and there are immigration and migration as given in Case 6 of Example 6.9. Then, it was shown in Example 6.9 that to the order of $O(N^{-2})$, the frequency $Y(t)$ of the A allele is a diffusion process with state space $[0, 1]$ and with diffusion coefficients

$$\left\{ m(x) = \gamma_2(1-x) - \gamma_1 x - \frac{x^2}{1+x}, \quad v(x) = x(1-x) \right\}.$$

Further, the states 0 and 1 are regular boundaries. Show that the density $\phi(x)$ of the stationary distribution is given by:

$$\phi(x) = Cx^{2\gamma_2-1}(1-x)^{2\gamma_1-1}(1-x^2)\,,$$

where C is a normalizing constant satisfying $\int_0^1 \phi(x)dx = 1$.

Exercise 7.3. Consider the Wright model in Exercise 7.1 and assume that $\alpha_1 = 2\alpha_2 = 2Z$. (This is the case of additive selection; see [5]). Then, during the time period $[t, t+\Delta t)$, the change of gene frequency of A allele given $Y(t) = x$ can be represented by $\Delta Y(t) = Y(t+\Delta t) - Y(t) = x(1-x)Z + [\gamma_2(1-x) - \gamma_1 x]\Delta t + \epsilon(t)$, where Z is the random variable representing the intensity of selection with mean $EX = s\Delta t + o(\Delta t)$ and with variance $\mathrm{Var}(Z) = \sigma^2 \Delta t + o(\Delta t)$ and where $\epsilon(t)$ is the random disturbance with mean 0 and variance $x(1-x)\Delta t$. Under this condition, to the order of $O(N^{-2})$, the frequency $Y(t)$ of the A allele at time t ($t \geq 0$) is a diffusion process with state space $[0, 1]$ and with diffusion coefficients

$$\{m(x) = sx(1-x) - \gamma_1 x + \gamma_2(1-x), v(x) = \sigma^2 x^2(1-x)^2 + x(1-x)\}\,.$$

This is the model describing random variation of selection first considered by Kimura [6]. In this case, the states 0 and 1 are regular boundary points.

Let $W = \frac{1}{\sigma^2}$ and put $\lambda_1 = \frac{1}{2}(1 + \sqrt{1+4W})$ ($\lambda_1 > 1$) and $\lambda_2 = \frac{1}{2}(1 - \sqrt{1+4W})$ ($\lambda_2 < 0$). Then $v(x) = \sigma^2(\lambda_1 - x)(x - \lambda_2)$. Show that the density of the stationary distribution is given by

$$\phi(x) = Cx^{2\gamma_2-1}(1-x)^{2\gamma_1-1}(\lambda_1 - x)^{W_1-1}(x - \lambda_2)^{W_2-1}, \quad 0 \leq x \leq 1\,,$$

where

$$W_1 = -\frac{2}{\lambda_1 - \lambda_2}(sW + \gamma_1\lambda_1 - \gamma_2\lambda_2)\,,$$

$$W_2 = -\frac{2}{\lambda_1 - \lambda_2}(sW + \gamma_1\lambda_2 - \gamma_2\lambda_1)\,,$$

and C is a normalizing constant satisfying the condition $\int_0^1 \phi(x)dx = 1$.

Exercise 7.4. Prove Eq. (7.31).

Exercise 7.5. Consider the Wright modwel given in Exercise 7.2. Assume that $\{\gamma_i = 0, i = 1, 2\}$ so that there are no mutations and no immigration and migration. In this case the states 0 and 1 are absorbing barriers.

(a) Obtain the ultimate absorptions into 0 and 1.

(b) Derive the mean and the variance of first absorption times.

Exercise 7.6.   Let $\{X(t), t \geq 0\}$ be a logistic birth-death process with state space $S = (0, 1, \ldots, M)$ and with birth rate $b_i(t) = ib(1 - i/M)$ and death rate $d_i(t) = id(1 - i/M)$. Then, it is shown in Example 6.9 that to the order of $O(M^{-2})$, $\{Y(t), t \geq 0\}$ is a diffusion process with state space $[0, 1]$ and with diffusion coefficients

$$\{m(x) = \epsilon x(1 - x), v(x) = \omega x(1 - x)\},$$

where $\{\epsilon = b - d, \omega = b + d\}$. In this case the states 0 and 1 are absorbing barriers.

(a) Obtain the probabilities of ultimate absorptions into 0 and 1, respectively.

(b) Derive the mean and the variance of first absorption times.

References

[1] T. Kato, *Perturbation Theory For Linear Operators*, Springer-Verlag, Berlin (1966).

[2] M. Kimura, *Diffusion Models in Population Genetics, Methuen's Monographs on Applied Probability and Statistics*, Vol. 2, Methuen & Co, LTD, London (1964).

[3] W. Y. Tan and M. Tiku, *Sampling Distributions in Terms of Lauguerre Polynomials and Application*, New Age International Publisher, New Delhi, India (1999).

[4] W. Y. Tan, *On the absorption probability and absorption time to finite homogeneous birth-death processes by diffusion approximation*, Metron **33** (1975) 389–401.

[5] J. F. Crow and M. Kimura, *An Introduction to Population Genetic Theory*, Harper and Row, New York (1972).

[6] M. Kimura, *Process leading to quasi-fixation of genes in natural populations due to random fluctuation of selection intensities*, Genetics **39** (1954) 280–295.

Chapter 8

State Space Models and Some Examples from Cancer and AIDS

As defined in Definition 1.4, state space models (Kalman filter models) of stochastic systems are stochastic models consisting of two sub-models: The stochastic system model which is the stochastic model of the system and the observation model which is a statistical model based on some data from the system. That is, the state space model adds one more dimension to the stochastic model and to the statistical model by combining both of these models into one model. It takes into account the basic mechanisms of the system and the random variation of the system through its stochastic system model and incorporate all these into the observed data from the system; furthermore, it validates and upgrades the stochastic model through its observation model and the observed data of the system. It is advantageous over both the stochastic model and the statistical model when used alone since it combines information and advantages from both of these models. Given below is a brief summary of the advantages over the statistical model or stochastic model used alone.

(1) The statistical model alone or the stochastic model alone very often are not identifiable and cannot provide information regarding some of the parameters and variables. For example, the backcalculation method (a statistical model of HIV epidemic) in AIDS research is not identifiable so that one cannot estimate simultaneously the HIV infection and the HIV incubation distribution; see [1] and [2, Chap. 5]. By using state space model, this difficulty is easily solved; see [2–4].

(2) State space model provides an optimal procedure to updating the model by new data which may become available in the future. This is the smoothing step of the state space models; see [5–7].

(3) The state space model provides an optimal procedure via Gibbs sampling to estimate simultaneously the unknown parameters and the state variables of interest; see [3, 4]. For example, by using the AIDS incidence data for the observation model in the state space model, the author and his associates were able to estimate the recruitment and immigration rates as well as the retirement and death rates besides the HIV infection, the HIV incubation, the numbers of S people, I people and AIDS incidence in the Swiss populations of IV drug users and homosexual/bisexual men [4]. This is not possible by other models used alone.

(4) The state space model provides an avenue to combine information from various sources. For example, this author and his associates have attempted to link the molecular events at the molecular level with the critical events at the population level in carcinogenesis via the multi-level Gibbs sampling method [8].

The state space model was originally proposed by Kalman and his associates in the early 60's for engineering control and communication [9]. Since then it has been successfully used as a powerful tool in aero-space research, satellite research and military missile research. It has also been used by economists in econometrical research [10] and by mathematician and statisticians in time series research [11] for solving many difficult problems which appear to be extremely difficult from other approaches. Since 1995, this author and his associates have attempted to apply the state space model and method to AIDS research and to cancer research; see [2–4, 12–23]. Because of its importance, in this chapter and the next chapter we will illustrate how to construct state space models for some problems in cancer and AIDS and demonstrate its applications to these areas.

8.1.　Some HIV Epidemic Models as Discrete-Time Linear State Space Models

In the first application of state space models including those by Kalman, time is discrete and both the stochastic system model and the observation model are linear functions of the state variables. These are the discrete-time linear state space models. In general, these state space models can be expressed as:

(A) Stochastic System Model:

$$\underset{\sim}{X}(t+1) = F(t+1,t)\underset{\sim}{X}(t) + \underset{\sim}{\epsilon}(t+1);$$

(B) Observation Model:

$$\underset{\sim}{Y}(t+1) = H(t+1)\underset{\sim}{X}(t+1) + \underset{\sim}{e}(t+1),$$

where $\underset{\sim}{\epsilon}(t+1)$ is the vector of random noises associated with $\underset{\sim}{X}(t+1)$ and $\underset{\sim}{e}(t+1)$ the vector of random measurement errors associated with measuring the observed vector $\underset{\sim}{Y}(t+1)$ and where $F(t+1,t)$ and $H(t+1)$ are transition matrices whose elements are deterministic (non-stochastic) functions of time t. In this model, the usual assumptions are:

(1) The elements of $\underset{\sim}{\epsilon}(t)$'s are independently distributed random noises with means 0 and with covariance matrix $\mathrm{Var}[\underset{\sim}{\epsilon}(t)] = V(t)$.

(2) The elements of $\underset{\sim}{e}(t)$'s are independently distributed random errors with means 0 and with covariance matrix $\mathrm{Var}[\underset{\sim}{e}(t)] = \Sigma(t)$.

(3) The $\underset{\sim}{\epsilon}(t)$'s are independently distributed of the $\underset{\sim}{e}(\tau)$'s for all $t \geq t_0$ and $\tau \geq t_0$.

(4) The $\underset{\sim}{\epsilon}(t)$'s and the $\underset{\sim}{e}(t)$'s are un-correlated with the state variables $\underset{\sim}{X}(\tau)$ for all $t \geq t_0$ and $\tau \geq t_0$.

In Example 1.1, we have illustrated how a hidden Markov model of HIV epidemic as proposed by Satten and Longini [24] can be expressed as a discrete-time linear state space model. In Subsec. 2.8.1 and in Example 2.17, we have demonstrated that for finite Markov chains, if some data sets are available from the system, then it can be expressed as discrete-time linear state space models. The following two examples show that some of the HIV transmission models can be expressed in terms of discrete-time linear state space models. These are the models first proposed in [12, 14, 17, 23]. Notice that in the example in Subsec. 8.1.1, the dimension of $\underset{\sim}{X}(t)$ increases with time t. This case has been referred to as an expanding state space model by Liu and Chen [25].

8.1.1.   *A state space model with variable infection for HIV epidemic in homosexual populations*

In Subsec. 2.8.2, we have developed a stochastic model for the HIV epidemic in homosexual populations or populations of IV drug users. In this model, the state variables are $\underset{\sim}{X}(t) = \{S(t), I(r,t), r = 0, 1, \ldots, t, A(t)\}$ with t denoting calendar time. When some observed data are available from the system, then one can develop a state space model for this system. Using the stochastic process from Subsec. 2.9.2, the stochastic system model of this state space model is given by the stochastic Eqs. (2.33)–(2.36) in Subsec. 2.9.2. Let $Y(t)$ be the observed number of AIDS cases at time t. Then the equation of the observation model of the state space model is given by:

$$Y(t) = A(t) + e(t),$$

where $e(t)$ is the measurement error in reporting AIDS cases at time t.

The $e(t)$'s are usually associated with under-reporting, reporting delay and/or other errors in reporting AIDS cases. When such reporting errors have been corrected, one may assume that the $e(t)$'s have expected values 0 and have covariances $\text{Cov}[e(t), e(\tau)] = \delta_{t\tau}\sigma_A^2(t)$, where δ_{ij} is the Kronecker's δ. These random errors are also un-correlated with the random noises $\underset{\sim}{\epsilon}(t) = \{\epsilon_S(t), \epsilon_u(t), u = 0, \ldots, t, \epsilon_A(t)\}'$ defined in Subsec. 2.8.2. In this model, if one replaces $p_S(t)$ by its estimates respectively, then one has a discrete-time linear state space model as given above. This model was first proposed by Tan and Xiang [17] (see also [3, 4]) for the HIV epidemic in homosexual populations.

To present the state space model in matrix form, denote $\{\phi_S(t) = 1 - p_S(t), \phi_i(t) = 1 - \gamma_i(t), i = 0, 1, \ldots, t\}$ and put;

$$F(t+1, t) = \begin{bmatrix} \phi_S(t) & 0 & 0 & \cdots & 0 & 0 \\ p_S(t) & 0 & 0 & \cdots & 0 & 0 \\ 0 & \phi_0(t) & 0 & \cdots & 0 & 0 \\ 0 & 0 & \phi_1(t) & \cdots & 0 & 0 \\ \vdots & \vdots & \vdots & \ddots & \vdots & \vdots \\ 0 & 0 & 0 & \cdots & \phi_t(t) & 0 \\ 0 & \gamma_0(t) & \gamma_1(t) & \cdots & \gamma_k(t) & 0 \end{bmatrix}.$$

Then,

$$\underset{\sim}{X}(t+1) = F(t+1,t)\underset{\sim}{X}(t) + \underset{\sim}{\epsilon}(t+1).$$

The expected value of $\underset{\sim}{\epsilon}(t)$ is a vector of zero's and the variances and covariances of elements of $\underset{\sim}{\epsilon}(t)$ are given in Table 8.1.

8.1.2. *A staged state-space model for HIV epidemic in homosexual populations*

In Subsec. 2.8.2, we have developed a staged-model for the HIV epidemic in homosexual populations. In this model, the state variables are $\{S(t), I(r,t), r = 1, \ldots, k, A(t)\}$. The stochastic equations for these state variables are given by Eqs. (2.39)–(2.43). By using these equations as the equations for the stochastic system model and by using Eq. (8.1) as the equation for the observation model we have a staged state space model for the HIV epidemic in homosexual populations. In this model, if estimates of $p_S(t)$ are available, then by substituting these probabilities by its estimates respectively, we have a discrete-time linear state space model as given above. This state space model was first proposed in [12, 14, 23] for the HIV epidemic in homosexual populations.

Table 8.1. The variances and covariances of the random noises for the chain binomial model of HIV epidemic.

$C_{SS}(t+1, \tau+1) = \text{Cov}[\epsilon_S(t+1), \epsilon_S(\tau+1)] = \delta_{t\tau} V_S(t+1) = \delta_{t\tau} E\{S(t)p_S(t)[1 - p_S(t)]\};$

$C_{Su}(t+1, \tau+1) = \text{Cov}[\epsilon_S(t+1), \epsilon_u(\tau+1)] = -\delta_{t\tau}\delta_{u0} V_S(t+1); u = 0, \ldots, t+1;$

$C_{SA}(t+1, \tau+1) = \text{Cov}[\epsilon_S(t+1), \epsilon_A(\tau+1)] = 0.$

For $u, v = 0, \ldots, t, t \geq 0$ and $\tau \geq 0$,

$$C_{uv}(t+1, \tau+1) = \text{Cov}[\epsilon_{u+1}(t+1), \epsilon_{v+1}(\tau+1)]$$

$$= \delta_{t\tau}\delta_{uv}\{\delta_{0u} V_S(t+1) + (1 - \delta_{0u})V_I(u+1, t+1)\},$$

where, for $u = 0, \ldots, t$,

$$V_I(u+1, t+1) = E[I(u,t)]\gamma_u(t)[1 - \gamma_u(t)].$$

$C_{AA}(t+1, \tau+1) = \text{Cov}[\epsilon_A(t+1), \epsilon_A(\tau+1)] = \delta_{t\tau} \sum_{u=0}^{t} V_I(u+1, t+1).$

For $u = 0, \ldots, t+1$,

$C_{Au}(t+1, \tau+1) = \text{Cov}[\epsilon_u(t+1), \epsilon_A(\tau+1)] = -\delta_{t\tau}(1 - \delta_{u0})V_I(u, t+1).$

To present the stochastic dynamic model in matrix form, put $\psi_S(t) = 1+\nu_S(t)-p_S(t)-d_S(t)$ and $\{\psi_r(t) = 1+\nu_r(t)-(1-\delta_{rk})\gamma_r(t)-(1-\delta_{r1})\beta_r(t)-\omega_r(t)-d_r(t), r = 1, \ldots, k\}$ and put:

$$
F(t+1,t) =
\begin{bmatrix}
\psi_S(t) & 0 & & & & & \\
p_S(t) & \psi_1(t) & \beta_2(t) & & & & \\
 & \gamma_1(t) & \psi_2(t) & \beta_3(t) & & & \\
 & & & & & & \\
 & & & & \gamma_{k-2}(t) & \psi_{k-1}(t) & \beta_k(t) \\
 & & & & & \gamma_{k-1}(t) & \psi_k(t) & 0 \\
0 & \omega_1(t) & & & & & \omega_k(t) & 1
\end{bmatrix}.
$$

Then, the stochastic system model is

$$
\underset{\sim}{X}(t+1) = F(t+1,t)\,\underset{\sim}{X}(t) + \underset{\sim}{\epsilon}(t+1),
$$

where $\underset{\sim}{\epsilon}(t)$ is the vector of random noises given by:

$$
\underset{\sim}{\epsilon}(t) = [\epsilon_S(t), \epsilon_1(t), \ldots, \epsilon_k(t), \epsilon_A(t)]'.
$$

The expected value of the random noise $\underset{\sim}{\epsilon}(t)$ is zero and the variances and covariances of these random noises are given in Table 8.2.

In [23], an additional observation model has been given by the total population size. Let $Y_1(t)$ be the total number of AIDS cases at time t including those who died from AIDS during $[t-1,t)$ and $Y_2(t)$ the sum of the total population size in the system at time t and the number of people who died from AIDS during $[t-1,t)$. Let $e_1(t)$ and $e_2(t)$ be the measurement errors for $Y_1(t)$ and $Y_2(t)$ respectively. Then,

$$
Y_1(t) = A(t) + e_1(t),
$$

$$
Y_2(t) = S(t) + \sum_{r=1}^{5} I_r(t) + A(t) + e_2(t).
$$

In the above equations, $Y_1(t)$ can be obtained from the surveillance reports from CDC of the United States or the World Health Organization (WHO); $Y_2(t)$ is usually available from population surveys and AIDS surveillance reports. In general, $e_1(t)$ is associated with AIDS report errors and $e_2(t)$ with

Table 8.2. The variances and covariances of the random noises for the staged model of HIV epidemic.

$C_{SS}(t+1, \tau+1) = \text{Cov}[\epsilon_S(t+1), \epsilon_S(\tau+1)] = \delta_{t\tau}\text{Var}[\epsilon_S(t+1)]$, where
$\text{Var}[\epsilon_S(t+1)] = E\{S(t)[p_S(t) + \mu_S(t)][1 - p_S(t) - \mu_S(t)]\} + \sigma_S^2(t)$ with
$\text{Var}[R_S(t)] = \sigma_S^2(t)$;

$C_{Su}(t+1, \tau+1) = \text{Cov}[\epsilon_S(t+1), \epsilon_u(t+1)] = -\delta_{u1}\delta_{t\tau}E\{S(t)p_S(t)[1 - p_S(t) - \mu_S(t)]\}$
for $u = 1, \ldots, k$;

$C_{SA}(t+1, \tau+1) = \text{Cov}[\epsilon_S(t+1), \epsilon_A(\tau+1)] = 0$ for all t, τ;

$C_{uv}(t+1, \tau+1) = \text{Cov}[\epsilon_u(t+1), \epsilon_v(\tau+1)] = \delta_{t\tau}\{\delta_{uv}\text{Var}[\epsilon_u(t+1)] + \delta_{v,u+1}\text{Cov}[\epsilon_u(t+1),$
$\epsilon_{u+1}(t+1)] + \delta_{v,u+2}\text{Cov}[\epsilon_u(t+1), \epsilon_{u+2}(t+1)] + \delta_{v,u-1}\text{Cov}[\epsilon_{u-1}(t+1), \epsilon_u(t+1)]\}$
for all $u, v = 1, \ldots, k$,
where
$V_u(t+1) = \text{Var}[\epsilon_u(t+1)] = \delta_{1u}E\{S(t)p_S(t)[1 - p_S(t)]\} + (1 - \delta_{1u})u_I(u - 1, t)\gamma_{u-1}(t)$
$[1 - \gamma_{u-1}(t)] + (1 - \delta_{uk})u_I(u + 1, t)\beta_{u+1}(t)[1 - \beta_{u+1}(t)] + u_I(u, t)\{p(I_u; t)[1 - p(I_u; t)]\} +$
$\sigma_u^2(t)$, with $p(I_u; t) = (1 - \delta_{uk})\gamma_u(t) + (1 - \delta_{1u})\beta_u(t) + \omega_u(t) + \mu_u(t)$ for $u = 1, \ldots, k$,
where $u_I(v, t) = EI(v, t), \sigma_r^2(t) = \text{Var}[R_I(r, t)]$ and $\beta_1(t) = 0$,

$C_{v,v+1}(t+1) = \text{Cov}[\epsilon_v(t+1), \epsilon_{v+1}(t+1)] = -u_I(v, t)\gamma_v(t)\{1 - \gamma_v(t) - (1 - \delta_{1v})\beta_v(t) -$
$\omega_v(t) - \mu_v(t)\} - u_I(v + 1, t)\beta_{v+1}(t)\{1 - \beta_{v+1}(t) - \omega_{v+1}(t) - (1 - \delta_{v+1,k})\ \gamma_{v+1}(t) -$
$\mu_{v+1}(t)\}$, for $v = 1, \ldots, k - 1$,

$C_{v,v+2}(t+1) = \text{Cov}[\epsilon_v(t+1), \epsilon_{v+2}(t+1)] = -u_I(v + 1, t)\beta_{v+1}(t)\gamma_{v+1}(t)$, for
$v = 1, \ldots, k - 2$;

$C_{Av}(t+1, \tau+1) = \text{Cov}[\epsilon_v(t+1), \epsilon_A(\tau+1)] = -\delta_{t\tau}u_I(v, t)\omega_v(t)[1 - \omega_v(t) - (1 - \delta_{vk})$
$\gamma_v(t) - \beta_v(t) - \mu_v(t)]$, for $v = 1, \ldots, k$;

$C_{AA}(t+1, \tau+1) = \text{Cov}[\epsilon_A(t+1), \epsilon_A(\tau+1)] = \delta_{t\tau}\{\sum_{v=1}^{k} u_I(v; t)\omega_v(t)[1 - \omega_v(t)]\}$.

---

population survey errors. Let $\underset{\sim}{Y}(t) = [Y_1(t), Y_2(t)]'$ and $\underset{\sim}{e}(t) = [e_1(t), e_2(t)]'$. Then in matrix notation, the observation equation is given by the stochastic equation

$$\underset{\sim}{Y}(t) = H(t+1)\underset{\sim}{X}(t) + \underset{\sim}{e}(t),$$

where

$$H(t) = \begin{bmatrix} 0 & 0 & 0 & 0 & 0 & 0 & 1 \\ 1 & 1 & 1 & 1 & 1 & 1 & 1 \end{bmatrix}.$$

8.2. Some State Space Models with Continuous-Time Stochastic System Model

In many state space models, the stochastic system model is usually based on continuous time. This is true in HIV pathogenesis as well as in carcinogenesis. In this section we give some examples from cancer treatment and HIV pathogenesis.

8.2.1. *A state space model for drug resistance in cancer chemotherapy*

In Example 1.1, we have considered a drug resistance model in cancer treatment. In this model, the state variables are the number of sensitive cancer tumor cells $(X_1(t))$ and resistant cancer tumor cells $(X_2(t))$ over time. Suppose that the total number of tumor cells have been counted at times $t_j, j = 1, \ldots, n$. Then we can develop a state space model with stochastic system model given by Eqs. (8.1)–(8.2) and with observation model given by Eq. (8.3). This is the state space model first proposed by Tan and his associates in [26].

8.2.1.1. *The stochastic system model*

As shown in [27], one may assume that the Type-i tumor cells proliferate by following a stochastic birth and death process with birth rate $b_i(t)$ and death rate $d_i(t)$. Also, many biological studies have shown that resistant tumor cells arise from sensitive tumor cells by mutation with rate $\alpha(t)$ and that one may ignore back mutation from resistant cells to sensitive tumor cells. To develop the stochastic system model, we thus denote by:

$B_S(t)(B_R(t))$ = Number of birth of sensitive (resistant) tumor cells during $[t, t + \Delta t)$,

$D_S(t)(D_R(t))$ = Number of death of sensitive (resistant) tumor cells during $[t, t + \Delta t)$,

$M_S(t)$ = Number of mutations from sensitive tumor cells to resistant tumor cells during $[t, t + \Delta t)$.

Then, as shown in Sec. 4.7, the conditional distributions of $\{B_S(t), D_S(t), M_S(t)\}$ given $X_1(t)$ and of $\{B_R(t), D_R(t)\}$ given $X_2(t)$ are multinomial distributions with parameters $\{X_1(t), b_1(t)\Delta t, d_1(t)\Delta t, \alpha(t)\Delta t\}$ and

$\{X_2(t), b_2(t)\Delta t, d_2(t)\Delta t\}$, respectively. That is,

$$\{B_S(t), D_S(t), M_S(t)\}|X_1(t) \sim ML\{X_1(t); b_1(t)\Delta t, d_1(t)\Delta t, \alpha(t)\Delta t\}$$

and

$$\{B_R(t), D_R(t)\}|X_2(t) \sim ML\{X_2(t); b_2(t)\Delta t, d_2(t)\Delta t\} \, .$$

Let

$$\epsilon_1(t)\Delta t = [B_S(t) - X_1(t)b_1(t)\Delta t] - [D_S(t) - X_1(t)d_1(t)\Delta t]$$

and

$$\epsilon_2(t)\Delta t = [M_S(t) - X_1(t)\alpha(t)\Delta t] + [B_R(t) - X_2(t)b_2(t)\Delta t]$$
$$- [D_R(t) - X_2(t)d_2(t)\Delta t] \, .$$

Then we have the following stochastic differential equations for the state variables $\{X_1(t), X_2(t)\}$ respectively:

$$dX_1(t) = X_1(t + \Delta t) - X_1(t) = B_S(t) - D_S(t)$$
$$= X_1(t)\gamma_1(t)\Delta t + \epsilon_1(t)\Delta t \, , \tag{8.1}$$
$$dX_2(t) = X_2(t + \Delta t) - X_2(t) = M_S(t) + B_R(t) - D_R(t)$$
$$= \{X_1(t)\alpha(t) + X_2(t)\gamma_2(t)\}\Delta t + \epsilon_2(t)\Delta t \, , \tag{8.2}$$

where $\gamma_i(t) = b_i(t) - d_i(t), i = 1, 2$.

In Eqs. (8.1)–(8.2) the expected values of the random noises $\{\epsilon_i(t), i = 1, 2\}$ are zero; further, these random noises are un-correlated with the state variables $\underset{\sim}{X}(t) = \{X_1(t), X_2(t)\}$. Also, to order of $o(\Delta t)$, the variances and the covariances of $\{\epsilon_i(t)\Delta t, i = 1, 2\}$ are given by:

$$\text{Cov}\{\epsilon_1(t_1)\Delta t, \epsilon_1(t)\Delta t\} = \delta_{t_1,t}E[X_1(t)]\{b_1(t) + d_1(t)\}\Delta t + o(\Delta t) \, ,$$
$$\text{Cov}\{\epsilon_2(t_1)\Delta t, \epsilon_2(t)\Delta t\} = \delta_{t_1,t}\{E[X_1(t)]\alpha(t) + [b_2(t)$$
$$+ d_2(t)]E[X_2(t)]\}\Delta t + o(\Delta t) \, ,$$
$$\text{Cov}\{\epsilon_1(t_1)\Delta t, \epsilon_2(t_2)\Delta t\} = o(\Delta t) \, .$$

8.2.1.2. *The observation model*

Let $Y(i,j)$ denote the observed total number of cancer tumor cells of the ith individual at time t_j with $(i = 1,\ldots,k, j = 1,\ldots,n)$. To describe the observation model, we assume that with $X(t) = \sum_{i=1}^{2} X_i(t)$, $e(i,j) = \{Y(i,j) - X(t_j)\}/\sqrt{X(t_j)}$ is distributed as normal with mean 0 and variance σ_i^2 independently for $i = 1,\ldots,k, j = 1,\ldots,n$. Then, the observation model is given by the following statistical equations (stochastic equations):

$$Y(i,j) = X(t_j) + \sqrt{X(t_j)}e(i,j), \quad i = 1,\ldots,k, \ j = 1,\ldots,n, \qquad (8.3)$$

where $e(i,j)$ is the random measurement error associated with measuring $Y(i,j)$ and $e(i,j) \sim N(0,\sigma_i^2)$ independently and are un-correlated with the random noises in the stochastic system equations.

8.2.2. *A state space model of HIV pathogenesis*

Consider a HIV-infected individual. For the HV pathogenesis in this individual, there are four types of cells in the blood: The normal uninfected $CD4^+$ T cells (to be denoted by T_1 cells), the latently HIV-infected $CD4^+$ T cells (to be denoted by T_2 cells), the actively HIV-infected $CD4^+$ T cells (to be denoted by T_3 cells) and the free HIV. Free HIV can infect T_1 cells as well as the precursor stem cells in the bone marrow and thymus. When a resting T_1 cell is infected by a free HIV, it becomes a T_2 cell which may either revert back to a T_1 cell or be activated at some time to become a T_3 cell. On the other hand, when a dividing T_1 cell is infected by a free HIV, it becomes a T_3 cell which will release free HIV when it dies. T_2 cells will not release free HIV until being activated to become T_3 cells. Further T_1 cells are generated by precursor stem cells in the bone marrow and mature in the thymus; the matured T_1 cells then move to the blood stream. For this stochastic dynamic system, Tan and Wu [28] have developed a stochastic model in terms of stochastic differential equations. To estimate the numbers of CD4 T cells and free HIV in the blood, Tan and Xiang [15] have developed a state space model for this system. For this state space model, the stochastic system model is expressed as stochastic differential equations given by (8.4)–(8.7) below; the observation model of this state space model is given by the statistical model given by Eq. (8.8) below.

8.2.2.1. *Stochastic system model*

Let $T_i(t), i = 1, 2, 3$ and $V(t)$ be the numbers of $T_i, i = 1, 2, 3$ cells and free HIV at time t respectively. Then $\underset{\sim}{X}(t) = \{T_i(t), i = 1, 2, 3, V(t)\}$ is a four-dimensional stochastic process. To derive stochastic differential equations for these state variables, consider the time interval $[t, t + \triangle t)$ and denote by:

(1) $S(t) =$ Number of T_1 cells per mm^3 blood generated stochastically by the precursor stem cells in the bone marrow and thymus during $[t, t + \triangle t)$;

(2) $G_1(t) =$ Number of T_1 cells per mm^3 generated by stochastic logistic growth of T_1 cells during $[t, t + \triangle t)$ through stimulation by free HIV and existing antigens;

(3) $F_1(t) =$ Number of T_1 cells infected by free HIV during $[t, t + \triangle t)$;

(4) $G_2(t) =$ Number of T_2 cells among the HIV-infected T_1 cells during $[t, t + \triangle t)$;

(5) $F_2(t) =$ Number of T_2 cells activating to become T_3 cells during $[t, t + \triangle t)$;

(6) $D_i(t) =$ Number of deaths of T_i cells during $[t, t + \triangle t), i = 1, 2, 3$;

(7) $D_V(t) =$ Number of free HIV which have lost infectivity, or die, or have been removed during $[t, t + \triangle t)$;

(8) $N(t) =$ Average number of free HIV released by a T_3 cell when it dies at time t.

Let k_1 be the HIV infection rate of T_1 cells, k_2 the rate of T_2 cells being activated, μ_i $(i = 1, 2, 3)$ the death rate of T_i cells $(i = 1, 2, 3)$ and μ_V the rate by which free HIV are being removed, die, or have lost infectivity. Let γ be the rate of proliferation of T_1 cells by stimulation by HIV and antigens, $\omega(t)$ the probability that an infected T_1 cell is a T_2 cell at time t and $s(t)$ the rate by which T_1 cells are generated by precursor stem cells in the bone marrow and thymus at time t. Then the conditional probability distributions of the above variables given $\underset{\sim}{X}(t)$ are specified as follows:

- $S(t)|V(t) \sim$ Poisson with mean $s(t)\triangle t$,
- $[G_1(t), D_1(t)]|T_1(t) \sim$ Multinomial $[T_1(t); b_T(t)\triangle t, \mu_1\triangle t]$,
- $[F_1(t), D_V(t)]|[V(t), T_1(t)] \sim$ Multinomial $[V(t); k_1 T_1(t)\triangle t, \mu_V\triangle t]$,
- $[F_2(t), D_2(t)]|T_2(t) \sim$ Multinomial $[T_2(t); k_2\triangle t, \mu_2\triangle t]$,

- $G_2(t)|F_1(t) \sim \text{Binomial}\,[F_1(t); \omega(t)]$,
- $D_3(t)|T_3(t) \sim \text{Binomial}\,[T_3(t); \mu_3 \triangle t]$,

where $b_T(t) = \gamma[1 - \sum_{j=1}^{3} T_j(t)/T_{\max}]$.

Given $\underset{\sim}{X}(t)$, conditionally $S(t), [G_1(t), D_1(t)], [F_1(t), D_V(t)], [F_2(t), D_2(t)]$, and $D_3(t)$ are independently distributed of one another; given $F_1(t)$, conditionally $G_2(t)$ is independently distributed of other variables.

Let $\epsilon_i(t), i = 1, 2, 3, 4$ be defined by:

$$\epsilon_1(t)dt = [S(t) - s(t)dt] + [G_1(t) - b_T(t)T_1(t)dt]$$
$$- [F_1(t) - k_1 T_1(t)V(t)dt] - [D_1(t) - \mu_1 T_1(t)dt]\,,$$

$$\epsilon_2(t)dt = [G_2(t) - \omega(t)k_1 V(t)T_1(t)dt] - [F_2(t) - k_2 T_2(t)dt]$$
$$- [D_2(t) - \mu_2 T_2(t)dt]\,,$$

$$\epsilon_3(t)dt = \{[F_1(t) - k_1 T_1(t)V(t)\triangle t] - [G_2(t) - \omega(t)k_1 T_1(t)V(t)dt]\}$$
$$+ [F_2(t) - k_2 T_2(t)dt] - [D_3(t) - \mu_3(t)T_3(t)dt]\,,$$

$$\epsilon_4(t)dt = N(t)[D_3(t) - T_3(t)\mu_3 dt] - [F_1(t) - k_1 T_1(t)V(t)dt]$$
$$- [D_V(t) - \mu_V V(t)dt]\,.$$

Using these distribution results, we obtain the following stochastic differential equations for $T_i(t), i = 1, 2, 3, V(t)$:

$$dT_1(t) = T_1(t + dt) - T_1(t) = S(t) + G_1(t) - F_1(t) - D_1(t)$$
$$= \{s(t) + b_T(t)T_1(t) - \mu_1 T_1(t) - k_1 V(t)T_1(t)\}dt + \epsilon_1(t)dt\,, \tag{8.4}$$

$$dT_2(t) = T_2(t + dt) - T_2(t) = G_2(t) - F_2(t) - D_2(t)$$
$$= \{\omega(t)k_1 V(t)T_1(t) - \mu_2 T_2(t) - k_2 T_2(t)\}dt + \epsilon_2(t)dt\,, \tag{8.5}$$

$$dT_3(t) = T_3(t + dt) - T_3(t) = [F_1(t) - G_2(t)] + F_2(t) - D_3(t)$$
$$= \{[1 - \omega(t)]k_1 V(t)T_1(t) + k_2 T_2(t) - \mu_3 T_3(t)\}dt + \epsilon_3(t)dt\,, \tag{8.6}$$

$$dV(t) = V(t + dt) - V(t) = N(t)D_3(t) - F_1(t) - D_V(t)$$
$$= \{N(t)\mu_3 T_3(t) - k_1 V(t)T_1(t) - \mu_V V(t)\}dt + \epsilon_4(t)dt\,. \tag{8.7}$$

In Eqs. (8.4)–(8.7), the random noises $\epsilon_j(t)$, $j = 1, 2, 3, 4$ have expectation zero. The variances and covariances of these random variables are easily obtained as $\mathrm{Cov}[\epsilon_i(t)dt, \epsilon_j(\tau)dt] = \delta(t - \tau)Q_{ij}(t)dt + o(dt)$, where

$$Q_{11}(t) = \mathrm{Var}[\epsilon_1(t)] = \mathrm{E}\{s(t) + b_T(t)T_1(t) + [\mu_1 + k_1 V(t)]T_1(t)\}\,,$$

$$Q_{12}(t) = \mathrm{Cov}[\epsilon_1(t), \epsilon_2(t)] = -k_1\omega(t)\mathrm{E}[V(t)T_1(t)]\,,$$

$$Q_{13}(t) = \mathrm{Cov}[\epsilon_1(t), \epsilon_3(t)] = -k_1[1 - \omega(t)]\mathrm{E}[V(t)T_1(t)]\,,$$

$$Q_{14}(t) = \mathrm{Cov}[\epsilon_1(t), \epsilon_4(t)] = k_1\mathrm{E}[V(t)T_1(t)]\,,$$

$$Q_{22}(t) = \mathrm{Var}[\epsilon_2(t)] = \mathrm{E}[k_1\omega(t)V(t)T_1(t) + (\mu_2 + k_2)T_2(t)]\,,$$

$$Q_{23}(t) = \mathrm{Cov}[\epsilon_2(t), \epsilon_3(t)] = -\mathrm{E}\{\omega(t)[1 - \omega(t)]k_1 V(t)T_1(t) + k_2 T_2(t)\}\,,$$

$$Q_{24}(t) = \mathrm{Cov}[\epsilon_2(t), \epsilon_4(t)] = -k_1\omega(t)\mathrm{E}[V(t)T_1(t)]\,,$$

$$Q_{33}(t) = \mathrm{Var}[\epsilon_3(t)] = \mathrm{E}\{[1 - \omega(t)]k_1 V(t)T_1(t) + k_2 T_2(t) + \mu_3 T_3(t)\}\,,$$

$$Q_{34}(t) = \mathrm{Cov}[\epsilon_3(t), \epsilon_4(t)] = -[1 - \omega(t)]k_1\mathrm{E}[V(t)T_1(t)] - N(t)\mu_3\mathrm{E}T_3(t)\,,$$

$$Q_{44}(t) = \mathrm{Var}[\epsilon_4(t)] = \mathrm{E}\{N^2(t)\mu_3 T_3(t) + [\mu_V + k_1 T_1(t)]V(t)\}\,.$$

By using the formula $\mathrm{Cov}(X, Y) = E\{\mathrm{Cov}[(X, Y)|Z]\} + \mathrm{Cov}[E(X|Z), E(Y|Z)]$, it can be shown easily that the random noises $\epsilon_j(t)$ are un-correlated with the state variables $T_i(t)$, $i = 1, 2, 3$ and $V(t)$. Since the random noises $\epsilon_j(t)$ are random variables associated with the random transitions during the interval $[t, t + \Delta t)$, one may also assume that the random noises $\epsilon_j(t)$ are un-correlated with the random noises $\epsilon_l(\tau)$ for all j and l if $t \neq \tau$.

8.2.2.2. *The observation model*

Let Y_j be the log of the observed total number of CD4$^+$ T-cell counts at time t_j. Then the observation model, based on the CD4$^+$ T-cell counts, is given by:

$$Y_j = \log\left[\sum_{i=1}^{3} T_i(t_j)\right] + e_j\,, \quad j = 1, \ldots, n\,, \tag{8.8}$$

where e_j is the random error associated with measuring y_j.

In Eq. (8.8), one may assume that e_j has expected value 0 and variance σ_j^2 and are un-correlated with the random noises of Eqs. (8.4)–(8.7) of the previous section. As shown in [28], one may also assume that e_j is un-correlated with e_u if the absolute value of $t_j - t_u$ is greater than 6 months.

8.3. Some State Space Models in Carcinogenesis

In the past 5 years, this author and his associates have developed some state space models for carcinogenesis. In this section we will thus present some of these models.

8.3.1. *The state space model of the extended multi-event model of carcinogenesis*

In Example 1.5, we have presented the k-stage ($k \geq 2$) multi-event model first proposed by Chu [29] in 1985; see also [30–35]. In human beings, these models are useful to describe the cascade of carcinogenesis of colon cancer, lung cancer and breast cancer which involve many oncogenes and suppressor genes; see [21, 36, 37]. A serious shortcoming of this model is that it has ignored the cancer progression by assuming that each primary cancer tumor cell grows instantaneously into a malignant cancer tumor; see Remark 8.1. In many practical situations, however, this assumption usually does not hold and may lead to confusing results; see [38]. Because, as shown by Yang and Chen [39], each cancer tumor develops by clonal expansion from a primary cancer tumor cell, Tan and Chen [21] have resolved this difficulty by assuming that each cancer tumor develops by following a stochastic birth-death process from a primary k-stage initiated cell. This model has been referred to by Tan and Chen as the extended k-stage multi-event model. A state space model for this model has been given by Tan, Chen and Wang [22]. In this state space model, the stochastic system model is represented by the stochastic differential equations given by (8.9)–(8.10) given below; the observation model is represented by a statistical model given in Subsec. 8.3.1.2 below involving data on the number of intermediate foci per individual and/or the number of detectable malignant tumors per individual.

Remark 8.1. For the k-stage multi-event model as described in [30–32], it is assumed that each primary I_k cell grows instantaneously into a malignant cancer tumor in which case one may consider each I_k cell as a cancer tumor. To relax this assumption and for modeling generation of malignant cancer tumors, we define the I_j cells which arise directly from I_{j-1} cells by mutation or genetic changes as primary I_j cells. Thus, I_j cells which arise from other I_j cells by cell division are referred to as secondary I_j cells. Thus, in the extended k-stage multi-event model, malignant cancer tumors derive from primary I_k cells by following a stochastic birth-death process.

8.3.1.1. *The stochastic system model*

The extended k-stage multi-event model views carcinogenesis as the end point of k ($k \geq 2$) discrete, heritable and irreversible events (mutations or genetic changes) and malignant cancer tumors arise from the primary I_k cells through clonal expansion. An important feature of this model is that it assumes that normal stem cells and intermediate-stage cells undergo stochastic cell proliferation (birth) and differentiation (death). Let $N = I_0$ denote normal stem cells, I_j the jth stage initiated cells arising from the $(j-1)$th stage initiated cells ($j = 1, \ldots, k$) by mutation or some genetic changes and T the cancer tumors arising from the primary I_k cells through clonal expansion by following some stochastic birth-death processes. Then the model assumes $N \to I_1 \to I_2 \to \cdots \to I_k$ with the N cells and the I_j cells subject to stochastic proliferation (birth) and differentiation (death); see Fig. 8.1. Let $I_j(t), j = 0, 1, \ldots, k - 1$ denote the number of I_j cells at time t and $T(t)$ the number of malignant cancer tumors at time t respectively. To model the stochastic process $\underset{\sim}{U}(t) = \{I_j(t), j = 0, 1, \ldots, k - 1, T(t)\}$, let $\{b_j(t), d_j(t), \alpha_j(t)\}$ denote the birth rate, the death rate and the mutation rate from $I_j \to I_{j+1}$ at time t for $j = 0, 1, \ldots, k - 1$. That is, during the

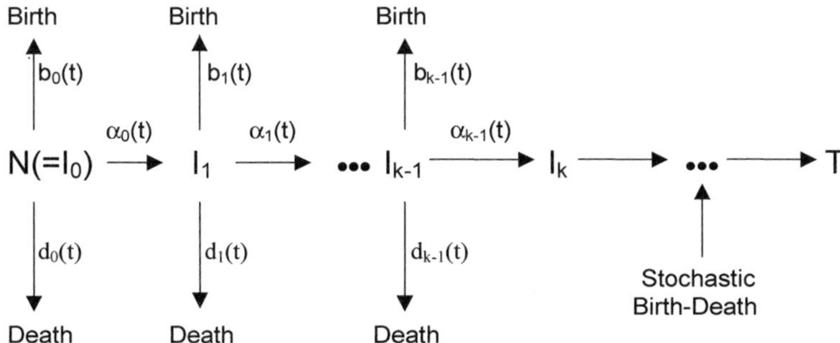

Fig. 8.1. A schematic representation of the extended multi-event model of carcinogenesis.

time interval $[t, t + \Delta t)$, the probabilities that an I_j cell at time t will yield two I_j cells, zero I_j cells, and one I_j cell and one I_{j+1} cell at time $t + \Delta t$ are given by $b_j(t)\Delta t + o(\Delta t)$, $d_j(t)\Delta t + o(\Delta t)$ and $\alpha_j(t)\Delta t + o(\Delta t)$ respectively. To model the generation of malignant cancer tumors, assume that each malignant cancer tumor develops from a primary I_k cell by following a stochastic birth and death process with birth rate $b_k(s, t)$ and death rate $d_k(s, t)$, where s is the onset time of the primary I_k cell. (Because growth of cancer tumor cells basically follow Gompertz growth [40], Tan and Chen [21] have assumed $b_k(s, t) = b_T \exp\{-\delta_T(t - s)\}$ and $d_k(s, t) = d_T \exp\{-\delta_T(t - s)\}$, where b_T, d_T and δ_T are non-negative numbers.)

8.3.1.2. *Stochastic differential equations for I_j cells, $j = 0, 1, \ldots, k - 1$*

To derive stochastic differential equations for these cells, let $\{B_j(t), D_j(t), M_j(t)\}$ denote the numbers of birth, of death and of mutation from $I_j \to I_{j+1}$ for $j = 0, 1, \ldots, k - 1$ during the time interval $[t, t + \Delta t)$. Then, as shown in Sec. 4.7, the conditional probability distribution of $\{B_j(t), D_j(t), M_j(t)\}$ given $I_j(t)$ for $j = 0, 1, \ldots, k - 1$ is multinomial with parameters $\{I_j(t), b_j(t)\Delta t, d_j(t)\Delta t, \alpha_j(t)\Delta t\}$ respectively:

$$\{B_j(t), D_j(t), M_j(t)\}|I_j(t) \sim ML\{I_j(t); b_j(t)\Delta t, d_j(t)\Delta t, \alpha_j(t)\Delta t\}\,,$$

$$j = 0, 1, \ldots, k - 1\,.$$

Using these distribution results, we obtain the following stochastic differential equations for $I_j(t), j = 0, 1, \ldots, k - 1$:

$$dI_0(t) = I_0(t + \Delta t) - I_0(t) = B_0(t) - D_0(t)$$

$$= I_0(t)\gamma_0(t)\Delta t + \epsilon_0(t)\Delta t\,, \tag{8.9}$$

$$dI_j(t) = I_j(t + \Delta t) - I_j(t) = M_{j-1}(t) + B_j(t) - D_j(t)$$

$$= \{I_{j-1}(t)\alpha_{j-1} + I_j(t)\gamma_j(t)\}\Delta t + \epsilon_j(t)\Delta t, j = 1, \ldots, k - 1, \tag{8.10}$$

where $\gamma_j(t) = b_j(t) - d_j(t), j = 0, 1, \ldots, k - 1$.

In Eqs. (8.9)–(8.10), the random noises are given by:

$$\epsilon_0(t)\Delta t = [B_0(t) - I_0(t)b_0(t)\Delta t] - [D_0(t) - I_0(t)d_0(t)\Delta t]$$

and for $j = 1, \ldots, k - 1$,

$$\epsilon_j(t)\Delta t = [M_{j-1}(t) - I_{j-1}(t)\alpha_{j-1}(t)\Delta t] + [B_j(t) - I_j(t)b_j(t)\Delta t]$$

$$- [D_j(t) - I_j(t)d_j(t)\Delta t].$$

From the above definition, the random noises $\{\epsilon_j(t), j = 0, 1, \ldots, k - 1\}$ have expectation zero conditionally. It follows that $E e_j(t) = 0$ for $j = 0, 1, \ldots, k - 1$. Using the basic formulae $\mathrm{Cov}(X, Y) = E\{\mathrm{Cov}[(X, Y)|Z]\} + \mathrm{Cov}[E(X|Z), E(Y|Z)]$, it is also obvious that $\epsilon_j(t)$'s are un-correlated with elements of $\underset{\sim}{X}(t)$. Further, using the distribution results given above, the variances and covariances of the $\epsilon_j(t)$'s are easily obtained as, to order $o(\Delta t)$, $\mathrm{Cov}(\epsilon_i(t)\Delta t, \epsilon_i(\tau)\Delta t) = \delta(t - \tau)C_{ij}(t)\Delta t + o(\Delta t)$, where

$$C_{jj}(t) = \mathrm{Var}[\epsilon_j(t)] = E\{[1 - \delta_{j0}]I_{j-1}(t)\alpha_{j-1}(t)$$

$$+ I_j(t)[b_j(t) + d_j(t)]\}, \quad (j = 0, 1, \ldots, k - 1) \tag{8.11}$$

and

$$C_{i,j}(t) = \mathrm{Cov}[\epsilon_i(t), \epsilon_j(t)] = 0 \quad \text{for } i \neq j.$$

8.3.1.3. *The probability distribution of $T(t)$*

To model the generation of malignant cancer tumors, let $P_T(s, t)$ be the probability that a primary I_k cell arising at time s will develop a detectable cancer tumor by time t. Assume that a cancer tumor is detectable only if it contains at least N_T cancer tumor cells. Then, given an I_k cell arising from an I_{k-1} cell at time s, by using results from Example 4.8 and [31, 40], the probability that this I_k cell will yield j I_k cells at time t is given by:

$$P_M(j) = \begin{cases} 1 - (h(t-s) + g(t-s))^{-1}, & \text{if } j = 1, \\ \left(\dfrac{g(t-s)}{h(t-s) + g(t-s)}\right)^{j-1} \dfrac{h(t-s)}{(h(t-s) + g(t-s))^2}, & \text{if } j > 1, \end{cases}$$

where

$$h(t-s) = \exp\left\{-\int_s^t [b_k(y-s) - d_k(y-s)]dy\right\}$$

$$= \exp\{-(\epsilon_T/\delta_T)[1 - \exp(-\delta_T(t-s))]\}$$

and

$$g(t - s) = \int_s^t b_k(y - s)h(y - s)dy = (b_T/\epsilon_T)[1 - h(t - s)].$$

Then $P_T(s, t)$ is given by:

$$P_T(s, t) = \sum_{j=N_T}^{\infty} P_M(j) = \frac{1}{h(t - s) + g(t - s)} \left(\frac{g(t - s)}{h(t - s) + g(t - s)} \right)^{N_T - 1}.$$

$$(8.12)$$

Theorem 8.1.   **(Probability Distribution of $T(t)$).** *Given the above specifications, the conditional probability distribution of $T(t)$ given $\{I_{k-1}(u), u \leq t\}$ is Poisson with parameters*

$$\lambda(t) = \int_{t_0}^t I_{k-1}(x)\alpha_{k-1}(x)P_T(x, t)dx.$$

That is,

$$T(t)|\{I_{k-1}(u), t_0 \leq u \leq t\} \sim \text{Poisson}\{\lambda(t)\}. \tag{8.13}$$

Proof.   To obtain the probability distribution of $T(t)$, let $t - t_0 = n\Delta t$ and let $C(s, t)$ be the total number of detectable cancer tumors at time t from primary I_k cells arising from the I_{k-1} cells by mutation during the time interval $[s, s + \Delta t)$. Then, to order $o(\Delta t)$, the probability distribution of $C(s, t)$ given $M_k(s)$ primary I_k cells at time s is binomial with parameters $\{M_k(s), P_T(s, t)\}$. Further, one has:

$$T(t) = \lim_{\Delta t \to 0} \sum_{j=1}^{n} C[t_0 + (j - 1)\Delta t, t]. \tag{8.14}$$

Now, given $I_{k-1}(s)$ I_{k-1} cells at time s, the number of I_k cells which are generated by mutation from these I_{k-1} cells during $[s, s + \Delta t)$ is, to order $o(\Delta t)$, a binomial variable with parameters $\{I_{k-1}(s), \alpha_{k-1}(s)\Delta t\}$. By using the moment generation function method, it is then easy to show that the probability distribution of $C(s, t)$ given $I_{k-1}(s)$ I_{k-1} cells at time s is, to order $o(\Delta t)$, binomial with parameters $\{I_{k-1}(s), \alpha_{k-1}(s)P_T(s, t)\Delta t\}$.

Hence, the conditional PGF of $T(t)$ given $\{I_{k-1}(x), t_0 \le x \le t_j\}$ is obtained as:

$$\phi(z,t) = \lim_{\Delta t \to 0} \prod_{j=1}^{n} E\{z^{C(t_0+(j-1)\Delta t,t)} | I_{k-1}(t_0 + (j-1)\Delta t)\}$$

$$= \lim_{\Delta t \to 0} \prod_{j=1}^{n} \{1 + (z-1)\alpha_{k-1}[t_0 + (j-1)\Delta t]$$

$$\times P_T[t_0 + (j-1)\Delta t, t]\Delta t\}^{I_{k-1}(t_0+(j-1)\Delta t)}$$

$$= \lim_{\Delta t \to 0} \prod_{j=1}^{n} \exp\{(z-1)I_{k-1}[t_0 + (j-1)\Delta t]\alpha_{k-1}[t_0 + (j-1)\Delta t]$$

$$\times P_T[t_0 + (j-1)\Delta t, t]\Delta t + o(\Delta t)\}$$

$$= \exp\left\{(z-1)\int_{t_0}^{t} I_{k-1}(x)\alpha_{k-1}(x)P_T(x,t)dt\right\}. \tag{8.15}$$

The above PGF is that of a Poisson distribution with parameter $\lambda(t)$ and hence the theorem is proved. $\qquad\square$

Let $Q_T(t_{j-1}, t_j)$ be the conditional probability of yielding a detectable cancer tumor during $[t_{j-1}, t_j)$ given $\{I_{k-1}(x), t_0 \le x \le t_j\}$. Theorem 8.1 then gives:

$$Q_T(t_{j-1}, t_j) = \exp\left\{-\int_{t_0}^{t_{j-1}} I_{k-1}(x)\alpha_{k-1}(x)P_T(x, t_{j-1})dx\right\}$$

$$- \exp\left\{-\int_{t_0}^{t_j} I_{k-1}(x)\alpha_{k-1}(x)P_T(x, t_j)dx\right\}. \tag{8.16}$$

8.3.1.4. *The probability distribution of intermediate foci in carcinogenesis studies*

In many carcinogenesis studies, very often the experimenters may have data on the number of intermediate foci per individual. For example, in animal carcinogenicity studies on skin cancer, one may have data on the number of papillomas per animal and number of carcinoma per animal over time. Under the MVK two-stage model, papillomas are derived by clonal expansion from

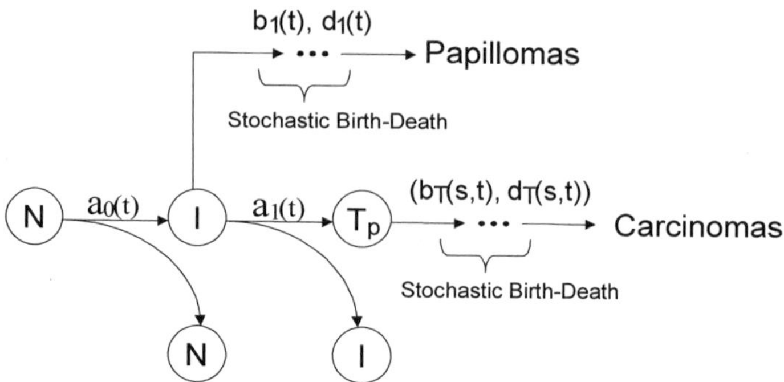

Fig. 8.2. A two-stage (multi-event with $k = 2$) model carcinogenesis in initiation-promotion experiments.

primary I_1 cells and carcinoma are derived from primary I_2 cells. This is represented schematically by Fig. 8.2. In the k-stage extended multi-event model, let T_j denote the Type-j intermediate foci and $T_j(t)$ the number of T_j foci per individual at time t. Then the T_j foci are derived from primary I_j cells. Using exactly the same procedures as in Subsec. 8.3.1.3, one may derive the probability distribution of $T_j(t)$.

To derive the probability distribution of $T_j(t)$, let $P_j(s,t)$ be the probability that a primary I_j cell arising at time s will develop a detectable Type-j foci by time t. Assume that a Type-j focus is detectable only if it contains at least N_j I_j cells. Then, using exactly the same procedure as those in Subsec. 8.3.1.3, we obtain:

$$P_j(s,t) = \frac{1}{h_j(s,t) + g_j(s,t)} \left(\frac{g_j(s,t)}{h_j(s,t) + g_j(s,t)} \right)^{N_j - 1}, \qquad (8.17)$$

where

$$h_j(s,t) = \exp\left\{ -\int_s^t [b_j(y) - d_j(y)]dy \right\}$$

and

$$g_j(s,t) = \int_s^t b_j(y)h_j(s,y)dy .$$

Further, the conditional probability distribution of $T_j(t)$ given $\{I_{j-1}(u), u \leq t\}$ is Poisson with parameters $\lambda_j(t) = \int_{t_0}^t I_{j-1}(x)\alpha_{j-1}(x)P_j(x,t)dx$.

That is,

$$T_j(t)|\{I_{j-1}(u), t_0 \le u \le t\} \sim \text{Poisson}\{\lambda_j(t)\}. \qquad (8.18)$$

Let $Q_i(t_{j-1}, t_j)$ be the conditional probability of yielding a Type-i intermediate focus during $[t_{j-1}, t_j)$ given $\{I_{i-1}(x), t_0 \le x \le t_j\}$. Then,

$$Q_i(t_{j-1}, t_j) = \exp\left\{-\int_{t_0}^{t_{j-1}} I_{i-1}(x)\alpha_{i-1}(x)P_i(x, t_{j-1})dx\right\}$$

$$- \exp\left\{-\int_{t_0}^{t_j} I_{i-1}(x)\alpha_{i-1}(x)P_i(x, t_j)dx\right\}. \qquad (8.19)$$

The proofs of the above results are exactly the same as that in Subsec. 8.3.1.3 above and hence is left as an exercise; see Exercise 8.8.

8.3.1.5. *The observation model*

The observation model depends on the type of data available. In the animal carcinogenicity studies on skin cancer, suppose that data on the number of papillomas per animal over time are available. Let $Y_1(i, j)$ be the observed number of papillomas for the ith animal during the time interval $[t_{i,j-1}, t_{ij}), j = 1, \ldots, k$ with $t_{i0} = t_0$ and $t_{ik} = \infty$. Then, the observation model is given by:

$$Y_1(i, j) = T_{1i}(t_j) + e_1(i, j), \quad j = 1, \ldots, k, \qquad (8.20)$$

independently for $i = 1, \ldots, n$, where $e_1(i, j)$ is the measurement error associated with counting $T_{1i}(t_j)$, the number of papillomas in the ith animal at time t_j. One may assume that $e_1(i, j) \sim N(0, \sigma_1^2)$ independently of $T_{1i}(t_j)$ in Eqs. (8.20).

Let n_j be the number of animals with intermediate foci being developed during $[t_{j-1}, t_j)$ with $\sum_{j=1}^{k} n_j \le n$. Then, starting with n animals with no intermediate foci, $\{n_j, j = 1, \ldots, k\}$ is distributed as a k-dimensional multinomial vector with parameters $\{n; Q_i(t_{j-1}, t_j), j = 1, \ldots, k\}$. That is,

$$(n_j, j = 1, \ldots, k) \sim ML\{n; Q_i(t_{j-1}, t_j), \quad j = 1, \ldots, k\}.$$

Suppose that data on the number of carcinomas per animal over time are also available. Let $Y_2(i, j)$ be the observed number of carcinomas for the ith animal during the time interval $[t_{i,j-1}, t_{ij}), j = 1, \ldots, k$ with $t_{i0} = t_0$ and

$t_{ik} = \infty$. Then, an additional equation for the observation model is

$$Y_2(i,j) = T_{2i}(t_j) + e_T(i,j), \quad j = 1, \ldots, k, \tag{8.21}$$

independently for $i = 1, \ldots, n$, where $e_T(i,j)$ is the measurement error associated with measuring $T_{2i}(t_j)$, the number of carcinomas in the ith animal at time t_j. One may assume that $e_T(i,j) \sim N(0, \sigma_T^2)$ independently of $e_1(i_1, j_1)$ for all $\{i_1, j_1\}$ and $T_{2i}(t_j)$ in Eqs. (8.21).

Let m_j be the number of animals with detectable tumors being developed during $[t_{j-1}, t_j)$ with $\sum_{j=1}^{k} m_j \leq n$. Then, starting with n animals with no tumor, $\{m_j, j = 1, \ldots, k\}$ is distributed as a k-dimensional multinomial vector with parameters $\{n; Q_T(t_{j-1}, t_j), j = 1, \ldots, k\}$, see equation (8.16). That is,

$$(m_j, j = 1, \ldots, k) \sim ML\{n; Q_T(t_{j-1}, t_j), j = 1, \ldots, k\}.$$

8.3.2. *A state space model for extended multiple pathways models of carcinogenesis*

While the multi-event models assume that cancer tumors develop from a single pathway through a multi-stage stochastic process, it has long been recognized that the same cancer might arise from different carcinogenic processes (See [31, Chap. 4], [41]). For example, in the animal studies of rats liver cancer by DeAngelo [41], at least four pathways are involved for the DCA-induced hepatocellular carcinomas in B6C3F1 mice; that is, a carcinoma may be preceded directly by hyperplastic nodules, adenomas, or dysplastic foci, and through an indirect pathway, carcinomas may also evolve from adenomas that have evolved from hyperplastic nodules. Let $I_0, I_j, j = 1, 2, 3$ and T_p denote the normal stem cell, the Type-j initiated cell $j = 1, 2, 3$ and the carcinoma proceeding cell, respectively. Then the model is represented schematically by Fig. 8.3. A state space model for this cascade of carcinogenesis has been provided by Tan, Chen and Wang [22]. In this state space model, the stochastic system model is represented by the stochastic differential Eqs. (8.22)–(8.26) given below and the observation model is represented by Eqs. (8.30)–(8.32) given below.

8.3.2.1. *The stochastic system model*

Let $I_0(t)$ and $I_j(t), j = 1, 2, 3\}$ denote the numbers of I_0 cells and I_j cells, $j = 1, 2, 3$ at time t respectively. Let $\{T(t), T_j(t), j = 1, 2, 3\}$ denote the number of carcinoma per animal, and the numbers of hyperplastic nodules per

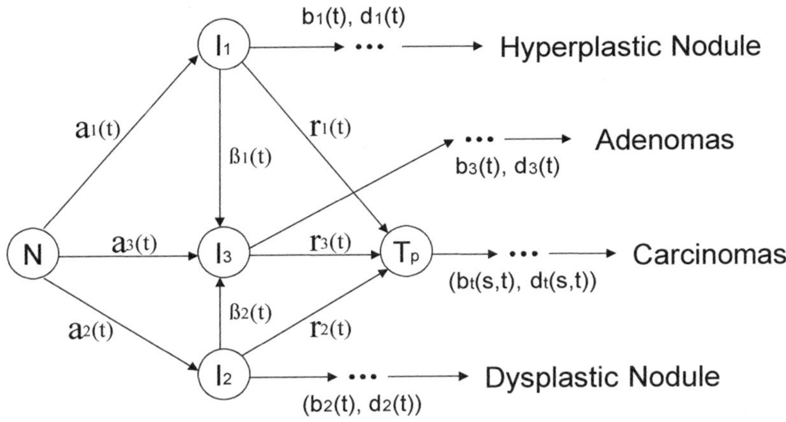

Fig. 8.3. A multiple pathways model for hepatocellular carcinomas induced by dichloroacetic acid (DCA).

animal, of adenomas per animal, and of dysplastic foci per animal at time t, respectively.

Let $\{b_j(t), d_j(t)\}(j = 0, 1, 2, 3)$ be the birth rate and death rate of I_j cells $(j = 0, 1, 2, 3)$, respectively and $\{b_T(s, t), d_T(s, t)\}$ the birth rate and death rate of W cells at time t which arise from initiated cells at time s. Let the mutation rates of $I_0 \to I_j(j = 1, 2, 3)$, $I_u \to I_3(u = 1, 2)$ and $I_j \to W(j = 1, 2, 3)$ be given by $\alpha_j(t)(j = 1, 2, 3)$, $\beta_u(t)(u = 1, 2)$ and $\gamma_j(t)(j = 1, 2, 3)$ at time t, respectively. Then, by using exactly the same approaches given in Subsecs. 5.4.2–5.4.3, we obtain the following stochastic differential equations for $I_j(t), j = 0, 1, 2, 3$:

$$\Delta I_0(t) = I_0(t + \Delta t) - I_0(t) = I_0(t)\gamma_0(t)\Delta t + \epsilon_0(t)\Delta t\,, \tag{8.22}$$

$$\Delta I_i(t) = I_i(t + \Delta t) - I_i(t) = I_0(t)\alpha_i(t)\Delta t$$
$$+ I_i(t)\gamma_i(t)\Delta t + \epsilon_i(t)\Delta t\,, \quad i = 1, 2\,, \tag{8.23}$$

$$\Delta I_3(t) = \left\{ I_0(t)\alpha_3(t) + \sum_{i=1}^{2} I_i(t)\beta_i(t) + I_3(t)\gamma_3(t) \right\} \Delta t + \epsilon_3(t)\Delta t\,, \tag{8.24}$$

where, $\gamma_j(t) = b_j(t) - d_j(t), j = 0, 1, 2, 3$.

The random noises $\epsilon_j(t)$'s have expected value 0 and are un-correlated with the $I_j(t)$'s. Further, to order $o(\Delta t)$, $\mathrm{Cov}[\epsilon_i(t)\Delta t, \epsilon_j(\tau)\Delta t] = \delta(t - \tau)Q_{ij}(t)\Delta t,$

where

$$Q_{00}(t) = E[I_0(t)][b_0(t) + d_0(t)], \quad Q_{0i}(t) = 0, \quad i = 1, 2, 3,$$

$$Q_{ii}(t) = E[I_0(t)]\alpha_i(t) + E[I_i(t)][b_i(t) + d_i(t)], \quad Q_{i3} = Q_{12} = 0, \quad i = 1, 2;$$

$$Q_{33}(t) = E[I_0(t)]\alpha_3(t) + \sum_{i=1}^{2} E[I_i(t)]\beta_i(t) + E[I_3(t)][b_3(t) + d_3(t)].$$

By using exactly the same approach as in Subsec. 8.3.1.3 and in Theorem 8.1, the conditional probability distributions of $T(t)$ given $\{I_i(t), i = 1, 2, 3\}$ and of $T_j(t)$ given $I_j(t)$ are given respectively by:

$$T(t)|\{I_i(u), t_0 \le u \le t, i = 1, 2, 3\} \sim \text{Poisson}\{\lambda(t)\}, \tag{8.25}$$

$$T_j(t)|\{I_j(u), t_0 \le u \le t\} \sim \text{Poisson}\{\lambda_j(t)\}, \quad j = 1, 2, 3, \tag{8.26}$$

where

$$\lambda(t) = \sum_{j=1}^{3} \int_{t_0}^{t} I_j(x)\gamma_j(x)P_j(x, t)dx,$$

$$\lambda_j(t) = \int_{t_0}^{t} I_0(x)\alpha_j(x)P_j(x, t)dx, \quad j = 1, 2,$$

$$\lambda_3(t) = \int_{t_0}^{t} \left[I_0(x)\alpha_3(x) + \sum_{i=1}^{2} I_i(x)\beta_i(x) \right] P_3(x, t)dx.$$

8.3.2.2. *The observation model*

Denote by:

(1) $Y_0(i, j)$ = Observed number of detectable cancer tumors for the ith animal at time t_{ij}, $j = 1, \ldots, n$, where $t_{i0} = t_0$.

(2) $Y_1(i, j)$ = Observed number of hyperplastic nodule for the ith animal at time t_{ij}, $j = 1, \ldots, n$, where $t_{i0} = t_0$.

(3) $Y_2(i, j)$ = Observed number of dysplastic nodule for the ith animal at time t_{ij}, $j = 1, \ldots, n$, where $t_{i0} = t_0$.

(4) $Y_3(i, j)$ = Observed number of adenomas for the ith animal at time t_{ij}, $j = 1, \ldots, n$, where $t_{i0} = t_0$.

The observation model based on data $\{Y_0(i,j), Y_r(i,j), r = 1,2,3, i = 1,\ldots,n, j = 1,\ldots,k\}$ are given by:

$$Y_0(i,j) = T_i(t_j) + e_T(i,j)\,, \quad j = 1,\ldots,k\,, \tag{8.27}$$

$$Y_r(i,j) = T_{ri}(t_j) + e_r(i,j)\,, \quad j = 1,\ldots,k,\ r = 1,2\,, \tag{8.28}$$

$$Y_3(i,j) = T_{3i}(t_j) + e_3(i,j)\,, \quad j = 1,\ldots,k\,, \tag{8.29}$$

where $\{e_T(i,j), e_r(i,j), r = 1,2,3\}$ are the measurement errors associated with measuring $\{T_i(t_j), T_{ri}(t_j), r = 1,2,3\}$ respectively. One may assume that these measurement errors are independently distributed as normal variables with means 0 and variances $\{\sigma_T^2, \sigma_r^2, r = 1,2,3\}$ respectively.

Let $\{Q_T(t_{j-1}, t_j), Q_i(t_{j-1}, t_j), i = 1,2,3\}$ denote the probabilities of yielding a detectable tumor, a detectable hyperplastic nodule, a dysplastic nodule and a adenomas during $[t_{j-1}, t_j)$ respectively. Then,

$$Q_T(t_{j-1}, t_j) = \exp\left\{-\sum_{i=1}^{3}\int_{t_0}^{t_{j-1}} I_i(x)\omega_i(x)P_T(x, t_{j-1})dx\right\}$$

$$- \exp\left\{-\sum_{i=1}^{3}\int_{t_0}^{t_j} I_i(x)\omega_i(x)P_T(x, t_j)dx\right\}\,, \tag{8.30}$$

$$Q_i(t_{j-1}, t_j) = \exp\left\{-\int_{t_0}^{t_{j-1}} I_0(x)\alpha_i(x)P_i(x, t_{j-1})dx\right\}$$

$$- \exp\left\{-\int_{t_0}^{t_j} I_0(x)\alpha_i(x)P_i(x, t_j)dx\right\}\,, \quad i = 1,2\,, \tag{8.31}$$

and

$$Q_3(t_{j-1}, t_j) = \exp\left\{-\int_{t_0}^{t_{j-1}} \left\{I_0(x)\alpha_3(x) + \sum_{i=1}^{2} I_i(x)\beta_i(x)\right\} P_3(x, t_{j-1})dx\right\}$$

$$- \exp\left\{-\int_{t_0}^{t_j} \left\{I_0(x)\alpha_3(x) + \sum_{i=1}^{2} I_i(x)\beta_i(x)\right\} P_i(x, t_j)dx\right\}\,. $$

$$\tag{8.32}$$

8.4. Some Classical Theories of Discrete and Linear State Space Models

In the classical analysis of state space models, the parameter values are assumed known or have been estimated from other sources. The main theories are to derive optimal estimates or predicted values of the state variables $\underset{\sim}{X}(t)$. These are the main results in most of standard texts on state space models (see for example [5–7]). In this section, we briefly summarize these results for discrete-time linear state space models and illustrate its applications using examples from AIDS epidemiology. Results for continuous-time state space models will be given in Sec. 9.1.

To proceed, we thus consider the state space model given by Eqs. (8.33) and (8.34), where the random noises $\underset{\sim}{\epsilon}(t)$ and the measurement errors $\underset{\sim}{e}(t)$ are assumed to be distributed independently of one another with expected values $\{E\underset{\sim}{\epsilon}(t+1) = \underset{\sim}{0}, E\underset{\sim}{e}(t) = \underset{\sim}{0}\}$ and covariance matrices $\{\mathrm{Var}[\underset{\sim}{\epsilon}(t)] = V(t), \mathrm{Var}[\underset{\sim}{e}(t)] = \Sigma(t)\}$:

$$\underset{\sim}{X}(t+1) = F(t+1,t)\underset{\sim}{X}(t) + \underset{\sim}{\epsilon}(t+1), \tag{8.33}$$

$$\underset{\sim}{Y}(t+1) = H(t+1)\underset{\sim}{X}(t+1) + \underset{\sim}{e}(t+1), \quad t = 1,\ldots,n. \tag{8.34}$$

In the classical analysis, it is assumed that the transition matrices $F(t+1,t)$ and $H(t)$ and the covariance matrices $\{V(t), \Sigma(t)\}$ are given matrices of real numbers. The main problem is to estimate and predict $\underset{\sim}{X}(t)$ based on data $\boldsymbol{D}(n) = \{Y(j), j = 1, 2, \ldots, n\}$. To this end, define:

- $\hat{\underset{\sim}{X}}(t|k) = $ Estimator (if $t \leq k$) or predictor (if $t > k$) of $\underset{\sim}{X}(t)$ given data $\boldsymbol{D}(k) = \{\underset{\sim}{Y}(j), j = 1, \ldots, k\}$,
- $\hat{\underset{\sim}{Y}}(k+r|k) = $ Predictor of $\underset{\sim}{Y}(k+r)$ given data $\boldsymbol{D}(k)$ for $r > 0$,
- $\hat{\underset{\sim}{\epsilon}}_X(t|k) = \hat{\underset{\sim}{X}}(t|k) - \underset{\sim}{X}(t) = $ Residual of estimating (if $t \leq k$) or predicting (if $t > k$) $\underset{\sim}{X}(t)$ by $\hat{\underset{\sim}{X}}(t|k)$,
- $\hat{\underset{\sim}{\epsilon}}_Y(k+r|k) = \hat{\underset{\sim}{Y}}(k+r|k) - \underset{\sim}{Y}(k+r) = $ Residual of predicting $\underset{\sim}{Y}(k+r)$ by $\hat{\underset{\sim}{Y}}(k+r|k)$,

- $P(t|k) = \text{Var}[\hat{\underset{\sim}{\epsilon}}(t|k)] = \text{Covariance matrix of the residual } \hat{\underset{\sim}{\epsilon}}_X(t|k)$ and $P_Y(k+r|k) = \text{Var}[\hat{\underset{\sim}{\epsilon}}_Y(k+r|k)] = \text{Covariance matrix of the residual } \hat{\underset{\sim}{\epsilon}}_Y(k+r|k)$.

In the above notation, $\hat{\underset{\sim}{X}}(t|k)$ has been referred to in the literature as a $(t-k)$-step predictor of $\underset{\sim}{X}(t)$ if $t > k$, a forward filter of $\underset{\sim}{X}(t)$ if $t = k$ and a smoother (or backward filter) of $\underset{\sim}{X}(t)$ if $k > t$.

To derive $\hat{\underset{\sim}{X}}(t|k)$, the standard least square method is to derive estimates by minimizing the residual sum of squares subjecting to the condition of unbiasedness. This is equivalent to seeking a linear estimator $\hat{\underset{\sim}{X}}(t|k) = R(0) + \sum_{i=1}^{k} R(i)\underset{\sim}{Y}(i)$ of $\underset{\sim}{X}(t)$ satisfying the two conditions:

(1) $E[\hat{\underset{\sim}{X}}(t|k)|\underset{\sim}{X}(t)] = \underset{\sim}{X}(t)$, and

(2) $\text{tr}\{D[\hat{\underset{\sim}{X}}(t|k) - \underset{\sim}{X}(t)][\hat{\underset{\sim}{X}}(t|k) - \underset{\sim}{X}(t)]'\}$ is minimized for any positive definite matrix D.

By the Gauss–Markov theorem of least square method (see [42]), this procedure then gives the linear, minimum varianced, unbiased estimator of $\underset{\sim}{X}(t)$ given data $\boldsymbol{D}(k)$ or the BLUE (best, linear and unbiased estimator) of $\underset{\sim}{X}(t)$ given data $\boldsymbol{D}(k)$, see Remark 8.2. These results are summarized in the following two theorems.

Remark 8.2. We define $\hat{\underset{\sim}{X}}(t|k)$ as an unbiased estimator (or predictor) of $\underset{\sim}{X}(t)$ given data $\boldsymbol{D}(k)$ if $E[\hat{\underset{\sim}{\epsilon}}(t|k)] = 0$. Define $\hat{\underset{\sim}{X}}(t|k)$ as the unbiased and minimum varianced estimator (or predictor) or BLUE of $\underset{\sim}{X}(t)$ given data $\boldsymbol{D}(k)$ if $\hat{\underset{\sim}{X}}(t|k)$ satisfies the following two conditions:

(1) $\hat{\underset{\sim}{X}}(t|k)$ is an unbiased estimator (or predictor) of $\underset{\sim}{X}(t)$ given data $\boldsymbol{D}(k)$.

(2) For any other unbiased estimator (or predictor) $\hat{\underset{\sim}{X}}^{(*)}(t|k)$ of $\underset{\sim}{X}(t)$ given data $\boldsymbol{D}(k)$ with residual $\hat{\underset{\sim}{\epsilon}}^{(*)}(t|k) = \hat{\underset{\sim}{X}}^{(*)}(t|k) - \underset{\sim}{X}(t)$ and with $P^{(*)}(t|k) = \text{Var}[\hat{\underset{\sim}{\epsilon}}^{(*)}(t|k)]$, $P^{(*)}(t|k) - P(t|k)$ is a non-negative definite matrix.

(That is, $\underset{\sim}{x}'[P^{(*)}(t|k) - P(t|k)]\underset{\sim}{x} \geq 0$, for any vector $\underset{\sim}{x}$ of real numbers with at least one non-zero element.)

8.4.1. *Some general theories*

Given the above specification, in this section we give some general theories which are the gold standard methods for discrete-time linear state space models in most texts of state space models.

Theorem 8.2. *Given the estimator of $\underset{\sim}{X}(0)$ as $\hat{\underset{\sim}{X}}(0|0)$ with $E[\hat{\underset{\sim}{\epsilon}}(0|0)] = 0$ and $\mathrm{Var}[\hat{\underset{\sim}{\epsilon}}(0|0)] = P(0|0)$, the linear, unbiased and minimum varianced estimator (filter) or BLUE $\hat{\underset{\sim}{X}}(j|j)$ of $\underset{\sim}{X}(j)$ given data $\boldsymbol{D}(j)$ are given by the following recursive equations:*

(1) Under the assumption, the estimator of $\underset{\sim}{X}(0)$ is $\hat{\underset{\sim}{X}}(0|0)$ with $E[\hat{\underset{\sim}{\epsilon}}(0|0)] = 0$ and $\mathrm{Var}[\hat{\underset{\sim}{\epsilon}}(0|0)] = P(0|0)$.

(2) For $j = 0, 1, \ldots, n = t_M$,

$$\hat{\underset{\sim}{X}}(j+1|j) = F(j+1, j)\,\hat{\underset{\sim}{X}}(j|j)\,, \tag{8.35}$$

$$\hat{\underset{\sim}{X}}(j+1|j+1) = \hat{\underset{\sim}{X}}(j+1|j) + K_{j+1}[\underset{\sim}{Y}(j+1)$$

$$- H(j+1)\,\hat{\underset{\sim}{X}}(j+1|j)]\,, \tag{8.36}$$

where

$$K_{j+1} = P(j+1|j)H'(j+1)[\Sigma_Y(j+1)]^{-1}\,,$$

and where $\Sigma_Y(j+1) = H(j+1)P(j+1|j)H'(j+1) + \Sigma(j+1)$ is the covariance matrix of $\hat{\underset{\sim}{\epsilon}}_Y(j+1|j) = H(j+1)\hat{\underset{\sim}{X}}(j+1|j) - \underset{\sim}{Y}(j+1) = H(j+1)\hat{\underset{\sim}{\epsilon}}_X(j+1|j) - \underset{\sim}{e}(j+1)$.

The covariance matrices of the residuals $\hat{\underset{\sim}{\epsilon}}_X(j+1|j)$ and $\hat{\underset{\sim}{\epsilon}}_X(j+1|j+1)$ are given respectively by:

$$P(j+1|j) = F(j+1, j)P(j|j)F'(j+1, j) + V(j+1)\,, \tag{8.37}$$

and

$$P(j+1|j+1) = [I - K_{j+1}H(j+1)]P(j+1|j)\,. \tag{8.38}$$

In the literature, the matrix K_{j+1} has been referred to as the Kalman gain matrix due to addition of the observation $\underset{\sim}{Y}(j+1)$. In the above theorem,

$\hat{X}(j+1|j)$ is a linear combination of $Y(1), \ldots, Y(j)$ and $\hat{X}(j+1|j+1)$ a linear combination of $Y(1), \ldots, Y(j+1)$. Also, since $E[\hat{X}(0|0)] = E[X(0)]$, both $\hat{X}(j+1|j)$ and $\hat{X}(j+1|j+1)$ are unbiased for $X(j+1)$; see Exercise 8.11.

Equations (8.35)–(8.38) have been referred to as the forward filtering.

Theorem 8.3. *Given the estimator of $X(0)$ as $\hat{X}(0|0)$ with $E[\hat{\epsilon}(0|0)] = 0$, the linear, unbiased and minimum varianced estimator or BLUE $\hat{X}(j|n)$ of $X(j)$ given data $\mathbf{D}(n)$ with $(n > j)$ is given by the following recursive equations:*

$$\hat{X}(j|n) = \hat{X}(j|j) + A_j\{\hat{X}(j+1|n) - \hat{X}(j+1|j)\}, \qquad (8.39)$$

where

$$A_j = P(j|j)F'(j+1,j)P^{-1}(j+1|j);$$

$$P(j|n) = P(j|j) - A_j\{P(j+1|j) - P(j+1|n)\}A'_j, \qquad (8.40)$$

for $j = 1, \ldots, n-1$.

Obviously, $\hat{X}(j|n)$ is a linear combination of $Y(1), \ldots, Y(n)$ and $E[\hat{X}(j|n)] = E[X(j)]$. Further, by mathematical induction it can easily be shown that $P(j|j) - P(j|m)$ is positive semi-definite if $m > j$ (Exercise 8.12). Hence the variances of $\hat{X}(j|j)$ are greater than or equal to the variances of the corresponding elements of $\hat{X}(j|m)$ if $m > j$ respectively.

Equations (8.39)–(8.40) have been referred to as backward filters.

Proof of Theorems 8.2 and 8.3. Theorems 8.2 and 8.3 can be proved by standard least square theories; see [5–7]. Since the least square estimators are equivalent to the maximum likelihood estimators (MLE) under the assumption of normal distributions for the random noises and random measurement errors, we prove Theorems 8.2 and 8.3 by assuming normality for the random noises and for the random measurement errors. We will prove this by using a basic result from multivariate normal distributions. This basic result is:

Let the $p \times 1$ random vector X be normal with mean u and covariance matrix Σ. That is, $X \sim N(u, \Sigma)$. Partition X, u and Σ by $X' =$

$(\underset{\sim}{X}{}'_1, \underset{\sim}{X}{}'_2)$, $\underset{\sim}{u}{}' = (\underset{\sim}{u}{}'_1, \underset{\sim}{u}{}'_2)$ and

$$\Sigma = \begin{bmatrix} \Sigma_{11} & \Sigma_{12} \\ \Sigma_{21} & \Sigma_{22} \end{bmatrix},$$

where $\underset{\sim}{X}_1$ and $\underset{\sim}{u}_1$ are $p_1 \times 1$ with $(1 \le p_1 < p)$ and Σ_{11} is $p_1 \times p_1$. Then, $\underset{\sim}{X}_1 \sim N(\underset{\sim}{u}_1, \Sigma_{11})$, $\underset{\sim}{X}_2 \sim N(\underset{\sim}{u}_2, \Sigma_{22})$, $\underset{\sim}{X}_1 | \underset{\sim}{X}_2 \sim N\{\underset{\sim}{u}_1 + \Sigma_{12}\Sigma_{22}^{-1}(\underset{\sim}{X}_2 - \underset{\sim}{u}_2), \Sigma_{11.2}\}$ and $\underset{\sim}{X}_2 | \underset{\sim}{X}_1 \sim N\{\underset{\sim}{u}_2 + \Sigma_{21}\Sigma_{11}^{-1}(\underset{\sim}{X}_1 - \underset{\sim}{u}_1), \Sigma_{22.1}\}$, where $\Sigma_{ii.j} = \Sigma_{ii} - \Sigma_{ij}\Sigma_{jj}^{-1}\Sigma_{ji}$ for $i \ne j$.

To prove Theorems 8.2 and 8.3, let $\hat{\underset{\sim}{X}}(t|j) = \mathrm{E}\{\underset{\sim}{X}(t)|D(j)\}$, $\hat{\underset{\sim}{\epsilon}}_X(t|j) = \hat{\underset{\sim}{X}}(t|j) - \underset{\sim}{X}(t)$ and

$$P(t|j) = E\{[\hat{\underset{\sim}{\epsilon}}_X(t|j)][\hat{\underset{\sim}{\epsilon}}_X(t|j)]'\},$$

where $D(j) = \{\underset{\sim}{Y}(1), \ldots, \underset{\sim}{Y}(j)\}$. Then, from the stochastic system equation, $\hat{\underset{\sim}{X}}(j+1|j) = E\{\underset{\sim}{X}(j+1)|D(j)\} = F(j+1,j)\hat{\underset{\sim}{X}}(j|j)$ and $\hat{\underset{\sim}{\epsilon}}_X(j+1|j) = \hat{\underset{\sim}{X}}(j+1|j) - \underset{\sim}{X}(j+1) = F(j+1,j)\hat{\underset{\sim}{\epsilon}}_X(j|j) - \underset{\sim}{\epsilon}(j+1)$ so that $P(j+1|j) = F(j+1,j)P(j|j)F(j+1,j)' + V(j+1)$. If the random noises $\underset{\sim}{\epsilon}(t)$ are normally distributed, then

$$\underset{\sim}{X}(j+1)|D(j) \sim N\{\hat{\underset{\sim}{X}}(j+1|j), P(j+1|j)\}.$$

Similarly, from the observation equation, we have:
$\hat{\underset{\sim}{Y}}(j+1|j) = E\{\underset{\sim}{Y}(j+1)|D(j)\} = H(j+1)\hat{\underset{\sim}{X}}(j+1|j)$ and $\hat{\underset{\sim}{\epsilon}}_Y(j+1|j) = \hat{\underset{\sim}{Y}}(j+1|j) - \underset{\sim}{Y}(j+1) = H(j+1)\hat{\underset{\sim}{\epsilon}}_X(j+1|j) - \underset{\sim}{e}(j+1)$ so that $\Sigma_Y(j+1|j) = E\{[\hat{\underset{\sim}{\epsilon}}_Y(j+1|j)][\hat{\underset{\sim}{\epsilon}}_Y(j+1|j)]'\} = H(j+1)P(j+1|j)H(j+1)' + \Sigma(j+1)$. Furthermore, the conditional covariance between $\underset{\sim}{X}(j+1)$ and $\underset{\sim}{Y}(j+1)$ given $D(j)$ is

$$\Sigma_{YX}(j+1|j) = \mathrm{Cov}[\underset{\sim}{Y}(j+1), \underset{\sim}{X}(j+1)|D(j)]$$

$$= H(j+1)\,\mathrm{Var}[\underset{\sim}{X}(j+1)|D(j)] = H(j+1)P(j+1|j).$$

If the random noises $\underset{\sim}{\epsilon}(t)$ and the random errors $\underset{\sim}{e}(t)$ are normally distributed, then

$$\underset{\sim}{Y}(j+1)|D(j) \sim N\{H(j+1)\hat{\underset{\sim}{X}}(j+1|j) = H(j+1)F(j+1,j)\hat{\underset{\sim}{X}}(j|j), \Sigma_Y(j+1|j)\}.$$

Thus the conditional mean of $\underset{\sim}{X}(j+1)$ given $\{D(j), \underset{\sim}{Y}(j+1)\} = D(j+1)$ is

$$\hat{\underset{\sim}{X}}(j+1|j+1) = \hat{\underset{\sim}{X}}(j+1|j) + \Sigma'_{YX}(j+1|j)\Sigma_Y^{-1}(j+1|j)[\underset{\sim}{Y}(j+1) - \hat{\underset{\sim}{Y}}(j+1|j)]$$

$$= \hat{X}(j+1|j) + K_{j+1}[\underset{\sim}{Y}(j+1) - H(j+1)\hat{\underset{\sim}{X}}(j+1|j)],$$

where

$$K_{j+1} = P(j+1|j)H'(j+1)[H(j+1)P(j+1|j)H(j+1)' + \Sigma(j+1)]^{-1}$$

is the Kalman gain matrix.

The covariance matrix of the residual $\hat{\underset{\sim}{\epsilon}}_X(j+1|j+1)$ is

$$P(j+1|j+1) = P(j+1|j) - \Sigma'_{YX}(j+1|j)\Sigma_Y^{-1}(j+1|j)\Sigma_{YX}(j+1|j)$$

$$= \{I - K_{j+1}H(j+1)\}P(j+1|j).$$

It follows that

$$\underset{\sim}{X}(j+1)|D(j+1) \sim N\{\hat{\underset{\sim}{X}}(j+1|j+1), P(j+1|j+1)\}.$$

Notice that $\hat{\underset{\sim}{X}}(j+1|j+1)$ is the MLE which is the forward filter of $\underset{\sim}{X}(j+1)$ as given in Theorem 8.2. Notice also that from Eqs. (8.35)–(8.36),

$$\hat{\underset{\sim}{X}}(j+1|j+1) - \hat{\underset{\sim}{X}}(j+1|j) = K_{j+1}[\underset{\sim}{Y}(j+1) - H(j+1)\hat{\underset{\sim}{X}}(j+1|j)]$$

and from Eqs. (8.37)–(8.38),

$$P(j+1|j) - P(j+1|j+1) = K_{j+1}H(j+1)P(j+1|j)$$

$$= P(j+1|j)H'(j+1)\Sigma_Y^{-1}(j+1|j)H(j+1)P(j+1|j)$$

which is positive semi-definite.

To prove Theorem 8.3, recall that if the random noises and random errors are normally distributed, then

$$\underset{\sim}{X}(j)|D(j) \sim N\{\hat{\underset{\sim}{X}}(j|j), P(j|j)\}.$$

Hence the conditional covariance matrix between

$$\underset{\sim}{X}(j) \text{ and } \underset{\sim}{Y}(j+1) = H(j+1)F(j+1,j)\underset{\sim}{X}(j) + [H(j+1)\underset{\sim}{\epsilon}(j+1) + \underset{\sim}{e}(j+1)]$$

given $D(j)$ is

$$\Sigma_{XY}(j) = P(j|j)F'(j+1,j)H'(j+1) = A(j)P(j+1|j)H'(j+1)\,,$$

where $A(j) = P(j|j)F'(j+1,j)P^{-1}(j+1|j)$.

Thus the conditional mean of $\underset{\sim}{X}(j)$ given $D(j+1) = \{D(j), \underset{\sim}{Y}(j+1)\}$ is

$$\hat{\underset{\sim}{X}}(j|j+1) = \hat{\underset{\sim}{X}}(j|j) + \Sigma_{XY}(j)\Sigma_Y^{-1}(j+1|j)[\underset{\sim}{Y}(j+1)$$

$$- H(j+1)F(j+1,j)\hat{\underset{\sim}{X}}(j|j)]$$

$$= \hat{\underset{\sim}{X}}(j|j) + A(j)K_{j+1}[\underset{\sim}{Y}(j+1) - H(j+1)\hat{\underset{\sim}{X}}(j+1|j)]$$

$$= \hat{\underset{\sim}{X}}(j|j) + A(j)\{\hat{\underset{\sim}{X}}(j+1|j+1) - \hat{\underset{\sim}{X}}(j+1|j)\}\,.$$

The covariance matrix of the residual $\hat{\underset{\sim}{\epsilon}}_X(j|j+1)$ is

$$P(j|j+1) = P(j|j) - \Sigma_{XY}(j)\Sigma_Y^{-1}(j+1|j)\Sigma'_{XY}(j)$$

$$= P(j|j) - A(j)K_{j+1}H(j+1)P(j+1|j)A'(j)$$

$$= P(j|j) - A(j)\{P(j+1|j+1) - P(j+1|j)\}A'(j)\,.$$

It follows that

$$\underset{\sim}{X}(j)|D(j+1) \sim N\{\hat{\underset{\sim}{X}}(j|j+1), P(j|j+1)\}\,.$$

By following this approach and using mathematical induction (cf. Exercise 8.13), we obtain:

$$\underset{\sim}{X}(j)|D(n) \sim N\{\hat{\underset{\sim}{X}}(j|n), P(j|n)\}\,,$$

where

$$\hat{\underset{\sim}{X}}(j|n) = \hat{\underset{\sim}{X}}(j|j) + A(j)\{\hat{\underset{\sim}{X}}(j+1|n) - \hat{\underset{\sim}{X}}(j+1|j)\}$$

and

$$P(j|n) = P(j|j) - A(j)\{P(j+1|n) - P(j+1|j)\}A'(j)$$

for $n > j$ are as given in Theorem 8.3.

To implement the above procedure for deriving the Kalman filter estimates of the state variables for given initial distribution of $\underset{\sim}{X}(t)$ at time 0, one first derives results by using formulas in Theorem 8.2. Then one goes backward from n to 1 by applying formulas in Theorem 8.3. $\qquad\square$

8.4.2. *Alternative representation of Kalman filters and smoothers*

Using the matrix result $(I + AB)^{-1} = I - A(I + BA)^{-1}B$, it can easily be shown that $P(j + 1|j + 1) = \{P^{-1}(j + 1|j) + H'(j + 1)\Sigma_{j+1}^{-1}H(j + 1)\}^{-1}$ and $K_{j+1} = P(j + 1|j + 1)H'(j + 1)\Sigma_{j+1}^{-1}$; see Exercise 8.14. It follows that $P(j+1|j) - P(j+1|j+1)$ is a positive semi-definite matrix so that the variances of elements of $\hat{\underset{\sim}{X}}(j + 1|j)$ are greater than or equal to the variances of the corresponding elements of $\hat{\underset{\sim}{X}}(j + 1|j + 1)$ respectively. Furthermore, $\hat{\underset{\sim}{X}}(j + 1|j + 1)$ can be expressed as (Exercise 8.15):

$$\hat{\underset{\sim}{X}}(j + 1|j + 1) = \{P^{-1}(j + 1|j) + H'(j + 1)\Sigma_{j+1}^{-1}H(j + 1)\}^{-1}$$

$$\times \{P^{-1}(j + 1|j)\hat{\underset{\sim}{X}}(j + 1|j)$$

$$+ H'(j + 1)\Sigma_{j+1}^{-1}H(j + 1)\hat{\underset{\sim}{X}}(j + 1|Y)\}, \qquad (8.41)$$

where

$$\hat{\underset{\sim}{X}}(j + 1|Y) = [H'(j + 1)\Sigma_{j+1}^{-1}H(j + 1)]^{-1}H'(j + 1)\Sigma_{j+1}^{-1}\underset{\sim}{Y}(j + 1)$$

is the least square solution of the linear model

$$\underset{\sim}{Y}(j + 1) = H(j + 1)\underset{\sim}{X}(j + 1) + \underset{\sim}{e}(j + 1).$$

Similarly, by using the matrix results $(I + AB)^{-1} = I - A(I + BA)^{-1}B$, it can be shown easily that $\hat{\underset{\sim}{X}}(j|n)$ can be expressed as:

$$\hat{\underset{\sim}{X}}(j|n) = \{P^{-1}(j|j) + F'(j + 1, j)V_{j+1}^{-1}F(j + 1, j)\}^{-1}\{P^{-1}(j|j)\hat{\underset{\sim}{X}}(j|j)$$

$$+ F'(j + 1, j)V_{j+1}^{-1}F(j + 1, j)\hat{\underset{\sim}{X}}(j|B, n)\}, \qquad (8.42)$$

where

$$\hat{X}\,(j|B,n) = [F'(j+1,j)V_{j+1}^{-1}F(j+1,j)]^{-1}F'(j+1,j)V_{j+1}^{-1}\,\hat{X}\,(j+1|n)$$

is the least equation solution of $X\,(t)$ from $X\,(j+1) = F(j+1,j)\,X\,(j)+\epsilon\,(j+1)$ with $\hat{X}\,(j+1|n)$ replacing $X\,(j+1)$; see Exercise 8.16.

8.4.3.   *Some classical theories for discrete-time linear state space models with missing data*

In the above model, it is assumed that data can be collected at each time point. In many practical situations, however, data are only collected at time point $t_j, j = 1,\ldots,k$ with $t_k \leq n$. Denote by $\{Y\,(t_j) = Z\,(j), e\,(t_j) = \xi\,(j)\}$. The state space model then becomes:

$$X\,(t+1) = F(t+1,t)\,X\,(t) + \epsilon\,(t+1), \quad t = 1,\ldots,n, \tag{8.43}$$

$$Z\,(j) = H(j)\,X\,(t_j) + \xi\,(j), \quad j = 1,\ldots,k. \tag{8.44}$$

To derive the BLUE of $X\,(t)$ when there are missing data, the basic trick is to reduce the model to that in Subsec. 8.4.1 at time points t_j $(j = 1,\ldots,k)$ when data are available; one uses the stochastic system equation at time points t $(t \neq t_j, j = 1,\ldots,k)$ when there are no data.

To derive the BLUE of state variables, define the following matrices:

$$F(j,i) = \prod_{r=i}^{j-1} F(r+1,r), \quad \text{for } j \geq i, \text{ with } F(i,i) = \prod_{r=i}^{i-1} F(r+1,r) = I_p,$$

$$\zeta\,(j+1,i) = \sum_{r=i}^{j} F(j+1,r+1)\,\epsilon\,(r+1), \quad \text{for } j \geq i,$$

$$G(j+1,j) = F(t_{j+1},t_j), \quad j = 1,\ldots,k,$$

$$\eta\,(j+1) = \zeta(t_{j+1},t_j), \quad j = 1,\ldots,k.$$

Then the expected values of $\xi\,(j)$, $\zeta\,(r+1,i)\}$ and $\eta\,(j)$ are vectors of zero's. The covariance matrices of these random vectors are given by $\mathrm{Var}[\xi\,(j)] =$

$$\Sigma(t_j) = \Omega(j),$$

$$\mathrm{Var}[\underset{\sim}{\zeta}(j+1,i)] = \sum_{r=i}^{j} F(j+1,r+1)V(r+1)F'(j+1,r+1)$$

$$= \Lambda(j+1,i), \quad j \geq i,$$

and

$$\mathrm{Var}[\underset{\sim}{\eta}(j+1)] = \Lambda(t_{j+1},t_j) = \Psi(j+1)$$

respectively. Furthermore, the $\underset{\sim}{\eta}(j)$ are independently distributed of one another and are un-correlated with the state variables.

Denote by $\underset{\sim}{X}(t_j) = \underset{\sim}{u}(j)$ and $\boldsymbol{D}(j) = \{Y(t_i), i = 1,\ldots,j\}$. Then, from Eq. (8.43), we have:

$$\underset{\sim}{X}(t) = F(t,t-1)\underset{\sim}{X}(t-1) + \underset{\sim}{\epsilon}(t)$$

$$= F(t,t-2)\underset{\sim}{X}(t-2) + \underset{\sim}{\zeta}(t,t-2)$$

$$= \cdots = F(t,t_j)\underset{\sim}{X}(t_j) + \underset{\sim}{\zeta}(t,t_j)$$

$$= F(t,t_j)\underset{\sim}{u}(j) + \underset{\sim}{\zeta}(t,t_j), \quad \text{for } t \geq t_j, \tag{8.45}$$

and for $j = 1,\ldots,k$,

$$\underset{\sim}{X}(t_{j+1}) = \underset{\sim}{u}(j+1) = G(j+1,j)\underset{\sim}{u}(j) + \underset{\sim}{\eta}(j+1), \tag{8.46}$$

$$\underset{\sim}{Z}(j+1) = H(j+1)\underset{\sim}{u}(j+1) + \underset{\sim}{\xi}(j+1), \quad t \geq t_j. \tag{8.47}$$

Using the state space model with stochastic system model given by Eq. (8.46) and with the observation model given by Eq. (8.47), by Theorems 8.2 and 8.3, the BLUE of $\underset{\sim}{u}(j) = \underset{\sim}{X}(t_j)$ given data $\boldsymbol{D}(j)$ and given data $\boldsymbol{D}(n)$ are given respectively by the following recursive equations:

(1) The BLUE $\underset{\sim}{\hat{X}}(t_j|j) = \underset{\sim}{\hat{u}}(j|j)$ of $\underset{\sim}{X}(t_j)$ Given $\boldsymbol{D}(j)$.

(a) By assumption, the unbiased estimator of $\underset{\sim}{X}(0)$ is $\underset{\sim}{\hat{u}}(0|0)$ with $\mathrm{Var}[\underset{\sim}{\hat{\epsilon}}(0|0)] = P(0|0) = P_u(0|0).$

(b) For $j = 0, 1, \ldots, k$,

$$\hat{\underset{\sim}{u}}(j+1|j) = G(j+1,j)\hat{\underset{\sim}{u}}(j|j) \, ,$$

$$\hat{\underset{\sim}{u}}(j+1|j+1) = \hat{\underset{\sim}{u}}(j+1|j) + K_Z(j+1)[\underset{\sim}{Z}(j+1) - H(j+1)\hat{\underset{\sim}{u}}(j+1|j)] \, ,$$

where

$$K_Z(j+1) = P_u(j+1|j)H'(j+1)[\Sigma_Z(j+1)]^{-1} \, ,$$

and where $\Sigma_Z(j+1) = H(j+1)P_u(j+1|j)H'(j+1) + \Omega(j+1)$ is the covariance matrix of $\hat{\underset{\sim}{\epsilon}}_Z(j+1|j) = H(j+1)\hat{\underset{\sim}{u}}(j+1|j) - \underset{\sim}{Z}(j+1) = H(j+1)\hat{\underset{\sim}{\epsilon}}_u(j+1|j) - \underset{\sim}{\xi}(j+1)$.

(c) The covariance matrices of the residuals $\hat{\underset{\sim}{\epsilon}}_u(j+1|j)$ and $\hat{\underset{\sim}{\epsilon}}_u(j+1|j+1)$ are given respectively by:

$$P_u(j+1|j) = G(j+1,j)P_u(j|j)G'(j+1,j) + \Psi(j+1) \, ,$$

and

$$P_u(j+1|j+1) = [I - K_Z(j+1)H(j+1)]P_u(j+1|j) \, .$$

(2) The BLUE $\hat{\underset{\sim}{X}}(t_j|k) = \hat{\underset{\sim}{u}}(j|k)$ of $\underset{\sim}{X}(t_j)$ Given $\boldsymbol{D}(k)$.

$$\hat{\underset{\sim}{u}}(j|k) = \hat{\underset{\sim}{u}}(j|j) + A_u(j)\{\hat{u}(j+1|k) - \hat{\underset{\sim}{u}}(j+1|j)\} \, ,$$

where

$$A_u(j) = P_u(j|j)G'(j+1,j)P_u^{-1}(j+1|j) \, ;$$

$$P_u(j|k) = P_u(j|j) - A_u(j)\{P_u(j+1|j) - P_u(j+1|k)\}A_u'(j) \, ,$$

for $j = 1, \ldots, k$.

To derive the BLUE of $\underset{\sim}{X}(t)$ for $t_j \leq t < t_{j+1}$, notice that from Eq. (8.45), linear unbiased estimators of $\underset{\sim}{X}(t)$ can be expressed as $F(t,t_j)\hat{\underset{\sim}{X}}(t_j)$, where $\hat{\underset{\sim}{X}}(t_j)$ is an unbiased linear estimator of $\underset{\sim}{X}(t_j) = \underset{\sim}{u}(j)$. Since the BLUE of $\underset{\sim}{u}(j)$ given data $\boldsymbol{D}(j)$ and given data $\boldsymbol{D}(k)$ are given respectively by $\hat{\underset{\sim}{u}}(j|j)$ and $\hat{\underset{\sim}{u}}(j|k)$, hence, for $t_j \leq t < t_{j+1}$, the BLUE of $\underset{\sim}{X}(t)$ given data $\boldsymbol{D}(j)$ and given data $\boldsymbol{D}(n)$ are given by $\hat{\underset{\sim}{X}}(t|j) = F(t,t_j)\hat{\underset{\sim}{u}}(j|j)$ and $\hat{\underset{\sim}{X}}(t|k) = F(t,t_j)\hat{\underset{\sim}{u}}(j|k)$, respectively. For $t_j \leq t < t_{j+1}, j = 1, \ldots, k-1$, the covariance matrices of the residuals $\hat{\underset{\sim}{\epsilon}}_X(t|j) = \hat{\underset{\sim}{X}}(t|j) - \underset{\sim}{X}(t)$ and $\hat{\underset{\sim}{\epsilon}}_X(t|k) = \hat{\underset{\sim}{X}}(t|k) - \underset{\sim}{X}(t)$ are given respectively by:

$$P_X(t|j) = F(t, t_j)P_u(j|j)F'(t, t_j) + \Lambda(t, t_j),$$

and

$$P_X(t|k) = F(t, t_j)P_u(j|k)F'(t, t_j) + \Lambda(t, t_j).$$

8.5. Estimation of HIV Prevalence and AIDS Cases in the San Francisco Homosexual Population

As an application of the discrete-time linear state space model, in this section we proceed to estimate and predict the HIV prevalence and the AIDS cases in the San Francisco homosexual population. Given in Table 8.3 is the monthly AIDS incidence data available from the gopher server of CDC. As shown in the example in Subsec. 8.1.2, this data set, together with the total population size and numbers of people who died from AIDS $(Y_2(t))$, will be used to construct the observation model of the state space model. (To avoid the problem of changing AIDS definition by CDC in 1993, we have used the AIDS incidence data only up to December 1992; see [43].) Since the sum of the population size

Table 8.3. San Francisco AIDS case report for 1981–1994 by month of primary diagnosis.

Year	Jan	Feb	Mar	Apr	May	Jun	Jul	Aug	Sep	Oct	Nov	Dec
81	1	3	2	1	1	3	3	3	5	3	3	8
82	6	5	0	6	6	15	12	10	10	14	20	12
83	24	19	31	25	19	21	27	35	28	31	26	31
84	46	32	39	43	39	48	67	59	70	54	60	56
85	77	62	72	77	73	80	94	89	76	89	70	89
86	104	93	114	100	103	109	121	138	112	149	99	142
87	136	137	142	130	149	149	150	148	161	140	123	131
88	156	144	184	141	130	155	139	136	156	117	132	155
89	159	141	179	197	163	201	169	160	138	155	138	142
90	200	174	191	156	180	173	176	195	156	175	182	150
91	220	195	195	190	208	193	229	243	220	303	227	241
92	286	303	299	215	218	241	267	244	249	240	196	226
93	243	234	229	188	190	224	235	179	182	162	162	161
94	213	164	162	131	90	38						

and the number of people who died from AIDS is quite stable, we follow Bailey [44] and Hethcote and Van Ark [45] to assume that the immigration rates of S people and I people equal to the death rates of these people respectively, i.e., $\mu_S(t) = d_S(t), \mu_r(t) = d_r(t)$, for $r = 1, 2, \ldots, 5$. Thus, based on estimates by Lemp *et al.* [46], $Y_2(t)$, is taken roughly as 58000 ± 5000. Because of the awareness of AIDS, one may also assume that there is no immigration for AIDS cases.

8.5.1.   *Estimation of parameter values in the San Francisco homosexual population*

In this application, we apply the staged model with $k = 6$ given in Example 8.2. We will use the estimates by Satten and Longini [24] for the forward, the backward and the direct transition rates (probabilities) of the infective stages under the assumption that these transition rates are constants and that there are no direct transition to AIDS for the first 3 infective stages. These estimates are given in Table 4.1. We will use the estimate $\mu_S(t) = \mu_u(t) = 0.000532(u = 1, \ldots, k)$ per month given in [45]. These estimates were obtained from the 1987 Bureau of Standard Statistics [47].

To implement the Kalman recursion, we need the estimates of the infection rate $p_S(t)$ of S people. This rate is the conditional probability that a S person contracts HIV for the first time to become an HIV carrier. Hence $p_S(j) = f_I(j)/[1 - F_I(j-1)]$, where $f_I(j)$ and $F_I(j)$ are the pdf and cdf of HIV infection, respectively.

To estimate $p_S(t)$, let $f_s(j)$ be the density of the seroconversion distribution and $f_w(t)$ the density of the window period which is defined as the random time between HIV infection and HIV seroconversion. Since the seroconversion distribution is a convolution of the HIV infection distribution and the distribution of the window period, one can readily estimate $f_I(j)$ from estimates of $f_s(j)$ and $f_w(t)$. Under the assumption that $f_w(t)$ is an exponential distribution with parameter θ, we have:

$$f_s(j) = \int_0^j f_I(x) f_w(j-x)dx = \sum_{i=1}^j f_I(i) \int_{i-1}^i \frac{1}{\theta} e^{-\frac{1}{\theta}(j-x)}dx$$

$$= \sum_{i=1}^j f_I(i) e^{-\frac{1}{\theta}(j-i)} (1 - e^{-\frac{1}{\theta}})$$

$$= f_I(j)(1 - e^{-\frac{1}{\theta}}) + e^{-\frac{1}{\theta}} \sum_{i=1}^{j-1} f_I(i)e^{-\frac{1}{\theta}(j-1-i)}(1 - e^{-\frac{1}{\theta}})$$

$$= f_I(j)(1 - e^{-\frac{1}{\theta}}) + e^{-\frac{1}{\theta}} f_s(j - 1).$$

Let $f_s(0) = 0$. Then we obtain

$$f_I(1) = f_s(1)(1 - e^{-\frac{1}{\theta}})^{-1},$$

$$f_I(j) = (1 - e^{-\frac{1}{\theta}})^{-1}[f_s(j) - e^{-\frac{1}{\theta}} f_s(j - 1)], \quad \text{for } j \geq 2. \tag{8.48}$$

By using the estimate of the $f_s(t)$ by Tan, Tang and Lee [48] and the estimates $\theta = 3.5$ month by Horsburgh *et al.* [49], we obtain the estimates of $f_I(t)$ as given in Fig. 8.4.

8.5.2. *The initial distribution*

The initial values of Kalman recursion, $\underset{\sim}{X}(0)$ and $P(0|0)$, are obtained by running only the dynamic models for 20 steps (months) starting at 10 infective people. That is, we are assuming that there were only 10 I_1 people in

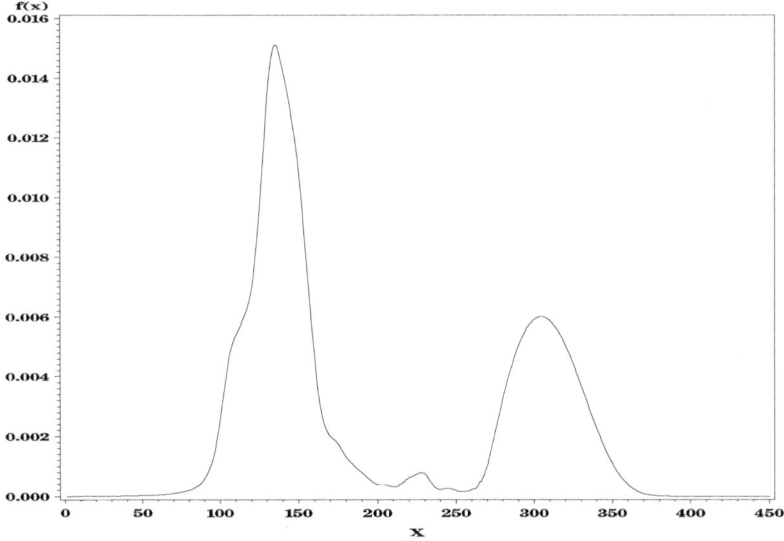

Fig. 8.4. Plot showing the estimates of HIV infection density.

May, 1979. We run our dynamic model (state equations) for 20 steps (up to December, 1980) with the random error being set as zero. Using $t = 0$ corresponding to December, 1980, the initial values of the states at 0 states are reported in Table 8.4. The result shows that in January, 1981 there would be approximately one AIDS case which is close to the observed number of A person given in Table 8.3.

8.5.3. *The variances and covariances of random noises and measurement error*

The variance-covariance matrix of the random noises of state variables is estimated by using formulas given in Table 8.2 with the mean of the

Table 8.4. The sources of parameters, initial values, observation and state variables.

Parameters			Sources
$\gamma_1(t) = 0.0441$	$\beta_2(t) = 0.0035$	$\omega_1(t) = \omega_2(t) = 0$	Satten and Longini [24]
$\gamma_2(t) = 0.0456$	$\beta_3(t) = 0.0121$	$\omega_3(t) = 0$	Satten and Longini [24]
$\gamma_3(t) = 0.0420$	$\beta_4(t) = 0.0071$	$\omega_4(t) = 0.0052$	Satten and Longini [24]
$\gamma_4(t) = 0.0404$	$\beta_5(t) = 0.0192$	$\omega_5(t) = 0.0474$	Satten and Longini [24]
$\mu_S(t) = d_S(t)$	$\mu_r(t) = d_r(t)$	for $r = 1, 2, \ldots, 5$	Assumption
$p_S(t)$	$t = 1, 2, \ldots, N$		Tan, Lee and Tang [48]

Initial Values	Sources
$\underset{\sim}{X}(0\vert 0) = [56392.5, 1247.7, 294.4, 57.3, 8.4, 0.95, 0.26]'$	dynamic models

$$P(0\vert 0) = \begin{bmatrix} 233.1 & -173.3 & 0 & 0 & 0 & 0 & 0 \\ -173.3 & 227.2 & -52.6 & 0 & 0 & 0 & 0 \\ 0 & -52.6 & 66.6 & -12.8 & 0 & 0 & 0 \\ 0 & 0 & -12.8 & 15.8 & -2.3 & 0 & 0 \\ 0 & 0 & 0 & -2.3 & 2.7 & -0.3 & -0.04 \\ 0 & 0 & 0 & 0 & -0.3 & 0.4 & -0.04 \\ 0 & 0 & 0 & 0 & -0.04 & -0.04 & 0.3 \end{bmatrix}$$

dynamic models

Variance-Covariance Matrix of Random Errors	Sources
$V(t)$	Table 8.2
$\Sigma(t) = \mathrm{diag}\{t, 6250000\}$	Assumption

Observation and State Variables	Sources
$Y_1(t)$	Table 8.3
$Y_2(t) = 58000 \pm 5000$	Survey [46]
$\underset{\sim}{X}(t) = [S(t), I_1(t), I_2(t), \ldots, I_5(t), A(t)]',\quad t = 1, 2, \ldots, N$	to be estimated by Kalman Recursion

state variables being replaced by the one-step ahead prediction of these variables.

In the observation models, we assume that the $e_1(t)$ and $e_2(t)$ are uncorrelated (i.e., $\Sigma(t)$ is a diagonal matrix, where the diagonal elements are the variances of $e_1(t)$ and $e_2(t)$). This is actually true since the observation variables $Y_1(t)$ and $Y_2(t)$ are obtained from two different sources. To obtain the variances of $e_1(t)$ and $e_2(t)$, notice that $e_1(t)$ is related to the error of reporting AIDS cases and $e_2(t)$ is related to the error of survey of the San Francisco homosexual population. Hence, the variance of $e_1(t)$ can be obtained from the estimates of the variance of reported AIDS cases. Because this estimate currently is not available to us, for illustration purpose, in [23] this variance is taken as a function of time, say, $\mathrm{Var}[e_1(t)] = t$. The variance of $e_2(t)$ can be easily obtained by assuming the error range 5000 to be two standard

Table 8.5. Estimated numbers of the infective people in the five stages for the years 1981–1992.

Time	I_1	I_2	I_3	I_4	I_5	Total
81.06	1280	690	279	76	11	2336
81.12	1519	911	529	212	46	3217
82.06	2080	1161	771	406	122	4540
82.12	3005	1537	1021	629	237	6429
83.06	4049	2071	1323	876	377	8696
83.12	5274	2739	1696	1156	548	11413
84.06	6621	3538	2146	1466	742	14513
84.12	8296	4495	2682	1809	927	18209
85.06	10229	5685	3320	2204	1133	22571
85.12	11580	7000	4061	2665	1385	26691
86.06	12146	8226	4855	3150	1665	30042
86.12	12040	9254	5659	3621	1909	32483
87.06	11134	9958	6442	4074	2098	33706
87.12	9694	10246	7158	4552	2260	33910
88.06	8187	10149	7745	5062	2399	33542
88.12	6785	9760	8154	5613	2706	33018
89.06	5551	9166	8383	6096	2780	31976
89.12	4549	8453	8413	6675	3138	31228
90.06	3755	7700	8238	7177	3511	30381
90.12	3104	6959	7880	7570	4215	29728
91.06	2578	6265	7407	7638	4902	28790
91.12	2153	5631	6924	7322	4830	26860
92.06	1883	5063	6481	6985	4240	24652
92.12	1724	4572	6054	6794	4030	23174

deviation. Then we have $\mathrm{Var}[e_2(t)] = 6250000$. This large variance downgrades the information from $Y_2(t)$.

All the parameters, initial values and assumptions are summarized in Table 8.4. Based on that information, we can apply the Kalman recursion in Sec. 8.2 to estimate the state variables $\underset{\sim}{X}(t) = [S(t), I_1(t), I_2(t), \ldots, I_5(t), A(t)]'$. The Kalman smoothing method is applied to the data (1981–1992) given in Table 8.3 and those of the total population size. Using these data sets we have obtained the estimates of the numbers of infective people in each of the five I stages and the cumulative numbers of AIDS cases for each month from January, 1981 to December, 1992. We have also computed the total number of infective people and the AIDS incidence. All these results are summarized

Table 8.6. Comparison of the estimated and the observed AIDS incidence and the AIDS cumulative numbers for the years 1981–1992.

Year and Month	AIDS incidence		AIDS cumulative	
	Estimated	Observed	Estimated	Observed
81.06	12	11	12	11
81.12	24	25	35	36
82.06	42	38	77	74
82.12	82	78	160	152
83.06	133	139	293	291
83.12	180	178	473	469
84.06	253	247	726	716
84.12	354	366	1080	1082
85.06	437	441	1517	1523
85.12	514	507	2031	2030
86.06	623	623	2654	2653
86.12	752	761	3406	3414
87.06	841	843	4247	4257
87.12	867	853	5114	5110
88.06	890	910	6005	6020
88.12	854	835	6859	6855
89.06	1011	1040	7869	7895
89.12	951	902	8821	8797
90.06	1047	1074	9868	9871
90.12	1064	1034	10932	10905
91.06	1203	1201	12134	12106
91.12	1463	1463	13597	13560
92.06	1540	1562	15138	15131
92.12	1416	1422	16553	16553

in Tables 8.5 and 8.6 by 6 months. For comparison purpose, the observed AIDS incidence and the observed cumulative numbers of AIDS cases are also included in Table 8.6.

8.5.4. *Estimation results*

From the results of Table 8.6, it is observed that the Kalman smoothing estimates trace the observed values surprisingly well. The estimated results reveal that the respective peaks of the numbers of infective people in the five stages are 12197, 10254, 8427, 7657, 4999 which are achieved in August 1986, January 1988, October 1989, April 1991, and September 1991 respectively; but the total number of infective people reaches its peak 33915 as early as November, 1987.

8.5.5. *Projection results*

To check the effectiveness of the Kalman projection, we have projected the AIDS incidence and the cumulative number for the years 1990–1992, and compared them with the observed values. These results are presented in Table 8.7. From the results of this table, it is observed that in general the projected values are close to the observed ones. However one may notice that as in other projections [45, 46] the results appear to be under-projected for the year 1991 and 1992, presumably due to the adjustment error for the new 1993 AIDS definition. It might be of interest to note that the projected results are at least as good (close to the observed values) as those of the other projections [45, 46].

In Table 8.8, we have listed the projected numbers of the infective people for each of the five stages, the total number of infective people, the AIDS incidence

Table 8.7. Comparison of the projected and the observed AIDS incidence and the AIDS cumulative numbers for 1990–1992 based on observations during 1981–1989.

Year and	AIDS incidence		AIDS cumulative	
Month	Predicted	Observed	Predicted	Observed
90.06	1005	1074	9812	9871
90.12	1103	1034	10915	10905
91.06	1186	1201	12101	12106
91.12	1251	1463	13353	13569
92.06	1298	1562	14650	15131
92.12	1326	1422	15977	16553

Table 8.8. Projected numbers of the infective people, the AIDS incidence and the AIDS cumulative number for 1993–2002 based on observations during 1981–1992.

Year and	Infective people						AIDS	
Month	I_1	I_2	I_3	I_4	I_5	Total	Incidence	Cumulative
93.06	1556	4145	5634	6615	4046	21996	1359	17912
93.12	1307	3754	5227	6393	4015	20696	1351	19264
94.06	1107	3383	4836	6140	3945	19411	1331	20594
94.12	977	3046	4461	5863	3845	18192	1299	21893
95.06	878	2746	4104	5571	3721	17020	1259	23152
95.12	936	2490	3771	5270	3577	16044	1211	24363
96.06	1479	2337	3468	4966	3420	15670	1160	25523
96.12	2656	2400	3224	4668	3254	16202	1104	26627
97.06	4103	2750	3084	4391	3085	17413	1047	27674
97.12	5220	3317	3084	4157	2920	18698	991	28666
98.06	5797	3945	3221	3985	2768	19716	939	29605
98.12	5930	4508	3459	3887	2639	20423	894	30499
99.06	5762	4939	3748	3863	2540	20852	859	31358
99.12	5414	5217	4043	3903	2476	21053	835	32193
00.06	4968	5347	4307	3989	2447	21058	822	33016
00.12	4481	5349	4516	4103	2448	20897	821	33837
01.06	3988	5247	4660	4227	2473	20595	828	34665
01.12	3515	5066	4734	4345	2525	20175	842	35507
02.06	3072	4826	4742	4445	2566	19651	858	36365
02.12	2669	4548	4688	4517	2617	19039	876	37240

and the AIDS cumulative numbers from 1993 to 2002 by every 6 months based on all the reliable data in 1981–1992. It is observed that the numbers of the infective people in each of the infective stages will have a second peak. This is caused by the bi-mode property of the estimated infection rate $p_S(t)$. The number of S people will dramatically decrease after 1995, and the total infective people will soon reach its second peak, 21077, in March, 2000.

8.6. Complements and Exercises

Exercise 8.1. In the models given in Secs. 8.1 and 8.2, show that if the random noises in the stochastic system equations have expected value 0, then the random noises are un-correlated with the state variables.

Exercise 8.2. Using the stochastic Eqs. (2.32)–(2.35) and the distribution results given in Subsec. 2.8.2, prove the covariance formula in Table 8.1.

Exercise 8.3. Using the stochastic Eqs. (2.40)–(2.44) and the distribution results given in Subsec. 2.8.2, prove the covariance formula between $I(i, t)$ and $I(j, \tau)$ and between $I(i, t)$ and $A(\tau)$ as given in Table 8.2.

Exercise 8.4. In the HIV pathogenesis model described in Subsec. 8.2.2, let $\{u_i(t) = E[T_i(t)], i = 1, 2, 3, u_V(t) = E[V(t)]\}$ denote the expected numbers of $\{T_i(t), i = 1, 2, 3, V(t)\}$ respectively. Using the stochastic Eqs. (8.4)–(8.7), show that these expected numbers satisfy the following system of differential equations:

$$\frac{d}{dt} u_1(t) = s(t) + \{E[b_T(t)] - \mu_1 - k_1 u_V(t)\} u_1(t) + \text{Cov}\{b_T(t), T_1(t)\}$$

$$- k_1 \text{ Cov}\{V(t), T_1(t)\},$$

$$\frac{d}{dt} u_2(t) = \omega(t) k_1 u_V(t) u_1(t) - [\mu_2 + k_2] u_2(t) + \omega(t) k_1 \text{ Cov}\{V(t), T_1(t)\},$$

$$\frac{d}{dt} u_3(t) = [1 - \omega(t)] k_1 u_V(t) u_1(t) + k_2 u_2(t) - \mu_3 u_3(t)$$

$$+ [1 - \omega(t)] k_1 \text{ Cov}\{V(t), T_1(t)\},$$

$$\frac{d}{dt} u_V(t) = N(t) \mu_3 u_3(t) - k_1 u_V(t) u_1(t) - \mu_V u_V(t) - k_1 \text{ Cov}\{V(t), T_1(t)\}.$$

Exercise 8.5. In the HIV pathogenesis model described in Subsec. 8.2.2, by using the distribution results in Subsec. 8.2.2, show that the covariances between the random noises are as given in Subsec. 8.2.2.

Exercise 8.6. For the multi-event model as described in Subsec. 8.3.1 and in Example 4.6, by using the stochastic Eqs. (8.9)–(8.10) and the multinomial distribution results for $\{B_j(t), D_j(t), M_j(t)\}$ given $I_j(t)$, show that the pgf of $\{I_j(t), j = 0, 1, \ldots, k - 1\}$ are given exactly by that in Part (b) of Exercise 4.8. This shows that the classical Kolmogorov equation method is equivalent to the stochastic equation representation and the conditional multinomial distributions for the associated random variables.

Exercise 8.7. Prove Eq. (8.11) for the variance of the random noises of Eqs. (8.9)–(8.10).

Exercise 8.8. Prove the distribution results given by Eqs. (8.18)–(8.19).

Exercise 8.9. For the multiple pathways model as described in Subsec. 8.3.2, by using the stochastic Eqs. (8.22)–(8.24) and the multinomial distribution results, show that the joint pgf $\phi(x_i, i = 0, 1, 2, 3; t) = \phi(\underset{\sim}{x}; t) = E\{\prod_{i=0}^{3} x_i^{I_i(t)} | I_0(0) = N_0\}$ of $\{I_i(t), i = 0, 1, 2, 3\}$ is given by:

$$\frac{\partial}{\partial t}\phi(\underset{\sim}{x}); t) = (x_0 - 1)\left\{ x_0 b_0(t) - d_0(t) + \sum_{i=1}^{3} x_i \alpha_i \right\} \frac{\partial}{\partial x_0}\phi(\underset{\sim}{x}; t)$$

$$+ \sum_{i=1}^{2} (x_i - 1)[x_i b_i(t) - d_i(t) + x_3 \beta_i]\frac{\partial}{\partial x_i}\phi(\underset{\sim}{x}; t)\}$$

$$+ (x_3 - 1)[x_3 b_3(t) - d_3(t)]\frac{\partial}{\partial x_3}\phi(\underset{\sim}{x}; t)\},$$

where the initial condition is $\phi(\underset{\sim}{x}; 0) = x_0^{N_0}$.

Exercise 8.10.   Prove the distribution results given by Eqs. (8.25)–(8.26).

Exercise 8.11.   Show that If $E\underset{\sim}{\hat{\epsilon}}(0|0) = \underset{\sim}{0}$, then $\underset{\sim}{\hat{X}}(j+1|j)$, $\underset{\sim}{\hat{X}}(j+1|j+1)$ and $\underset{\sim}{\hat{X}}(j|n)(n > j)$ are all unbiased for $\underset{\sim}{X}(t)$.

Exercise 8.12.   Using mathematical induction, show that in Theorem 8.3, $P(j|j) - P(j|m)$ is positive semi-definite for all $m \geq j$.

Exercise 8.13.   By using mathematical induction method, show that

$$\underset{\sim}{X}(j)|D(n) \sim N\{\underset{\sim}{\hat{X}}(j|n), P(j|n)\},$$

where $\{\underset{\sim}{\hat{X}}(j|n), P(j|n)\}$ are given in Theorem 8.3.

Exercise 8.14.   Show that the Kalman gain matrix K_{j+1} can be expressed as $P(j+1|j+1)H'(j+1)\Sigma_{j+1}^{-1}$. Hence show that $P(j+1|j) - P(j+1|j+1)$ is positive semi-definite.

Exercise 8.15.   Prove Eq. (8.41).

Exercise 8.16.   Prove Eq. (8.42).

Exercise 8.17.   Prove Eq. (8.48) under the assumptions:

(a) The HIV sero-conversion distribution is a convolution of HIV infection distribution and the window period.

(b) The window period follows an exponential distribution with parameter θ.

References

[1] R. Brookmeyer and M. Gail, *AIDS Epidemiology: A Quantitative Approach*, Oxford University Press, Oxford (1994).

[2] W. Y. Tan, *Stochastic Modeling of the AIDS Epidemiology and HIV Pathogenesis*, World Scientific, Singapore (2000).

[3] W. Y. Tan and Z. Z. Ye, *Estimation of HIV infection and HIV incubation via state space models*, Math. Biosciences **167** (2000) 31–50.

[4] W. Y. Tan and Z. Z. Ye, *Some state space models of HIV epidemic and applications for the estimation of HIV infection and HIV incubation*, Comm. Statistics (Theory and Methods) **29** (2000) 1059–1088.

[5] D. E. Catlin, *Estimation, Control and Discrete Kalman Filter*, Spring-Verlag, New York (1989).

[6] A. Gelb, *Applied Optimal Estimation*, M.I.T. Press, Cambridge, MA (1974).

[7] A. P. Sage and J. L. Melsa, *Estimation Theory with Application to Communication and Control*, McGraw-Hill Book Com., New York (1971).

[8] W. Y. Tan, W. C. Chen and W. Wang, *A generalized state space model of carcinogenesis*, in 2000 International Biometric Conference at UC Berkerly.

[9] R. E. Kalman, *A new approach to linear filter and prediction problems*, J. Basic Eng. **82** (1960) 35–45.

[10] A. C. Harvey, *Forecasting, Structural Time Series Models and the Kalman Filter*, Cambridge University Press, Cambridge (1994).

[11] M. Aoki, *State Space Modeling of Time Series*, Second edition, Spring-Verlag, Berlin (1990).

[12] H. Wu and W. Y. Tan, *Modeling the HIV epidemic: A state space approach*, in: *ASA 1995 Proceeding of the Epidemiology Section*, ASA, Alexdria (1995).

[13] W. Y. Tan and Z. H. Xiang, *A stochastic model for the HIV epidemic in homosexual populations: Estimation of parameters and validation of the model*, in: Simulation in the Medical Sciences, eds. J. G. Anderson and M. Katzper, The Society for Computer Simulation, San Diego (1996).

[14] W. Y. Tan and Z. H. Xiang, *State space models of the HIV epidemic in homosexual populations and some applications*, Math. Biosciences **152** (1998) 29–61.

[15] W. Y. Tan and Z. H. Xiang, *State Space Models for the HIV pathogenesis*, in: *Mathematical Models in Medicine and Health Sciences*, eds. M. A. Horn, G. Simonett and G. Webb, Vanderbilt University Press, Nashville (1998).

[16] W. Y. Tan and Z. H. Xiang, *Estimating and predicting the numbers of T cells and free HIV by non-linear Kalman filter*, in: *Artificial Immune Systems and Their Applications*, ed. DasGupta, Springer-Verlag, Berlin (1998).

[17] W. Y. Tan and Z. H. Xiang, *Modeling the HIV epidemic with variable infection in homosexual populations by state space models*, J. Statist. Inference and Planning **78** (1999) 71–87.

[18] W. Y. Tan and Z. H. Xiang, *A state space model of HIV pathogenesis under treatment by anti-viral drugs in HIV-infected individuals*, Math. Biosciences **156** (1999) 69–94.

[19] W. Y. Tan and W. C. Chen, *Stochastic models of carcinogenesis, Some new insight*, Math Comput. Modeling **28** (1998) 49–71.

[20] W. Y. Tan, C. W. Chen and W. Wang, *Some state space models of carcinogenesis*, in: *Simulation in Medical Sciences*, eds. J. G. Anderson and M. Katzper, The Society of Computer Simulation International, San Diego (1999).

[21] W. Y. Tan, C. W. Chen and W. Wang, *Some multiple pathways state space models of carcinogenesis*, in: *Simulation in Medical Sciences*, eds. J. G. Anderson and M. Katzper, The Society of Computer Simulation International, San Diego (2000).

[22] W. Y. Tan, C. W. Chen and W. Wang, *Stochastic modeling of carcinogenesis by state space models: A New approach*, Math and Computer Modeling **33** (2001) 1323–1345.

[23] H. Wu and W. Y. Tan, *Modeling the HIV epidemic: A state space approach*, Math. Compt. Modeling **32** (2000) 197–215.

[24] G. Satten and Ira M. Jr. Longini, *Markov Chain With Measurement Error: Estimating the "True" Course of Marker of the Progression of Human Immunodeficiency Virus Disease*, Appl. Statist. **45** (1996) 275–309.

[25] J. S. Liu and R. Chen, *Sequential Monte Carlo method for dynamic systems*, Jour. American Statist. Association **93** (1998) 1032–1044.

[26] W. Y. Tan, W. Wang and J. H. Zhu, *A State Space Model for Cancer Tumor Cells under Drug Resistance and Immunostimulation*, in: *Simulation in the Medical Sciences*, eds. J. G. Anderson and M. Katzper, The Society for Computer Simulation, San Diego (2001).

[27] W. Y. Tan and C. C. Brown, *A nonhomogeneous two stages model of carcinogenesis*, Math. Modeling **9** (1987) 631–642.

[28] W. Y. Tan and H. Wu, *Stochastic modeling of the dynamics of CD4+ T cell infection by HIV and some Monte Carlo studies*, Math. Biosciences **147** (1998) 173–205.

[29] K. C. Chu, *Multi-event model for carcinogenesis: A model for cancer causation and prevention, in: Carcinogenesis: A Comprehensive Survey Volume 8: Cancer of the Respiratory Tract-Predisposing Factors*, eds. M. J. Mass, D. G. Ksufman, J. M. Siegfied, V. E. Steel and S. Nesnow, Raven Press, New York (1985).

[30] K. C. Chu, C. C. Brown, R. E. Tarone and W. Y. Tan, *Differentiating between proposed mechanisms for tumor promotion in mouse skin using the multi-vent model for cancer*, Jour. Nat. Cancer Inst. **79** (1987) 789–796.

[31] W. Y. Tan, *Stochastic Models of Carcinogenesis*, Marcel Dekker, New York (1991).

[32] M. P. Little, *Are two mutations sufficient to cause cancer? Some generalizations of the two-mutation model of carcinogenesis of Moolgavkar, Venson and Knudson, and of the multistage model of Armitage and Doll*, Biometrics **51** (1995) 1278–1291.

[33] M. P. Little, *Generalizations of the two-mutation and classical multi-stage models of carcinogenesis fitted to the Japanese atomic bomb survivor data*, J. Radiol. Prot. **16** (1996) 7–24.

[34] M. P. Little, C. R. Muirhead, J. D. Boice Jr. and R. A. Kleinerman, *Using multistage models to describe radiation-induced leukaemia*, J. Radiol. Prot. **15** (1995) 315–334.

[35] M. P. Little, C. R. Muirhead and C. A. Stiller, *Modelling lymphocytic leukaemia incidence in England and Wales using generalizations of the two-mutation model of carcinogenesis of Moolgavkar, Venzon and Knudson*, Statistics in Medicine **15** (1996) 1003–1022.

[36] S. H. Moolgavkar, *A population perspective on multistage carcinogenesis*, in: *Multistage Carcinogenesis*, eds. C. C. Harris, S. Hirohashi, N. Ito, H. C. Pitot, T. Sugimura, M. Terada and J. Yokota, CRC Press, Florida (1992).

[37] K. M. Kinzler and B. Vogelstein, *Colorectal tumors, in The Genetic Basis of Human Cancer*, eds. B. Vogelstein and K. M. Kinzler, McGraw-Hill, New York (1998).

[38] A. Y. Yakovlev and A. D. Tsodikov, *Stochastic Models of Tumor Latency and Their Biostatistical Applications*, World Scientific, Singapore (1996).

[39] G. L. Yang and C. W. Chen, *A stochastic two-stage carcinogenesis model: A new approach to computing the probability of observing tumor in animal bioassays*, Math. Biosci. **104** (1991) 247–258.

[40] W. Y. Tan, *A Stochastic Gompertz birth-death process*, Statist.& Prob. Lett. **4** (1986) 25–28.

[41] A. DeAngelo, *Dichloroacetic acid case study, in: Expert Panel to Evaluate EPA's Proposed Guidelines for Cancer Risk Assessment Using Chloroform and Dichloroacetate as Case Studies Workshop*, ILSI Health and Environmental Sciences Institute, Washington, D.C (1996).

[42] W. Y. Tan, *Note on an extension of Gauss–Markov theorem to multivariate regression models*, SIAM J. Applied Mathematics **20** (1971) 24–29.

[43] CDC, *1993 Revised Classification System for HIV Infection and Expanded Surveillance Case Definition for AIDS Among Adolescents and Adults*, MMWR **41** (1992), No. RR17.

[44] N. T. J. Bailey, *Estimating HIV incidence & AIDS projections: Prediction and validation in the public health modelling of HIV/AIDS*, Statistics in Medicine **13** (1994) 1933–1944.

[45] H. W. Hethcote and J. W. Van Ark, *Modeling HIV transmission and AIDS in the United States*, Lecture Notes in Biomath. Springer-Verlag, Berlin (1992).

[46] G. F. Lemp, S. F. Payne, G. W. Rutherford, and et al., *Projections of AIDS morbidity and mortality in San Francisco*, Jour. of Amer. Medic. Assoc. **263** (1989) 1497–1501.

[47] U. S. Bureau of the Census, *Statistical Abstract of the United States: 108th edition*, Washington, D. C. (1987).

[48] W. Y. Tan, S. C. Tang and S. R. Lee, *Estimation of HIV Seroconversion and Effects of Age in San Francisco Homosexual Populations*, Jour. Applied Statistics **25** (1998) 85–102.

[49] C. R. Jr. Horsburgh, C. Y. Qu, I. M. Jason, *et al., Duration of human immuodeficiency virus infection before detection of antibody*, Lancet **2** (1989) 637–640.

Chapter 9

Some General Theories of State Space Models and Applications

In Chap. 8, I have illustrated how to develop state space models in some cancer and AIDS models. As a continuation, in this chapter I proceed to give some general theories and illustrate its application.

9.1. Some Classical Theories of Linear State Space Models with Continuous-Time Stochastic System Model

In the previous chapter I have given some general theories for discrete-time linear state space models. Because most of the state space models in HIV pathogenesis and in carcinogenesis have continuous-time for the stochastic system models, in this section I will present some general theories for the classical analysis of continuous-time state space models. Thus, we consider a state space model with stochastic system model given by Eq. (9.1) below and with observation model given by Eq. (9.2) below, where the matrices $\{F(t), H(j)\}$ are given non-stochastic transition matrices:

$$\frac{d}{dt}\underset{\sim}{X}(t) = F(t)\underset{\sim}{X}(t) + \underset{\sim}{\epsilon}(t), \quad t \geq 0; \tag{9.1}$$

$$\underset{\sim}{Y}(j) = H(j)\underset{\sim}{X}(t_j) + \underset{\sim}{e}(j), \quad j = 1,\ldots,n. \tag{9.2}$$

In the above equations, it is assumed that the elements of the random noises $\underset{\sim}{\epsilon}(t)$ and of the random measurement errors $\underset{\sim}{e}(j)$ have expected values

0 and are independently distributed of one another. The covariance matrices of $\underset{\sim}{\epsilon}(t)dt$ and $\underset{\sim}{e}(j)$ are given by $\mathrm{Cov}[\underset{\sim}{\epsilon}(t)dt, \underset{\sim}{\epsilon}(\tau)dt] = \delta(t-\tau)V(t)dt + o(dt)$ and $\mathrm{Var}[\underset{\sim}{e}(j)] = \Sigma(j)$, where $\delta(x)$ is the Dirac's δ function.

Let $\underset{\sim}{\hat{X}}(t|k)$ denote a predictor (or estimator) of $\underset{\sim}{X}(t)$ given data $D(k) = \{\underset{\sim}{Y}(u), u = 1, \ldots, k\}$ with residual $\underset{\sim}{\hat{\epsilon}}(t|k) = \underset{\sim}{\hat{X}}(t|k) - \underset{\sim}{X}(t)$, where $k = 1, \ldots, n$. Denote the covariance matrix of $\underset{\sim}{\hat{\epsilon}}(t|k)$ by $Q(t|k)$.

To derive the optimal $\underset{\sim}{\hat{X}}(t|k)$, write $\underset{\sim}{X}(t_j) = \underset{\sim}{u}(j)$ and $\underset{\sim}{\hat{X}}(t_j|k) = \underset{\sim}{\hat{u}}(j|k)$, $Q(t_j|k) = P(j|k)$ for $k = 1, \ldots, n$ and $Q(t_{j+1}|j) = P(j+1|j)$. Then, we have the following two theorems which provide the optimal estimator and predictor of $\underset{\sim}{X}(t)$ given $D(k) = \{\underset{\sim}{Y}(u), u = 1, \ldots, k\}$.

Theorem 9.1. *Assume that for $t \geq t_j$, there exists a non-singular matrix $R(t, t_j)$ satisfying the conditions $\lim_{t \to t_j} R(t, t_j) = I_p$ and $\frac{d}{dt}R(t, t_j) = -F(t)R(t, t_j) = -R(t, t_j)F(t)$; see Remark 9.1. Then, given $\underset{\sim}{\hat{u}}(0|0)$ as an unbiased estimator of $\underset{\sim}{X}(0)$ with $P(0|0)$ as the covariance matrix of $\underset{\sim}{\hat{\epsilon}}(0|0)$, for $t_j \leq t \leq t_{j+1}$, the BLUE $\underset{\sim}{\hat{X}}(t|j)$ of $\underset{\sim}{X}(t)$ given data $D(j)$ are given by the following recursive equations:*

(i) For $t_j \leq t < t_{j+1}$, $j = 0, 1, \ldots, n$ ($t_0 = 0, t_{n+1} = \infty$), $\underset{\sim}{\hat{X}}(t|j)$ satisfies the following differential equation with boundary condition $\lim_{t \to t_j} \underset{\sim}{\hat{X}}(t|j) = \underset{\sim}{\hat{u}}(j|j)$:

$$\frac{d}{dt}\underset{\sim}{\hat{X}}(t|j) = F(t)\underset{\sim}{\hat{X}}(t|j), \tag{9.3}$$

where for $j > 0$, $\underset{\sim}{\hat{u}}(j|j)$ is given in (iii).

(ii) For $t_j \leq t < t_{j+1}$, $j = 0, 1, \ldots, n$, $Q(t|j)$ satisfies the following differential equation with boundary condition $\lim_{t \to t_j} Q(t|j) = P(j|j)$:

$$\frac{d}{dt}Q(t|j) = F(t)Q(t|j) + Q(t|j)F(t)' + V(t), \tag{9.4}$$

where for $j > 0$, $P(j|j)$ is given in (iii).

(iii) For $j = 0, 1, \ldots, n$, put $G(j+1, j) = R^{-1}(t_{j+1}, t_j)$. Then, $\underset{\sim}{\hat{u}}(j+1|j), P(j+1|j), \underset{\sim}{\hat{u}}(j+1|j+1)$ and $P(j+1|j+1)$ are given by the following recurssive equations (see Remark 9.2):

$$\underset{\sim}{\hat{u}}(j+1|j) = G(j+1, j)\underset{\sim}{\hat{u}}(j|j), \tag{9.5}$$

$$\hat{\underset{\sim}{u}}\,(j+1|j+1) = \hat{\underset{\sim}{u}}\,(j+1|j) + K_{j+1}\{\underset{\sim}{Y}\,(j+1) - H(j+1)\hat{\underset{\sim}{u}}\,(j+1|j)\}, \quad (9.6)$$

$$P(j+1|j) = G(j+1,j)P(j|j)G'(j+1,j) + V_\xi(j+1), \quad (9.7)$$

and

$$P(j+1|j+1) = [I - K_{j+1}H(j+1)]P(j+1|j), \quad (9.8)$$

where

$$V_\xi(j+1) = \int_{t_j}^{t_{j+1}} R(t_{j+1},x)V(x)R'(t_{j+1},x)dx,$$

and

$$K_{j+1} = P(j+1|j)H(j+1)'[H(j+1)P(j+1|j)H'(j+1) + \Sigma(j+1)]^{-1}.$$

In the literature the procedures in Theorem 9.1 have been referred to as the forward filtering.

Remark 9.1. In most cases, there exists a non-singular $p \times p$ matrix $R(t,t_j)$ satisfying the conditions $\lim_{t\to t_j} R(t,t_j) = I_p$ and $\frac{d}{dt}R(t,t_j) = -F(t)R(t,t_j) = -R(t,t_j)F(t)$ for all $t \geq t_j$. Observe that if $F(t) = F$, then

$$R(t,t_j) = \exp[-F(t-t_j)],$$

where

$$\exp[-F(t-t_j)] = \sum_{i=0}^{\infty} \frac{1}{i!}(t-t_j)^i(-F)^i$$

provided that the series converges.

If $F(t) = F_i$ for $s_i, \leq t < s_{i+1}$, then

$$R(t,t_j) = \exp[-F_k(t-s_k)]\left\{ \prod_{l=u+1}^{k-1} \exp(-F_{k-l+u}\tau_{k-l+u}) \right\} \exp[-F_u(s_{u+1} - t_j)]$$

for $s_k \leq t < s_{k+1}$ and $s_u \leq t_j < s_{u+1}$ where $\tau_l = s_{l+1} - s_l$.

In general, if elements of $F(t)$ are bounded, then one may write $R(t,t_j)$ as

$$R(t,t_j) = \lim_{\Delta\to 0} \prod_{i=0}^{n_j-1} [I_p - F(t_j + i\Delta t)\Delta t],$$

where for fixed $\Delta > 0$, n_j is the largest integer satisfying $(n_j + 1)\Delta > t - t_j \geq n_j\Delta$.

Theorem 9.2. *Suppose that for $t_j \leq t < t_{j+1}$ with $0 \leq j \leq n$, there exists a non-singular matrix $R(t, t_j)$ satisfying the conditions $\lim_{t \to t_j} R(t, t_j) = I_p$ and $\frac{d}{dt} R(t, t_j) = -F(t)R(t, t_j) = -R(t, t_j)F(t)$. Then, given $\hat{\underset{\sim}{X}}(0|0) = \hat{\underset{\sim}{u}}(0|0)$ as an unbiased estimator of $\underset{\sim}{X}(0)$ with $P(0|0)$ as the covariance matrix of the residual $\hat{\underset{\sim}{\epsilon}}(0|0)$, the BLUE $\hat{\underset{\sim}{X}}(t|n)$ of $\underset{\sim}{X}(t)$ given data $D(n)$ are given by the following recursive equations:*

(i) For $t_j \leq t < t_{j+1}$, $j = 0, 1, \ldots, n$, $\hat{\underset{\sim}{X}}(t|n)$ satisfies the following differential equation with boundary condition $\lim_{t \to t_j} \hat{\underset{\sim}{X}}(t|n) = \underset{\sim}{u}(j|n)$:

$$\frac{d}{dt} \hat{\underset{\sim}{X}}(t|n) = F(t) \hat{\underset{\sim}{X}}(t|n),$$ (9.9)

where $\hat{\underset{\sim}{u}}(j|n)$ is given in (iii).

(ii) For $t_j \leq t < t_{j+1}$, $j = 0, 1, \ldots, n$, $Q(t|n)$ satisfies the following differential equation with boundary condition $\lim_{t \to t_j} Q(t|n) = P(j|n)$:

$$\frac{d}{dt} Q(t|n) = F(t)Q(t|n) + Q(t|n)F(t)' + V(t),$$ (9.10)

where $P(j|n)$ is given in (iii).

(iii) For $j = 0, 1, \ldots, n$, put $G(j + 1, j) = R^{-1}(t_{j+1}, t_j)$. Then, $\hat{\underset{\sim}{u}}(j|n)$ and $P(j|n)$ are given by the following recursive equations respectively (see Remark 9.2):

$$\hat{\underset{\sim}{u}}(j|n) = \hat{\underset{\sim}{u}}(j|j) + A_j\{\hat{\underset{\sim}{u}}(j + 1|n) - \hat{\underset{\sim}{u}}(j + 1|j)\}$$ (9.11)

and

$$P(j|n) = P(j|j) - A_j\{P(j + 1|j) - P(j + 1|n)\}A_j',$$ (9.12)

where

$$A_j = P(j|j)G(j + 1, j)'P^{-1}(j + 1|j).$$

In the literature, the procedures in Theorem 9.2 have been referred to as the smoothing procedures. This has also been referred to as the backward filtering.

Notice that results of the above theorems are basically results from linear least square methods. Hence, results of the above two theorems may be considered as extensions of the Gauss-Markov theorems in linear least square models; see [1].

Proof of Theorems 9.1 and 9.2. As in Subsec. 8.4.3, the basic trick to prove Theorems 9.1 and 9.2 is that at the time points $t = t_j$ $(j = 0, 1, \ldots, n)$, we reduce the model to a standard discrete-time linear Kalman filter model and then apply results from Theorems 8.2 and 8.3; for time t in $t_j < t < t_{j+1}$, since observed data are not available, we use stochastic system equations and take conditional expectations given data.

Now, if $R(t, t_j)$ is a $p \times p$ matrix satisfying the conditions $\lim_{t \to t_j} R(t, t_j) = I_p$ and $\frac{d}{dt} R(t, t_j) = -F(t) R(t, t_j) = -R(t, t_j) F(t)$ for all $t \geq t_j$. Then in terms of mean squared error, the solution of Eq. (9.1) is

$$\underset{\sim}{X}(t) = R^{-1}(t, t_j) \underset{\sim}{X}(t_j) + \underset{\sim}{\eta}(t, t_j), \tag{9.13}$$

where $\underset{\sim}{\eta}(t, t_j)$ is given by:

$$\underset{\sim}{\eta}(t, t_j) = R^{-1}(t, t_j) \int_{t_j}^{t} R(x, t_j) \underset{\sim}{\epsilon}(x) dx.$$

Also, the above solution of Eq. (9.1) is unique almost surely (see Remark 9.2).

Obviously, $E[\underset{\sim}{\eta}(t, t_i)] = 0$ and the covariance matrix of $\underset{\sim}{\eta}(t, t_j)$ is

$$V_0(t, t_j) = R^{-1}(t, t_j) \left\{ \int_{t_j}^{t} R(x, t_j) V(x) R'(x, t_j) dx \right\} \{R^{-1}(t, t_j)\}'.$$

Furthermore, the $\underset{\sim}{\eta}(t, t_i)$'s are independently distributed of the measurement error $\underset{\sim}{e}(j)$'s and are uncorrelated with $\underset{\sim}{X}(t)$.

Put $\underset{\sim}{X}(t_j) = \underset{\sim}{u}(j), R^{-1}(t_{j+1}, t_j) = G(j + 1, j), \underset{\sim}{\eta}(t_{j+1}, t_j) = \underset{\sim}{\xi}(j + 1)$ and $V_0(t_{j+1}, t_j) = V_\xi(j + 1)$. Obviously, $\underset{\sim}{\xi}(j + 1)$ are uncorrelated with $\underset{\sim}{\xi}(i + 1)$ if $i \neq j$. Then, we have the following state space model for $\underset{\sim}{u}(j)$, $j = 0, \ldots, n - 1$:

$$\underset{\sim}{u}(j + 1) = G(j + 1, j) \underset{\sim}{u}(j) + \underset{\sim}{\xi}(j + 1) \tag{9.14}$$

and

$$\underset{\sim}{Y}(j+1) = H(j+1)\underset{\sim}{u}(j+1) + \underset{\sim}{e}(j+1). \tag{9.15}$$

The above is a standard linear discrete-time Kalman filter model as given in Subsec. 8.4.1. Part (iii) of Theorems 9.1 and 9.2 then follow from basic results as given in Theorems 8.2 and 8.3 respectively.

To prove (i) of Theorems 9.1 and 9.2, we put $\underset{\sim}{\hat{X}}(t|j) = R^{-1}(t,t_j)\underset{\sim}{\hat{u}}(j|j)$ $(t \geq t_j)$ and $\underset{\sim}{\hat{X}}(t|n) = R^{-1}(t,t_j)\underset{\sim}{\hat{u}}(j|n)$ $(0 < j \leq n)$. Then, since $\frac{d}{dt}R(t,t_j) = -F(t)R(t,t_j) = -R(t,t_j)F(t)$, obviously, $\underset{\sim}{\hat{X}}(t|j)$ and $\underset{\sim}{\hat{X}}(t|n)$ satisfy the equation in (i) of Theorems 9.1 and 9.2 respectively. To prove (i) of Theorems 9.1 and 9.2, it remains to show that these estimators are the optimal estimators. Writing the residual $\underset{\sim}{\hat{\epsilon}}(t|j) = \underset{\sim}{\hat{X}}(t|j) - X(t)$ as

$$\underset{\sim}{\hat{\epsilon}}(t|j) = R^{-1}(t,t_j)\underset{\sim}{\hat{\epsilon}}_u(j|j) - \underset{\sim}{\eta}(t,t_j),$$

where $\underset{\sim}{\hat{\epsilon}}_u(j|j) = \underset{\sim}{\hat{u}}(j|j) - u(j)$, the covariance matrix of $\underset{\sim}{\hat{\epsilon}}(t|j)$ is

$$Q(t|t_j) = R^{-1}(t,t_j)P(j|j)\{R^{-1}(t,t_j)\}' + V_0(t,t_j).$$

Let $\underset{\sim}{\hat{X}}^{(*)}(t|j)$ be any other linear unbiased estimator of $\underset{\sim}{X}(t)$ given data $D(j)$, where $t_j < t < t_{j+1}$. Then, by Eq. (9.13), $\underset{\sim}{\hat{X}}^{(*)}(t|j) = R^{-1}(t,t_j)\underset{\sim}{\hat{u}}^{(*)}(j|j)$, where $\underset{\sim}{\hat{u}}^{(*)}(j|j)$ is an unbiased estimator of $\underset{\sim}{X}(t_j) = \underset{\sim}{u}(j)$ and is a linear function of elements of $D(j)$. Let $Q^{(*)}(t|j)$ be the covariance matrix of $\underset{\sim}{\hat{\epsilon}}^{(*)}(t|j) = \underset{\sim}{\hat{X}}^{(*)}(t|j) - \underset{\sim}{X}(t)$ and $P^{(*)}(j|j)$ the covariance matrix of $\underset{\sim}{\hat{\epsilon}}^{(*)}_u(j|j) = \underset{\sim}{\hat{u}}^{(*)}(j|j) - \underset{\sim}{u}(j)$. Then,

$$Q^{(*)}(t|j) = R^{-1}(t,t_j)P^{(*)}(j|j)\{R^{-1}(t,t_j)\}' + V_0(t,t_j).$$

But by (iii) of Theorem 9.1, $P^{(*)}(j|j) - P(j|j)$ is positive semi-definite. It follows that $Q^{(*)}(t|j) - Q(t,t_j) = R^{-1}(t,t_j)\{P^{(*)}(j|j) - P(j|j)\}\{R^{-1}(t,t_j)\}'$ is positive semi-definite. This shows that for $t \geq t_j$, $\underset{\sim}{\hat{X}}(t|j)$ is the BLUE of $\underset{\sim}{X}(t)$ given $D(j)$. This proves (i) of Theorem 9.1. Similarly one proves (i) of Theorem 9.2

To proves (ii) of Theorems 9.1 and 9.2, notice that the covariance matrix of the residuals $\underset{\sim}{\hat{\epsilon}}(t|j) = \underset{\sim}{\hat{X}}(t|j) - \underset{\sim}{X}(t)$ and $\underset{\sim}{\hat{\epsilon}}(t|n) = \underset{\sim}{\hat{X}}(t|n) - \underset{\sim}{X}(t)$ are given by the following two formula respectively:

$$Q(t|j) = R^{-1}(t,t_j)P(j|j)\{R^{-1}(t,t_j)\}' + V_0(t,t_j)\,,$$

$$Q(t|n) = R^{-1}(t,t_j)P(j|n)\{R^{-1}(t,t_j)\}' + V_0(t,t_j)\,.$$

Noting that $R(t,t) = I_p$ and $\frac{d}{dt}R(t,t_i) = -F(t)R(t,t_i) = -R(t,t_i)F(t)$, by taking derivatives on both sides of the above equation, it is obvious that (ii) of Theorems 9.1 and 9.2 hold. This proves (ii) of Theorems 9.1 and 9.2.

Remark 9.2. It can be shown that if the solutions of Eq. (9.1) exist for each given $\underset{\sim}{\epsilon}(t)$, then the solution is unique almost surely (i.e. with probability one). It follows that the estimators $\hat{\underset{\sim}{u}}(j|k)$ are unique almost surely. Because the proof is quite complicated, we will not present it here. We only note that the least square solutions are unique almost surely.

9.2. The Extended State Space Models with Continuous-Time Stochastic System Model

As shown in examples in Secs. 8.2 and 8.3, in many state space models of HIV pathogenesis and carcinogenesis, either the stochastic system model or the observation model or both are non-linear models. In these cases, one may express the stochastic system equations and the observation equation respectively as:

$$d\underset{\sim}{X}(t) = \underset{\sim}{f}[\underset{\sim}{X}(t)]dt + \underset{\sim}{\epsilon}(t)dt, \quad t_j \le t < t_{j+1}, j = 0, 1, \ldots, n; \quad (9.16)$$

$$\underset{\sim}{Y}(j) = \underset{\sim}{h}[\underset{\sim}{X}(t_j)] + \underset{\sim}{e}_j, \quad j = 1, \ldots, n. \quad (9.17)$$

Given that the above state space model can be closely approximated by the extended Kalman filter model, then one may use the procedures given in Theorems 9.1 and 9.2 to derive approximate optimal estimators of $\underset{\sim}{X}(t)$. The basic idea is to expand $\underset{\sim}{f}[\underset{\sim}{X}(t)]$ and $\underset{\sim}{h}[\underset{\sim}{X}(t_j)]$ in Taylor series with respect to the optimal estimates up to the first order and then take conditional expectation given data. For state space models of HIV pathogenesis, it has been shown by Tan and Xiang [2–4] through Monte carlo simulation studies that this approximation is indeed very close. This is expected and justified by the observation that the HIV infection rates are usually very small.

Notice that if one expands $\underset{\sim}{f}[\underset{\sim}{X}(t)]$ and $\underset{\sim}{h}[\underset{\sim}{X}(t_j)]$ in Taylor series with respect to the optimal estimates up to the first order and then take conditional expectation given data, the models reduce to linear models given in the previous section. Then one may use results of the previous two theorems to derive optimal estimates. Since the basic procedures are practically the same as those in the above two theorems, we summarize without proof the results in the next two theorems. We will leave the proofs of these theorems as an exercise; see Exercise 9.1.

Theorem 9.3. *Suppose that elements of $\underset{\sim}{h}[\underset{\sim}{X}(t)]$ and elements of $\underset{\sim}{f}[\underset{\sim}{X}(t)]$ have continuous and bounded first derivatives and can be closely approximated by Taylor series expansion up to first order. Then, given $\hat{\underset{\sim}{X}}(0|0) = \hat{\underset{\sim}{u}}(0|0)$ as an unbiased estimator of $\underset{\sim}{X}(0)$ with $P(0|0)$ as the covariance matrix of the residual $\hat{\underset{\sim}{\epsilon}}(0|0)$, for $t_j \leq t$, the BLUE $\hat{\underset{\sim}{X}}(t|j)$ of $\underset{\sim}{X}(t)$ given data $D(j)$ are closely approximated by the following recursive equations:*

(i) For $t_j \leq t < t_{j+1}$, $j = 0, 1, \ldots, n$ $(t_0 = 0, t_{n+1} = \infty)$, $\hat{\underset{\sim}{X}}(t|j)$ satisfies the following differential equation with boundary condition $\lim_{t \to t_j} \hat{\underset{\sim}{X}}(t|j) = \underset{\sim}{u}(j|j)$:

$$\frac{d}{dt}\hat{\underset{\sim}{X}}(t|j) = \underset{\sim}{f}[\hat{\underset{\sim}{X}}(t|j)], \qquad (9.18)$$

where for $j > 0$, $\hat{\underset{\sim}{X}}(t_j|j) = \hat{\underset{\sim}{u}}(j|j)$ is given in (iii).

(ii) For $t_j \leq t < t_{j+1}$, $j = 0, 1, \ldots, n$, $Q(t|j)$ satisfies the following differential equation with boundary condition $\lim_{t \to t_j} Q(t|j) = P(j|j)$:

$$\frac{d}{dt}Q(t|j) = F(t|j)Q(t|j) + Q(t|j)F(t|j)' + V(t), \qquad (9.19)$$

where $F(t|j) = \left(\dfrac{\partial}{\partial \underset{\sim}{X}(t)}\underset{\sim}{f}[\underset{\sim}{X}(t)]\right)_{\underset{\sim}{X}(t) = \hat{\underset{\sim}{X}}(t|j)}$ with $\lim_{t \to t_j} Q(t|j) = P(j|j)$ being given in (iii).

(iii) Denote by $\lim_{t \to t_{j+1}} \hat{\underset{\sim}{X}}(t|j) = \hat{\underset{\sim}{u}}(j+1|j)$ and $\lim_{t \to t_{j+1}} Q(t|j) = P(j+1|j)$. Let $H_0(j+1) = \left(\dfrac{\partial}{\partial \underset{\sim}{X}(t)}\underset{\sim}{h}[\underset{\sim}{X}(t)]\right)_{\underset{\sim}{X}(t) = \hat{\underset{\sim}{u}}(j|j)}$. Then $\hat{\underset{\sim}{u}}(j+1|j+1)$ and $P(j+1|j+1)$ are given by the following recursive equations respectively:

$$\hat{\underset{\sim}{u}}(j+1|j+1) = \hat{\underset{\sim}{u}}(j+1|j) + K_{j+1}\{\underset{\sim}{Y}(j+1) - \underset{\sim}{h}[\hat{\underset{\sim}{u}}(j+1|j)]\}, \qquad (9.20)$$

and

$$P(j+1|j+1) = [I - K_{j+1}H_0(j+1)]P(j+1|j), \qquad (9.21)$$

and

$$K_{j+1} = P(j+1|j)H_0(j+1)'[H_0(j+1)P(j+1|j)H_0(j+1)' + \Sigma(j+1)]^{-1}.$$

To implement the procedures in Theorems 9.1 and/or 9.3, one starts with $\hat{\underset{\sim}{X}}(0|0) = \hat{\underset{\sim}{u}}(0|0)$ and $P(0|0)$. Then by (i) and (ii) of the respective theorems, one derive $\hat{\underset{\sim}{X}}(t|0)$ and $Q(t|0)$ for $t_0 \leq t \leq t_1$ and derives $\hat{\underset{\sim}{u}}(1|1)$ and $P(1|1)$ by (iii) of the respective theorems. Repeating these procedures one may derive $\hat{\underset{\sim}{X}}(t|j)$ and $Q(t|j)$ for $t_j \leq t < t_{j+1}$, $j = 0, 1, \ldots, n$. These procedures are referred to as forward filtering procedures.

Theorem 9.4.   *Suppose that elements of $\underset{\sim}{h}[\underset{\sim}{X}(t)]$ and elements of $\underset{\sim}{f}[\underset{\sim}{X}(t)]$ have continuous and bounded first derivatives and can be closely approximated by Taylor series expansions up to first order. Then, given $\hat{\underset{\sim}{X}}(0|0) = \hat{\underset{\sim}{u}}(0|0)$ as an unbiased estimator of $\underset{\sim}{X}(0)$ with $P(0|0)$ as the covariance matrix of the residual $\hat{\underset{\sim}{\epsilon}}(0|0)$, for $t_j \leq t < t_{j+1}$ with $0 \leq j \leq n$, the BLUE $\hat{\underset{\sim}{X}}(t|n)$ of $\underset{\sim}{X}(t)$ given data $D(n)$ are closely approximated by the following recursive equations respectively:*

(i) For $t_j \leq t < t_{j+1}$, $j = 0, 1, \ldots, n, j \leq n$ $(t_0 = 0, t_{n+1} = \infty)$, $\hat{\underset{\sim}{X}}(t|n)$ satisfies the following differential equation with boundary condition $\lim_{t \to t_j} \hat{\underset{\sim}{X}}(t|n) = \underset{\sim}{u}(j|n)$:

$$\frac{d}{dt}\hat{\underset{\sim}{X}}(t|n) = \underset{\sim}{f}[\hat{\underset{\sim}{X}}(t|n)], \qquad (9.22)$$

where for $j > 0$, $\hat{\underset{\sim}{X}}(t_j|n) = \hat{\underset{\sim}{u}}(j|n)$ is given in (iii).

(ii) For $t_j \leq t < t_{j+1}$, $j = 0, 1, \ldots, n, j \leq n$, $Q(t|j)$ satisfies the following differential equation with boundary condition $\lim_{t \to t_j} Q(t|n) = P(j|n)$:

$$\frac{d}{dt}Q(t|n) = F(t|n)Q(t|n) + Q(t|n)F(t|n)' + V(t), \qquad (9.23)$$

where $F(t|n) = \left(\frac{\partial}{\partial \hat{\underset{\sim}{X}}(t)}\underset{\sim}{f}[\underset{\sim}{X}(t)]\right)_{\underset{\sim}{X}(t)=\hat{\underset{\sim}{X}}(t|n)}$ with $\lim_{t \to t_j} Q(t|n) = P(j|n)$ being given in (iii).

(iii) *For $j = 1, \ldots, n$, $\hat{\underset{\sim}{u}}(j|n)$ and $P(j|n)$ are given by the following recursive equations respectively:*

$$\hat{\underset{\sim}{u}}(j|n) = \hat{\underset{\sim}{u}}(j|j) + A_j\{\hat{\underset{\sim}{u}}(j+1|n) - \hat{\underset{\sim}{u}}(j+1|j)\} \tag{9.24}$$

and

$$P(j|n) = P(j|j) - A_j\{P(j+1|j) - P(j+1|n)\}A'_j, \tag{9.25}$$

where, with $(n_j + 1)\Delta > t_{j+1} - t_j \geq n_j\Delta$, $G(j+1,j) = \lim_{\Delta \to 0} \prod_{s=0}^{n_j-1}[I + F(t_j + s\Delta|j)\Delta]$, and

$$A_j = P(j|j)G(j+1,j)'P^{-1}(j+1|j)$$

and where $P(j|j)$ and $P(j+1|j)$ are given in Theorem 9.3. (To compute $G(j+1,j)$, we assume Δ to be a very small number such as one hour and then approximate $G(j+1,j)$ by $G(j+1,j) \cong \prod_{s=0}^{n_j-1}[I + F(t_j + s\Delta|j)\Delta]$.)

To implement the procedures in Theorems 9.2 or 9.4 to derive $\hat{\underset{\sim}{X}}(t|n)$ for $t_j \leq t < t_{j+1}$, one first derives $\hat{\underset{\sim}{X}}(t|j)$ for $t_j \leq t$ by using formulas in Theorems 9.1 or 9.3 (forward filtering). Then one goes backward from n to 1 to derive the estimates given data $D(n)$ for $t_j \leq t < t_{j+1}$ by using formulas in Theorems 9.2 or 9.4. These are the backward filtering procedures.

9.3.   Estimation of CD4$^{(+)}$ T Cell Counts and Number of HIV in Blood in HIV-Infected Individuals

As an application of Theorems 9.1–9.4, we consider a hemophilia patient from NCI/NIH Multi-Center Hemophilia Cohort Studies. This patient contracted HIV at age 11 through a blood transfusion. The CD4$^{(+)}$ T-Cell counts for this patient were taken at 16 occasions and are given in Table 9.1. (Notice that $|t_j - t_u| > 6$ months for $j \neq u$, see Table 9.1.)

For this individual, the estimates of the parameters are: $s = 10/\text{day}/\text{mm}^3$, $\theta = 0.7571/\text{day}/\text{mm}^3$, $\gamma = 0.03/\text{day}/\text{mm}^3$, $\mu_1 = \mu_2 = 0.02/\text{day}$, $\mu_3 = 0.3/\text{day}$, $\mu_V = 4.6/\text{day}$, $T_{\max} = 1656.16/\text{day}/\text{mm}^3$, $N_0 = .3351$, $\beta_1 = 0.794378E - 05$, $\beta_2 = 49.9061E + 03/$, $k_1 = 1.3162 \times 10^{-6}/\text{day}/\text{mm}^3$, $k_2 = 7.9873 \times 10^{-3}/\text{day}/\text{mm}^3$, $\omega = 0.2$, where $N(t) = N_0 \exp(-\beta_1 t) + \beta_2(1 - \exp(-\beta_1 t))$ and $s(t) = s/[\theta + V(t)]$. In these estimates the values of the parameters

$s, \gamma, \mu_i, i = 1, 2, 3, \mu_V, \omega$ were taken from the literature [5–9] but the values of the parameters $k_1, k_2, N_0, \beta_1, \beta_2, \theta, T_{\max}$ were estimated by nonlinear least square procedures [10, 11].

Using these estimates and the methods given in Theorems 9.3–9.4, we have estimated the number of free HIV, the total number of CD4$^{(+)}$ T cells and

Table 9.1. The Observed Total Numbers of CD4$^{(+)}$ T Cells of an HIV Infected Patient.

Days after Infection	T Cell Counts (mm^3)	Days after Infection	T Cell Counts (mm^3)
1152	1254	2952	686
1404	1005	3096	357
1692	1022	3312	440
1872	1105	3456	584
2124	372	3708	508
2376	432	3852	583
2592	520	4032	328
2736	660	4212	345

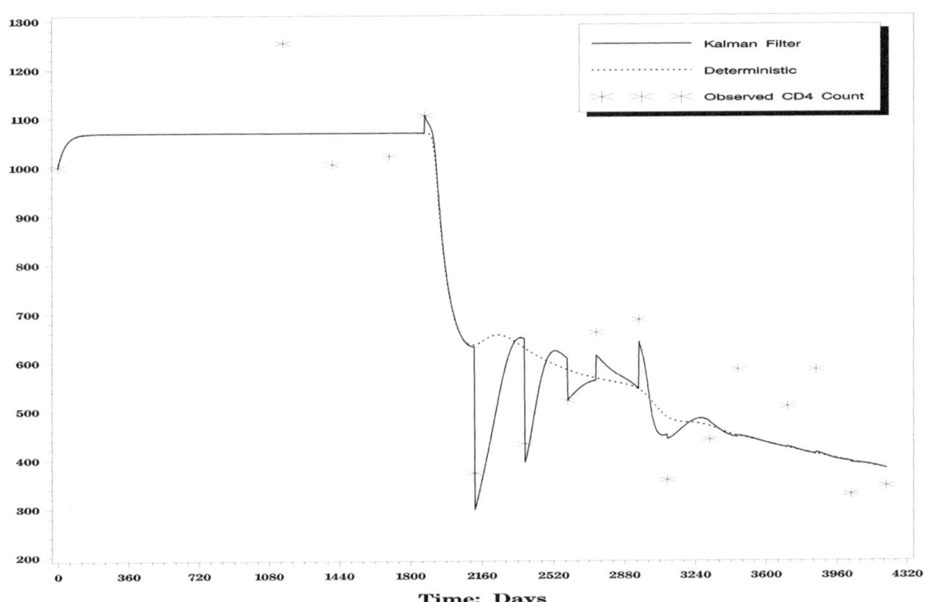

Fig. 9.1.   Plots showing the observed total number of CD4$^{(+)}$ T cell counts and the estimates by the Kalman filter method and by deterministic model.

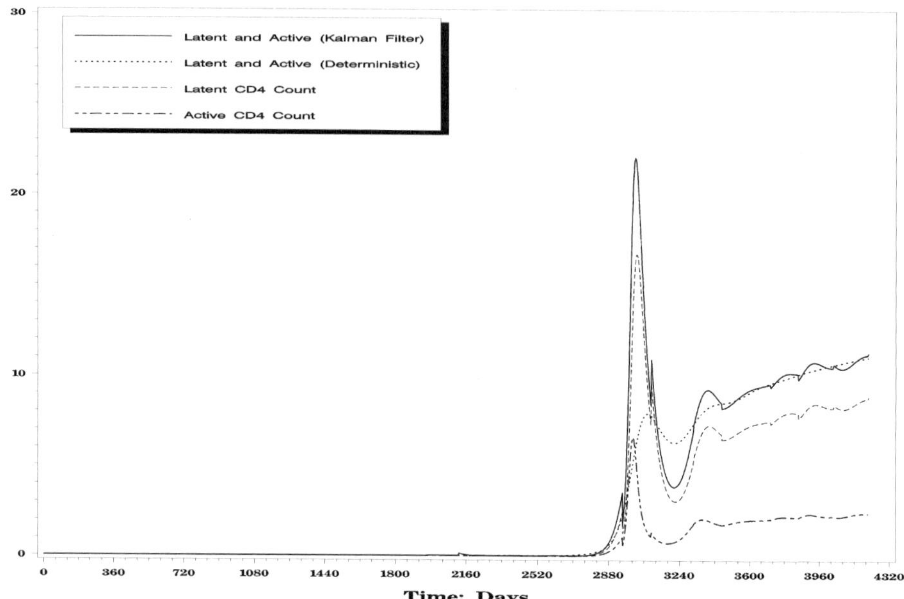

Fig. 9.2. Plots showing the Kalman filter estimates of the numbers of HIV infected T cells and the estimates by the deterministic model.

the numbers of T_i cells $(i = 1, 2, 3)$ at different times. Plotted in Fig. 9.1 are the estimated total number of $CD4^{(+)}$ T-cell counts together with the observed $CD4^{(+)}$ T-cell counts. Plotted in Figs. 9.2–9.3 are the estimated numbers of infected $CD4^{(+)}$ T cells and free HIV respectively.

From Fig. 9.1, we observed that the Kalman filter estimates tried to trace the observed numbers. On the other hand, if we ignore the random variations and simply use the deterministic model to fit the data, then the fitted numbers would draw a smooth line between the observed numbers. This is not surprising since the deterministic model would yield continuous solutions.

From Figs. 9.2–9.3, we observed that before days 2,800 since HIV infection, the estimated numbers of the HIV infected T cells and free HIV are very small (almost zero) by both the Kalman filter methods and the deterministic model. After days 2800 since HIV infection, however, there are significant differences between the Kalman filtering estimates and the estimates by the deterministic model. It appears that for the Kalman filter estimates, both the numbers of the HIV-infected $CD4^{(+)}$ T cells and free HIV first increase very sharply, reach a maximum at about 3,000 days since HIV infection and then decrease

Number of Free HIV

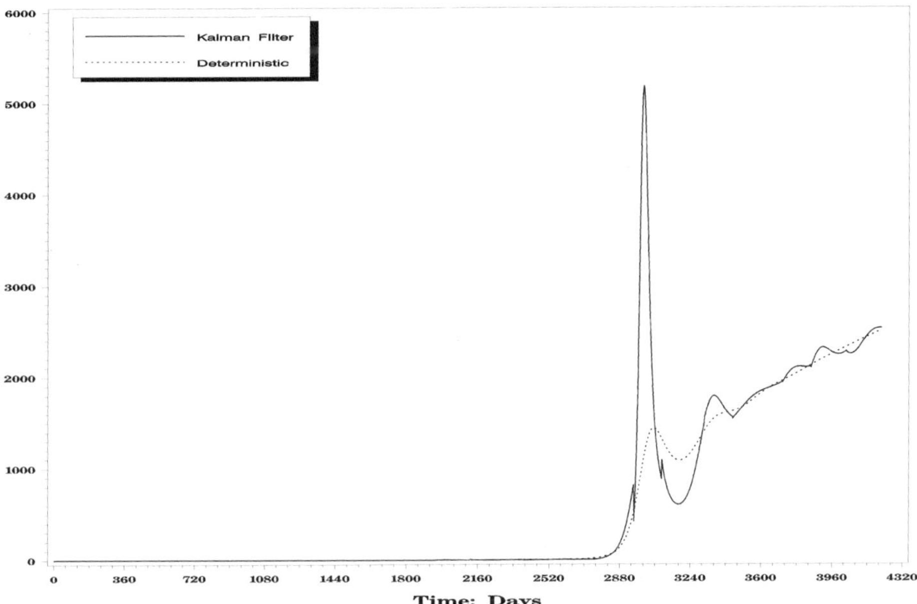

Fig. 9.3.   Plots showing Kalman filter estimates of the number of free HIV together with the estimates by the deterministic model.

to a very lower level around 3240 days since infection; after that the curves increase steadily. These results appear to be consistent with the observation by Phillips [8], suggesting an acute infectivity phase early on. On the other hand, the deterministic results show only moderately increasing curves without an initial high peak; also no decreasing pattern has been observed. From Figs. 9.2–9.3, we also observed that the estimated curves of the infected T cells and free HIV show similar pattern, suggesting a positive correlation between infected $CD4^{(+)}$ T cells and free HIV. This is not surprising since free HIV are released mainly from the death of T_3 cells.

9.4.   A General Bayesian Procedure for Estimating the Unknown Parameters and the State Variables by State Space Models Simultaneously

To apply the theorems of Chap. 8 and Secs. 9.1–9.2 to estimate the state variables, one would need to assume the parameter values as known or given. In many problems, however, the parameters may be unknown. In such cases, one would need to estimate simultaneously the unknown parameters and the state variables. In this section, we introduce some of these procedures for the model given in Secs. 8.1 and 8.2 through the multi-level Gibb's sampling method. The basic procedure was first proposed by this author and his associates in AIDS research and has been referred to as a generalized Bayesian method; see [12, 13] and [14, Chaps. 6 and 9]. The multi-level Gibbs sampling method is an extension of the Gibb's sampler method to multivariate cases first proposed by Sheppard [15], as described in Subsec. 3.3.1. Theoretical studies of this extension have been given in [15–17].

To proceed, let $\boldsymbol{X} = \{\underset{\sim}{X}(1), \ldots, \underset{\sim}{X}(t_M)\}$ be the collection of all state variables, Θ the collection of all unknown parameters and $\boldsymbol{Y} = \{\underset{\sim}{Y}(t_1), \ldots, \underset{\sim}{Y}(t_k)\}$ $(0 \le t_1 < \cdots < t_k \le t_M)$ the collection of all vectors of observed data sets. Let $P(\Theta)$ be the prior distribution of the parameters Θ, $P(\boldsymbol{X}|\Theta)$ the conditional probability density of $\boldsymbol{X}$ given the parameters Θ and $P(\boldsymbol{Y}|\boldsymbol{X}, \Theta)$ the conditional probability density of $\boldsymbol{Y}$ given $\boldsymbol{X}$ and Θ. In these distributions, the prior density $P(\Theta)$ can be constructed from previous studies. (This is the so-called *empirical Bayesian procedure*; see [18].) If there are no prior information or our prior knowledge is vague and imprecise, then one may adopt an uniform prior or non-informative prior. The conditional probability density $P(\boldsymbol{X}|\Theta)$ represents the stochastic system model and may be derived theoretically from the stochastic system equations. Given $\boldsymbol{X}$, the probability density $P(\boldsymbol{Y}|\Theta, \boldsymbol{X})$ has usually been referred to as the likelihood function of the parameters. Based on the type of probability distributions being used, the standard inference in the literature may be classified as:

(1) The Sampling Theory Inference: Given $\boldsymbol{X}$, inference about Θ is derived only from the likelihood function $P(\boldsymbol{Y}|\boldsymbol{X}, \Theta)$. For example, one derives estimate of Θ by maximizing $P(\boldsymbol{Y}|\boldsymbol{X}, \Theta)$; these are the MLE (maximum likelihood estimator) of Θ.

(2) The Bayesian Inference: Given $\boldsymbol{X}$, the Bayesian inference about Θ is derived from the posterior distribution of Θ which is proportional to the product of $P(\Theta)$ and $P(\boldsymbol{Y}|\boldsymbol{X}, \Theta)$. For example, one may use the posterior mean $E\{\Theta|\boldsymbol{X}, \boldsymbol{Y}\}$ of Θ given $\{\boldsymbol{X}, \boldsymbol{Y}\}$ or the posterior mode of Θ given $\{\boldsymbol{X}, \boldsymbol{Y}\}$ as an estimate of Θ. These are the Bayesian estimate of Θ.

(3) The Classical Kalman Filter: The classical theories of Kalman filter (see Sec. 8.4 and Secs. 9.1 and 9.2) derive optimal estimators or predictors of the state variables $\boldsymbol{X}$ by using $P(\boldsymbol{X}|\Theta)P(\boldsymbol{Y}|\boldsymbol{X}, \Theta)$ with Θ being assumed as

given or known. These are the procedures given in almost all of the texts on Kalman filters published to date; see for example [19–21].

In the above, notice that in the sampling theory inference, the prior information about Θ and the information about $\boldsymbol{X}$ from the stochastic system model are completely ignored; in the Bayesian inference, the information from the stochastic system model have been ignored. In the classical Kalman filter theories, the parameters Θ are assumed known. Thus, in each of these cases, some information have been lost or ignored. In this section we proceed to derive a general procedure to estimate simultaneously the unknown parameters and the state variables by taking into account information from all three sources. For this purpose, notice first that the joint probability density function of $(\boldsymbol{X}, \boldsymbol{Y}, \Theta)$ is $P(\Theta, \boldsymbol{X}, \boldsymbol{Y}) = P(\Theta)P(\boldsymbol{X}|\Theta)P(\boldsymbol{Y}|\boldsymbol{X}, \Theta)$. From this, one derives the conditional probability density function $P(\boldsymbol{X}|\Theta, \boldsymbol{Y})$ of $\boldsymbol{X}$ given $(\Theta, \boldsymbol{Y})$ and the conditional probability density function $P(\Theta|\boldsymbol{X}, \boldsymbol{Y})$ of Θ given $(\boldsymbol{X}, \boldsymbol{Y})$ respectively as:

$$P(\boldsymbol{X}|\Theta, \boldsymbol{Y}) \propto P(\boldsymbol{X}|\Theta)P(\boldsymbol{Y}|\boldsymbol{X}, \Theta) \,, \tag{9.26}$$

and

$$P(\Theta|\boldsymbol{X}, \boldsymbol{Y}) \propto P(\Theta)P(\boldsymbol{X}|\Theta)P(\boldsymbol{Y}|\boldsymbol{X}, \Theta) \,. \tag{9.27}$$

Given these conditional distributions, one may then use the multi-level Gibb's sampler method [15–16] to estimate simultaneously Θ and $\boldsymbol{X}$. The multi-level Gibb's sampler method is a Monte Carlo method to estimate $P(\boldsymbol{X}|\boldsymbol{Y})$ (the conditional density function of $\boldsymbol{X}$ given $\boldsymbol{Y}$) and $P(\Theta|\boldsymbol{Y})$ (the posterior density function of Θ given $\boldsymbol{Y}$) through a sequential procedure by drawing from $P(\boldsymbol{X}|\Theta, \boldsymbol{Y})$ and $P(\Theta|\boldsymbol{X}, \boldsymbol{Y})$ alternatively and sequentially. As proposed by Sheppard [15] and used by Liu and Chen [16], the algorithm of this method iterates through the following loop:

(1) Given $\Theta^{(*)}$ and $\boldsymbol{Y}$, generate $\boldsymbol{X}^{(*)}$ from $P(\boldsymbol{X}|\boldsymbol{Y}, \Theta^{(*)})$.

(2) Generate $\Theta^{(*)}$ from $P(\Theta|\boldsymbol{Y}, \boldsymbol{X}^{(*)})$ where $\boldsymbol{X}^{(*)}$ is the value obtained in (1).

(3) Using $\Theta^{(*)}$ obtained from (2) as initial values, go back to (1) and repeat the (1)–(2) loop until convergence.

At convergence, the above procedure then leads to random samples of $\boldsymbol{X}$ from the conditional density $P(\boldsymbol{X}|\boldsymbol{Y})$ of $\boldsymbol{X}$ given $\boldsymbol{Y}$ independently of Θ and to random samples of Θ from the posterior density $P(\Theta|\boldsymbol{Y})$ of Θ independently of $\boldsymbol{X}$. The proof of the convergence of the multi-level Gibbs sampling method is easily derived by extending the proof given in Subsec. 3.3.1; see Exercise 9.2.

9.4.1. *Generating data from $P(\boldsymbol{X}|\boldsymbol{Y}, \Theta)$*

Because in practice it is often very difficult to derive $P(\boldsymbol{X}|\boldsymbol{Y}, \Theta)$ whereas it is easy to generate $\boldsymbol{X}$ from $P(\boldsymbol{X}|\Theta)$, Tan and Ye [12, 13] have developed an indirect method by using the weighted bootstrap method due to Smith and Gelfand [22] to generate $\boldsymbol{X}$ from $P(\boldsymbol{X}|\boldsymbol{Y}, \Theta)$ through the generation of $\boldsymbol{X}$ from $P(\boldsymbol{X}|\Theta)$. The algorithm of the weighted bootstrap method is given by the following steps. The proof of this algorithm has been given in Subsec. 3.3.2:

(a) Given $\Theta^{(*)}$ and $\underset{\sim}{X}(l)$ $(0 \leq l \leq j)$, generate a large random sample of size N on $\underset{\sim}{X}(j+1)$ by using $P\{\underset{\sim}{X}(j+1)|\underset{\sim}{X}(j)\}$ from the stochastic system model; denote it by: $\{\underset{\sim}{X}^{(1)}(j+1), \ldots, \underset{\sim}{X}^{(N)}(j+1)\}$.

(b) Compute $w_k = P\{\underset{\sim}{Y}(j+1)|\underset{\sim}{X}(s), s = 0, 1, \ldots, j, \underset{\sim}{X}^{(k)}(j+1), \Theta^{(*)}\}$ and $q_k = w_k / \sum_{i=1}^{N} w_i$ for $k = 1, \ldots, N$. (The observation model facilitates the computation of w_k.)

(c) Construct a population Π with elements $\{E_1, \ldots, E_N\}$ and with $P(E_k) = q_k$. (Note $\sum_{i=1}^{N} q_i = 1$) Draw an element randomly from Π. If the outcome is E_k, then $\underset{\sim}{X}^{(k)}(j+1)$ is the element of $\underset{\sim}{X}(j+1)$ generated from the conditional distribution of $\boldsymbol{X}$ given the observed data and the parameter values.

(d) Start with $j = 1$ and repeat (a)–(c) until $j = t_M$ to generate a random sample from $P(\boldsymbol{X}|\boldsymbol{Y}, \Theta^{(*)})$.

9.4.2. *Generating Θ from $P(\Theta|\boldsymbol{Y}, \boldsymbol{X})$*

To generate Θ from $P(\Theta|\boldsymbol{X}, \boldsymbol{Y})$, very often it is convenient to partition Θ into non-disjoint subsets $\Theta = \{\Theta_1, \ldots, \Theta_l\}$ and apply multi-level Gibbs sampling method to these subsets. With no loss of generality we illustrate the method with $l = 3$ and write $\Theta = \{\Theta_1, \Theta_2, \Theta_3\}$. Let the conditional posterior distribution of Θ_i given $\{\boldsymbol{Y}, \boldsymbol{X}, \Theta_j, \text{ all } j \neq i\}$ be $P(\Theta_i|\boldsymbol{Y}, \boldsymbol{X}, \Theta_j, \text{ all } j \neq i)$. Then the algorithm goes through the following loop:

(1) Given $\Theta = \Theta^{(*)}$ and $\boldsymbol{X} = \boldsymbol{X}^{(*)}$, generate Θ_1 from $P(\Theta_1|\boldsymbol{Y}, \boldsymbol{X}^{(*)}, \Theta_2^{(*)}, \Theta_3^{(*)})$, and denote it by $\Theta_1^{(1)}$.

(2) Generate Θ_2 from $P(\Theta_2|\boldsymbol{Y}, \boldsymbol{X}^{(*)}, \Theta_1^{(1)}, \Theta_3^{(*)})$, where $\Theta_1^{(1)}$ is the value obtained in (1), and denote it by $\Theta_2^{(1)}$.

(3) Generate Θ_3 from $P(\Theta_3|\boldsymbol{Y}, \boldsymbol{X}^{(*)}, \Theta_1^{(1)}, \Theta_2^{(1)})$, where $\Theta_1^{(1)}$ is the value obtained in (1) and $\Theta_2^{(1)}$ the value obtained in (2), and denote it by $\Theta_3^{(1)}$.

(4) Using $\{\Theta_j^{(1)}, j = 1, 2, 3\}$ obtained from (1)–(3) as initial values, go back to steps in Subsec. 9.4.1 to generate $\boldsymbol{X}$ from $P(\boldsymbol{X}|\boldsymbol{Y}, \Theta)$.

Starting with $j = 0$ and continuing until $j = t_M$, by combining the above iterative procedures from Subsecs. 9.4.1 and 9.4.2, one can readily generate a random sample for $\boldsymbol{X}$ from $P(\boldsymbol{X}|\boldsymbol{Y})$ and a random sample for Θ from $P(\Theta|\boldsymbol{Y})$. From these generated samples, one may use the sample means as estimates of $\boldsymbol{X}$ and Θ and use the sample variances and sample covariances as the estimates of the variances and covariances of the estimates of the parameters and state variables.

In the next two sections we illustrate the application of the theories of this section by two examples: One from the discrete-time state space model and the other from the continuous-time state space model.

9.5. Simultaneous Estimation in the San Francisco Population

In this section, we illustrate the application of the theories in the previous section to estimate the unknown parameters and the state variables simultaneously in the San Francisco homosexual population. We will adopt the state space model given by the example in Sec. 8.1.2. In this example, we assume only one sexual activity level by visualizing $p_S(t)$ as a mixture from different sexual activity levels; we note, however, that this will not affect the results because the probabilities of HIV infection of S people are functions of time.

9.5.1. *A state space model for the San Francisco homosexual population*

To develop a state space model for the San Francisco homosexual population, we make the following assumptions:

(1) Based on [23, 24], we assume that $p_S(t)$ and $\gamma(u,t) = \gamma(t-u)$ are deterministic functions.

(2) Because of the awareness of AIDS, we assume that there are no immigrants and recruitment of AIDS cases. Because the total population size changes very little over time, for S people and I people we also assume that the numbers of immigrants and recruitment are balanced out by death and retirement. As shown by Hethcode and Van Ark [25], this is approximately true for the San Francisco homosexual population. It follows that for the S people and I people, one may ignore immigration and death; see Remark 9.3.

(3) As in the literature, we assume that there are no reverse transition from I to S and from A to I; see [23, 24].

Remark 9.3.   In [12], it is shown that assumption (2) has little impacts on the estimation of the HIV infection distribution and the HIV incubation distribution. However, it does have some impacts on the estimation of the numbers of S people and I people. To correct for this we have thus figured in a 1% increase per year in estimating the numbers of S people and I people in the San Francisco homosexual population; for more detail, see [12].

9.5.1.1. *The stochastic system model*

Under the above assumptions, the stochastic model is the chain binomial model given in Subsec. 2.9.2. In this model, if we ignore $A(t)$, the state variables at time t are $\underset{\sim}{X}(t) = \{S(t), I(u,t), u = 0, 1, \ldots, t\}$ and the parameters are $\Theta = \{p_S(t), \gamma(t), t = 1, \ldots, t_M\}$. Let $\boldsymbol{X} = \{\underset{\sim}{X}(1), \ldots, \underset{\sim}{X}(t_M)\}$, where t_M is the last time point. Then $\boldsymbol{X}$ is the collection of all the state variables and Θ the collection of all the parameters. For this model, the conditional probability distribution $\Pr\{\boldsymbol{X}|\underset{\sim}{X}(0)\}$ of $\boldsymbol{X}$ given $\underset{\sim}{X}(0)$ is available from Subsec. 2.9.2 and is given by

$$P\{\boldsymbol{X}|\underset{\sim}{X}(0)\} = \prod_{j=0}^{t_M-1} P\{\underset{\sim}{X}(j+1)|\underset{\sim}{X}(j), \Theta\},$$

where

$$\Pr\{\underset{\sim}{X}(j+1)|\underset{\sim}{X}(j),\Theta\} = \begin{pmatrix} S(t) \\ I(0,t+1) \end{pmatrix} [p_S(t)]^{I(0,t+1)}[1-p_S(t)]^{S(t)-I(0,t+1)}$$

$$\times \prod_{u=0}^{t} \begin{pmatrix} I(u,t) \\ I(u,t)-I(u+1,t+1) \end{pmatrix}$$

$$\times [\gamma(u)]^{I(u,t)-I(u+1,t+1)}[1-\gamma(u)]^{I(u+1,t+1)}.$$

$$(9.28)$$

For generating X, we copy the stochastic equations from Subsec. 2.9.2 to give:

$$S(t+1) = S(t) - F_S(t), \qquad (9.29)$$

$$I(0,t+1) = F_S(t), \qquad (9.30)$$

$$I(u+1,t+1) = I(u,t) - F_I(u,t), u = 0,\ldots,t, \qquad (9.31)$$

where $F_S(t)|S(t) \sim B\{S(t),p_S(t)\}$ and $F_I(u,t)|I(u,t) \sim B\{I(u,t),\gamma(u)\}$.

9.5.1.2. *The observation model*

Let $Y(j+1)$ be the observed AIDS incidence during $[j,j+1)$. Then the stochastic equation for the observation model is

$$Y(j+1) = A(j+1) + e(j+1),$$

where $A(j+1) = \sum_{u=0}^{j}[I(u,t) - I(u+1,t+1)]$ and $e(t+1)$ is the random measurement error associated with observing $Y(j+1)$. Assuming that the $e(j)$ are independently distributed as normal with means 0 and variance σ_j^2, then the likelihood function $L_A = L(\Theta|\underset{\sim}{y})$ given the state variables is

$$L_A \propto \prod_{j=1}^{t_M} \sigma_j^{-1} \exp\left\{ -\frac{1}{2\sigma_j^2}[Y(j) - A(j)]^2 \right\}. \qquad (9.32)$$

9.5.2. *The initial distribution*

Since the average AIDS incubation period is around 10 years and since the first AIDS cases were discovered in 1981, we assume January 1, 1970 as $t_0 = 0$.

It is also assumed that at time 0 there are no AIDS cases and no HIV infected people with infection duration $u > 0$; but to start the HIV epidemic, some HIV were introduced into the population at time 0. (That is, there are some I people with zero infection duration at time 0.) Tan, Tang and Lee [26], and Tan and Xiang [27, 28] have shown that the results are very insensitive to the choice of time origin.

Based on the estimate by Tan and Xiang [27, 28], we assume $S(0) = 40,000$; see Remark 9.4. Since the total number of AIDS cases during 1981 in the city of San Francisco was observed as 36, we assume $I(0,0) = 36$ at time 0. The initial variances and covariances of these initial numbers are given by the variances and covariances of $\epsilon_S(0)$ and $\epsilon_0(0)$. That is, we have:

$$\hat{\underset{\sim}{X}}(0|0) = [\hat{S}(0) = 40,000, \hat{I}(0,0) = 36]$$

and

$$P(0|0) = \begin{bmatrix} \text{Var}[\epsilon_S(0)] & \text{Cov}[\epsilon_S(0), \epsilon_0(0)] \\ \text{Cov}[\epsilon_S(0), \epsilon_0(0)] & \text{Var}[\epsilon_0(0)] \end{bmatrix}.$$

Remark 9.4. This number is the number of S people who were at risk for AIDS at $t_0 = 0$. This does not include those S people who would not be infected or would not contribute to the total number of AIDS cases in his/or her life time. Recently, molecular biologists [29] have found that besides the CD4 receptor, successful infection of T cells by HIV also requires an CC-chemokine receptor. For M-tropic HIV, this CC-chemokine receptor is CCR-5. Since during the long asymptomatic period, M-tropic HIV appears to be the dominant type [30], if the CCR-5 gene has mutated to deplete the function of the CCR-5 receptor, then the infection of T cells by HIV can not be completed in which case the individual would not be infected by HIV nor will he/she proceed to the AIDS stage. Let C denote the wild CCR-5 gene and c the mutated form of CCR-5. Let q be the frequency of the c gene in the population. Then the combined frequency of the CC type and the Cc type is $q^2 + 2q(1-q)$ in the population. Based on the estimate $\hat{q} = 0.01$ by [29] of the frequency of the CCR-5 mutant gene in the population, then there are about 20% ($q^2 + 2q(1-q) = 0.199$) people in the population with at least one mutant gene. It follows that the estimated total size of S people at $t_0 = 0$ is around $40,000 + 10,000 = 50,000$. With a 1% population increase as estimated by the Census survey [31], this leads to the population size of $58048 = 50,000 \times (1.01)^{15}$ in 1985 which is very close to the estimate 58500 of the size of the San Francisco homosexual population in 1985 by Lemp *et al.* [32].

9.5.3. *The prior distribution*

In the above model, there are two sets of parameters: (1) The probabilities $\{p_S(j), j = 1, 2, \ldots\}$ of the HIV infection of S people, and (2) the transition rates $\{\gamma(j), j = 1, 2, \ldots\}$ of $I(j) \to A$. The $p_S(j)$ are the hazard functions of the HIV infection distribution and the $\gamma(j)$ the hazard functions of the HIV incubation distribution. That is, the pdf's of the HIV infection and the HIV incubation are given respectively by $f_I(j) = p_S(j) \prod_{i=1}^{j-1}(1 - p_S(i)), j > 0$ and $g(s, t) = g(t - s) = \gamma(t - s) \prod_{i=1}^{t-s-1}(1 - \gamma(i)), t \geq s$. In[33, 34], the prior distributions of these parameters (i.e. $\{p_S(j), \gamma(j), j = 1, \ldots t_M\}$) are specified by the following procedures:

(1) It is assumed that **a priori** the $p_S(i)$'s are independently distributed of the $\gamma(j)$'s.

(2) The Prior Distribution of $p_S(i)$.

Since $0 \leq p_S(j) \leq 1$, and since the complete likelihood function of the hazard function of the HIV sero-conversion in the EM-algorithm can be expressed as a product of beta densities [26, 35], a natural conjugate prior of $p_S(j)$ is:

$$P\{p_S(i), i = 1, \ldots) \propto \prod_{i \geq 1}\left\{[p_S(i)]^{a_1(i)-1}[1 - p_S(i)]^{a_2(i)-1}\right\}, \qquad (9.33)$$

where $a_1(i)$ and $a_2(i)$ are determined by some prior studies. Notice that if $a_1(i) = a_2(i) = 1$, the above prior is an uniform prior which corresponds to no prior information.

To determine the prior parameters from some prior data, one may specify $a_1(j) - 1$ as the number of HIV infected individuals with HIV infection in $[j - 1, j)$ in the prior data. If there are n_0 individuals in the prior study, then the $a_1(j) - 1$ are distributed as multinomial random variables with parameters $\{n_0, f_I(t_0, j), j = 1, \ldots.\}$, where $f_I(t_0, j)$ is the probability density distribution (pdf) of the HIV infection at time j for S people at $t_0 = 0$. It follows that $\hat{a}_1(j) = 1 + n_0 \hat{f}_I(t_0, j)$ and $\hat{a}_2(j) = 1 + n_0 \sum_{l \geq j+1} \hat{f}_I(t_0, l)$, where $\hat{f}_I(t_0, j)$ is the estimate of $f_I(t_0, j)$ and n_0 the sample size in the previous study.

(3) The prior distribution of $\gamma(j)$.

Table 9.2. Prior Information for Infection Distribution.

Time	Jun. 77	Dec. 77	Jun. 78	Dec. 78	Jun. 79	Dec. 79	Jun. 80	Dec. 80
$a_1(t)$	4.03	5.92	7.42	8.39	11.15	15.61	15.50	14.02
$a_2(t)$	1060.42	1033.92	1001.62	957.60	903.19	829.79	733.92	650.26

Time	Jun. 81	Dec. 81	Jun. 82	Dec. 82	Jun. 83	Dec. 83	Jun. 84	Dec. 84
$a_1(t)$	13.53	10.25	6.67	4.95	3.40	2.45	2.03	1.85
$a_2(t)$	574.57	511.87	469.05	441.70	424.51	413.57	406.48	401.15

Time	Jun. 85	Dec. 85	Jun. 86	Dec. 86	Jun. 87	Dec. 87	Jun. 88	Dec. 88
$a_1(t)$	1.52	1.52	1.28	1.31	1.61	1.87	1.47	1.25
$a_2(t)$	397.31	394.28	391.98	390.37	387.22	383.12	377.77	375.83

Time	Jun. 89	Dec. 89	Jun. 90	Dec. 90	Jun. 91	Dec. 91	Jun. 92	Dec. 92
$a_1(t)$	1.41	1.18	1.22	1.69	5.09	9.33	10.25	9.04
$a_2(t)$	373.92	372.56	371.38	368.87	355.00	313.97	259.07	207.49

Since $0 \leq \gamma(j) \leq 1$ and since the complete likelihood function of the incubation hazard functions is a product of beta densities, a natural conjugate prior for $\gamma(j)$ is,

$$P(\gamma(j), j = 1, 2, \ldots) \propto \prod_{j \geq 1} [\gamma(j)]^{b_1(j)-1} [1 - \gamma(j)]^{b_2(j)-1}, \qquad (9.34)$$

where the prior parameters $b_1(j)$ and $b_2(j)$ are estimated from some prior studies on the HIV incubation distribution. As above, if $\hat{g}(j)$ is an estimate of the HIV incubation density $g(j)$ from some study with sample size n, then an estimate of $b_1(j)$ and $b_2(j)$ are given by $\hat{b}_1(j) = 1 + n\hat{g}(j)$ and $\hat{b}_2(j) = 1 + \sum_{i \geq j+1} \hat{b}_1(i)$.

Given in Tables 9.2 and 9.3 are the estimated hyperparameters of the prior distributions for $p_S(t)$ and $\gamma(t)$ respectively.

To compare effects of different prior distributions, we also assume that our prior knowledge about the parameters are vague and imprecise so that we assume uniform prior for the parameters. That is, we take $a_i(t) = b_i(t) = 1$ for $i = 1, 2$. We will compare results from different prior distributions.

9.5.4.   *Generating X from the conditional density $P(X|\Theta, Y)$*

We use the weighted bootstrap method as described in Sec. 9.4 to generate X from $P(X|\Theta)$ through the stochastic Eqs. (9.29)–(9.32) given above. Thus,

Table 9.3. Prior Information for Incubation Distribution.

Time	Jun. 77	Dec. 77	Jun. 78	Dec. 78	Jun. 79	Dec. 79	Jun. 80	Dec. 80
$b_1(t)$	5.34	5.46	5.51	5.52	5.48	5.40	5.29	5.15
$b_2(t)$	528.82	502.34	475.37	448.24	421.23	394.61	368.58	343.32

Time	Jun. 81	Dec. 81	Jun. 82	Dec. 82	Jun. 83	Dec. 83	Jun. 84	Dec. 84
$b_1(t)$	4.99	4.82	4.63	4.44	4.25	4.05	3.86	3.67
$b_2(t)$	318.96	295.62	273.36	252.24	232.27	213.48	195.84	179.34

Time	Jun. 85	Dec. 85	Jun. 86	Dec. 86	Jun. 87	Dec. 87	Jun. 88	Dec. 88
$b_1(t)$	3.49	3.31	3.15	2.99	2.84	2.69	2.56	2.44
$b_2(t)$	163.95	149.63	136.34	124.02	112.62	102.10	92.41	83.48

Time	Jun. 89	Dec. 89	Jun. 90	Dec. 90	Jun. 91	Dec. 91	Jun. 92	Dec. 92
$b_1(t)$	2.32	2.21	2.11	2.02	1.93	1.85	1.78	1.71
$b_2(t)$	75.28	67.74	60.83	54.50	48.70	43.39	38.54	34.11

given $\underset{\sim}{X}(j) = \{S(j), I(u,j), u = 0, 1, \ldots, j\}$ and given the parameter values, we use the binomial generator to generate $F_S(t)$ and $F_I(u,t)$ through the conditional binomial distributions $F_S(t)|S(t) \sim B\{S(t), p_S(t)\}$ and $F_I(u,t)| I(u,t) \sim B\{I(u,t), \gamma(u)\}$. These lead to

$$S(t+1) = S(t) - F_S(t), I(0, t+1) = F_S(t), I(u+1, t+1)$$

$$= I(u,t) - F_I(u,t), \quad u = 0, 1, \ldots, t$$

and

$$A(t+1) = \sum_{u=0}^{t} F_I(u,t).$$

The binomial generator is readily available from the IMSL subroutines [36] or other software packages such as SAS. With the generation of X from $P(X|\Theta)$, one may then apply the weighted bootstrap method to generate X from $P(X|Y, \Theta)$.

9.5.5. *Generating Θ from the conditional density $P(\Theta|X, Y)$*

Using Eq. (9.28) given in Subsec. 9.5.1, and the prior distribution from Subsec. 9.5.3, we obtain

$$P(\Theta|\boldsymbol{X},\boldsymbol{Y}) \propto \prod_{t=0}^{t_M-1} \{[p_S(t)]^{I(0,t+1)+a_1(t)-1}[1-p_S(t)]^{S(t+1)+a_2(t)-1}\}$$

$$\times \prod_{u=1}^{t_M-1} \{[\gamma(u)]^{c_1(u)+b_1(u)-1}[1-\gamma(u)]^{c_2(u)+b_2(u)-1}\}, \qquad (9.35)$$

The above equation shows that the conditional distribution of $p_S(t)$ given $\boldsymbol{X}$ and given $\boldsymbol{Y}$ is a Beta distribution with parameters $\{I(0,t+1)+a_1(t), S(t+1)+a_2(t)\}$. Similarly, the conditional distribution of $\gamma(t)$ given $\boldsymbol{X}$ and given $\boldsymbol{Y}$ is a Beta distribution with parameters $\{c_1(u)+b_1(u), c_2(u)+b_2(u)\}$. Since generating a large sample from the Beta distribution to give sample means are numerically identical to compute the mean values from the Beta distribution, the estimates of $p_S(t)$ and $\gamma(t)$ are then given by:

$$\hat{p}_S(t) = \frac{I(0,t+1)+a_1(t)}{I(0,t+1)+S(t+1)+a_1(t)+a_2(t)},$$

$$\hat{\gamma}(u) = \frac{c_1(u)+b_1(u)}{\sum_{i=1}^{2}[c_i(u)+b_i(u)]}.$$

We will use these estimates as the generated sample means.

Using the above approach, one can readily estimate simultaneously the numbers of S people, I people and AIDS cases as well as the parameters $\{p_S(t),\gamma(t)\}$. With the estimation of $\{p_S(t),\gamma(t)\}$, one may readily estimate the HIV infection distribution $f_I(t)$ and the HIV incubation distribution $g(t)$ through the formula $f_I(t) = p_S(t)\prod_{i=1}^{t-1}(1-p_S(i))$ and $g(t) = \gamma(t)\prod_{i=1}^{t-1}(1-\gamma(i))$. For the San Francisco homosexual population, these estimates are plotted in Figs. 9.4–9.6. Comparing results from Sec. 8.5 and with the corresponding results from this section, one notices that these two approaches gave similar results and hence almost the same conclusions for the San Francisco homosexual population. Given below we summarize the basic findings:

(a) From Fig. 9.4, the estimated density of the HIV infection clearly showed a mixture of distributions with two obvious peaks.

(b) From Fig. 9.5, the estimated density of the HIV incubation distribution appeared to be a mixture of distributions with two obvious peaks.

(c) From Fig. 9.6(a), we observe that the estimates of the AIDS incidence by the Gibbs sampler are almost identical to the corresponding observed AIDS incidence respectively. These results are almost identical to the estimates by the method of Sec. 8.5; see Sec. 8.5.

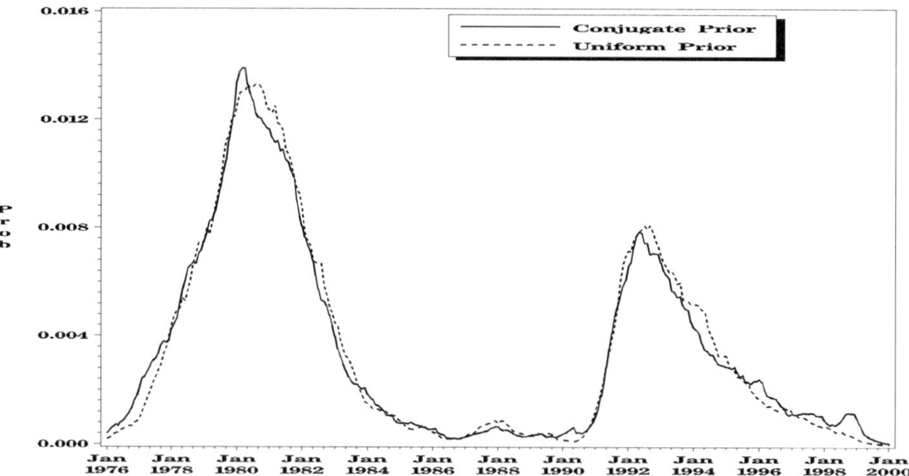

Fig. 9.4.   Plots of the estimated HIV infection distribution.

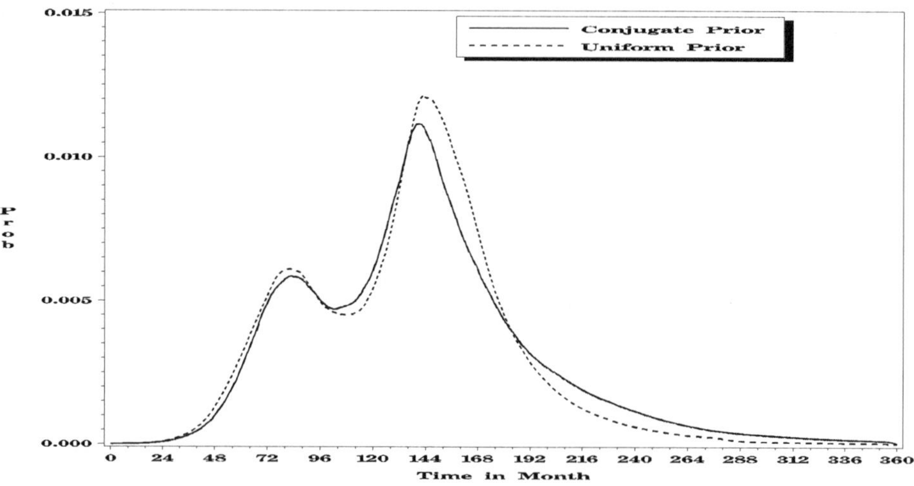

Fig. 9.5.   Plots of the estimated HIV incubation distribution.

(d) To assess influence of prior information on $\{p_S(t), \gamma(t)\}$, we plot in Figs. 9.4–9.5 the estimates of the HIV infection and the HIV incubation under both with and without (i.e. non-informative uniform prior) prior information. The results show clearly that the prior information seem to have little effects, especially in the case of HIV infection.

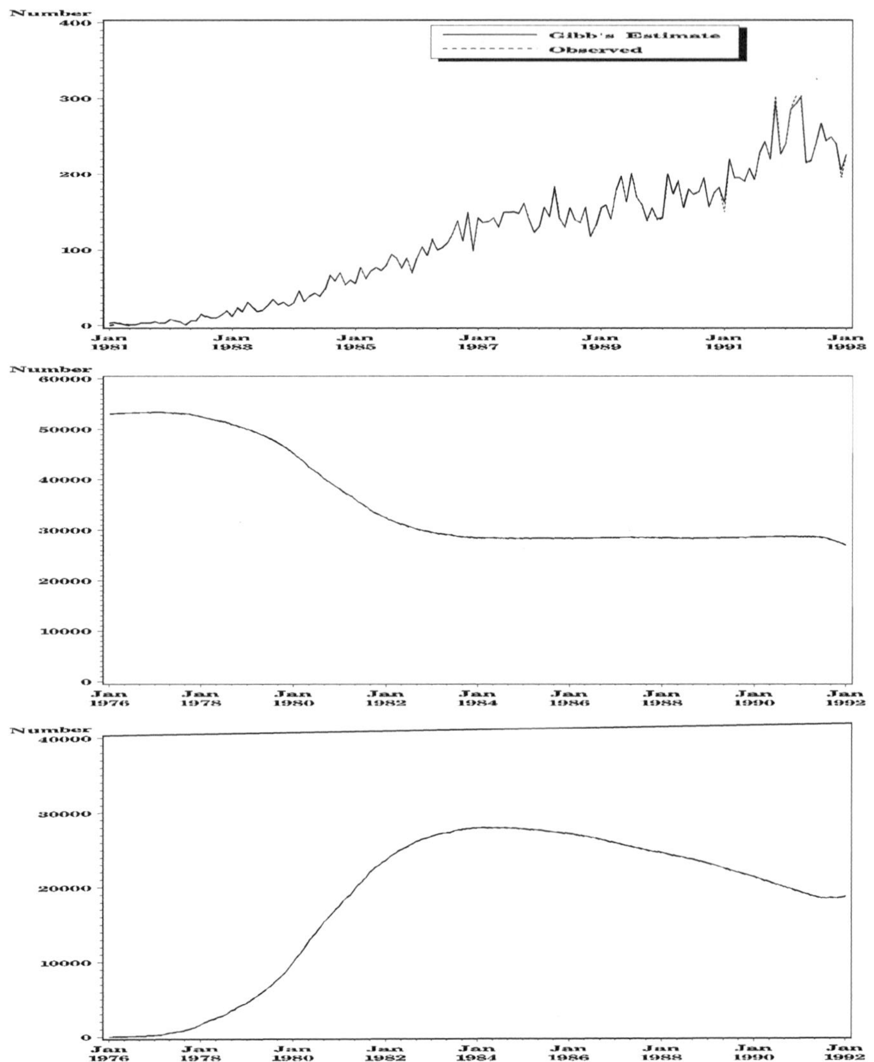

Fig. 9.6.   Plots of the observed AIDS incidence, the Gibb's sampler estimate and the estimated numbers of susceptible and infected people.

9.6.   Simultaneous Estimation in the Cancer Drug-Resistant Model

In Subsec. 8.2.1, we have considered the cancer drug-resistant model under chemotherapy and immuno-stimulation. In this model, the state variables are the numbers of sensitive cancer tumor cells $(X_1(t))$ and resistant cancer tumor cells $(X_2(t))$. The parameters are the birth rates $(b_i(t),\ i = 1, 2)$, the death rates $(d_i(t),\ i = 1, 2)$ and the mutation rates $\alpha(t)$ from sensitive tumor cells to resistant tumor cells. If some data are available from this system so that one can develop a state space model, then one may use the basic procedures in the previous section to estimate simultaneously the state variables and the parameter values.

To serve as an example, in this section we will use the model in Subsec. 8.2.1 to generate some Monte Carlo data; by using the generated data we then proceed to derive estimates of the unknown parameters and the state variables.

9.6.1. *Data generation*

As in [37], we use the model given in Subsec. 8.2.1 to generate some Monte Carlo data over a period of 60 days. We partition the time interval $[0, 60]$ of 60 days into 5 sub-intervals:

$$[0, 10), [10, 20), [20, 30), [30, 40), [40, 60].$$

Chemotherapy is applied to sub-intervals $[10, 20)$ and $[30, 40)$ but immuno-stimulation is applied to all sub-intervals. For the birth rates and death rates, if chemotherapy is not applied, we assume $b_1(t) = b_1 = 0.055$ and $d_1(t) = d_1 = 0.035$ to yield $\gamma_1(t) = b_1(t) - d_1(t) = 0.02$ which corresponding to a doubling time of 35 days for cancer tumor cells; if chemotherapy is applied, then we assume $b_1(t) = b_1 = 0.055$ but $d_1(t) = \delta_S = 0.535$. For resistant cancer tumor cells, because of resistance, we assume $b_2(t) = b_2 = 0.045$ and $d_2(t) = d_2 = 0.025$ to reflect the observation that resistant tumor cells in general have smaller fitness than sensitive tumor cells. (Notice that $b_2(t) - d_2(t) = 0.02$ so that the doubling time is again 35 days for resistant tumor cells; see [37].) For the mutation rates from sensitive tumor cells to resistant tumor cells, if chemotherapy is not applied, we take $\alpha(t) = \alpha_0 = 10^{-7}$ (Notice that this is the spontaneous mutation rate.); if chemotherapy is applied, then we take $\alpha(t) = \alpha_1 = 10^{-3}$. In this model, therefore, the parameters to be estimated are: $\Theta_1 = \{b_1, d_1, \delta_S\}, \Theta_2 = \{b_2, d_2\}$ and $\Theta_3 = \{\alpha_0, \alpha_1\}$.

Using these parameter values and the model in Sec. 9.3, we have generated 20 observed total number of cancer tumor cells every 3 days over a period of 60 days $(k = 1, n = 20)$. In the generation of these observed numbers, we assume

Table 4. Generated Observed And Fitted Numbers.

Time (days)	Fitted Numbers	Observed Numbers
3	$1.0613E + 07$	$1.0614E+07$
6	$1.1264E + 07$	$1.1258E+07$
9	$1.1955E + 07$	$1.1940E+07$
12	6488948.00	6475337.00
15	938810.00	932789.60
18	158220.00	156488.50
21	49455.00	48068.66
24	52387.00	50940.54
27	55492.00	54344.58
30	58815.00	57735.37
33	44822.00	43981.10
36	41900.00	41307.60
39	43489.00	42858.89
42	45951.00	45424.34
45	48597.00	48512.52
48	51373.00	51272.01
51	54331.00	54666.74
54	57447.00	58480.43
57	60754.00	61710.66
60	64222.00	65850.72

that the Gaussian errors have mean 0 and variance $\sigma_i^2 = 1$. These generated observed numbers are given in Table 9.4.

9.6.2. *The state space model*

Using the state space model given in Subsec. 8.2.1, one may derive the probability density $P\{\boldsymbol{X}|\Theta\}$ and the likelihood function of the parameters $P\{\boldsymbol{Y}|\boldsymbol{X},\Theta\}$. Combining with the prior distribution of the parameters, one may then use the generalized Bayesian method of Sec. 9.4 to estimate the unknown parameters and the state variables.

9.6.2.1. *The probability distribution of the state variables*

Let $\boldsymbol{X} = \{\underset{\sim}{X}(1), \ldots, \underset{\sim}{X}(t_M)\}$, where t_M is the time of termination of the study and $\underset{\sim}{X}(t) = \{X_1(t), X_2(t)\}'$. Then, by the Markov condition, the conditional probability density $P\{\boldsymbol{X}|\Theta\}$ of $\boldsymbol{X}$ given Θ is

$$P\{\boldsymbol{X}|\Theta\} = P\{\underset{\sim}{X}(0)|\Theta\} \prod_{t=1}^{t_M} P\{\underset{\sim}{X}(t)|\underset{\sim}{X}(t-1),\Theta\}\,.$$

To derive the conditional density $P\{\underset{\sim}{X}(t)|\underset{\sim}{X}(t-1),\Theta\}$, denote by:

$$m_1(i,t) = X_1(t) + i - X_1(t+1)\,, \qquad n_1(i,j,t) = X_1(t+1) - 2i - j\,,$$
$$m_2(k,j,t) = X_2(t) + k + j - X_2(t+1)\,, \quad n_2(k,j,t) = X_2(t+1) - 2k - j\,.$$

Then, using the stochastic equations given by Eqs. (8.1)–(8.2) and noting that the probability distributions of $\{B_S(t), D_S(t), M_S(t)\}$ and $\{B_R(t), D_R(t)\}$ are multinomial, we obtain:

$$P\{\underset{\sim}{X}(t+1)|\underset{\sim}{X}(t),\Theta\} = \sum_{i=0}^{X_1(t)} \sum_{j=0}^{X_1(t)-i} \binom{X_1(t)}{i} \binom{X_1(t)-i}{j} \binom{X_1(t)-i-j}{m_1(i,t)}$$

$$\times [b_1(t)]^i [\alpha(t)]^j [d_1(t)]^{m_1(i,t)} [1-b_1(t)-d_1(t)-\alpha(t)]^{n_1(i,j,t)}$$

$$\times \sum_{k=0}^{X_2(t)} \binom{X_2(t)}{k} \binom{X_2(t)-k}{m_2(k,j,t)}$$

$$\times [b_2(t)]^k [d_2(t)]^{m_2(k,j,t)} [1-b_2(t)-d_2(t)]^{n_2(k,j,t)}\,. \quad (9.36)$$

The above distribution is quite complicated, to implement the multi-level Gibbs sampling method, we thus introduce the augmented data $\boldsymbol{U} = \{\underset{\sim}{U}(j), j = 1,\ldots,t_M\}$, where $\underset{\sim}{U}(j) = \{B_S(j), M_S(j), B_R(j)\}$. Then, by using Eqs. (8.1)–(8.2) and the multinomial distributions for the augmented variables, one can generate $\boldsymbol{X}$ given $\boldsymbol{U}$ and Θ; similarly, one can generate $\boldsymbol{U}$ given $\boldsymbol{X}$ and Θ. The conditional density $P\{\boldsymbol{X}|\boldsymbol{U},\Theta\}$ of $\boldsymbol{X}$ given $\boldsymbol{U}$ and Θ is,

$$P\{\boldsymbol{X}|\boldsymbol{U},\Theta\} = P\{\underset{\sim}{X}(0)|\Theta\} \prod_{t=1}^{t_M} P\{\underset{\sim}{X}(t)|\underset{\sim}{X}(t-1),\underset{\sim}{U}(t-1),\Theta\}\,,$$

$$P\{\underset{\sim}{X}(t+1)|\underset{\sim}{X}(t),\underset{\sim}{U}(t)\} \propto C_1(t)C_2(t)[b_1(t)]^{B_S(t)}[\alpha(t)]^{M_S(t)}[d_1(t)]^{m_S(t)}$$

$$\times [1-b_1(t)-d_1(t)-\alpha(t)]^{n_S(t)}[b_2(t)]^{B_R(t)}$$

$$\times [d_2(t)]^{m_R(t)}[1-b_2(t)-d_2(t)]^{n_R(t)}\,, \quad (9.37)$$

where

$$C_1(t) = \begin{pmatrix} X_1(t) \\ B_S(t) \end{pmatrix} \begin{pmatrix} X_1(t) - B_S(t) \\ M_S(t) \end{pmatrix} \begin{pmatrix} X_1(t) - B_S(t) - M_S(t) \\ m_S(t) \end{pmatrix},$$

and

$$C_2(t) = \begin{pmatrix} X_2(t) \\ B_R(t) \end{pmatrix} \begin{pmatrix} X_2(t) - B_R(t) \\ m_R(t) \end{pmatrix}$$

with

$$m_S(t) = X_1(t) - X_1(t+1) + B_S(t),$$

$$n_S(t) = X_1(t+1) - 2B_S(t) - M_S(t),$$

$$m_R(t) = X_2(t) - X_2(t+1) + B_R(t) + M_S(t),$$

and

$$n_R(t) = X_2(t+1) - 2B_R(t) - M_S(t).$$

9.6.2.2.   *The conditional likelihood function*

Write $X(t) = \sum_{i=1}^{2} X_i(t)$. Under the assumption that $e(1, j) = \{Y(1, j) - X(t_j)\}/\sqrt{X(t_j)}$ is distributed as normal with mean 0 and variance σ^2 independently, the conditional likelihood function of the parameters given $\boldsymbol{X}$ is

$$P\{\boldsymbol{Y}|\boldsymbol{X}, \Theta\} \propto \sigma^{-n} \prod_{j=1}^{n} \frac{1}{\sqrt{X(t_j)}} \exp\left\{-\frac{1}{2\sigma^2 X(t_j)}[Y(1, j) - X(t_j)]^2\right\}. \quad (9.38)$$

9.6.3.   **The prior distribution and the conditional posterior distributions of the parameters**

To assign the prior distribution for the parameters, we assume that **a priori** the $\{\Theta_i, i = 1, 2, 3\}$ are independently distributed of one another. Then the prior distribution of $\{\Theta_i, i = 1, 2, 3\}$ is

$$P\{\Theta_i, i = 1, 2, 3\} = \prod_{i=1}^{3} P\{\Theta_i\}, \quad (9.39)$$

where $P\{\Theta_i\}$ is the prior density of Θ_i.

To derive the conditional posterior distribution of the parameters, let $\{L_{1i}, i = 1, \ldots, k_1\}$ ($k_1 = 3$ in the above) denote the sub-intervals in which chemotherapy has not been applied and $\{L_{2i}, i = 1, \ldots, k_2\}$ ($k_2 = 2$ in the above) the sub-intervals in which chemotherapy has been applied. Denote by:

$$S_1 = \sum_{i=1}^{k_1} \sum_{t \in L_{1i}} X_1(t), \; S_2 = \sum_{i=1}^{k_2} \sum_{t \in L_{2i}} X_1(t), \quad \text{and} \quad N_R = \sum_{t} X_2(t),$$

$$n_S(1) = \sum_{i=1}^{k_1} \sum_{t \in L_{1i}} B_S(t), \; n_S(2) = \sum_{i=1}^{k_2} \sum_{t \in L_{2i}} B_S(t), \; m_S(1) = \sum_{i=1}^{k_1} \sum_{t \in L_{1i}} D_S(t),$$

$$m_S(2) = \sum_{i=1}^{k_2} \sum_{t \in L_{2i}} D_S(t), \; k_S(1) = \sum_{i=1}^{k_1} \sum_{t \in L_{1i}} M_S(t), \; k_S(2) = \sum_{i=1}^{k_2} \sum_{t \in L_{2i}} M_S(t),$$

and

$$n_R = \sum_{t} B_R(t), \qquad m_R = \sum_{t} D_R(t).$$

Then, using the probability densities of the state variables given above, we obtain the conditional posterior distribution of Θ_i given the other parameters and given $\{\boldsymbol{X}, \boldsymbol{U}, \boldsymbol{Y}\}$ as:

$$P(\Theta_1 | \Theta_2, \Theta_3, \boldsymbol{Y}, \boldsymbol{X}, \boldsymbol{U}) \propto P(\Theta_1) b_1^{n_S(1)+n_S(2)} [d_1]^{m_S(1)} [\delta_S]^{m_S(2)}$$

$$\times (1 - b_1 - d_1)^{S_1 - n_S(1) - m_S(1)}$$

$$\times (1 - b_1 - \delta_s)^{S_2 - n_S(2) - m_S(2)}, \tag{9.40}$$

$$P(\Theta_2 | \Theta_1, \Theta_3, \boldsymbol{Y}, \boldsymbol{X}.\boldsymbol{U}) \propto P(\Theta_2) (b_2)^{n_R}$$

$$\times (d_2)^{m_R} (1 - b_2 - d_2)^{N_R - n_R - m_R}, \tag{9.41}$$

$$P(\Theta_3 | \Theta_1, \Theta_2, \boldsymbol{Y}, \boldsymbol{X}, \boldsymbol{U}) \propto P(\Theta_3) \left(\frac{\alpha_0}{1 - b_1 - d_1} \right)^{k_S(1)}$$

$$\times \left(1 - \frac{\alpha_0}{1 - b_1 - d_1} \right)^{S_1 - n_S(1) - m_S(1) - k_S(1)}$$

$$\times \left(\frac{\alpha_1}{1 - b_1 - \delta_S} \right)^{k_S(2)}$$

$$\times \left(1 - \frac{\alpha_1}{1 - b_1 - \delta_S} \right)^{S_2 - n_S(2) - m_S(2) - k_S(2)} . \qquad (9.42)$$

To implement the multi-level Gibbs sampling method, because we have no prior information about the parameters, we assume non- informative prior to reflect that our knowledge about the parameters are vague and imprecise. Notice that if we assume a non-informative prior, generating a large sample of Θ_1 from the posterior distribution of Eq. (9.40) and taking sample mean is numerically equivalent to estimating $\{b_1, d_1, \delta_S\}$ by

$$\hat{b}_1 = \frac{n_S(1) + n_S(2)}{S_1 + S_2}, \quad \hat{d}_1 = \frac{(1 - \hat{b}_1)m_S(1)}{S_1 - n_S(1)}, \quad \text{and} \quad \hat{\delta}_S = \frac{(1 - \hat{b}_1)m_S(2)}{S_2 - n_S(2)} .$$

Similarly, if one uses a non-informative prior, generating a large sample of Θ_2 from the posterior distribution of Eq. (9.41) and taking sample mean is numerically equivalent to estimating $\{b_2, d_2\}$ by

$$\hat{b}_2 = \frac{n_R}{N_R}, \quad \text{and} \quad \hat{d}_2 = \frac{m_R}{N_R} ;$$

generating a large sample of Θ_3 from the posterior distribution of Eq. (9.42) and taking sample mean is equivalent numerically to estimating $\{\alpha_0, \alpha_1\}$ by

$$\hat{\alpha}_0 = \frac{(1 - \hat{b}_1 - \hat{d}_1)k_S(1)}{S_1 - n_S(1) - m_S(1)}, \quad \hat{\alpha}_1 = \frac{(1 - \hat{b}_1 - \hat{\delta}_S)k_S(2)}{S_2 - n_S(2) - m_S(2)} .$$

9.6.4. *The multi-level Gibbs sampling procedure*

Using results from Subsecs. 9.6.2–9.6.3, the multi-level Gibbs sampling method to estimate Θ and the state variables are then given by the following loop:

(1) Combining a large sample from $P\{U|X, \Theta\}$ for given X with $P\{Y|\Theta, X\}$ through the weighted Bootstrap method due to Smith and Gelfant [21], we generate U (denote the generated sample $U^{(*)}$) from $P\{U|\Theta, X, Y\}$ although the latter density is unknown.

(2) Combining a large sample from $P\{X|U^{(*)}, \Theta\}$ with $P\{Y|\Theta, X\}$ through the weighted Bootstrap method due to Smith and Gelfant [21], we

generate $\boldsymbol{X}$ (denote the generated sample by $\boldsymbol{X}^{(*)}$) from $P\{\boldsymbol{X}|\Theta, \boldsymbol{U}^{(*)}, \boldsymbol{Y}\}$ although the latter density is unknown.

(3) On substituting $\{\boldsymbol{U}^{(*)}, \boldsymbol{X}^{(*)}\}$ which are generated numbers from the above two steps and assuming non-informative uniform priors, generate Θ from the conditional density $P\{\Theta|\boldsymbol{X}^{(*)}.\boldsymbol{U}^{(*)}, \boldsymbol{Y}\}$ given by Eqs. (9.22)–(9.24) in Subsec. 9.6.3.

(4) On substituting $\boldsymbol{X}^{(*)}$ generated from Step (2) above and with Θ being generated from Step (3) above, go back to Step (1) and repeat the above (1)–(3) loop until convergence.

At convergence, one then generates a random sample of $\boldsymbol{X}$ from the conditional distribution $P\{\boldsymbol{X}|\boldsymbol{Y}\}$ of $\boldsymbol{X}$ given $\boldsymbol{Y}$, independent of $\boldsymbol{U}$ and Θ and a random sample of Θ from the posterior distribution $P\{\Theta|\boldsymbol{Y}\}$ of Θ given $\boldsymbol{Y}$, independent of $\{\boldsymbol{X}, \boldsymbol{U}\}$. Repeat these procedures we then generate a random sample of size n of $\boldsymbol{X}$ and a random sample of size m of Θ. One may then use the sample means to derive the estimates of $\boldsymbol{X}$ and Θ and use the sample variances as the variances of these estimates. The convergence of these procedures are proved by using the basic theory of homogeneous Markov chains given in Subsec. 3.3.1.

9.6.5. *Estimated results*

Using procedures given in Subsec. 9.6.4 and the data in Table 9.4, we have estimated the unknown parameters and the state variables. Given in Figs. 9.7–9.8 are the estimated values of the $X_i(t)$'s over time. Given in Table 9.4 are the estimated total numbers of tumor cells over time. The estimates of the parameters together with the estimated standard errors by the Bootstrap method are given as follows:

$$\hat{b}_1 = 5.5006038E - 02 \pm 6.6009081E - 05\,,$$

$$\hat{d}_1 = 3.4966093E - 02 \pm 1.3446721E - 04\,,$$

$$\hat{\delta}_S = 0.5343983 \pm 1.1880239E - 03\,,$$

$$\hat{b}_2 = 4.1315455E - 02 \pm 1.0685023E - 02\,,$$

$$\hat{d}_2 = 2.3084166E - 02 \pm 5.0952979E - 03\,,$$

$$\hat{\alpha}_0 = 5.0931044E - 07 \pm 1.5631909E - 06\,,$$

$$\hat{\alpha}_1 = 1.0274288E - 03 \pm 5.3189488E - 05\,.$$

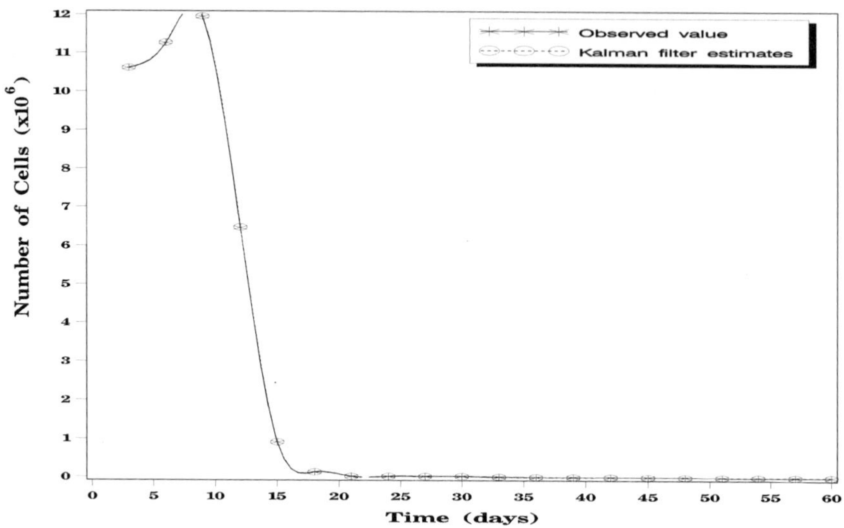

Fig. 9.7.   Plots showing the number of sensitive tumor cells.

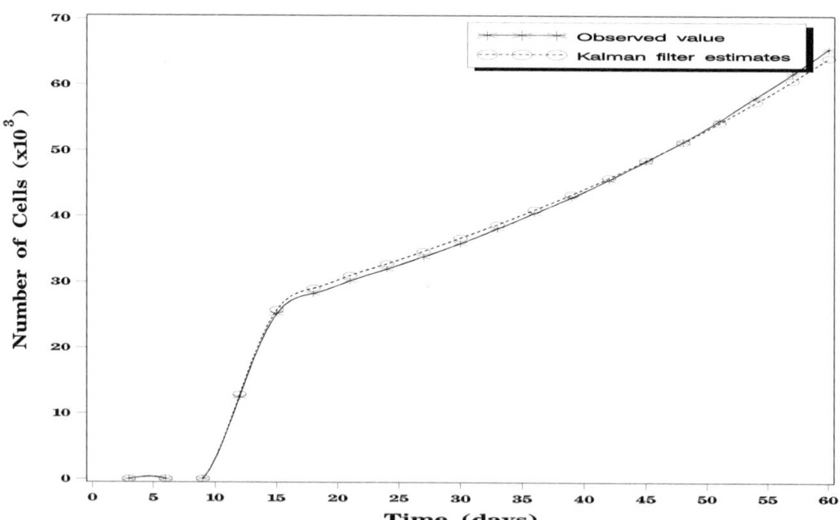

Fig. 9.8.   Plots showing the number of resistant tumor cells.

From the above estimates and the fitted results given in Table 9.4, it is clear that the methods appear to be quite promising. The Kalman filter estimates of the numbers of $X_i(t)$ also appear to be very close to the generated true numbers. Notice that if one uses the classical least square method to estimate the unknown parameters, the problem is not identifiable in the sense that one can not estimate the birth rates and the death rates separately but only the difference between the birth rate and the death rate.

9.7. Complements and Exercises

Exercise 9.1. Prove Theorem 9.3.

Exercise 9.2. Prove Theorem 9.4.

Exercise 9.3. Prove the convergence of the multi-level Gibbs sampling procedures in Sec. 9.4.

Exercise 9.4. Using the stochastic Eqs. (9.29)–(9.31), derive the distribution result in (9.28).

Exercise 9.5. Using the density given by (9.35), derive the expected values and variance of $p_S(t)$ and $\gamma(t)$.

Exercise 9.6. Using the stochastic Eqs. (8.1) and (8.2), derive the distribution results in Eqs. (9.36) and (9.37).

Exercise 9.7. Using the densities given by (9.40)–(9.42), derive the expected values and variance of b_i, d_i and α_i.

References

[1] W. Y. Tan, *Note on an extension of Gauss-Markov theorem to multivariate regression models*, SIAM J. Applied Mathematics **20** (1971) 24–29.
[2] W. Y. Tan and Z. H. Xiang, *State Space Models for the HIV pathogenesis*, in *Mathematical Models in Medicine and Health Sciences*, eds. M. A. Horn, G. Simonett and G. Webb, Vanderbilt University Press, Nashville (1998).
[3] W. Y. Tan and Z. H. Xiang, *Estimating and predicting the numbers of T cells and free HIV by non-linear Kalman filter*, in *Artificial Immune Systems and Their Applications*, ed. DasGupta, Springer-Verlag, Berlin (1998).

[4] W. Y. Tan and Z. H. Xiang, *A state space model of HIV pathogenesis under treatment by anti-viral drugs in HIV-infected individuals*, Math. Biosciences **156** (1999) 69–94.

[5] D. D. Ho, A. U. Neumann, A. S. Perelson, W. Chen, J. M. Leonard and M. Markowitz, *Rapid turnover of plasma virus and CD4 lymphocytes in HIV-1 infection*, Nature **373** (1995) 123–126.

[6] X. Wei, S. K. Ghosh, M. E. Taylor, V. A. Johnson, E. A. Emini, P. Deutsch, J. D. Lifson, S. Bonhoeffer, M. A. Nowak, B. H. Hahn, M. S. Saag and G. M. Shaw, *Viral dynamics in human immunodeficiency virus type 1 infection*, Nature **373** (1995) 117–122.

[7] A. S. Perelson, A. U. Neumann, M. Markowitz, J. M. Leonard and D. D. Ho, *HIV-1 dynamics in vivo: Virion clearance rate, infected cell life-span, and viral generation time*, Science **271** (1996) 1582–1586.

[8] A. N. Phillips, *Reduction of HIV Concentration During Acute Infection: Independence from a Specific Immune Response*, Science **271** (1996) 497–499.

[9] D. Schenzle, *A Model for AIDS Pathogenesis*, Statistics in Medicine **13** (1994) 2067–2079.

[10] W. Y. Tan and H. Wu, *A stochastic model for the pathogenesis of HIV at the cellular level and some Monte Carlo studies*, in *Simulation in the Medical Sciences*, eds. J. G. Anderson and M. Katzper, The Society for Computer Simulation, San Diego (1997).

[11] W. Y. Tan and H. Wu, *Stochastic modeling of the dynamics of $CD4^{(+)}$ T cells by HIV infection and some Monte Carlo studies*, Math. Biosciences **147** (1998) 173–205.

[12] W. Y. Tan and Z. Z. Ye, *Estimation of HIV infection and HIV incubation via state space models*, Math. Biosciences **167** (2000) 31–50.

[13] W. Y. Tan and Z. Z. Ye, *Some state space models of HIV epidemic and applications for the estimation of HIV infection and HIV incubation*, Comm. Statistics (Theory and Methods) **29** (2000) 1059–1088.

[14] W. Y. Tan, *Stochastic Modeling of AIDS Epidemiology and HIV Pathogenesis*, World Scientific, Singapore (2000).

[15] N. Shephard, *Partial non-Gaussian state space*, Biometrika **81** (1994) 115–131.

[16] J. S. Liu and R. Chen, *Sequential Monte Carlo method for dynamic systems*, Jour. Amer. Statist. Association **93** (1998) 1032–1044.

[17] G. Kitagawa, *A self organizing state space model*, Jour. American Statist. Association **93** (1998) 1203–1215.

[18] B. P. Carlin and T. A. Louis, *Bayes and Empirical Bayes Methods for Data Analysis*, Chapman and Hall, London (1996).

[19] D. E. Catlin, *Estimation, Control and Discrete Kalman Filter*, Spring-Verlag, New York (1989).

[20] A. Gelb, *Applied Optimal Estimation*, M.I.T. Press, Cambridge (1974).

[21] A. P. Sage and J. L. Melsa, *Estimation Theory With Application to Communication and Control*, McGraw-Hill Book Com., New York (1971).

[22] A. F. M. Smith and A. E. Gelfand, *Bayesian statistics without tears: A sampling-resampling perspective*, American Statistician **46** (1992) 84–88.

[23] W. Y. Tan, *On the chain multinomial model of HIV epidemic in homosexual populations and effects of randomness of risk factors*, in *Mathematical population Dynamics* 3, eds. O. Arino, D. E. Axelrod and M. Kimmel, Wuerz Publishing Ltd., Winnipeg, Manitoba (1995).

[24] W. Y. Tan, S. C. Tang and S. R. Lee, *Effects of Randomness of Risk Factors on the HIV Epidemic in Homo- sexual Populations*, SIAM Jour. Appl. Math. **55** (1995) 1697–1723.

[25] H. W. Hethcote and J. W. Van Ark, *Modeling HIV transmission and AIDS in the United States*, Lecture Notes in Biomath, Springer-Verlag, Berlin (1992).

[26] W. Y. Tan, S. C. Tang and S. R. Lee, *Estimation of HIV Seroconversion and Effects of Age in San Francisco Homosexual Populations*, Jour. Applied Statistics **25** (1998) 85–102.

[27] W. Y. Tan and Z. H. Xiang, *State space models of the HIV epidemic in homosexual populations and some applications*, Math. Biosciences **152** (1998) 29–61.

[28] W. Y. Tan and Z. H. Xiang, *Modeling the HIV epidemic with variable infection in homosexual populations by state space models*, J. Statist. Inference and Planning **78** (1999) 71–87.

[29] C. M. Hill and D. R. Littman, *Natural resistance to HIV*, Nature **382** (1996) 668–669.

[30] R. I. Connor, H. Mohri, Y. Cao and D. D. Ho, *Increased viral burden and cytopathicity correlate temporally with $CD4^{(+)}$ T-lymphocyte decline and clinical progression in human immunodeficiency virus type 1-infected individuals*, J. Virology **678** (1993) 1772–1777.

[31] U. S. Bureau of the Census, *Statistical Abstract of the United States: 108th edition*, Washington, D. C. (1987).

[32] G. F. Lemp, S. F. Payne, G. W. Rutherford *et al.*, *Projections of AIDS morbidity and mortality in San Francisco*, Jour. of Amer. Medic. Assoc. **263** (1989) 1497–1501.

[33] Z. H. Xiang, *Modeling the HIV Epidemic: Part I. Bayesian Estimation of the HIV Infection and Incubation Via Backcalculation. Part II. The State Space Model of the HIV Epidemic in Homosexual Populations*, Ph. D Thesis, Department of Mathematical Sciences, the University of Memphis (1997).

[34] W. Y. Tan and Z. H. Xiang, *Bayesian estimation of the HIV infection and incubation via the backcalculation method*, Invited paper at the IMS Asian and Pacific Region and ICSA meeting, July 7-9, 1997.

[35] P. Bacchetti, *Estimating the incubation period of AIDS comparing population infection and diagnosis pattern*, Jour. Amer. Statist. Association **85** (1990) 1002–1008.

[36] IMSL. *MATH/LIBRARY User's Manual*, IMSL, Houston, Texas (1989).

[37] W. Y. Tan and C. C. Brown, *A stochastic model for drug resistance and immunization. I. One drug case*, Math. Biosciences **97** (1989) 145–160.

Chapter 10

Stochastic Models of Carcinogenesis

In previous chapters, we have developed basic theories of some important stochastic models. To further illustrate applications of these models in medical research, in this chapter we will develop some important biologically supported stochastic models of carcinogenesis and illustrate its applications.

10.1. The Extended Multi-event Model of Carcinogenesis

Results of cancer molecular biology (see McDonald *et al.* [1], Tan [2], Tan *et al.* [3]–[5]; Weinberg [6]) imply that carciniogenesis is a stochastic multi-stage model with cancer tumors being developed from stem cells which have sustained a finite number of genetic and/or epigenetic changes and with intermediate-stage cells subjecting to stochastic cell proliferation and differentiation. For a single pathway, this model has been referred to by Tan and Chen ([7]–[8]) as an extended multi-event model of carcinogenesis. This is the most general model of carcinogenesis for a single pathway and is an extension of the multi-event model first proposed by Chu [9] and studied by Tan [2] and Little [10]. In this section we will develop some general theories for this model. In Secs. (10.2) and (10.3), we will give some examples to illustrate some applications of this model. As in Section (8.3.1.) in Chapter 8, let $N = I_0$ denote normal stem cells, T the cancer tumors and I_j the $j-$th stage initiated cells arising from the $(j-1)-$th stage initiated cells $(j = 1, \ldots, k)$ by some genetic and/or epigenetic changes. Then the stochastic extended k-stage multi-event model of carcinogenesis is characterized by the following postulates:

(a) As shown in Figure (8.1) in Chapter 8, the model assumes that cancer tumors develop from normal stem cells ($N = I_0$ cells) by the pathway $N(I_0) \rightarrow I_1 \rightarrow \ldots I_k \rightarrow Tumor$ by genetic and/or epigenetic changes.

(b) The I_j cells ($j = 0,, \ldots, k$) are subjected to stochastic proliferation (i.e. birth) and differentiation (i.e. death).

(c) Define I_k cells generated from I_{k-1} cells directly as primary I_k cells. Then cancer tumors develop from primary I_k cells by clonal expansion (i.e. stochastic birth-death process); see Yang and Chen [11] and **Remark (10.1)**.

(d) All cells develop independently of one another.

Because the total number of I_k cells at any time is the total number of primary I_k cells plus I_k cells generated by cell proliferation of other exiting I_k cells, postulate (c) implies that at any time the number of cancer tumors is much smaller than the total number of I_k cells. It follows that the I_k stage is transitory and the identifiable stages in the process of carcinogenesis are I_j ($j = 0, 1, \ldots, k - 1$) cells and cancer tumor T. For modeling the process of carcinogenesis given in Figure (8.1) in Chapter 8, we will therefore be entertaining a k-dimensional stochastic process $Z(t) = \{I_j(t), j = 0, 1, \ldots, k - 1, T(t)\}'$ ($t \geq t_0$), where $I_j(t)$ is the number of I_j cells at time t and $T(t)$ the number of tumors at time t.

As examples, in Figure (10.1) we present a multi-stage pathway for the squamous NSCLC (Non-Small Cell Lung cancer) as proposed by Osaka and Takahashi [12] and Wituba $et\ al.$ [13]. Similarly, given in Figure (10.2) is the multi-stage model for the $APC - \beta - Catenin - Tcf$ pathway for human colon cancer (Tan $et\ al.$ [14], Tan and Yan [15]). This is a major pathway which accounts for 80% of all human colon cancers. In Figure (10.3), we present another multistage model for human colon cancer involving mis-match repair genes (mostly hMLH1 and hMSH2) and the Bax gene in chromosome 19q and the $TGF\beta R_{II}$ gene in chromosome 3p (Tan $et\ al.$ [14], Tan and Yan [15]). This pathway accounts for about 15% of all human colon cancer.

To derive mathematical theories for the Stochastic Process $\{Z(t), t \geq t_0\}$, denote by $X(t) = (I_j(t), j = 0, 1, \ldots, k - 1)'$. Then one may readily assume that $X(t)$ is a Markov process even though $T(t)$ is in general not Markov (see Tan $et\ al.$ [3]–[5], [16]–[18]). Under this assumption, one can readily develop stochastic differential equations for the state variables $I_j(t)$ in

$\underset{\sim}{X}(t) = (I_j(t), j = 0, 1, \ldots, k-1)'$. This is given in equations (8.9)–(8.10) in Section (8.3.1) of Chapter 8. In Chapter 8, we have also derived the conditional probability distributions of $T(t)$ given $\{I_{k-1}(s), t_0 \leq s \leq t)$ (see equations (8.13) and (8.15) in Chapter 8). In these equations, $b_j(t), d_j(t)$ and $\alpha_j(t)$ are the birth rate, the death rate and the mutation rate of the I_j cells at time t respectively; $\{B_i(t), D_i(t), M_i(t)\}$ $(i = 0, 1, \ldots, k-1)$ are the numbers of birth and death of I_i cells and the number of mutations from $I_i \to I_{i+1}$ during $[t, t + \Delta t)$ respectively. Under the assumption that all cells develop by following the same biological mechanism independently of one another, we have that to the order of $o(\Delta t)$, the conditional probability distribution of $\{B_j(t), D_j(t), M_j(t)\}$ given $I_j(t)$ I_j cells at time t is multinomial with parameters $\{I_j(t), b_j(t)\Delta t, d_j(t)\Delta t, \alpha_j(t)\Delta t\}$. That is, to the order of $o(\Delta t)$, $\{B_j(t), D_j(t), M_j(t)\} | I_j(t) \sim \text{Multinomial}\{I_j(t); b_j(t)\Delta t, d_j(t)\Delta t, \alpha_j(t)\Delta t\}$.

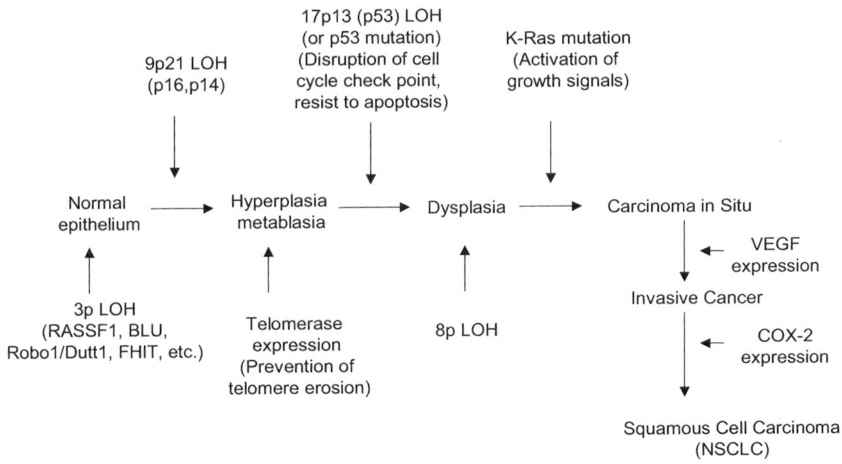

Fig. 10.1.   Genetic pathway of squamous cell carcinoma of non-small cell lung ancer

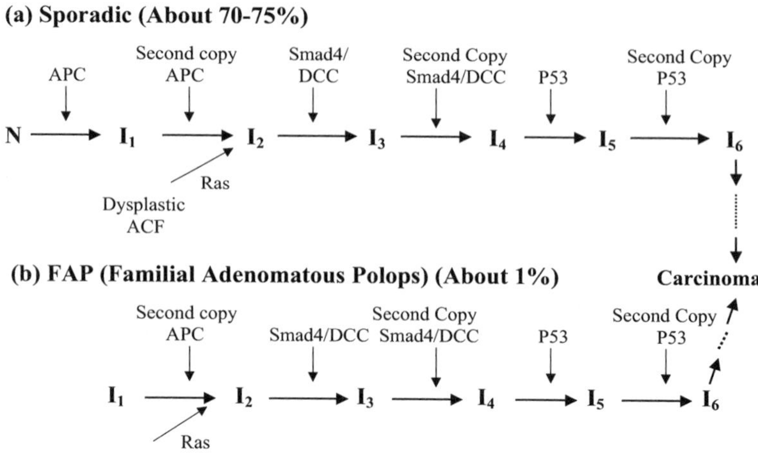

Fig. 10.2.   The CIN pathway of human colon cancer

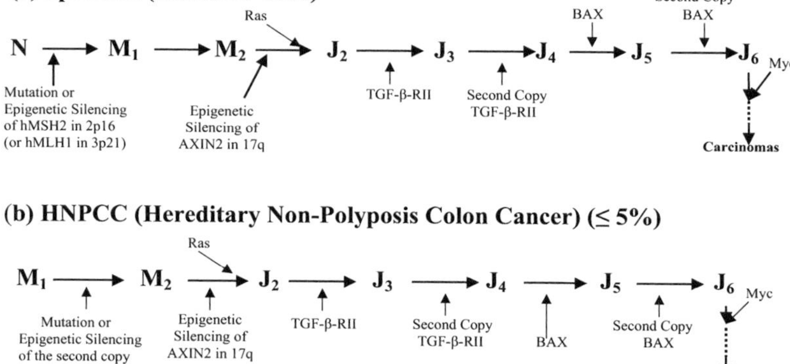

Fig. 10.3.   The MSI pathway of human colon cancer

Remark (10.1): To develop stochastic models of carcinogenesis, in the literature ([2], [19], [20]), it is conveniently assumed that the last stage cells (i.e. I_k cells) develop instantaneously into cancer tumors as soon as they are generated. In this case, one may identity I_k cells as cancer tumors so that $T(t)$

is Markov. However, because tumors develop from primary I_k cells by clonal expansion, for continuous time $T(t)$ can hardly be Markov ([3]–[5], [16]–[18]). On the other hand, for human cancers, if one let one time unit (i.e. $\Delta t = 1$) corresponding to 3 months or six months or one year, then because the growth of I_k cells is very rapid, practically one may assume that with probability close to one a primary I_k cell at time t would develop into a detectable tumor by time $t + 1$.

10.1.1. *The stochastic differential equations for $\underset{\sim}{X}(t)$*

Denote by $\underset{\sim}{e}(t) = \{e_j(t), j = 0, 1, \ldots, k-1\}'$. Put $\gamma_j(t) = b_j(t) - d_j(t)$ and let $A(t) = (a_{ij}(t))$ be the $k \times k$ matrix with the (i,j)–th element $a_{ij}(t)$ given by:

$$a_{ij}(t) = \gamma_j(t) \ \text{if} \ i = j, i = 1, \ldots, k$$
$$= \alpha_{i-1}(t) \ \text{if} \ j = i - 1, j = 2, \ldots, k$$
$$= 0 \ \text{if} \ j < i - 1 \ \text{or} \ j > i.$$

Then, in matrix notation the stochastic differential equations for $\underset{\sim}{X}(t)$ in equations (8.9)–(8.10) of Section (8.3.1) in Chapter 8 are expressed as:

$$\frac{d}{dt}\underset{\sim}{X}(t) = A(t)\underset{\sim}{X}(t) + \underset{\sim}{e}(t). \tag{10.1}$$

If the model is time homogeneous so that $(b_j(t) = b_j, d_j(t) = d_j, \alpha_j(t) = \alpha_j)$ for all $j = 0, 1, \ldots, k-1$, then $A(t) = A$ is independent of time t. In this case, the solution of equation (1) is, for $(t > t_0)$:

$$\underset{\sim}{X}(t) = e^{A(t-t_0)}\underset{\sim}{X}(t_0) + \underset{\sim}{\eta}(t), \tag{10.2}$$

where $\underset{\sim}{\eta}(t) = \int_{t_0}^{t} e^{A(t-y)}\underset{\sim}{e}(y)dy$, and where $e^{A(t-x)}$ is a matrix exponential function of t defined by $e^{A(t-x)} = \sum_{i=0}^{\infty} \frac{1}{i!}A^i(t-x)^i$ (see Section (4.4) in Chapter 4).

In equation (10.2), the expected value $E\underset{\sim}{\eta}(t)$ of $\underset{\sim}{\eta}(t)$ are zero's (i.e. $E\underset{\sim}{\eta}(t) = \underset{\sim}{0}_k$) and the covariance matrix $\Sigma_e(t)$ of $\underset{\sim}{\eta}(t)$ is $\Sigma_e(t) = \int_{t_0}^{t} e^{A(t-y)}V_e(y)e^{A'(t-y)}dy$, where A' is the transpose of A. Hence, the expected value $E[\underset{\sim}{X}(t)]$ and the covariance matrix $\Sigma_X(t)$ of $\underset{\sim}{X}(t)$ are given respectively by:

$$E[\underset{\sim}{X}(t)] = e^{A(t-t_0)}E[\underset{\sim}{X}(t_0)], \text{ and, } \Sigma_X(t) = e^{A(t-t_0)}\Sigma_X(t_0)e^{A'(t-t_0)} + \Sigma_e(t).$$

Note that the diagonal elements γ_j $(j = 0, 1, \ldots, k-1)$ of the A matrix in the above equations are eigenvalues of A. (Note that $\sum_{j=1}^{m} E_j = I_{k-1}$, $E_j^2 = E_j$ and $E_i E_j = 0$ if $i \neq j$; see Appendix (2.11.3) in Chapter 1.) Using these spectral expansions, the solutions in equation (10.12) become:

$$\underset{\sim}{X}(t) = \sum_{j=1}^{m} e^{\omega_j(t-t_0)} E_j \underset{\sim}{X}(t_0) + \sum_{j=1}^{m} \underset{\sim}{\eta}_j(t), \tag{10.3}$$

where $\underset{\sim}{\eta}_j(t) = \int_{t_0}^{t} e^{\omega_j(t-y)} E_j \underset{\sim}{e}(y)dy$.

If all the γ_j's are distinct, then all eigenvalues of A are distinct. In this case, $m = k$ and $\omega_j = \gamma_{j-1}, j = 1, \ldots, k$; the right eigenvector of A for ω_j is $\underset{\sim}{B}_j = (B_{j1}, \ldots, B_{jk})'$ and the left eigenvector of A for ω_j is $\underset{\sim}{C}_j = (C_{1j}, \ldots, C_{kj})'$, where

$$B_{jr} = \prod_{u=j}^{r-1} [\alpha_{u-1}/(\gamma_{j-1} - \gamma_u)] \text{ if } j < r$$
$$= 1 \text{ if } j = r$$
$$= 0 \text{ if } j > r,$$

and

$$C_{rj} = \prod_{v=r-1}^{j-2} [\alpha_v/(\gamma_{j-1} - \gamma_v)] \text{ if } r < j$$
$$= 1 \text{ if } r = j,$$
$$= 0 \text{ if } r > j$$

In the spectral expansions of A and $e^{A(t-t_0)}$, if the γ_j are all distinct, then $m = k, \omega_j = \gamma_{j-1}$ and $E_j = \underset{\sim}{B}_j \underset{\sim}{C}_j'$. Note that if all the γ_j's are distinct, for normal people at time t_0, the expected number of $I_j(t)$ is, for $j = 1, \ldots, k-1$:

$$E[I_j(t)] = u_j(t) = \frac{1}{\alpha_j} \sum_{v=0}^{j} \theta_j(v) e^{\gamma_v(t-t_0)} \text{ if } b_0 - d_0 \neq 0, \tag{10.4}$$

$$= \frac{1}{\alpha_j} \sum_{v=1}^{j} \theta_j(v)[e^{\gamma_v(t-t_0)} - 1] \quad \text{if } b_0 - d_0 = 0, \tag{10.5}$$

where $\theta_j(v) = I_0(t_0)\{\prod_{r=0}^{j} \alpha_r\} A_{0j}(v)$ with $A_{0j}(v) = \{\prod_{u=0, u \neq v}^{j}(\gamma_v - \gamma_u)\}^{-1}$. In particular,

$$E[I_{k-1}(t)] = u_{k-1}(t) = \frac{1}{\alpha_{k-1}} \sum_{v=0}^{k-1} \theta_{k-1}(v) e^{\gamma_v(t-t_0)} \quad \text{if } b_0 - d_0 \neq 0$$

$$= \frac{1}{\alpha_{k-1}} \sum_{v=1}^{k-1} \theta_{k-1}(v)[e^{\gamma_v(t-t_0)} - 1] \quad \text{if } b_0 - d_0 = 0.$$

10.1.2. *The probability distributions of the state variables $\underset{\sim}{X}(t)$ and dummy state variables*

To derive the transition probabilities $P\{\underset{\sim}{X}(t + \Delta t) = \underset{\sim}{v} \,|\, \underset{\sim}{X} = \underset{\sim}{u}\} = P\{I_j(t + \Delta t) = v_j, j = 0, 1, \ldots, k - 1 | I_j(t) = u_j, j = 0, 1, \ldots, k - 1\}$, denote by $h(x; \lambda)$ the probability density at x of a Poisson random variable $X \sim Poisson(\lambda)$, $f(x; n, p)$ the probability density at x of a Binomial random variable $X \sim Binomial(n; p)$ and $g(x, y; n, p_1, p_2)$ the probability density at (x, y) of a bivariate multinomial random variables $(X, Y) \sim Multinomial(n; p_1, p_2)$. Also, note the following results:

(a) If $\{B_j(t), D_j(t), M_j(t)\}|I_j(t) \sim \text{Multinomial}\{I_j(t); b_j(t)\Delta t, d_j(t)\Delta t, \alpha_j(t)\Delta t\}$, then $\{B_j(t), D_j(t)\}|I_j(t) \sim Multinomial\{I_j(t); b_j(t)\Delta t, d_j(t)\Delta t\}$ and $D_j(t)|\{I_j(t), B_j(t)\} \sim Binomial\{I_j(t) - B_j(t), \frac{d_j(t)\Delta t}{1-b_j(t)\Delta t}\}$. By Lemma (10.1), to order of $o(\Delta t)$, $M_j(t)|I_j(t) \sim Binomial\{I_j(t), \alpha_j(t)\Delta t\}$, independently of $\{B_j(t), D_j(t)\}$.

(b) Since $I_0(t + \Delta t) = I_0(t) + B_0(t) - D_0(t)$, $I_0(t + \Delta t)$ depends only on $I_0(t)$ and is independent of $I_j(t)$ $(j = 1, \ldots, k - 1)$; further, given $(I_0(t) = u_0, B_0(t) = r_0)$ $I_0(t + \Delta t) = v_0$ if and only if $D_0(t) = u_0 - v_0 + r_0$. Thus,

$$P\{I_0(t + \Delta t) = v_0 \,\mid\, I_i(t) = u_i, i = 0, 1, \ldots, k - 1\}$$
$$= P\{I_0(t + \Delta t) = v_0 | I_0(t) = u_0\}$$

and

$$P\{I_0(t + \Delta t) = v_0 \,\mid\, I_0(t) = u_0, B_0(t) = r_0\}$$

$$= P\{D_0(t) = u_0 - v_0 + r_0 | I_0(t) = u_0, B_0(t) = r_0\}$$

$$= f\left(u_0 - v_0 + r_0; u_0 - r_0, \frac{d_0(t)\Delta t}{1 - b_0(t)\Delta t}\right).$$

(c) Since $I_j(t + \Delta t) = I_j(t) + M_{j-1}(t) + B_j(t) - D_j(t)$ for $j = 1, \ldots, k-1$, $I_j(t + \Delta t)$ depends only on $\{I_{j-1}(t), I_j(t)\}$, and is independent of $I_r(t)$ for $r \neq j$ and $r \neq j - 1$. Further, given $\{I_j(t) = u_j, B_j(t) = r_j, D_j(t) = m_j\}$, $I_j(t + \Delta t) = v_j$ if and only if $M_{j-1}(t) = v_j - u_j - r_j + m_j$. Thus, for $j > 1$,

$$P\{I_j(t + \Delta t) = v_j \mid I_i(t) = u_i, i = 0, 1, \ldots, k-1\}$$
$$= P\{I_j(t + \Delta t) = v_j | I_j(t) = u_j, I_{j-1}(t) = u_{j-1}\},$$

and

$$P\{I_j(t + \Delta t) = v_j \mid I_{j-1}(t) = u_{j-1}, I_j(t) = u_j, B_j(t) = r_j, D_j(t) = m_j\}$$
$$= P\{M_{j-1}(t) = v_j - u_j - r_j + m_j | I_{j-1}(t) = u_{j-1}\}$$
$$= f(v_j - u_j - r_j + m_j; u_{j-1}, \alpha_{j-1}(t)\Delta t).$$

From results (b)–(c) given above, we have:

$$P\{I_j(t + \Delta t) = v_j, \, j = 0, 1, \ldots, k-1 | I_j(t) = u_j, j = 0, 1, \ldots, k-1\}$$
$$= P\{I_0(t + \Delta t) = v_0 | I_0(t) = u_0\}$$
$$\times \prod_{j=1}^{k-1} P\{I_j(t + \Delta t) = v_j | I_{j-1}(t) = u_{j-1}, I_j(t) = u_j\}. \quad (10.6)$$

From result (a) above, we have for $j = 1, \ldots, k-1$:

$$P\{B_j(t) = r_j, D_j(t) = m_j | I_j(t) = u_j\} = g(r_j, m_j; u_j, b_j(t)\Delta t, d_j(t)\Delta t).$$

From result (c) above and noting that $0 \leq v_j - u_j - r_j + m_j \leq u_{j-1}$ if and only $u_j + r_j - v_j \leq m_j \leq u_{j-1} + u_j + r_j - v_j$ and $0 \leq m_j \leq u_j - r_j$ given $(I_j(t) = u_j, B_j(t) = r_j)$, we obtain for $j = 1, \ldots, k-1$:

$$P\{I_j(t + \Delta t) = v_j \mid I_{j-1}(t) = u_{j-1}, I_j(t) = u_j\}$$
$$= \sum_{r_j=0}^{u_j} \sum_{m_j=a_{1j}}^{a_{2j}} g\{r_j, m_j; u_j, b_j(t)\Delta t, d_j(t)\Delta t\}$$
$$\times f\{v_j - u_j - r_j + m_j; u_{j-1}, \alpha_{j-1}(t)\Delta t\}, \quad (10.7)$$

where $a_{1j} = Max(0, u_j + r_j - v_j)$ and $a_{2j} = Min(u_j - r_j, u_{j-1} + u_j + r_j - v_j)$.

Similarly, $P\{B_0(t) = r_0|I_0(t) = u_0\} = f(r_0; u_0, b_0(t)\Delta t)$. From result (b) given above and noting that $0 \leq r_0 \leq u_0$ given $I_0(t) = u_0$ and $0 \leq u_0 - v_0 + r_0 \leq u_0 - r_0$ if and only if $v_0 - u_0 \leq r_0$ and $2r_0 \leq v_0$, we obtain:

$$P\{I_0(t + \Delta t) = v_0|I_0(t) = u_0\} = \sum_{r_0 = Max(0, v_0 - u_0)}^{Min(u_0, [\frac{v_0}{2}])} f(r_0; u_0, b_0(t)\Delta t)$$

$$\times f\left(u_0 - v_0 + r_0; u_0 - r_0, \frac{d_0(t)\Delta t}{1 - b_0(t)\Delta t}\right)$$

$$= \sum_{r_0 = Max(0, v_0 - u_0)}^{Min(u_0, [\frac{v_0}{2}])} g(r_0, u_0 - v_0 + r_0; u_0, b_0(t)\Delta t, d_0(t)\Delta t), \quad (10.8)$$

where $[\frac{v_0}{2}]$ is the largest integer $\leq \frac{v_0}{2}$.

For applying the above probabilities to fit cancer data, expand the model by defining the dummy state variables $\underset{\sim}{U}(t) = \{B_0(t), B_i(t), D_i(t), i = 1, \ldots, k - 1\}$. Put $\underset{\sim}{W}(t + \Delta t) = \{\underset{\sim}{U}(t)', \underset{\sim}{X}(t + \Delta t)'\}'$. Then, $\{\underset{\sim}{W}(t), t \geq t_0\}$ is Markov with transition probability $P\{\underset{\sim}{W}(t + \Delta t)|\underset{\sim}{W}(t)\}$ given by:

$$P\{\underset{\sim}{W}(t + \Delta t)|\underset{\sim}{W}(t)\} = P\{\underset{\sim}{X}(t + \Delta t)|\underset{\sim}{U}(t), \underset{\sim}{X}(t)\}P\{\underset{\sim}{U}(t)|\underset{\sim}{X}(t)\}, \quad (10.9)$$

where

$$P\{\underset{\sim}{U}(t) \mid \underset{\sim}{X}(t)\} = f(B_0(t); I_0(t), b_0(t)\Delta t)$$

$$\times \prod_{r=1}^{k-1} g\{B_r(t), D_r(t); I_r(t), b_r(t)\Delta t, d_r(t)\Delta t\}; \quad (10.10)$$

and

$$P\{\underset{\sim}{X}(t + \Delta t) \mid \underset{\sim}{U}(t), \underset{\sim}{X}(t)\}$$

$$= f\left\{I_0(t) - I_0(t + \Delta t) + B_0(t); I_0(t) - B_0(t), \frac{d_0(t)\Delta t}{1 - b_0(t)\Delta t}\right\}$$

$$\times \prod_{r=1}^{k-1} f\{M_r(t); I_{r-1}(t), \alpha_{r-1}(t)\Delta t\}, \quad (10.11)$$

where $M_r(t) = I_r(t + \Delta t) - I_r(t) - B_r(t) + D_r(t)$ for $r = 1, \ldots, k - 1$.

Lemma (10.1). Let

$$\{B_j(t), D_j(t), M_j(t)\}|I_j(t) \sim \text{Multinomial}\{I_j(t); b_j(t)\Delta t, d_j(t)\Delta t, \alpha_j(t)\Delta t\}.$$

Then to the order of $o(\Delta t)$, $\{B_j(t), D_j(t)\}|I_j(t) \sim \text{Multinomial}\{I_j(t); b_j(t)\Delta t, d_j(t)\Delta t\}$ and $M_j(t)\}|I_j(t) \sim \text{Binomial}\{I_j(t); \alpha_j(t)\Delta t\}$ independently of $\{B_j(t), D_j(t)\}$.

Proof: By using the method of probability generating function (pgf), the proof is straightforward. (See Exercise 10.4)

10.1.3. *The probability distribution of the number of detectable tumors and an approximation*

In Section (8.3.1) of Chapter 8, it is shown that given $\{I_{k-1}(s)\ (t_0 \leq s \leq t)\}$, the conditional probability that a normal individual at time t_0 would develop cancer tumors during the time period $[t_{j-1}, t_j)\ (t_j > t_{j-1})$ is $Q(t_{j-1}, t_j)$, where

$$Q(t_{j-1}, t_j) = e^{-\int_{t_0}^{t_{j-1}} I_{k-1}(s)\alpha_{k-1}(s)P_T(s,t_{j-1})ds} - e^{-\int_{t_0}^{t_j} I_{k-1}(s)\alpha_{k-1}(s)P_T(s,t_j)ds},$$

and where $P_T(s, t)$ is the probability that a primary I_k cell at time s develops into a detectable cancer tumor by time t (see equation (8.12) in Chapter 8).

It follows that for people who are normal at time t_0, the probability that these people develop cancer tumors during the time period $[t_{j-1}, t_j)\ (t_j > t_{j-1})$ is $Q_T(j) = EQ(t_{j-1}, t_j)$, the expected value of $Q(t_{j-1}, t_j)$.

To derive an approximation to $Q_T(j)$, observe that in many human cancers the mutation rate $\alpha_{k-1}(t)$ is very small (e.g. $10^{-4} \to 10^{-8}$) and that one may practically assume $\alpha_{k-1}(t) = \alpha_{k-1}$. In this case, we have:

$$Q_T(j) = E\{e^{-\alpha_{k-1}\int_{t_0}^{t_{j-1}} I_{k-1}(s)P_T(s,t_{j-1})ds} - e^{-\alpha_{k-1}\int_{t_0}^{t_j} I_{k-1}(s)P_T(s,t_j)ds}\}$$

$$= \{e^{-\alpha_{k-1}\int_{t_0}^{t_{j-1}} E[I_{k-1}(s)]P_T(s,t_{j-1})ds} - e^{-\alpha_{k-1}\int_{t_0}^{t_j} E[I_{k-1}(s)]P_T(s,t_j)ds}\}$$

$$+ o(\alpha_{k-1}). \tag{10.12}$$

The result in equation (10.12) shows that to order of $o(\alpha_{k-1})$, $Q_T(j)$ depends on the state variables only through the expected numbers of $I_{k-1}(s)$ $(t_0 \leq s \leq t)$.

10.1.4. *A statistical model based on cancer incidence*

For studying cancer models, the data available usually are cancer incidence data which give numbers of new cancer cases during different time periods given numbers of people at risk for cancer. For the lung cancer example, the data given in Table (10.1) are given by $Y = \{y_i(j), n_i(j), i = 1, \ldots, m, j = 1, \ldots, n\}$, where $y_i(j)$ is the observed number of cancer cases developed during the time period $[t_{j-1}, t_j)$ under exposure level s_i of smoking and where $n_i(j)$ is the number of normal people at t_0 from whom $y_i(j)$ of them developed cancer during $[t_{j-1}, t_j)$.

To find the probability distribution of $y_i(j)$ given $n_i(j)$ and given the state variables, let $I_{k-1}(t; i)$ denote the number of I_{k-1} cells at time t in individuals who are exposed to the risk factor with dose level s_i. Let $Q_i(t_{j-1}, t_j)$ denote the conditional probability given $\{I_{k-1}(s; i), s \leq t_j\}$ that people who are normal at t_0 and who are exposed to the risk factor with dose level s_i develop tumors during $[t_{j-1}, t_j)$. Then

$$Q_i(t_{j-1}, t_j) = e^{-\int_{t_0}^{t_{j-1}} I_{k-1}(s;i)\alpha_{k-1}(s;i)P_T(s,t_{j-1})ds} - e^{-\int_{t_0}^{t_j} I_{k-1}(s;i)\alpha_{k-1}(s;i)P_T(s,t_j)ds},$$

where $\alpha_{k-1}(t; i)$ is the $\alpha_{k-1}(t)$ under exposure to the risk factor with dose level s_i.

For people who are normal at time t_0 and who are exposed to the risk factor with dose level s_i, the probability that these people develop tumors during the j–th age period is $Q_T(j; i) = E[Q_i(t_{j-1}, t_j)]$. Assuming that all people under exposure to risk factor with level s_i develop cancer by following the same biological mechanism independently of one another, then the conditional probability distribution of $y_i(j)$ given $n_i(j)$ is binomial with parameters $\{n_i(j), Q_T(j; i)\}$. Because $n_i(j)$ is very large and $Q_T(j; i)$ is very small, we have:

$$y_j(i)|n_i(j) \sim \text{Poisson}\{n_i(j)Q_T(j; i)\},$$

Or equivalently,

$$P\{y_i(j)|n_i(j)\} = h\{y_i(j); n_i(j)Q_T(j; i)\}. \tag{10.13}$$

Let $\alpha_{k-1}(t;i) = \alpha_{k-1}(i)$ be the rate of transition $I_{k-1} \to I_k$ under exposure to the risk factor with dose level s_i. Then, if $\alpha_{(k-1)}(i)$ is very small,

$$Q_T(j;i) = \left\{ e^{-\alpha_{k-1}(i) \int_{t_0}^{t_{j-1}} E[I_{k-1}(s;i)] P_T(s,t_{j-1}) ds} \right.$$
$$\left. - e^{-\alpha_{k-1}(i) \int_{t_0}^{t_j} E[I_{k-1}(s;i)] P_T(s,t_j) ds} \right\} + o(\alpha_{k-1}(i)). \qquad (10.14)$$

With $P\{y_i(j)|n_i(j)\}$ as given in equation (10.13), the joint density of $\boldsymbol{Y}$ given the state variables is then given by:

$$P\{\boldsymbol{Y}|\boldsymbol{X},\boldsymbol{U}\} = \prod_{i=1}^{m}\prod_{j=1}^{n} P\{y_i(j)|n_i(j)\} = \prod_{i=1}^{m}\prod_{j=1}^{n} h\{y_i(j); n_i(j)Q_T(j;i)\}.$$

$$(10.15)$$

10.1.5. *A state space model for the extended multi-event stochastic model of carcinogenesis*

State space model is a stochastic model which consists of two sub-models: The stochastic system model which is the stochastic model of the system and the observation model which is a statistical model based on available observed data from the system. For the extended multi-event model of carcinogenesis defined above, the stochastic system model is the stochastic process $\{\underset{\sim}{Z}(t), t \geq t_0\} = \{\underset{\sim}{X}(t), T(t), t \geq t_0\}$ whereas the observation model is a statistical model based on available data from the system. For the lung cancer example given in Table (10.1), the observation model is the statistical model for $\boldsymbol{Y} = \{y_i(j), i = 1, \ldots, mj = 1, \ldots, n\}$ given in Section (10.1.4) above.

By using this state space model, one can readily estimate the unknown genetic parameters, predict cancer incidence and check the validity of the model by using the generalized Bayesian inference and Gibbs sampling procedures as described in Chapter 9. For these purposes, we chose some fixed small time interval as 1 time unit (i.e. $\Delta t \sim 1$). For animal models, this time unit may be 1 day or 3 days; for human beings, this time unit may be 3 months or six months or longer. This would change the model to a model with discrete time $(t = t_0, t_0 + 1, \ldots, t_M)$.

10.1.5.1. *The stochastic system model, the augmented state variables and probability distributions*

Assuming discrete time, the state variables and the augmented state variables are then given by $\boldsymbol{X} = \{\underset{\sim}{X}(j), j = t_0, \ldots, t_M\}$ and $\boldsymbol{U} = \{\underset{\sim}{U}(j), j = t_0, \ldots, t_M - 1\}$ respectively. The probability distribution for the state variables $\boldsymbol{X}$ and the augmented state variables $\boldsymbol{U}$ are derived in equation (10.9). From equations (10.9)–(10.11), the joint density of $\{\boldsymbol{X}, \boldsymbol{U}\}$ is

$$P\{\boldsymbol{X}, \boldsymbol{U} | \Theta, \underset{\sim}{X}(t_0)\} = \prod_{t=t_0}^{t_M - 1} P\{\underset{\sim}{X}(t+1) | \underset{\sim}{X}(t), \underset{\sim}{U}(t)\} P\{\underset{\sim}{U}(t) | \underset{\sim}{X}(t)\},$$

$$(10.16)$$

where

$$P\{\underset{\sim}{U}(t) | \underset{\sim}{X}(t)\} = f(B_0(t); I_0(t), b_0(t)) \prod_{r=1}^{k-1} g\{B_r(t), D_r(t); I_r(t), b_r(t), d_r(t)\},$$

$$t = t_0, \ldots, T_M - 1;$$

and

$$P\{\underset{\sim}{X}(t+1) | \underset{\sim}{U}(t), \underset{\sim}{X}(t)\} = f\left\{ I_0(t) - I_0(t+1) + B_0(t); I_0(t) - B_0(t), \frac{d_0(t)}{1 - b_0(t)} \right\}$$

$$\times \prod_{r=1}^{k-1} f\{M_r(t); I_{r-1}(t), \alpha_{r-1}(t)\}, t = t_0, \ldots, t_M - 1$$

where $M_r(t) = I_r(t+1) - I_r(t) - B_r(t) + D_r(t)$ for $r = 1, \ldots, k - 1$.

This probability distribution provides information from the stochastic system model.

10.1.5.2. *The observation model and the probability distribution of cancer incidence*

The observation model is a statistical model based on data $\boldsymbol{Y}$ as given in Section (10.1.4). This statistical model depends on the stochastic model $\{\underset{\sim}{Z}(t), t \geq t_0\}$ through the probabilities $Q_T(j; i)$. To derive this probability under the assumption that one time unit (i.e. $\Delta t \sim 1$) corresponds to 3 months or 6 months or longer, note that one may practically assume $P_T(s, t) \approx 1$ if

$t - s > 1$ (see Tan *et al.* [3]–[5]). Then, as shown in (8.3.1.3) in Chapter 8 and in Section (10.1.3), $Q_T(j; i)$ is given by:

$$Q_T(j; i) = exp\left\{ -\alpha_{k-1}(i) \sum_{t=t_0}^{t_{j-1}-1} E[I_{k-1}(t; i)] \right\}$$
$$\times \left(1 - e^{-R(i;j)\alpha_{k-1}(i)} \right) + o(Min(\alpha_{k-1}(i), i = 1, \ldots, m)),$$

where $R(i; j) = \sum_{t=t_{j-1}}^{t_j - 1} E[I_{k-1}(t; i)]$.

From equation (10.15), the conditional likelihood of the parameters Θ given data $\boldsymbol{Y} = \{y_i(j), i = 1, \ldots, m, j = 1, \ldots, n\}$ and given the state variables is

$$L\{\Theta | \boldsymbol{Y}, \boldsymbol{X}, \boldsymbol{U}\} = \prod_{i=1}^{m} \prod_{j=1}^{n} h\{y_i(j); n_i(j)Q_T(j; i)\}. \tag{10.17}$$

From the above distribution results, the joint density of $\{\boldsymbol{X}, \boldsymbol{U}, \boldsymbol{Y}\}$ is

$$P\{\boldsymbol{X}, \boldsymbol{U}, \boldsymbol{Y}\} = \prod_{i=1}^{m} \prod_{j=1}^{n} h\{y_i(j); n_i(j)Q_T(j; i)\}$$
$$\times \prod_{t=t_{j-1}+1}^{t_j} P\{\underset{\sim}{X}(t) | \underset{\sim}{X}(t-1), \underset{\sim}{U}(t-1)\}$$
$$\times P\{\underset{\sim}{U}(t-1) | \underset{\sim}{X}(t-1)\}. \tag{10.18}$$

The above distribution will be used to derive the conditional posterior distribution of the unknown parameters Θ given $\{\boldsymbol{X}, \boldsymbol{U}, \boldsymbol{Y}\}$. Notice that because the number of parameters is very large, the classical sampling theory approach by using the likelihood function $P\{\boldsymbol{Y} | \boldsymbol{X}, \boldsymbol{U}\}$ alone is not possible without making assumptions about the parameters; however, this problem can easily be avoided by new information from the stochastic system model and the prior distribution of the parameters.

10.1.6. *Statistical inference for the multi-event model of carcinogenesis*

For fitting the multi-event model to cancer incidence data, one would need to estimate the unknown parameters and to predict cancer incidence. The most efficient inference procedure for this purpose is the generalized Bayesian and

Gibbs sampling procedures (Carlin and Louis [21]). The unknown parameters for the multi-event model are the birth rates, death rates and mutation rates which are in general functions of the dose levels of the risk factors to which individuals are exposed. To assess effects of dose level, let $x_i = \log(1 + s_i)$, and assume $\{\alpha_0(t; i) = \alpha_{00}(t)e^{\alpha_{01}(t)x_i}, \alpha_j(t; i) = \alpha_{j0}(t)e^{\alpha_{j1}(t)x_i}, b_j(t; i) = b_{j0}(t) + b_{j1}(t)x_i, d_j(t; i) = d_{j0}(t) + d_{j1}(t)x_i, j = 1, \ldots, k-1\}$ (Tan *et al.* [3]–[4], Tan [5]). Let $\lambda_i = N(t_0)\alpha_{00}e^{\alpha_{01}x_i}$. When the model is time homogeneous, then the set Θ of unknown parameters are $\Theta = \{\alpha_{0u}, \alpha_{ju}, b_{ji}, d_{ju}, u = 0, 1, j = 1, \ldots, k - 1\}$.

To estimate these unknown parameters and to predict the state variables, we will use the generalized Bayesian and Gibbs sampling procedures as described in Chapter 9. This method combines the prior distribution $P\{\Theta\}$ of the parameters with the joint density $P\{\boldsymbol{Y}, \boldsymbol{X}, \boldsymbol{U}|\Theta\}$. As such, the generalized Bayesian approach not only extract information from the data and the prior distribution of the parameters but also from the basic mechanism of the system through stochastic system model.

10.1.6.1.   *The prior distribution of the parameters*

Put $\Theta = \{\Theta_1, \Theta_2\}$, where $\Theta_2 = \{\alpha_{(k-1)u}, u = 0, 1\}$ and $\Theta_1 = \Theta - \Theta_2 = \{\alpha_{(j-1)u}, b_{ju}, d_{ju}, u = 0, 1, j = 1, \ldots, k - 1\}$. For the prior distributions of Θ, we assume that **a priori**, Θ_1 and Θ_2 are independently distributed of each other. Because biological information have suggested some lower bounds and upper bounds for the mutation rates and for the proliferation rates, we assume

$$P(\Theta_i) \propto c \, (c > 0)$$

where c is a positive constant if these parameters satisfy some biologically specified constraints; and equal to zero for otherwise. These biological constraints are:

(1) For the mutation rates of the I_i cells, $1 < N(t_0)\alpha_{00} < 1000 \ (N \to I_1)$, $-10^{-1} < \alpha_{01} < 10^2$, $10^{-6} < \alpha_{i0} < 10^{-4}, -10^{-1} < \alpha_{i1} < 10^2, i = 1, \ldots, k - 1$.

(2) For the proliferation rates of I_i cells, $0 < b_{i0}, b_{i1} < 0.5,$ and $- 10^{-2} < d_{i0}, d_{i1} < 0.5, i = 1, \ldots, k - 1$.

We will refer the above prior as a partially informative prior which may be considered as an extension of the traditional non- informative prior given in Box and Tiao [22].

10.1.6.2. *The posterior distribution of the unknown parameters*

To derive the posterior distribution of Θ given $\{\boldsymbol{X}, \boldsymbol{U}, \boldsymbol{Y}\}$, for the $i-$th dose level, denote the $\{\underset{\sim}{X}(t), \underset{\sim}{U}(t)\}$ by $\underset{\sim}{X}(t; i) = \{I_j(t; i), j = 1, \ldots, k-1\}$ and $\underset{\sim}{U}(t; i) = \{B_j(t; i), D_j(t; i), j = 1, \ldots, k-1\}$; also, denote the $M_j(t)$ as $M_j(t; i)$ under dose level s_i. Denote by $S_B(j; i) = \sum_{t=t_0}^{t_M-1} B_j(t; i)$, $S_D(j; i) = \sum_{t=t_0}^{t_M-1} D_j(t; i)$, $S_R(j; i) = \sum_{t=t_0}^{t_M-1}[I_j(t; i) - B_j(t; i) - D_j(t; i)]$, $S_M(j; i) = \sum_{t=t_0}^{t_M-1}[I_j(t+1; i) - I_j(t; i) - B_j(t; i) + D_j(t; i)]$, $S_A(j; i) = \sum_{t=t_0}^{t_M-1}[I_{j-1}(t; i) - I_j(t+1; i) + I_j(t; i) + B_j(t; i) - D_j(t; i)]$. From equations (10.10)-(10.11), the conditional posterior distribution of $\Theta_1 = \{\alpha_{(j-1)u}, b_{ju}, d_{ju}, u = 0, 1, j = 1, \ldots, k-1\}$ is:

$$P\{\Theta_1 | \boldsymbol{X}, \boldsymbol{U}, \boldsymbol{Y}, \alpha_{(k-1)u}, u = 0, 1\} \propto P\{\Theta_1\} \prod_{i=1}^{m} \lambda_i^{S_M(0;i)} e^{-(t_M-t_0)\lambda_i}$$

$$\times (b_1(i))^{S_B(1;i)} (d_1(i))^{S_D(1;i)} (1 - b_1(i) - d_1(i))^{S_R(1;i)}$$

$$\times \prod_{j=2}^{k-1} (b_j(i))^{S_B(j;i)} (d_j(i))^{S_D(j;i)} (1 - b_j(i) - d_j(i))^{S_R(j;i)}$$

$$\times \alpha_{j-1}(i)^{S_M(j;i)} (1 - \alpha_{j-1}(i))^{S_R(j;i)}. \tag{10.19}$$

From equation (10.14), the conditional posterior distribution of $\{\alpha_{(k-1)u}, u = 0, 1\}$ given $\{\boldsymbol{X}, \boldsymbol{U}, \boldsymbol{Y}, \Theta_1\}$ is

$$P\{\alpha_{(k-1)u}, u = 0, 1 | \boldsymbol{X}, \boldsymbol{U}, \boldsymbol{Y}, \Theta_1\} \propto P\{\Theta_2\} \prod_{i=1}^{m} \prod_{j=1}^{n} e^{-y_j(i)\alpha_{k-1}(i)A(i,j;t_{j-1})}$$

$$\times e^{-n_j(i)Q_T(j;i)} \left(1 - e^{-\alpha_{k-1}(i)R(j;i)}\right)^{y_j(i)}, \tag{10.20}$$

where $A(i, j; t) = \sum_{s=t_0}^{t_{j-1}-1} E[I_{k-1}(s; i)]$ for $i = 1, \ldots, m; j = 1, \ldots, n$, and $R(j; i) = \sum_{t=t_{j-1}}^{t_j-1} E[I_{k-1}(t; i)]$.

10.1.6.3. *The multi-level Gibbs sampling procedures*

Using the above distribution results, the multi-level Gibbs sampling procedures for estimating the unknown parameters Θ and the state variables $\boldsymbol{X}$ are given by the following loop:

(1) Given the parameter values, use the stochastic equation (10.1) (or

equations (8.9)–(8.10) in Section (8.3.1) in Chapter 8) and the associated probability distributions to generate a large sample of $\{X, U\}$. Then, by combining this large sample with $P\{Y|X, U\}$, select $\{X, U\}$ from this sample through the weighted Bootstrap method due to Smith and Gelfant [23]. This selected $\{X, U\}$ is then a sample generated from $P\{X, U|\Theta, Y\}$ although the latter density is unknown (for proof, see Chapter 3). Call the generated sample $\{X^{(*)}, U^{(*)}\}$.

(2). On substituting $\{U^{(*)}, X^{(*)}\}$ which are generated numbers from the above step, generate Θ_i from the conditional density $P\{\Theta|X^{(*)}.U^{(*)}, Y, \Theta_j\}(i \neq j)$ given by equations (10.19)–(10.20).

(3). With Θ being generated from Step 2 above, go back to Step 1 and repeat the above [1]–[2] loop until convergence.

The convergence of the above algorithm has been proved in Chapter 3. At convergence, one then generates a random sample of $\{X, U\}$ from the conditional distribution $P\{X, U|Y\}$ of $\{X, U\}$ given Y, independent of Θ and a random sample of Θ from the posterior distribution $P\{\Theta|Y\}$ of Θ given Y, independent of $\{X, U\}$. Repeat these procedures one then generates a random sample of size N of $\{X, U\}$ and a random sample of size M of Θ. One may then use the sample means to derive the estimates of $\{X, U\}$ and Θ and use the sample variances as the variances of these estimates. Alternatively, one may also use Efron's bootstrap method (Efron [24]) to derive estimates of the standard errors of the estimates.

10.2. Assessment of Effects of Smoking on Lung Cancer: An Example of Application of the Model in (10.1)

It has long been recognized that smoking can cause lung cancer (Osada and Takahashi [12], Wistuba *et al.* [13], Fong and Sekido [25]) in most cases. To reveal the basic mechanisms of how tobacco nicotine cause lung cancer, in this section we will apply the multi-event model in Section (10.1) to analyze the British physician smoking data given in Doll and Peto [26]. Given in Table (10.1) is the British physician data extracted from the paper by Doll and Peto [26]. In this data set, we have included only the age groups between 40 years old and 80 years old because in this data set, lung cancer incidence are non-existent before 40 years old and are also rare among people who are older than 80 years old.

Table (10.1) British Physician Lung Cancer Data with Smoking Information

Age Group (years)	CIGARETTES/DAY(RANGE AND MEAN)								
	0 — 0	1-4 — 2.7	5-9 — 6.6	10-14 — 11.3	15-19 — 16	20-24 — 20.4	25-29 — 25.4	30-34 — 30.2	35-40 — 38
40-44	17846.5^{1}	1216	2041.5	3795.5	4824	7046	2523	1715.5	892.5
	0^{2}	0	0	1	0	1	0	1	0
	0^{3}	0	0	0	0	1	0	1	0
45-49	15832.5	1000.5	1745	3205	3995	6460.5	2565.5	2123	1150
	0	0	0	1	1	1	2	2	0
	0	0	0	1	1	2	1	2	1
50-54	12226	853.5	1562.5	2727	3278.5	5583	2620	2226.5	1281
	1	0	0	2	4	6	3	3	3
	1	0	0	1	2	4	3	3	2
55-59	8905.5	625	1355	2288	2466.5	4357.5	2108.5	1923	1063
	2	1	0	1	0	8	5	6	4
	1	0	0	2	2	6	4	5	4
60-64	6248	509.5	1068	1714	1829.5	2863.5	1508.5	1362	826
	0	1	1	1	2	13	4	11	7
	0	0	0	2	2	9	5	7	6
65-69	4351	392.5	843.5	1214	1237	1930	974.5	763.5	515
	0	0	1	2	2	12	5	9	9
	0	0	0	3	2	10	6	7	7
70-74	2723.5	242	696.5	862	683.5	1055	527	317.5	233
	1	1	2	4	4	10	7	2	5
	1	0	0	4	2	8	7	4	5
75-79	1772	208.5	517.5	547	370.5	512	209.5	130	88.5
	2	0	0	4	5	7	4	2	2
	0	0	0	4	2	7	4	3	3

Notes: [1]–population [2]–observed lung cancer incidence [3]–predicted lung cancer incidence based on 4 stage homogeneous model

For data in Table (10.1), observe that there are 8 dose levels represented by the number of cigarettes smoked per day and there are 8 age groups each with a period of 5 years. To implement the Bayesian and Gibbs sampling procedures, we let $\Delta t \sim 1$ correspond to a period of 3 months.

To analyze data given in Table (10.1), we would use the state space model in Section (10.1.5) with the observation model being given by the number $(y_i(j))$ of total lung cancer incidence. For the stochastic system model we

Table (10.2) BIC, AIC and Loglikelihood Values for 2, 3 and 4 Stages Models

Model	BIC	AIC	Log of Likelihood Values $=\log L(\hat{\Theta}|Y)$	P-value for Goodness of fit
Homogeneous 2-stage	188.93	182.07	-87.04	0.10
None-Homogeneous 2-stage	177.48	167.20	-77.60	0.42
Homogeneous 3-stage	151.42	139.42	-62.71	0.98
Homogeneous 4-stage	150.76	133.61	-56.81	0.99
Homogeneous 5-stage	211.67	189.38	-81.69	0.06

Notes: (1)The p-value is based on $\sum_{i=1}^{m}\sum_{j=1}^{n}\{y_j(i)-\hat{y}_j(i)\}^2/\hat{y}_j(i) \sim \chi^2(72-k)$, (k = the number of parameters under the model)

(2) $\log L(\hat{\Theta}|Y) = \sum_{i=1}^{m}\sum_{j=1}^{n}\{-\hat{y}_j(i)+y_j(i)\log\hat{y}_j(i)-\log(y_j(i)!)\}, \hat{y}_j(i) = n_{ij}Q_T(j;i)$

would entertain four extended k-stage multi-event models: (a) A time non-homogeneous 2-stage model, (b) a time homogeneous 2-stage model, (c) a time homogeneous 3-stage model, (d) a time homogeneous 4-stage model and (e) a time homogeneous 5-stage model. In models in (b)–(e), the mutation rates, the birth rates and the death rates are assumed to be independent of time. In model (a), while the mutation rates are independent of time, we assume that the birth rate and the death rate on initiated cells are 2-step piece-wise homogeneous with $t_1 = 60$ years old as the cut-off time point.

Applying the procedures in Section (10.1.6), we have estimated the parameters and fitted the data in Table (10.1). The AIC and BIC values of all models as well as the p-values for testing goodness of fit of the models are given in Table (10.2). The p-values are computed using the approximate probability distribution results $\sum_{i=1}^{m}\sum_{j=1}^{n}(y_i(j)-\hat{y}_i(j))^2/\hat{y}_i(j) \sim \chi^2(mn-k)$, for large mn, where k is the number of parameters estimated under the model. The estimates for the unknown parameters given the 4-stage model are given in Table (10.3). From the p-values of the models, apparently all models except the 5-stage model fit the data well, but the values of AIC and BIC suggested that the 4-stage model is more appropriate for the data. This 4-stage model seems to fit the following molecular biological model for squamous cell lung carcinoma proposed recently by Wistuba *et al.* [13]: Normal epithelium $\rightarrow$ Hyperplasia (3p/9p LOH, Genomic Instability) $\rightarrow$ Dysplasia (Telomerase dysregulation) $\rightarrow$ In situ Carcinoma (8p LOH, FHIT gene inactivation, gene methylation) $\rightarrow$ Invasive Carcinoma (p53 gene inactivation, k-ras mutation). This model is represented schematically in Figure (10.1).

Table (10.3) Estimates of Parameters for the 4 stage homogeneous Model with predictive

Parameters	CIGARETTES/DAY (RANGE AND MEAN)				
	0	1-4	5-9	10-14	15-19
	0	2.7	6.6	11.3	16
$\lambda_0(i)$	218.03	247.95	286.41	285.07	295.57
	± 12.81	± 15.79	± 16.79	± 16.04	± 16.73
$b_1(i)$	$5.93E-02$	$6.72E-02$	$7.17E-02$	$7.47E-02$	$7.66E-02$
	$\pm 8.77E-05$	$\pm 5.40E-05$	$\pm 3.43E-05$	$\pm 1.46E-05$	$1.16E-05$
$d_1(i)$	$6.18E-02$	$6.02E-02$	$5.99E-02$	$5.95E-02$	$5.93E-02$
	$\pm 9.68E-05$	$\pm 4.78E-05$	$\pm 2.90E-05$	$\pm 1.06E-05$	$1.50E-05$
$b_1(i)-d_1(i)$	$-1.31E-03$	$1.78E-03$	$3.95E-03$	$5.21E-03$	$6.21E-03$
	$\pm 5.60E-05$	$\pm 6.07E-05$	$\pm 6.64E-05$	$\pm 4.12E-05$	$3.17E-05$
$\alpha_1(i)$	$5.29E-05$	$6.74E-05$	$5.48E-05$	$5.68E-05$	$5.52E-05$
	$\pm 2.67E-06$	$\pm 1.67E-06$	$\pm 1.12E-06$	$\pm 3.96E-07$	$\pm 3.65E-07$
$b_2(i)$	$8.08E-02$	$8.92E-02$	$8.97E-02$	$9.14E-02$	$9.34E-02$
	$\pm 1.59E-03$	$\pm 7.34E-04$	$\pm 5.04E-04$	$\pm 2.10E-04$	$2.62E-04$
$d_2(i)$	$8.84E-02$	$8.63E-02$	$9.00E-02$	$9.19E-02$	$9.37E-02$
	$\pm 1.99E-03$	$\pm 6.82E-04$	$\pm 5.52E-04$	$\pm 2.15E-04$	$2.15E-04$
$b_2(i)-d_2(i)$	$-7.60E-03$	$2.90E-03$	$-3.00E-04$	$-5.00E-04$	$-3.00E-04$
	$\pm 2.55E-03$	$\pm 1.00E-03$	$\pm 7.47E-03$	$\pm 3.01E-04$	$3.39E-04$
$\alpha_2(i)$	$4.01E-05$	$5.55E-06$	$3.28E-06$	$3.84E-05$	$1.61E-05$
	$\pm 3.86E-05$	$\pm 4.99E-06$	$\pm 3.70E-06$	$\pm 4.75E-06$	$\pm 3.37E-06$
$b_3(i)$	$1.08E-02$	$9.32E-02$	$4.79E-02$	$9.84E-02$	$1.01E-01$
	$\pm 1.07E-02$	$\pm 2.53E-04$	$\pm 1.65E-02$	$\pm 7.63E-04$	$4.89E-04$
$d_3(i)$	$7.18E-02$	$6.27E-02$	$1.10E-01$	$8.59E-02$	$8.16E-02$
	$\pm 2.45E-02$	$\pm 1.68E-04$	$\pm 2.45E-02$	$\pm 5.60E-04$	$4.49E-04$
$b_3(i)-d_3(i)$	$-2.21E-03$	$7.88E-03$	$1.46E-02$	$1.97E-02$	$2.11E-02$
	$\pm 2.42E-03$	$\pm 2.71E-03$	$\pm 1.84E-03$	$\pm 4.78E-04$	$5.79E-04$
$\alpha_3(i)=\alpha_3$	$9.54E-06 \pm 2.51E-08$				

From results given in Tables (10.3), we observe the following interesting results:

(1). The estimates of $\lambda_0(i)$ increase as the dose level increases. This indicates that the tobacco nicotine is an initiator. From molecular biological studies, this initiation process may either be associated with the LOH (loss of heterozygosity) of some suppressor genes from chromosomes 3p or silencing of these genes by epigenetic actions (Jones and Baylin [27], Baylin and Ohm [28]).

(2). The estimates of the mutation rates $\alpha_j(i) = \alpha_j$ for $j = 1, 2, 3$ are in

(Table (10.3) continued)

Parameters	CIGARETTES/DAY(RANGE AND MEAN)			
	20-24 20.4	25-29 25.4	30-34 30.2	35-40 38
$\lambda_0(i)$	313.02 ± 16.86	318.24 ± 16.40	330.79 ± 18.61	341.41 ± 19.33
$b_1(i)$	$7.81E-02$ $\pm 7.85E-06$	$7.94E-02$ $\pm 8.83E-06$	$8.04E-02$ $\pm 9.26E-06$	$8.18E-02$ $\pm 7.29E-06$
$d_1(i)$	$5.92E-02$ $\pm 7.83E-06$	$5.91E-02$ $\pm 7.30E-06$	$5.90E-02$ $\pm 7.99E-06$	$5.89E-02$ $\pm 6.56E-06$
$b_1(i)-d_1(i)$	$6.79E-03$ $\pm 2.09E-05$	$7.30E-03$ $\pm 2.66E-05$	$7.86E-03$ $\pm 2.15E-05$	$8.30E-03$ $\pm 3.21E-05$
$\alpha_1(i)$	$5.52E-05$ $\pm 2.03E-07$	$5.44E-05$ $\pm 2.24E-07$	$5.56E-05$ $\pm 2.28E-07$	$5.51E-05$ $\pm 1.85E-07$
$b_2(i)$	$9.49E-02$ $\pm 1.49E-04$	$9.61E-02$ $\pm 1.34E-04$	$9.70E-02$ $\pm 1.30E-04$	$9.80E-02$ $\pm 1.26E-04$
$d_2(i)$	$9.42E-02$ $\pm 1.54E-04$	$9.40E-02$ $\pm 1.42E-04$	$9.46E-02$ $\pm 1.67E-04$	$9.53E-02$ $\pm 1.24E-04$
$b_2(i)-d_2(i)$	$7.00E-04$ $\pm 2.13E-04$	$2.10E-03$ $\pm 1.95E-04$	$2.40E-03$ $\pm 2.12E-04$	$2.70E-03$ $\pm 1.77E-04$
$\alpha_2(i)$	$1.42E-05$ $\pm 2.12E-06$	$7.39E-06$ $\pm 1.16E-06$	$1.48E-05$ $\pm 2.11E-06$	$1.20E-05$ $\pm 1.50E-06$
$b_3(i)$	$1.01E-01$ $\pm 9.55E-04$	$1.05E-01$ $\pm 4.39E-04$	$1.04E-01$ $\pm 6.81E-04$	$1.07E-01$ $\pm 2.70E-04$
$d_3(i)$	$8.16E-02$ $\pm 8.36E-04$	$7.90E-02$ $\pm 4.00E-04$	$7.99E-02$ $\pm 6.56E-04$	$8.02E-02$ $\pm 2.06E-04$
$b_3(i)-d_3(i)$	$2.48E-02$ $\pm 2.58E-04$	$2.69E-02$ $\pm 2.98E-04$	$2.74E-02$ $\pm 1.85E-04$	$2.98E-02$ $\pm 3.85E-04$
$\alpha_3(i)=\alpha_3$	$9.54E-06 \pm 2.51E-08$			

Notes:

$\lambda_{00} = 215.17 \pm 4.90 \qquad \lambda_{01} = 0.12 \pm 7.90\text{E-}03$

$b_{10} = 0.0592 \pm 0.0001 \qquad b_{11} = 0.0062 \pm 0.0001$

$d_{10} = 0.0615 \pm 0.0002 \qquad d_{11} = -0.0008 \pm 0.0001$

$\alpha_{10} = 5.82\text{E-}05 \pm 3.63\text{E-}06 \qquad \alpha_{11} = -0.01 \pm 0.02$

$b_{20} = 0.0814 \pm 0.0008 \qquad b_{21} = 0.0044 \pm 0.0003$

$d_{20} = 0.0862 \pm 0.0012 \qquad d_{21} = 0.0024 \pm 0.0005$

$\alpha_{20} = 3.30\text{E-}05 \pm 1.16\text{E-}05 \qquad \alpha_{21} = -0.30 \pm 0.18$

$b_{30} = 0.0265 \pm 0.0153 \qquad b_{31} = 0.0239 \pm 0.0057$

$d_{30} = 0.0752 \pm 0.0107 \qquad d_{31} = 0.0025 \pm 0.0040$

general independent of the dose level x_i. These estimates are of order 10^{-5} and do not differ significantly from one another.

(3). The results in Table (10.3) indicate that for non-smokers, the death rates $d_j(0)$ ($j = 1, 2, 3$) are slightly greater than the birth rates $b_j(0)$

so that the proliferation rates $\gamma_j(0) = b_j(0) - d_j(0)$ are negative. For smokers, however, the proliferation rates $\gamma_j(i) = b_j(i) - d_j(i)(i > 0)$ are positive and increases as dose level increases (the only exception is $\gamma_2(3)$). This is not surprising since most of the genes are tumor suppressor genes which are involved in cell differentiation and cell proliferation and apoptosis (e.g., p53) (see references [12]–[13], [25]).

(4). From Table (10.3), we observed that the estimates of $\gamma_3(i)$ were of order 10^{-2} which are considerably greater than the estimates of $\gamma_1(i)$ respectively. The estimates of $\gamma_1(i)$ are of order 10^{-3} and are considerably greater than the estimates of $\gamma_2(i)$ respectively. The estimates of $\gamma_2(i)$ are of order 10^{-4} and $\gamma_2(3)$ assumes negative value. One may explain these observations by noting the results: (1) Significant cell proliferation may trickle apoptosis leading to increased cell death unless the apoptosis gene (p53) has been inactivated and (2) the inactivation of the apoptosis gene (p53) occurred in the very last stage; see Osaka and Takahashi [12] and Wistuba *et al.* [13].

From the above analysis, we make the following conclusions: (a) The best fitted 4-stage model appears to be consistent with the molecular biological model of squamous cell lung carcinoma proposed by Osaka and Takahashi [12] and Wistuba *et al.* [13]. (b) The tobacco nicotine is both an initiator and promoter. (c) The mutation rates can best be described by the Cox regression model so that $\alpha_j(i,t) = \alpha_{j0}(t)\exp\{\alpha_{j1}(t)x_i\}, j = 0, 1$; similarly, the birth rates $b_j(i,t)$ and the death rates $d_j(i,t)$ can best be described by linear regression models. (d) The estimates of the mutation rates and proliferation rates appear to be consistent with biological observations.

10.3. Cancer Risk Assessment of Environmental Agents by Stochastic Models of Carcinogenesis

Human beings are constantly exposed to environmental agents such as arsenic in drinking water. For environmental health and public safety, it is important to assess if these agents can cause cancer and/or other serious diseases. To answer these questions, statisticians have employed different statistical models and methods to analyze exposure data from human beings; see for example, Morales *et al.* [29]. As illustrated in Morales *et al.*[29], these methods have ignored information from cancer biology. Furthermore, given that these carcinogens can cause cancer, purely statistical analysis can not provide information about the mechanism on the action of the carcinogens. Van Ryzin [30] has shown that the same data can be fitted equally well by different models; yet different models gave very different results in risk assessment of environmental agents. It follows that to assess risk of environmental agents at low dose levels, it is important to use biologically supported stochastic models of carcinogenesis. This has led EPA to revise guidelines for assessing effects of environmental agents. This new 1996 EPA cancer guidelines (EPA, [31]) call for the use of biologically supported model if sufficient biological information is available to support such a model.

To serve as an example for the application of the extended multi-event model of carcinogenesis, in this section we apply the model and methods in Section (10.2) to the data in Table (10.4) to assess cancer risk of arsenic in drinking water. This analysis will demonstrate how the arsenic in drinking water would affect human health by inducing bladder cancer.

10.3.1. *Cancer risk of environmental agent*

The data given in Table (10.4) are data on person-years at risk by age, sex and arsenic level in drinking water with observed number of deaths from bladder cancer in Taiwan. For both males and females, the data are given by $\{y_i(j), n_i(j), i = 1, j =\}$, where $y_i(j)$ is the number of people die from bladder cancer during the j-th age group under exposure to arsenic with level s_i and $n_i(j)$ the number of people at risk for cancer incidence during the j-th age group under exposure level s_i. For this data set, note that the time period is 10 years for age groups $[20, 30]$ and 20 years for age groups $[30, 50]$ and $[50, 70]$. Thus, practically one may assume the number of people who die from bladder cancer during each age group as the number of people who develop bladder cancer during each age group.

To assess cancer risk of arsenic in drinking water, we will consider a state space model as given in Section (10.1.5) with the multi-event model as the stochastic system model and with a statistical model based on data in Table (10.4) as the observation model. Note that the data in Table (10.4) is similar to those in Table (10.1) so that the observation model is exactly the same as that in Section (10.1.5). For the stochastic system model, studies of recent molecular biology on human bladder cancer (see Baffa *et al.* [32], Cordon-

Cardo [33], Castillo-Martin *et al.* [34]) would suggest a three stage stochastic multi-stage model ($N \rightarrow I_1 \rightarrow I_2 \rightarrow I_3 \rightarrow Tumor$) by mean of which the bladder cancer develops. We thus assume a 3-stage extended multi-event model as in Section (10.1) as the stochastic system model. For this model, the genetic parameters are the mutation rates $\alpha_j(i)$ ($j = 0, 1, 2$) from $I_j \rightarrow I_{j+1}$ under exposure dose level s_i, the birth rate $b_j(i)$ and the death rate $d_j(i)$ of I_j ($j = 1, 2$) cells under exposure dose level s_i.

Table (10.4). Person-years at risk by age, sex and arsenic level with observed number of deaths from bladder cancer

Arsenic(ug/L)	Age (year)*				
	20-30	30-49	50-69	$\geq$70	Total
MALE					
< 100	35,818(0)	34,196(1)	21,040(6)	4,401(10)	95,455(17)
100-299	18,578(0)	16,301(0)	10,223(7)	2,166(2)	47,268(9)
300-599	27,556(0)	25,544(5)	15,747(15)	3,221(12)	72,068(32)
$\geq$ 600	16,609(0)	15,773(4)	8,573(15)	1,224(8)	42,179(27)
Total	98,561(0)	91,814(10)	55,583(43)	11,012(32)	256,970(85)
FEMALE					
< 100	27,901(0)	32,471(3)	21,556(9)	5,047(9)	86,975(21)
100-299	13,381(0)	15,514(0)	11,357(9)	2,960(2)	43,212(11)
300-599	19,831(0)	24,343(0)	16,881(19)	3,848(11)	64,903(30)
$\geq$ 600	12,988(0)	15,540(0)	9,084(21)	1,257(7)	38,869(28)
Total	74,101(0)	87,868(3)	58,878(58)	13,112(29)	233,959(90)

Notes: *–Values in parentheses are number of deaths from bladder

To fit the model to the data in Table (10.4), we assume 6 months as one time unit ($\Delta t = 1$) so that $P_T(s, t) \approx 1$ for $t - s \geq 1$. To assess effect of dose levels, as in Section (10.2), we define $x_i = \log(1 + s_i)$ and assume Cox regression models for mutation rates ($\alpha_j(i)$) and use linear model for the birth rates and the death rates. Thus, $\alpha_j(i) = \alpha_{j0}e^{\alpha_{j1}x_i}, j = 0, 1, 2, \lambda_i = N(t_0)\alpha_0(i)$ and $(b_j(i) = b_{j0} + b_{j1}x_i, d_j(i) = d_{j0} + d_{j1}x_i, j = 1, 2)$.

Applying the methods in Section (10.1.6) to the data in Table (10.4), we have estimated these unknown parameters. These estimates are given in Table (10.5) and Table 10.6 for males and females respectively. Based on these results, we have made the following observations:

(a) The estimates of $\lambda_0(i)$ indicates that the arsenic in drinking water is an initiator to initiate the carcinogenesis process. It increase linearly with dose. Comparing results between Table (10.5) and Table (10.6), it appeared that the rates in females were at least 5–7 times greater than those in males; further these differences increase with dose levels.

(b) The proliferation rates of the first stage(I_1) and second stage(I_2) initiated cells increase as dose level increases, albeit slowly. This indicates that the arsenic in drinking water is also a weak promoter for initiated stem cancer cells.

(c) The proliferation rates of I_2 cells appears to be smaller than those of I_1 cells, indicating that the rapid proliferation after two stages may have invoked the apoptosis process to slow down the proliferation.

(d) The estimates of the mutation rates of I_1 and I_2 cells indicate that the arsenic in drinking water would not affect the mutation process in intermediate initiated cells.

Table (10.5). Estimates of Parameters for the 3 Stage Homogeneous Model (male)

Parameters	dose			
	50	200	450	800
$\lambda_0(i)$	12.76	14.66	15.79	17.84
	± 3.37	± 3.74	± 3.28	± 4.07
$b_1(i)$	$2.32E-02$	$2.42E-02$	$2.47E-02$	$2.51E-02$
	$\pm 1.31E-05$	$\pm 2.21E-05$	$\pm 9.18E-06$	$\pm 8.89E-06$
$d_1(i)$	$5.01E-03$	$4.99E-03$	$5.00E-03$	$5.00E-03$
	$\pm 7.83E-06$	$\pm 8.33E-06$	$\pm 3.92E-06$	$\pm 3.92E-06$
$b_1(i)-d_1(i)$	$1.82E-02$	$1.92E-02$	$1.97E-02$	$2.01E-02$
	$\pm 1.53E-05$	$\pm 2.36E-05$	$\pm 9.98E-06$	$\pm 9.72E-06$
$\alpha_1(i)$	$6.52E-07$	$4.07E-07$	$6.40E-07$	$7.74E-07$
	$\pm 7.91E-08$	$\pm 9.56E-08$	$\pm 5.08E-8$	$\pm 4.72E-8$
$b_2(i)$	$3.88E-04$	$8.25E-04$	$5.24E-03$	$1.24E-02$
	$\pm 3.07E-04$	$\pm 8.67E-04$	$\pm 6.00E-04$	$\pm 8.61E-04$
$d_2(i)$	$8.01E-03$	$5.75E-03$	$4.64E-03$	$4.03E-03$
	$\pm 1.45E-03$	$\pm 2.36E-03$	$\pm 5.25E-04$	$\pm 4.57E-04$
$b_2(i)-d_2(i)$	$-7.63E-03$	$-4.93E-03$	$6.05E-04$	$8.41E-03$
	$\pm 1.48E-03$	$\pm 2.52E-03$	$\pm 7.97E-04$	$\pm 9.74E-04$
$\alpha_2(i)=\alpha_2$	$5.75E-06 \pm 6.42E-09$			

Notes:

$$\lambda_{00} = 7.86 \pm 0.75 \qquad \lambda_{01} = 0.12 \pm 1.64\text{E-}02$$
$$b_{10} = 0.0205 \pm 0.0001 \qquad b_{11} = 0.0007 \pm 0.0001$$
$$d_{10} = 0.005 \pm 0.0002 \qquad d_{11} = 0 \pm 0.0001$$
$$\alpha_{10} = 3.86\text{E-}07 \pm 3.19\text{E-}07 \qquad \alpha_{11} = 8.48\text{E-}02 \pm 1.43\text{E-}01$$
$$b_{20} = -0.0172 \pm 0.0097 \qquad b_{21} = 0.0040 \pm 0.0017$$
$$d_{20} = 0.0137 \pm 0.0005 \qquad d_{21} = -0.0015 \pm 0.0001$$

Table (10.6). Estimates of Parameters for the 3 Stage Homogeneous Model (female)

Parameters	dose			
	50	200	450	800
$\lambda_0(i)$	64.33	97.44	109.92	119.77
	± 7.44	± 9.84	± 9.30	± 11.32
$b_1(i)$	$1.76E - 02$	$1.92E - 02$	$2.01E - 02$	$2.08E - 02$
	$\pm 1.07E - 05$	$\pm 1.16E - 05$	$\pm 5.58E - 06$	$\pm 4.96E - 06$
$d_1(i)$	$4.99E - 03$	$5.00E - 03$	$5.00E - 03$	$5.00E - 03$
	$\pm 5.87E - 06$	$\pm 6.68E - 06$	$\pm 3.05E - 06$	$\pm 2.89E - 06$
$b_1(i) - d_1(i)$	$1.26E - 02$	$1.42E - 02$	$1.51E - 02$	$1.58E - 02$
	$\pm 1.22E - 05$	$\pm 1.33E - 05$	$\pm 6.36E - 06$	$\pm 5.74E - 06$
$\alpha_1(i)$	$2.59E - 07$	$2.00E - 07$	$2.54E - 07$	$2.81E - 07$
	$\pm 3.20E - 08$	$\pm 3.68E - 08$	$\pm 1.68E - 08$	$\pm 1.81E - 08$
$b_2(i)$	$1.74E - 02$	$1.88E - 02$	$1.91E - 02$	$1.86E - 02$
	$\pm 1.09E - 03$	$\pm 1.71E - 03$	$\pm 6.41E - 04$	$\pm 8.18E - 04$
$d_2(i)$	$4.10E - 03$	$5.11E - 03$	$5.42E - 03$	$5.48E - 03$
	$\pm 4.97E - 04$	$\pm 9.19E - 04$	$\pm 3.54E - 04$	$\pm 4.22E - 04$
$b_2(i) - d_2(i)$	$1.33E - 02$	$1.37E - 02$	$1.37E - 02$	$1.31E - 02$
	$\pm 1.20E - 03$	$\pm 1.94E - 03$	$\pm 7.33E - 04$	$\pm 9.21E - 04$
$\alpha_2(i) = \alpha_2$	$1.36E - 06 \pm 5.87E - 08$			

Notes:

$$\lambda_{00} = 30.00 \pm 5.69 \qquad \lambda_{01} = 0.21 \pm 3.18\text{E-}02$$
$$b_{10} = 0.0131 \pm 0.0001 \qquad b_{11} = 0.0012 \pm 0.0001$$
$$d_{10} = 0.005 \pm 0.0001 \qquad d_{11} = 0 \pm 0.0001$$
$$\alpha_{10} = 2.03\text{E-}07 \pm 9.18\text{E-}08 \qquad \alpha_{11} = 3.64\text{E-}02 \pm 7.96\text{E-}02$$
$$b_{20} = 0.0157 \pm 0.0015 \qquad b_{21} = 0.0005 \pm 0.0003$$
$$d_{20} = 0.0022 \pm 0.0005 \qquad d_{21} = 0.0005 \pm 0.0001$$

From the above analysis, we conclude that the arsenic in drinking water is an initiator to initiate the process of carcinogenesis but is a weak promoter. Further, it would not affect the mutation process of later stages. Notice that results in Table (10.4) of Morale *et al.* [29] revealed that during the dose level interval $(0, 300]$, the observed deaths from cancers decreased as the dose level increases; these observed results would suggest that the agent might be a complete carcinogen and that the last stage involves genes for apoptosis and /or immortalization.

Table (10.7). Estimates of Probabilities (male)

Age	dose			
	50	200	450	800
$20-30$	$2.79E-06$	$1.08E-05$	$1.81E-05$	$2.41E-05$
	$\pm 2.75E-06$	$\pm 7.18E-06$	$\pm 6.05E-06$	$\pm 9.83E-06$
$30-50$	$5.26E-05$	$8.59E-05$	$1.33E-04$	$2.22E-04$
	$\pm 9.55E-06$	$\pm 2.62E-05$	$\pm 2.20E-05$	$\pm 4.36E-05$
$50-70$	$4.61E-04$	$3.42E-04$	$8.07E-04$	$1.31E-03$
	$\pm 4.07E-05$	$\pm 4.89E-05$	$\pm 7.52E-05$	$\pm 1.41E-04$
>70	$2.52E-03$	$1.15E-03$	$4.69E-03$	$7.60E-03$
	$\pm 2.66E-04$	$\pm 3.10E-04$	$\pm 2.94E-04$	$\pm 4.40E-04$

Table (10.8). Estimates of Probabilities (female)

Age	dose			
	50	200	450	800
$20-30$	$3.84E-06$	$6.61E-06$	$8.74E-06$	$1.27E-05$
	$\pm 1.16E-06$	$\pm 8.07E-07$	$\pm 5.17E-07$	$\pm 9.25E-07$
$30-50$	$4.25E-05$	$6.85E-05$	$9.03E-05$	$1.36E-04$
	$\pm 1.22E-05$	$\pm 7.93E-06$	$\pm 5.29E-06$	$\pm 1.04E-05$
$50-70$	$2.05E-04$	$3.11E-04$	$4.15E-04$	$6.50E-04$
	$\pm 5.54E-05$	$\pm 3.42E-05$	$\pm 2.43E-05$	$\pm 5.31E-05$
>70	$7.71E-04$	$1.13E-03$	$1.56E-03$	$2.56E-03$
	$\pm 1.99E-04$	$\pm 1.20E-04$	$\pm 9.16E-05$	$\pm 2.20E-04$

10.3.2. *Developing dose-response curves of environmental agent*

In cancer risk assessment of environmental agents, one usually derives dose-response curves for the probability of developing cancer by the environmental agent. Thus, for each dose level s_i we derive estimates of $Q_T(j;i)$ and $P_i = \sum_{j=1}^{r} Q_T(j;i)$. Then we will use $Q_T(j;i)$ and P_i to assess how the dose level $d(i)$ affects $Q_T(j;i)$ and P_i. These are the so-called dose response curves and these type of curves reflect the basic mechanism by means of which the environmental agent affects the process of carcinogenesis. Given in Table 10.7 and Table 10.8 are the estimates of the probabilities to cancer tumor under different dose levels. Plotted in Figure (10.4) and Figure (10.5) are dose response curves for these probabilities.

The dose response curves in Figure (10.4) and Figure (10.5) clearly indicate that the arsenic in drinking water is carcinogenic although for males in higher age groups the curves may first decrease slightly due to competing death and possibly apoptosis. Combining these with estimates of the genetic parameters, one may conclude that the arsenic in drinking water is mainly acting

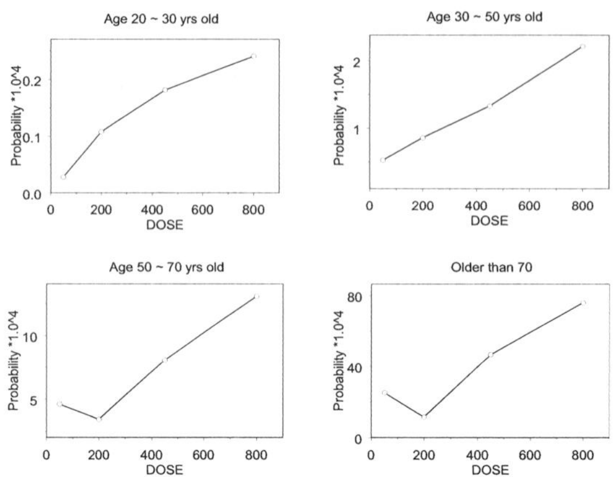

Fig. 10.4.   Plot of the Probabilities Estimates to Cancer Under Different Dose Levels (male)

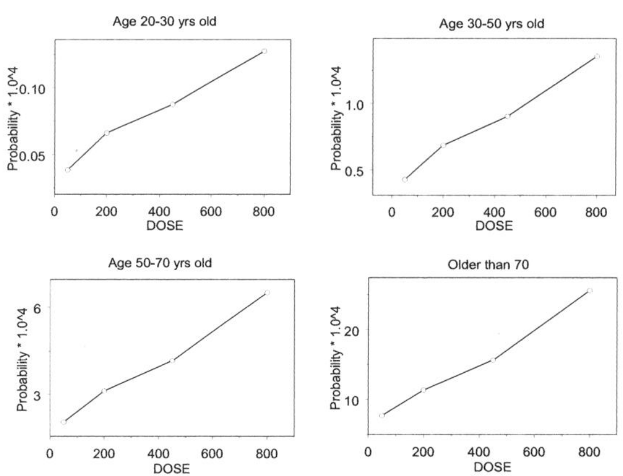

Fig. 10.5.   Plot of the Probabilities Estimates to Cancer Under Different Dose Levels (female)

as an initiator to initiate the cancer process and has little effects in cancer progression.

10.4.   Multiple Pathway Models of Carcinogenesis and Human Colon Cancer

While carcinogenesis is basically a stochastic multi-stage model involving a finite number of stages, depending on environmental conditions cancer tumors can be developed either through a single pathway or through several pathways. Specific examples involving multiple pathways in human cancers include colon cancer, liver cancer, melanoma and lung cancer among many others. Using colon cancer as an example, in this section we will illustrate how to develop biologically supported stochastic models of carcinogenesis involving multiple pathways. For this purpose, in Section (10.4.1) we first present a brief summary of most recent results of molecular biology of human colon cancer.

10.4.1.   *A brief summary of human colon cancer*

Genetic studies have indicated that there are two major avenues by means of which human colon cancer is developed: The Chromosomal Instability (CIN) and the Micro-Satellite Instability (MSI);see references [14]–[15], [35]–[43]. The first avenue is associated with the loss of heterozygosity (LOH) pathway involving the APC gene in chromosome 5q whereas the second avenue is associated with the micro-satellite pathway involving mis-match repair genes. The most important oncogene in these pathways is the β-Catenin gene in chromosome 3p22.

10.4.1.1.   *The CIN (LOH) pathway of human colon cancer*
(the APC-$\beta - catenin - Tcf - myc$ pathway)

The CIN pathway involves loss or inactivation of the tumor suppressor genes-the APC gene in chromosome 5q, the Smad-4 gene in chromosome 18q and the p53 gene in chromosome 17p; see **Remark (10.2)**. This pathway accounts for about 85% of all colon cancers. It has been referred to as the LOH pathway because it is characterized by aneuploidy /or loss of chromosome segments (chromosomal instability); see **Remark (10.3)**. This pathway has also been referred to as APC-$\beta - catenin - Tcf - myc$ pathway because it involves the destruction complex GSK-3β-Axin-APC which phosphorylates the β-Catenin protein leading to its degradation; when both copies of the APC gene are inactivated or mutated, the destruction complex is then inactive leading to accumulation of free β-Catenin protein in the cytoplasm which move to the nucleus to complex with Tcf/Lef transcription factor to activate and transcript oncogenes myc, cyclin D and CD44. (Free β-Catenin protein in the cytoplasm also binds with E-cadherin and α-Catenin to disrupt the gap junction between cells, leading to migration and metastasis of cancer tumors.)

Morphological studies have indicated that inactivation, or loss or mutation of APC creates dysplastic aberrant crypt foci (ACF) which grow into dysplastic adenomas. These adenomas grow to a maximum size of about 10 mm^3; further growth and malignancy require the abrogation of differentiation, cell cycle inhibition and apoptosis which are facilitated by the inactivation, or mutation or loss of Smad-4 gene in 18q and the p53 gene in 17p. The mutation or activation of the oncogene H-ras in chromosome 11p and /or mutation and/or activation of the oncogene src in chromosome 20q would speed up these transitions by promoting the proliferation rates of the respective intermediate initiated cells; see Jessup *et al.* [40]. This pathway is represented schematically by Figure (10.2).

The model in Figure (10.2) is a 6-stage model. However, because of the haplo-insufficiency of the Smad4 gene (see Alberici *et al.* [44]) and the haplo-insufficiency of the p53 gene (Lynch and Milner [45]), one may reduce this 6-stage model into a 4-stage model by combining the third stage and the fourth stage into one stage and by combining the fifth stage and the sixth stage into one stage. This may help explain why for single pathway models, the 4-stage model fits the human colon cancer better than other single pathway multi-stage models (see [46]). Recent biological studies by Green and Kaplan [38] and others have also shown that the inactivation or deletion or mutation of one copy of the APC gene in chromosome 5 can cause defects in microtubule plus-end attachment during mitosis dominantly, leading to aneuploidy and chromosome instability. This would speed up the mutation or inactivation of the second copy of the APC gene and increase fitness of the APC-carrying cells in the micro-evolution process of cancer progression. This could also help explain why the APC LOH pathway is more frequent than other pathways.

Remark (10.2): As observed by Sparks *et al.* [42], instead of the APC gene, this pathway can also be initiated by mutation of the oncogene β-catenin gene; however, the proportion of human colon cancer due to mutation of β-catenin is very small (less than 1%) as compared to the APC gene, due presumably to the contribution of the APC on chromosome instability ([38]). Similarly, the destruction complex can become inactive either by the inhibition of GSK-3β

through the Wnt signalling pathway (see [47]–[48]) or the inactivation or mutation of the Axin protein, leading to accumulation of the β- Catenin proteins in the cytoplasm; but the proportion of colon cancer caused by inhibition of GSK-3β is also very small as compared to the colon cancer cases caused by the CIN and the MSI pathways.

Remark (10.3): The APC gene in chromosome 5 acts both as a tumor suppressor gene and an oncogene in initiating and promoting colon carcinogenesis. As an oncogene, the APC gene acts dominantly in regulating microtubule plus-end attachment during mitosis ([38]). Thus, the inactivation or deletion or mutation of one copy of the APC gene in chromosome 5 can cause defects in microtubule plus-end attachment during mitosis, leading to aneuploidy and chromosome instability. This would speed up the mutation or inactivation of the second copy of the APC gene and increase fitness of the APC-carrying cells in the micro-evolution process of cancer progression. This could also help explain why the APC LOH pathway is more frequent than other pathways.

10.4.1.2. *The MSI (micro-satellite instability) pathway of human colon cancer*

This pathway accounts for about 15% of all colon cancers and appears mostly in the right colon. It has been referred to as the MSI pathway or the mutator phenotype pathway because it is initiated by the mutations or epigenetic methylation of the mis-match repair genes (mostly hMLH1 in chromosome 3p21 and hMSH2 in chromosome 2p16) creating a mutator phenotype to significantly increase the mutation rate of many critical genes 10 to 1000 times. Normally these critical genes are TGF-βR_{II}, Bax (the X protein of bcl-2 gene), IGF2R, or CDX-2. The mis-match repair genes are hMLH1, hMSH2, hPMS1, hPMS2, hMSH6 and hMSH3; mostly hMLH1 (50%) and hMSH2 (40%). This pathway is represented Schematically by Figure (10.3).

As in the LOH pathway, assuming haplo-insufficiency of tumor suppressor genes, one may approximate this pathway by a 5-stage model.

Morphologically, mutation or methylation silencing of the MMR gene hMLH1 or hMSH2 generates hyperplastic polyp which leads to the generation of serrated adenomas. These adenomas develop into carcinomas after the inactivation, or loss or mutations of the TGF-βR_{II} gene and the Bax gene, thus abrogating differentiation and apoptosis. (Bax is an anti-apoptosis gene.)

In what follows, we denote by J_i the i–th stage cells in the MSI pathways. Then for sporadic MSI, the model is $N \rightarrow J_1 \rightarrow J_2 \rightarrow J_3 \rightarrow J_4 \rightarrow J_5 \rightarrow$ *Cancer tumor*.

10.4.1.3. *The major signalling pathways for human colon cancer*

Recent biological studies ([47]–[48]) have shown that both the CIN and the MSI pathways involve the Wnt signalling pathway and the destruction complex (this complex is a downstream of the Wnt signalling pathway), the TGF-β inhibiting signalling pathway and the p53-Bax apoptosis signalling pathway but different genes in different pathways are affected in these signalling processes. In the CIN pathway, the affected gene is the APC gene in the Wnt signalling, the Smad4 in the TGF-β signalling and p53 gene in the p53-Bax signalling; on the other hand, in the MSI pathway, the affected gene is the Axin 2 gene in the Wnt signalling, the TGF-$\beta - R_{II}$ in the TGF-β signalling and the Bax gene in the p53-Bax signalling.

Because the probability of point mutation or genetic changes of genes are in general very small compared to epigenetic changes, one may speculate that colon cancer may actually be initiated with some epigenetic mechanisms ([47]–[50]).

Based on the above biological studies, we thus postulate that the incidence data of human colon cancer are described and generated by a multi-stage model involving 2 pathways as defined above. Because of haploid-insufficiency of the tumor suppressor genes $\{Smad_4, p53, Axin, Bax, TGF - \beta - ReceptorII\}$, the number of stages for the CIN pathway and MSI are assumed as 4 and 5 respectively.

10.4.2. *Stochastic multi-stage model of carcinogenesis for human colon cancer involving multiple pathways*

From results of Section (10.4.1), it follows that the stochastic multi-stage model for human colon cancer involves two pathways with pathway 1 as a k_1-stage multi-stage model involving I_i ($i = 1, \ldots, k_1$) cells and with pathway 2 as a k_2-stage multi-stage model involving J_j ($j = 1, \ldots, k_2$) cells. (For human colon cancer, $k_1 = 4, k_2 = 5$.) Assuming that the number $N(t)$ of normal stem cells is very large so that one may practically assume $N(t)$ as a deterministic function of time t, the state variables are then $\underset{\sim}{X}(t) = \{I_i(t), i = 1, \ldots, k_1 -$

$1, J_j(t), j = 1, \ldots, k_2 - 1\}$ and $T(t)$, where $T(t)$ denote the number of cancer tumors at time t and where $I_i(t)$ $(J_j(t))$ denote the number of the I_i (J_j) initiated cells for $\{i = 1, \ldots, k_1 - 1$ $(j = 1, \ldots, k_2 - 1)\}$ respectively. To develop a stochastic model for this process, notice that one may practically assume that the $\{\underset{\sim}{X}(t), t \geq t_0\}$ is a Markov process with continuous time; however $T(t)$ in general may not be Markov.

10.4.2.1. *The stochastic equation for state variables*

To derive stochastic equations for the state variables, let $B_i^{(I)}(t)$ $(B_j^{(J)}(t))$ be the number of birth of the I_i (J_j) initiated cells during $[t, t + \Delta t)$ $\{i = 1, \ldots, k_1 - 1$ $(j = 1, \ldots, k_2 - 1)\}$, $D_i^{(I)}(t)$ $(D_j^{(J)}(t))$ the number of death of the I_i (J_j) initiated cells during $[t, t + \Delta t)$ $\{i = 1, \ldots, k_1 - 1$ $(j = 1, \ldots, k_2 - 1)\}$ and $M_i^{(I)}(t)$ $(M_j^{(J)}(t))$ the number of mutation $(I_i \rightarrow I_{i+1})$ $(J_j \rightarrow J_{j+1})$ of I_i (J_j) cells during $[t, t + \Delta t)$ $\{i = 1, \ldots, k_1 - 1$ $(j = 1, \ldots, k_2 - 1)\}$. Also let $M_0^{(I)}(t)$ $(M_0^{(J)}(t))$ be the number of mutation of $N \rightarrow I_1$ $(N \rightarrow J_1)$ during $[t, t + \Delta t)$. By the conservation law, we have then the following stochastic equations for the state variables:

$$I_i(t + \Delta t) = I_i(t) + M_{i-1}^{(I)}(t) + B_i^{(I)}(t) - D_i^{(I)}(t), i = 1, \ldots, k_1 - 1, \quad (10.21)$$

$$J_j(t + \Delta t) = J_j(t) + M_{j-1}^{(J)}(t) + B_j^{(J)}(t) - D_j^{(J)}(t), j = 1, \ldots, k_2 - 1, (10.22)$$

Because the transition variables $\{M_i^{(I)}(t), M_j^{(J)}(t), B_i^{(I)}(t), D_i^{(I)}(t), B_j^{(J)}(t), D_j^{(J)}(t)\}$ are random variables, the above equations are stochastic equations. To derive the probability distributions of these variables and the staging variables, let $\{\alpha_i(t), \beta_j(t), b_i^{(I)}(t), b_j^{(J)}(t), d_i^{(I)}(t), d_j^{(J)}(t)\}$ denote the mutation rates, the birth rates and the death rates of $\{I_i, J_j\}$ cells as given in Table (10.9) respectively.

Then as in Section (8.3.1) of Chapter 8, we have that to the order of $o(\Delta t)$, the conditional probability distributions of $M_0^{(I)}(t)$ and $M_0^{(J)}(t)$ given $N(t)$ are Poisson with means $\lambda_I(t)\Delta t$ and $\lambda_I(t)\Delta t$ respectively whereas the conditional probability distributions of $\{B_i^{(I)}(t), D_i^{(I)}(t), M_i^{(I)}(t)\}$ given $I_i(t)$ and of $\{B_j^{(J)}(t), D_j^{(J)}(t), M_j^{(J)}(t)\}$ given $J_j(t)$ follow multinomial distributions independently. Thus, by Lemma (10.1), we have:

$$M_0^{(I)}(t)|N(t) \sim \text{Poisson}\{\lambda_I(t)\Delta t\}, \text{independently of } M_0^{(J)}(t); \quad (10.23)$$

$$M_0^{(J)}(t)|N(t) \sim \text{Poisson}\{\lambda_J(t)\Delta t\}, \text{independently of } M_0^{(I)}(t); \quad (10.24)$$

Table (10.9). Transition rates and transition probabilities for human colon carcinogenesis

1 N	$\longrightarrow$	1 N, 1 I_1	$\alpha_0(t)\Delta t$
1 N	$\longrightarrow$	1 N, 1 J_1	$\beta_0(t)\Delta t$
1 I_i	$\longrightarrow$	2 I_i	$b_i^{(I)}(t)\Delta t$
1 I_i	$\longrightarrow$	death	$d_i^{(I)}(t)\Delta t$
1 I_i	$\longrightarrow$	1 I_i,1 I_{i+1}	$\alpha_i(t)\Delta t$
	$i = 1, \ldots, k_1 - 1$		
1 J_j	$\longrightarrow$	2 J_j	$b_j^{(J)}(t)\Delta t$
1 J_j	$\longrightarrow$	death	$d_j^{(J)}(t)\Delta t$
1 J_j	$\longrightarrow$	1 J_j,1 J_{j+1}	$\beta_i(t)\Delta t$
	$j = 1, \ldots, k_2 - 1$		

for $i = 1, 2, \ldots, k_1 - 1$,

$$\{B_i^{(I)}(t), D_i^{(I)}(t)\}|I_i(t) \sim \text{Multinomial}\{I_i(t); b_i^{(I)}(t)\Delta t, d_i^{(I)}(t)\Delta t\}; \quad (10.25)$$

for $j = 1, \ldots, k_2 - 1$,

$$\{B_j^{(J)}(t), D_j^{(J)}(t)\}|J_j(t) \sim \text{Multinomial}\{J_j(t); b_j^{(J)}(t)\Delta t, d_j^{(J)}(t)\Delta t\}. \quad (10.26)$$

$$M_i^{(I)}(t)|I_i(t) \sim \text{Binomial}\{I_i(t); \alpha_i(t)\Delta t\}, i = 1, \ldots, k_1 - 1, \quad (10.27)$$

independently of $\{B_i^{(I)}(t), D_i^{(I)}(t)\}$ and other transition variables, and

$$M_j^{(J)}(t)|J_j(t) \sim \text{Binomial}\{J_j(t), \beta_j(t)\Delta t\}, j = 1, \ldots, k_2 - 1, \quad (10.28)$$

independently of $\{B_j^{(J)}(t), D_j^{(J)}(t)\}$ and other transition variables.

Using the above probability distributions and by subtracting from the transition variables the conditional expected values respectively, we have from equations (10.21)–(10.22) the following stochastic differential equations for the staging state variables:

$$
\begin{aligned}
dI_i(t) &= I_i(t + \Delta t) - I_i(t) = M_{i-1}^{(I)}(t) + B_i^{(I)}(t) - D_i^{(I)}(t) \\
&= \{I_{i-1}(t)\alpha_{i-1}(t) + I_i(t)\gamma_i^{(I)}(t)\}\Delta t + e_i^{(I)}(t)\Delta t, \\
&\quad i = 1, \ldots, k_1 - 1,
\end{aligned}
\quad (10.29)
$$

$$dJ_j(t) = J_j(t + \Delta t) - J_j(t) = M_{j-1}^{(J)}(t) + B_j^{(J)}(t) - D_j^{(J)}(t)$$

$$= \{J_{j-1}(t)\beta_{j-1}(t) + J_j(t)\gamma_j^{(J)}(t)\}\Delta t + e_j^{(J)}(t)\Delta t,$$

$$j = 1, \ldots, k_2 - 1, \tag{10.30}$$

where $\gamma_i^{(I)}(t) = b_i^{(I)}(t) - d_i^{(I)}(t)$, $\gamma_j^{(J)}(t) = b_j^{(J)}(t) - d_j^{(J)}(t)$.

In the above equations, the random noises $\{e_i^{(I)}(t)\Delta t, e_j^{(J)}(t)\Delta t\}$ are derived by subtracting the conditional expected numbers from the random transition variables respectively. Obviously, these random noises are linear combinations of Poisson and multinomial random variables. These random noises have expected value zero and are un-correlated with the state variables $\{I_i(t), i = 1, \ldots, k_1 - 1, J_j(t), j = 1, \ldots, k_2 - 1\}$. It can also be shown that to the order of $o(\Delta t)$, these random noises are uncorrelated with one another and have variances given by:

$$Var\{e_i^{(I)}(t)\Delta t\} = EI_{i-1}(t)\alpha_{i-1}(t)\Delta t + EI_i(t)[b_i^{(I)}(t) + d_i^{(I)}(t)]\Delta t + o(\Delta t),$$

$$\text{for } i = 1, \ldots, k_1 - 1,$$

$$Var\{e_j^{(J)}(t)\Delta t\} = EJ_{j-1}(t)\beta_{j-1}(t)\Delta t + EJ_j(t)[b_j^{(J)}(t) + d_j^{(J)}(t)]\Delta t + o(\Delta t),$$

$$\text{for } j = 1, \ldots, k_2 - 1,$$

where $I_0(t) = J_0(t) = N(t)$.

To solve the above equations under the initial condition $I_i(t_0) = J_j(t_0) = 0$ for $i = 1, \ldots, k_1 - 1$ and $j = 1, \ldots, k_2 - 1$, denote by $\eta_i^{(I)}(t) = \int_{t_0}^t e^{\int_x^t \gamma_i^{(I)}(y)dy} e_i^{(I)}(x)dx$, and $\eta_j^{(J)}(t) = \int_{t_0}^t e^{\int_x^t \gamma_j^{(J)}(y)dy} e_j^{(J)}(x)dx$. Then $E[\eta_i^{(I)}(t)] = E[\eta_j^{(J)}(t)] = 0$ for all $i = 1, \ldots, k_1 - 1$ and $j = 1, \ldots, k_2 - 1$. These random variables are uncorrelated with one another and have variances given by:

$$Var\{\eta_i^{(I)}(t)\} = \int_{t_0}^t e^{2\int_x^t \gamma_i^{(I)}(y)dy}\{\alpha_{i-1}(x)E[I_{i-1}(x)]$$

$$+[b_i^{(I)}(x) + d_i^{(I)}(x)]E[I_i(x)]\}dx,$$

$$\text{for } i = 1, \ldots, k_1 - 1,$$

$$Var\{\eta_j^{(J)}(t)\} = \int_{t_0}^t e^{2\int_x^t \gamma_j^{(J)}(y)dy}\{\beta_{j-1}(x)EJ_{j-1}(x)$$

$$+[b_j^{(J)}(x) + d_j^{(J)}(x)]E[J_j(x)]\}dx,$$

$$\text{for } j = 1, \ldots, k_2 - 1.$$

Then, for normal people at time t_0, the solutions of the equations (10.29)–(10.30) are given respectively by:

$$I_1(t) = \int_{t_0}^{t} e^{\int_x^t \gamma_1^{(I)}(y)dy} \lambda_I(x)dx + \eta_1(I)(t),$$

$$\text{with } \lambda_I(t) = N(t)\alpha_0(t); \tag{10.31}$$

and for $i = 2, \ldots, k_1 - 1$,

$$I_i(t) = \int_{t_0}^{t} e^{\int_x^t \gamma_i^{(I)}(y)dy} I_{i-1}(x)\alpha_{i-1}(x)dx + \eta_i(I)(t); \tag{10.32}$$

$$J_1(t) = \int_{t_0}^{t} e^{\int_x^t \gamma_1^{(J)}(y)dy} \lambda_J(x)dx + \eta_1(J)(t), \text{with}$$

$$\lambda_J(t) = N(t)\beta_0(t); \tag{10.33}$$

and for $j = 2, \ldots, k_2 - 1$,

$$J_j(t) = \int_{t_0}^{t} e^{\int_x^t \gamma_j^{(J)}(y)dy} J_{j-1}(x)\beta_{j-1}(x)dx + \eta_j(J)(t). \tag{10.34}$$

10.4.2.2. *The expected numbers*

Let $u_I(i,t) = E[I_i(t)]$ and $u_J(j,t) = E[J_j(t)]$ denote the expected numbers of $I_i(t)$ and $J_j(t)$ respectively and write $u_I(0,t) = u_J(0,t) = N(t)$. Using equations (10.31)–(10.34), we obtain these expected numbers as:

$$u_I(1,t) = \int_{t_0}^{t} \lambda_I(x)e^{\int_x^t \gamma_1^{(I)}(z)dz}dx,$$

$$u_J(1,t) = \int_{t_0}^{t} \lambda_J(x)e^{\int_x^t \gamma_1^{(J)}(z)dz}dx,$$

$$u_I(i,t) = \int_{t_0}^{t} u_I(i-1,x)e^{\int_x^t \gamma_i^{(I)}(z)dz}dx, \text{for } i = 2, \ldots, k_1 - 1,$$

$$u_J(j,t) = \int_{t_0}^{t} u_J(j-1,x)e^{\int_x^t \gamma_j^{(J)}(z)dz}dx, \text{for} j = 2, \ldots, k_2 - 1.$$

Because for normal stem cells it is expected that $b_0(t) = d_0(t)$ so that $N(t) = N(t_0)$, if the model is time homogeneous, then $\lambda_I(t) = \lambda_I, \lambda_J(t) = \lambda_J, \alpha_i(t) = \alpha_i, \gamma_i^{(I)}(t) = \gamma_i^{(I)}$ for $i = 1, \ldots, k_1 - 1$ and $\beta_j(t) = \beta_j, \gamma_j^{(J)}(t) =$

$\gamma_j^{(J)}$ for $j = 1, \ldots, k_2 - 1$. Thus, if the model is time homogeneous, if the proliferation rates of I_i cells and J_j cells are not zero and if $\gamma_i^{(I)} \neq \gamma_u^{(I)} \neq \gamma_j^{(J)} \neq \gamma_v^{(J)}$ for all $i \neq u$ and $j \neq v$, the above solutions reduce respectively to (Note $b_0 = d_0$ so that $\gamma_0 = b_0 - d_0 = 0$):

$$u_I(1, t) = \frac{\lambda_I}{\gamma_1^{(I)}}[e^{\gamma_1^{(I)}} - 1], u_J(1, t) = \frac{\lambda_J}{\gamma_1^{(J)}}[e^{\gamma_1^{(J)}} - 1],$$

$$u_I(i, t) = \sum_{u=1}^{i} A_{1i}(u)[e^{\gamma_u^{(I)}t} - 1], \text{ for } i = 2, \ldots, k_1 - 1;$$

$$u_J(j, t) = \sum_{u=1}^{j} B_{1j}(u)[e^{\gamma_u^{(J)}t} - 1], \text{ for } j = 2, \ldots, k_2 - 1$$

where $A_{1i}(u) = \frac{\lambda_I}{\gamma_u^{(I)}}\{\prod_{r=1}^{i-1} \alpha_r\}\{\prod_{\substack{v=1 \\ v \neq u}}^{i} (\gamma_u^{(I)} - \gamma_v^{(I)})\}^{-1}, i = 2, \ldots, k_1 - 1.$

$B_{1j}(u) = \frac{\lambda_J}{\gamma_u^{(J)}}\{\prod_{r=1}^{j-1} \beta_r\}\{\prod_{\substack{v=1 \\ v \neq u}}^{j} (\gamma_u^{(J)} - \gamma_v^{(J)})\}^{-1}, j = 2, \ldots, k_2 - 1.$

10.4.2.3. *The probability distribution of state variables and dummy state variables*

To derive the transition probability of $\{\underset{\sim}{X}(t), t \geq t_0\}$, as in Section (10.1), denote by $g(x, y : N, p_1, p_2)$ is the density at (x, y) of the multinomial distribution $ML(N; p_1, p_2)$ with parameters $(N; p_1, p_2)$, $f(x; N, p)$ the density at x of a binomial distribution Binomial(N, p) and $h(x : \lambda)$ the density at x of the Poisson distribution with mean λ. Then, using the probability distributions given by equations (10.23)–(10.28), the transition probability of this Markov process is, to order of $o(\Delta t)$:

$$P\{\underset{\sim}{X}(t + \Delta t) \mid \underset{\sim}{X}(t)\} = P\{I_1(t + \Delta t)|I_1(t)\} \prod_{i=2}^{k_1-1} P\{I_i(t + \Delta t)|I_{i-1}(t), I_i(t)\}$$

$$\times P\{J_1(t + \Delta t)|J_1(t)\} \prod_{j=2}^{k_2-1} P\{J_j(t + \Delta t)|J_{j-1}(t), J_j(t)\},$$

where with $m_I(i, u, v; t) = I_i(t + \Delta t) - I_i(t) - u + v, a_I(i, t; u) = Max(0, I_i(t) - I_i(t + \Delta t) + u), i = 1, \ldots, k - 1$ and $b_I(i, t; u) = Min(I_i(t) - u, I_{i-1}(t) + I_i(t) - I_i(t + \Delta t) + u), i = 2, \ldots, k - 1,$

$$P\{I_1(t+\Delta t)|I_1(t)\} = \left\{ \sum_{u=0}^{I_1(t)} \sum_{v=a_I(1,t;u)}^{I_1(t)-u} g\{u,v;I_1(t),b_1^{(I)}(t)\Delta t, d_1^{(I)}(t)\Delta t\} \right.$$

$$\left. \times\, h\{m_I(1,u,v;t); \lambda_I(t)\Delta t\} \right\},$$

$$P\{I_i(t+\Delta t)|I_{i-1}(t),I_i(t)\} = \left\{ \sum_{u=0}^{I_i(t)} \sum_{v=a_I(i,t;u)}^{b_I(i,t;u)} g\{u,v;I_i(t),b_i^{(I)}(t)\Delta t, d_i^{(I)}(t)\Delta t\} \right.$$

$$\left. \times\, f\{m_I(i,u,v;t); I_{i-1}(t),\alpha_{i-1}(t)\Delta t\} \right\},$$

and with $m_J(j,u,v;t) = J_j(t+\Delta t) - J_j(t) - u + v$, $a_J(j,t;u) = Max(0, J_j(t) - J_j(t+\Delta t) + u)$, $j = 1,\ldots,k_2 - 1$ and $b_J(j,t;u) = Min(J_j(t) - u, J_{j-1}(t) + J_j(t) - J_j(t+\Delta t) + u)$, $j = 2,\ldots,k_2 - 1$,

$$P\{J_1(t+\Delta t)|J_1(t)\} = \left\{ \sum_{u=0}^{J_1(t)} \sum_{v=a_J(1,t;u)}^{J_1(t)-u} g\{u,v;J_1(t),b_1^{(J)}(t)\Delta t, d_1^{(J)}(t)\Delta t\} \right.$$

$$\left. \times\, h\{m_J(1,u,v;t); \lambda_J(t)\Delta t\} \right\},$$

$$P\{J_j(t+\Delta t)|J_{j-1}(t),J_j(t)\} = \left\{ \sum_{u=0}^{J_j(t)} \sum_{v=a_J(j,t;u)}^{b_J(j,t;u)} g\{u,v;J_j(t),b_j^{(J)}(t)\Delta t, d_j^{(J)}(t)\Delta t\} \right.$$

$$\left. \times\, f\{m_J(j,u,v;t); J_{j-1}(t),\beta_{j-1}(t)\Delta t\} \right\},$$

For implementing the Gibbs sampling procedure to estimate parameters and to predict state variables, we define the dummy state variables $\underset{\sim}{U}(t) = \{B_i^{(I)}(t), D_i^{(I)}(t), i = 1,\ldots,k_1-1, B_j^{(J)}(t), D_j^{(J)}(t), j = 1,\ldots,k_2-1\}$. (In what follows we will refer these variables as the transition variables, unless otherwise stated.) In Bayesian inference, this is referred to as the data augmentation method.

Put $\underset{\sim}{Z}(t) = \{\underset{\sim}{X}(t)', \underset{\sim}{U}(t-\Delta t)'\}'$. Then $\{\underset{\sim}{Z}(t), t \geq t_0\}$ is Markov with continuous time. Using the probability distributions of the transition random variables given by equations (10.23)–(10.28), the transition probability

$P\{Z(t+\Delta t)|Z(t)\}$ is

$$P\{Z(t+\Delta t)|Z(t)\} = P\{X(t+\Delta t)|X(t), U(t)\}$$
$$\times\, P\{U(t)|X(t)\}, \tag{10.35}$$

where

$$P\{U(t)|X(t)\} = \prod_{i=1}^{k_1-1} f\{B_i^{(I)}(t), D_i^{(I)}(t); I_i(t), b_i^{(I)}(t)\Delta t, d_i^{(I)}(t)\Delta t\}$$
$$\times \prod_{j=1}^{k_2-1} f\{B_j^{(J)}(t), D_j^{(J)}(t); J_j(t), b_j^{(J)}(t)\Delta t, d_j^{(J)}(t)\Delta t\}; \tag{10.36}$$

and

$$P\{X(t+\Delta t)|U(t), X(t)\} = h(u_I(1,t); \lambda_I(t)\Delta t)$$
$$\times \prod_{i=2}^{k_1-1} f\{M_I(i,t); I_{i-1}(t), \alpha_{i-1}(t)\Delta t\}$$
$$\times h(v_J(1,t); \lambda_J(t)\Delta t) \prod_{j=2}^{k_2-1} f\{M_J(j,t); J_{j-1}(t), \beta_{j-1}(t)\Delta t\}, \tag{10.37}$$

where $M_I(i,t) = I_i(t+\Delta t) - I_i(t) - B_i^{(I)}(t) + D_i^{(I)}(t)$ for $i = 1, \ldots, k_1 - 1$ and $M_J(j,t) = J_j(t+\Delta t) - J_j(t) - B_j^{(J)}(t) + D_j^{(J)}(t)$ for $j = 1, \ldots, k_2 - 1$.

The probability distribution given by equation (10.35) will be used to derive estimates and predicted numbers of state variables. This is discussed in Section (10.4.5).

10.4.3. *A statistical model and the probability distribution of the number of detectable tumors*

The data available for modeling carcinogenesis are usually cancer incidence over different time periods. For example, the SEER data of NCI/NIH for human cancers are given by $\{(y_j, n_j), j = 1, \ldots, n\}$, where y_j is the number of cancer cases during the j−th age group and n_j is the number of normal people who are at risk for cancer and from whom y_j of them have developed cancer during the age group. Given in Table (10.10) are the SEER data for human colon cancer adjusted for genetic cancer cases.

10.4.3.1. *The probability distribution of the number of detectable tumors for colon cancer*

To derive the probability distribution of time to tumors, one needs the probability distribution of $T(t)$. For deriving this probability distribution, we observe that malignant cancer tumors arise by clonal expansion from primary I_{k_1} cells and primary J_{k_2} cells, where primary I_{k_1} cells are I_{k_1} cells derived from I_{k_1-1} cells by mutation of I_{k_1-1} cells and primary J_{k_2} cells are J_{k_2} cells derived from J_{k_2-1} cells by mutation of J_{k_2-1} cells.

Let $P_T^{(I)}(s,t)$ $(P_T^{(J)}(s,t))$ be the probability that a primary I_{k_1} (J_{k_2}) cancer cell at time s develops into a detectable cancer tumor at time t. Let $T_i(t)$ be the number of cancer tumors derived from the $i-$th pathway. Then, to order of $o(\Delta t)$, the conditional probability distribution of $T_1(t)$ given $I_{k_1-1}(s), \{s \leq t\}$ is Poisson with mean $\omega_1(t)$ independently of $T_2(t)$, where

$$\omega_1(t) = \int_{t_0}^t I_{k_1-1}(x)\alpha_{k_1-1}(x)P_T^{(I)}(x,t)dx.$$

Similarly, to order of $o(\Delta t)$, the conditional probability distribution of $T_2(t)$ given $J_{k_2-1}(s), \{s \leq t\}$ is Poisson with mean $\omega_2(t)$ independently of $T_1(t)$, where

$$\omega_2(t) = \int_{t_0}^t J_{k_2-1}(x)\beta_{k_2-1}(x)P_T^{(J)}(x,t)dx.$$

Let $Q_i(j)$ $(i = 1, 2)$ be defined by:

$$Q_i(j) = E\{e^{-\omega_i(t_{j-1})} - e^{-\omega_i(t_j)}\} = E\{e^{-\omega_i(t_{j-1})}(1 - e^{R_i(t_{j-1},t_j)})\},$$

where $R_i(t_{j-1}, t_j) = \omega_i(t_{j-1}) - \omega_i(t_j)$.

Then $Q_i(j)$ is the probability that cancer tumors would develop during the $j-$th age group by the $i-$th pathway. Since cancer tumors develop if and only if at least one of the two pathways yield cancer tumors, the probability that a normal person at time t_0 will develop cancer tumors during $[t_{j-1}, t_j)$ is given by $Q_T(j)$, where

$$Q_T(j) = 1 - [1 - Q_1(j)][1 - Q_2(j)] = Q_1(j) + Q_2(j) - Q_1(j)Q_2(j).$$

For practical applications, we observe that to order of $o(\alpha_{k_1-1}(t))$ and $o(\beta_{k_2-1}(t))$ respectively, the $\omega_i(t)$ in $Q_i(j)$ are approximated by

$$\omega_1(t) \sim \int_{t_0}^{t} E[I_{k_1-1}(s)]\alpha_{k_1-1}(s)P_T^{(I)}(s,t)ds,$$

$$\omega_2(t) \sim \int_{t_0}^{t} E[J_{k_2-1}(s)]\beta_{k_2-1}(s)P_T^{(J)}(s,t)ds.$$

Table (10.10). Colon Cancer Data from SEER (overall population)

Age Group	Number of People at Risk	Observed Colon Cancer Cases	Total Predicted Colon Cancer
0	9934747	1	0
0−4	38690768	2	0
5−9	48506058	2	6
10−14	49881935	35	44
15−19	50447512	104	164
20−24	51612785	337	370
25−29	54071811	847	965
30−34	54194486	1829	2080
35−39	50363957	3420	3534
40−44	46029771	6174	6698
45−49	40674188	10950	11072
50−54	36070434	18716	18256
55−59	31084543	27438	25875
60−64	26507762	37155	34867
65−69	22772688	47202	45156
70−74	18785224	53190	52810
75−79	14592602	52887	53479
80−84	9751212	42589	41517

Similarly, it can readily be shown that to the order of $Min\{o(\alpha_{k_1-1}(t))$, $o(\beta_{k_2-1}(t))\}$, $Q_T(t) \sim Q_1(t) + Q_2(t)$. (See Exercise 5.)

To further simplify the calculation of $Q_T(j)$, we observe that in studying human cancers, one time unit (i.e. $\Delta t = 1$) is usually assumed to be 3 months or 6 months or longer. In these cases, one may practically assume $P_T^{(I)}(s,t) \sim 1$ and $P_T^{(J)}(s,t) \sim 1$ if $t - s \geq 1$.

10.4.3.2.　*A statistical model for cancer incidence data*

Let y_j be the observed number of cancer cases during $[t_{j-1}, t_j)$ given n_j normal people at risk at t_0 for cancer. We assume that each individual develops colon cancer tumor by the same mechanism independently of one another. Then for each normal person at time t_0, the probability that this individual would develop colon cancer tumor during the j-th age group $[t_{j-1}, t_j)$ is given by $Q_T(j)$. It follows that the probability distribution of y_j given that n_j people are at risk for colon cancer at time t_0 is:

$$y_j \sim \text{Poisson}\{\tau_j\}, \tag{10.38}$$

where $\tau_j = n_j Q_T(j)$.

Notice that to the order of $Max\{o(\alpha_{k_1-1}(t)), o(\beta_{k_2-1}(t))\}$, τ_j (and hence y_j) depends on the stochastic model of colon carcinogenesis through the expected number $\{E[I_{k_1-1}(t)],$
$E[J_{k_2-1}(t)]\}$ of $\{I_{k_1-1}(t), J_{k_2-1}(t)\}$ and the parameters $\{\alpha_{k_1-1}(t), \beta_{k_2-1}(t)\}$ over the time period $[t_{j-1}, t_j)$.

10.4.4.　**The state space model of human colon cancer**

State space model is a stochastic model which consists of two sub-models: The stochastic system model which is the stochastic model of the system and the observation model which is a statistical model based on available observed data from the system. For human colon cancer, the stochastic system model of the state space model is the stochastic multiple pathways model consisting of 2 pathways with each pathway following a multi-stage model as described in Section 10.3; the observation model of this state space model is a statistic model based on the observed number of colon cancer cases as described in Section 10.4.

10.4.4.1.　*The stochastic system model and the state variables*

Putting $\Delta t = 1$ for some fixed small interval, then the staging variables are $\boldsymbol{X} = \{\underset{\sim}{X}(t), t = t_0, t_0 + 1, \ldots, t_M\}$ and the transition variables are

$U = \{U(t), t = t_0, t_0 + 1, \ldots, t_M - 1\}$. From results in Section (10.4.3), the joint probability distribution of $\{X, U\}$ given the parameters Θ is:

$$P\{X, U|\Theta\} = \prod_{t=t_0+1}^{t_M} P\{X(t)|X(t-1), U(t-1)\}$$
$$\times P\{U(t-1)|X(t-1)\}, \tag{10.39}$$

where $P\{U(t-1)|X(t-1)\}$ and $P\{X(t)|X(t-1), U(t-1)\}$ are given by equations (10.36) and (10.37) respectively and where $\Theta = \{\lambda_I, \lambda_J, \alpha_i(t), \beta_j(y), b_i^{(I)}(t), d_i(t)^{(I)}(t), b_j^{(J)}(t), d_j(t)^{(J)}(t), i = 1, \ldots, k_1 - 1, j = 1, \ldots, k_2 - 1\}$.

Notice that this probability distribution is basically a product of Poisson distributions and multinomial distributions.

10.4.4.2. *The observation model using SEER data*

Put $Y = (y_j, j = 1, \ldots, m)$. By the probability distribution given by equation (10.38), the conditional probability density of Y given $\{X, U, \Theta\}$ is:

$$P\{Y|X, U, \Theta\} = \prod_{j=1}^{m} h(y_j; \tau_j), \tag{10.40}$$

where $h(y_j : \tau_j)$ is the density at y_j of the Poisson distribution with mean τ_j.

The deviance from this density is given by:

$$Dev = \sum_{j=1}^{m} \left\{ \tau_j - y_j - y_j \log \frac{\tau_j}{y_j} \right\}.$$

From equations (10.39)–(10.40), we have for the joint density of (X, U, Y) given Θ:

$$P\{X, U, Y|\Theta\} = P\{X, U|\Theta\}P\{Y|X, U, \Theta\}. \tag{10.41}$$

To apply the above distribution to estimate unknown parameters and to fit real data, we also make the following assumptions: (a) From biological observations ([35]–[43]), one may practically assume that $\{\alpha_i(t) = \alpha_i, i = 0, 1, 2, 3; \beta_j(t) = \beta_j, j = 0, 1, 2, 3, 4, b_3^{(I)}(t) = b_3^{(I)}, d_3^{(I)}(t) = d_3^{(I)}, b_4^{(J)}(t) = b_4^{(J)}, d_4^{(J)}(t) = d_4^{(J)}\}$. (b) Because the colon polys are generated by proliferation of I_2 cells and J_3 cells and because the polys can only grow to a maximum size of about 10 mm^3, we assume that $\{b_2^{(I)}(t) = b_2^{(I)} e^{-\delta_1 t}, d_2^{(I)}(t) = d_2^{(I)} e^{-\delta_1 t}\}$ and $\{b_3^{(J)}(t) = b_3^{(J)} e^{-\delta_2 t}, d_3^{(J)}(t) = d_3^{(J)} e^{-\delta_2 t}\}$ for some small $(\delta_i > 0, i = 1, 2)$. (c) Because colon cell divisions are mainly due to action of the β-Catenin gene, one may also assume $\{\gamma_1^{(I)}(t) = \gamma_j^{(J)}(t) = 0, j = 1, 2\}$. In this case, one has approximately $I_1(t + \Delta t) = I_1(t) + M_0^{(I)}(t)$ and $J_j(t + \Delta t) = J_j(t) + M_{j-1}^{(J)}(t), j = 1, 2$. Under these assumptions, the unknown parameters of interest are $\Theta = \{\Theta_1, \Theta_2\}$, where $\Theta_1 = \{\lambda_I, \lambda_J, \beta_3, \alpha_i, \beta_j, b_{i+1}^{(I)}, d_{i+1}^{(I)}, b_{j+2}^{(J)}, d_{j+2}^{(J)}, i = 1, 2, j = 1, 2, \delta_l, l = 1, 2\}$ and $\Theta_2 = (\alpha_3, \beta_4)$.

10.4.5.   *The generalized Bayesian inference and the Gibbs sampling procedure*

To estimate the unknown parameters and to predict state variables, the most efficient inference procedure is the generalized Bayesian inference procedure. This procedure is based on the posterior distribution $P\{\Theta|\boldsymbol{X}, \boldsymbol{U}, \boldsymbol{Y}\}$ of Θ given $\{\boldsymbol{X}, \boldsymbol{U}, \boldsymbol{Y}\}$. This posterior distribution is derived by combining the prior distribution $P\{\Theta\}$ of Θ with the probability distribution $P\{\boldsymbol{X}, \boldsymbol{U}, \boldsymbol{Y}|\Theta\}$ given by equation (10.41). It follows that this inference procedure would combine information from three sources: (1) Previous information and experiences about the parameters in terms of the prior distribution $P\{\Theta\}$ of the parameters, (2) Biological information via the stochastic system equations of the stochastic system $(P\{\boldsymbol{X}, \boldsymbol{U}|\Theta\})$ and (3) Information from observed data via the statistical model from the system $(P\{\boldsymbol{Y}|\boldsymbol{X}, \boldsymbol{U}, \Theta\})$. Because of additional information from the stochastic system model, this inference procedure is advantageous over the standard Bayesian procedure in that it can avoid the identifiability problems associated with standard Bayesian method.

10.4.5.1.   *The prior distribution of the parameters*

For the prior distributions of Θ, because biological information have suggested some lower bounds and upper bounds for the mutation rates and for the proliferation rates, we assume

$$P(\Theta) \propto c \ (c > 0)$$

where c is a positive constant if these parameters satisfy some biologically specified constraints; and equal to zero for otherwise. These biological constraints are:

(i) For the mutation rates of the I_i cells in the LOH pathway, $1 < \lambda_I < 1000$ ($N \to I_1$), $10^{-6} < \alpha_i < 10^{-4}, i = 1, 2, 3$. For the proliferation rates of I_i cells in the LOH pathway, $\gamma_1(t) = 0, 0 < b_i^{(I)} < 0.5, i = 2, 3, \gamma_2(t) = \gamma_2 e^{-\delta_1 t}, 10^{-4} < \gamma_2 < 2 * 10^{-2}, 10^{-5} < \delta_1 < 5 * 10^{-3}, 10^{-2} < \gamma_3 < 0.5$.

(ii) For the mutation rates in the MSI pathway, $1 < \lambda_J < 1000$ ($N \to I_1$), $10^{-8} < \beta_1 < 10^{-5}$, $10^{-6} < \beta_j < 10^{-2}, j = 2, 3, 4$. For the proliferation rates in the MSI pathway, $\gamma_i^{(J)}(t) = 0, i = 1, 2, \gamma_3^{(J)}(t) = \gamma_3^{(J)} e^{-\delta_2 t}, 10^{-3} < \gamma_j^{(J)} < 0.5, j = 3, 4, 10^{-6} < \delta_2 < 10^{-4}, 0 < b_j^{(J)} < 0.5, j = 3, 4$.

We will refer the above prior as a partially informative prior which may be considered as an extension of the traditional non- informative prior given in Box and Tiao [22].

10.4.5.2. *The posterior distribution of the parameters given* $\{Y, X, U\}$

Combining the prior distribution given in (10.4.5.1) with the density of $P\{X, U, Y | \Theta\}$ given in equation (10.41), one can readily derive the conditional posterior distribution of Θ given $\{X, U, Y\}$. For ($i = 2, 3$), denote by: $N_{iI} = \sum_{t=1}^{t_M} I_i(t), B_{iI} = \sum_{t=1}^{t_M} B_i^{(I)}(t)$ and $D_{iI} = \sum_{t=1}^{t_M} D_i^{(I)}(t)$; similarly, for $j = 3, 4$, we define $\{B_{jJ}, D_{jJ}, N_{jJ}\}$ by replacing $(I_i(t), B_i^{(I)}(t), D_i^{(I)}\}$ by $(J_j(t), B_j^{(J)}(t), D_j^{(J)}\}$ respectively. Then, we have the following results for the conditional posterior distributions:

(i) The conditional posterior distributions of $\Theta_1(1) = \{\lambda_I, \lambda_J, \alpha_i, i = 1, 2, \beta_j, j = 1, 2, 3\}$ given $\{X, U, Y\}$ is:

$$P\{\Theta_1(1) | X, U, Y\} \propto P\{\Theta_1(1)\} e^{-(\lambda_I + \lambda_J)(t_M - t_0)} [\lambda_I]^{\sum_{t=t_0}^{t_M} M_0^{(I)}(t)}$$

$$\times [\lambda_J]^{\sum_{t=t_0}^{t_M} M_0^{(J)}(t)} \prod_{i=1}^{2} e^{-N_{iI}\alpha_i} [\alpha_i]^{\sum_{j=1}^{m} y_j}$$

$$\times \prod_{j=1}^{3} e^{-N_{jJ}\beta_j} [\beta_j]^{\sum_{j=1}^{m} y_j}.$$

(ii) The conditional posterior distributions of $\Theta_1(2) = \{b_3^{(I)}, d_3^{(I)}, b_4^{(J)}, d_4^{(J)}\}$ given $\{X, U, Y\}$ is:

$$P\{\Theta_1(2)|\boldsymbol{X},\boldsymbol{U},\boldsymbol{Y}\} \propto P\{\Theta_1(2)\}f(B_{3I},D_{3I};N_{3I},b_3^{(I)},d_3^{(I)})$$
$$f(B_{4J},D_{4J};N_{4J},b_4^{(J)},d_4^{(J)}).$$

(iii) The conditional posterior distribution of $\{\alpha_3,\beta_4\}$ given $\{\boldsymbol{X},\boldsymbol{U},\boldsymbol{Y}\}$ is:

$$P\{\alpha_3,\beta_4|\boldsymbol{X},\boldsymbol{U},\boldsymbol{Y}\} \propto P(\alpha_3,\beta_4)\prod_{j=1}^{m} e^{-\tau_j}(\tau_j)^{y_j}.$$

(vi) The conditional posterior distribution of $\{b_2^{(I)},d_2^{(I)},\delta_1\}$ given $\{\boldsymbol{X},\boldsymbol{U},\boldsymbol{Y}\}$ and the conditional posterior distribution of $\{b_3^{(J)},d_3^{(J)},\delta_2\}$ given $\{\boldsymbol{X},\boldsymbol{U},\boldsymbol{Y}\}$ are given respectively by:

$$P\{b_2^{(I)},d_2^{(I)},\delta_1|\boldsymbol{X},\boldsymbol{U},\boldsymbol{Y}\} \propto P(b_2^{(I)},d_2^{(I)},\delta_1)f(B_{2I},D_{2I};N_{2I},b_2^{(I)},d_2^{(I)})$$
$$\times \prod_{t=1}^{t_M}\left(1-\frac{(b_2^{(I)}+d_2^{(I)})(1-e^{-\delta_1 t})}{1-b_2^{(I)}-d_2^{(I)}}\right)^{I_2(t)-B_2^{(I)}(t)-D_2^{(I)}(t)},$$

$$P\{b_3^{(J)},d_3^{(J)},\delta_2|\boldsymbol{X},\boldsymbol{U},\boldsymbol{Y}\} \propto P(b_3^{(J)},d_3^{(J)},\delta_2)f(B_{3J},D_{3J};N_{3J},b_3^{(J)},d_3^{(J)})$$
$$\times \prod_{t=1}^{t_M}\left(1-\frac{(b_3^{(J)}+d_3^{(J)})(1-e^{-\delta_2 t})}{1-b_3^{(J)}-d_3^{(J)}}\right)^{J_3(t)-B_3^{(J)}(t)-D_3^{(J)}(t)}.$$

10.4.5.3. *The multi-level Gibbs sampling procedure for estimating parameters*

Given the above probability distributions, the multi-level Gibbs sampling procedure for deriving estimates of the unknown parameters are given by:

(a) Step 1: Generating $(\boldsymbol{X},\boldsymbol{U})$ Given $(\boldsymbol{Y},\Theta)$ (The Data-Augmentation Step):

Given $\boldsymbol{Y}$ and given Θ, use the stochastic equations (10.21)–(10.22) and the probability distributions given by equations (10.23)–(10.28) in Section (10.4.1.1) to generate a large sample of $(\boldsymbol{X},\boldsymbol{U})$. Then, by combining this sample with $P\{\boldsymbol{Y}|\boldsymbol{X},\boldsymbol{U},\Theta\}$ to select $(\boldsymbol{X},\boldsymbol{U})$ through the weighted bootstrap method due to Smith and Gelfant [23]. This selected $(\boldsymbol{X},\boldsymbol{U})$ is then a sample from $P\{\boldsymbol{X},\boldsymbol{U}|\boldsymbol{Y},\Theta\}$ even though the latter is unknown. Call the generated sample $(\hat{\boldsymbol{X}},\hat{\boldsymbol{U}})$.

(b) Step 2: Estimation of $\Theta = \{\Theta_1, \Theta_2\}$ Given $\{Y, X, U\}$:

Given Y and given $(X, U) = (\hat{X}, \hat{U})$ from Step 1, derive the posterior mode of the parameters by maximizing the conditional posterior distribution $P\{\Theta | \hat{X}, \hat{U}, Y\}$. Denote the generated mode as $\hat{\Theta}$.

(c) Step 3: Recycling Step.

With $\{(X, U) = (\hat{X}, \hat{U}), \Theta = \hat{\Theta}\}$ given above, go back to Step (a) and continue until convergence.

The proof of convergence of the above steps can be proved using procedure given in Chapter 3. At convergence, the $\hat{\Theta}$ are the generated values from the posterior distribution of Θ given Y independently of (X, U). Repeat the above procedures one then generates a random sample of Θ from the posterior distribution of Θ given Y; then one uses the sample mean as the estimates of (Θ) and use the sample variances and covarainces as estimates of the variances and covariances of these estimates.

10.5.   Analysis of Colon Cancer Incidence: Fitting of Colon Cancer Models to the SEER Data

In this section, we will apply the model in Section (10.4) to the NCI/NIH colon cancer data from the SEER project. Given in Table (10.10) are the numbers of people at risk and colon cancer cases in the age groups together with the predicted cases from the model. There are 18 age groups with each group spanning over 5 years.

To fit the data, we have assumed that $\gamma_1^{(I)} = \gamma_j^{(J)} = 0$ for $j = 1, 2$ because of the observation that uncontrolled cell division of colon stem cells is mainly initiated by the oncogene β-Catenin in 3p22. Given in Table (10.11) are the estimates of the mutation rates, the birth and death rates of the I_i cells and J_j cells.

From these results, we have made the following observations:

(a) As shown by results in Table (10.10), the predicted number of cancer cases are very close to the observed cases in all pathways. This indicates that the model fits the data well and that one can safely assume that the human colon cancer can be described by a model of 2 pathways. The AIC (Akaike Information Criterion) and the BIC (Bayesian Information Criterion) from the model are 55.96 and 81.30 which are smaller than the AIC of 816.0667 and the BIC value of 827.1513 from a single pathway 4-stage model respectively

(Luebeck and Moolgavkar [46]). This shows that the multiple pathway model fits better than the single pathway 4-stage model as proposed by Luebeck and Moolgavkar [46].

(b) From Table (10.10), it is observed that the largest number of cancer cases is in the age group between 70 and 75 years old. Comparing the values of $Q_i(j)$ between the CIN pathway ($i = 1$) and the MSI pathway ($i = 2$), it appears that the largest cancer cases is between the age group 65 and 70 years old for the CIN pathway and is between 85 and 90 years old for the MSI pathways. Presumably this might be due to the fact that the MSI pathway has one more stage than the CIN pathway.

(c) Reflecting the contribution of the APC gene on chromosomal instability, results in Table (10.11) showed that the mutation rates of the I_i cells from $I_1 \rightarrow I_2$ and from $I_2 \rightarrow I_3$ had increased about 100 times and 1000 times respectively than the mutation rate from $N \rightarrow I_1$ cells. Similarly, due to the contribution to genomic instability by the mis-match repair genes, the mutation rates from $J_1 \rightarrow J_2$, from $J_2 \rightarrow J_3$ and $J_3 \rightarrow J_4$ had increased about $5 * 10^2, 0.5 * 10^4$ and 10^4 times respectively. Notice also from Table (10.11) that the mutation rates from $J_1 \rightarrow J_2 \rightarrow J_3 \rightarrow J_4$ are about 2 to 3 times of those from $I_1 \rightarrow I_2 \rightarrow I_3$. It appears that these increases have speeded up the time to cancer in MSI pathway by about 5-10 years.

(d) Results in Table (10.11) showed that the mutation rates from $I_3 \rightarrow I_4$ and from $J_4 \rightarrow J_5$ are of the order 10^{-6} which were about $10^{-3} \rightarrow 10^{-2}$ times smaller than the mutation rates from $I_1 \rightarrow I_2 \rightarrow I_3$ and from $J_1 \rightarrow J_2 \rightarrow J_3 \rightarrow J_4$. These results might be the consequence that we had ignored the stages of vascular carcinogenesis (i.e. angiogenesis and metastasis; see Weinberg [6], Hanahan and Weinberg [51]–[52]) by merging these stages into the last stage. From Weinberg ([6], Chapters 13–14), notice that the angiogenesis and metastasis are also multi-stage processes.

(e) Results in Table (10.11) showed that the proliferation rates (birth rate - death rate) of the I_3 cells and the J_4 cells are of order 10^{-2} which much larger than the proliferation rates of the I_2 cells and the J_3 cells, due presumably to effects of the silencing or inactivation of the cell cycle inhibition genes (Smad4 and TGF-$\beta - R_{II}$) and the apoptosis inhibition gene (p53 and Bax). Notice from Table (10.11) that the estimates of the proliferation rates of the I_2 and I_3 cells are approximately equal to those of the J_3 and J_4 cells respectively. These results seemed to suggest that the genomic instabilities had little effects on cell proliferations.

Table (10.11). Estimates of Parameters for Each Pathway

LOH Pathway				
	I_0	I_1	I_2	I_3
Mutation Rate	1.4E-06	2.2E-04	3.2E-03	1.2E-06
	±1.69E-08	±1.32E-05	±3.33E-04	±2.06E-07
Proliferation Rate	0	0	3.6E-03	1.6E-02
	N/A	N/A	±1.12E-03	±4.78E-04
Birth Rate Para.	0	0	7.4E-03	1.9E-02
	N/A	N/A	±1.03E-03	±4.08E-04
Growth Limiting Para.	N/A	N/A	8.3E-05	N/A
	N/A	N/A	±1.4E-05	N/A

MSI Pathway					
	J_0	J_1	J_2	J_3	J_4
Mutation Rate	8.3E-07	3.5E-04	1.4E-03	9.3E-03	7.7E-06
	±1.38E-08	±1.89E-05	±8.57E-05	±1.22E-03	±1.7 9E-06
Proliferation Rate	0	0	0	2.8E-03	2.0E-02
	N/A	N/A	N/A	±7.01E-04	±3.31E-04
Birth Rate	0	0	0	9.6E-03	2.6E-02
	N/A	N/A	N/A	±6.08E-04	±2.88E-04
Growth Limiting Para.	N/A	N/A	N/A	1.6E-03	N/A
	N/A	N/A	N/A	±3.7E-04	N/A

10.6.   Complements and Exercises

Exercise 1.   The Multi-event Model as A Model of Multi-variate Stochastic Birth-Death Process

Let $\underset{\sim}{e}_j$ denote a $k \times 1$ column with 1 in the $j-$th position and 0 in other positions (i.e. $\underset{\sim}{e}_j{'} = (0,\ldots,0,1,0,\ldots,0)$).

(a) Show that the probability distribution results in Section (10.1.2) lead to the following transition probabilities during $[t, t + \Delta t)$:

$$P\{\underset{\sim}{X}(t+\Delta t) = \underset{\sim}{v} \mid \underset{\sim}{X}(t) = \underset{\sim}{u}\} = [u_j b_j(t) + (1 - \delta_{j0})u_{j-1}\alpha_{j-1}(t)]\Delta t + o(\Delta t)$$

$$\text{if } \underset{\sim}{v} = \underset{\sim}{u} + \underset{\sim}{e}_j, j = 0, 1, \ldots, k-1$$

$$= u_j d_j(t)\Delta t + o(\Delta t) \text{ if } \underset{\sim}{v} = \underset{\sim}{u} - \underset{\sim}{e}_j, j = 0, 1, \ldots, k-1$$

$$= o(\Delta t) \text{ if } |\underset{\sim}{1}_k{'}(\underset{\sim}{u} - \underset{\sim}{v})| \geq 2.$$

(b) Results in (a) show that $X(t)$ is a k- dimensional birth-death process with birth rates $\{jb_i(t), i = 0, 1, \ldots, k - 1\}$, death rates $\{jd_i(t), i = 0, 1, \ldots, k - 1\}$ and cross transition rates $\{\alpha_{i,i+1}(j, t) = j\alpha_i(t), i = 0, 1, \ldots, k - 1, \alpha_{u,v}(j, t) = 0$ if $v \neq u + 1\}$. (See Definition (4.1) in Chapter 4.). Using these results, show that the Kolmogorov forward equation for the probabilities $P\{I_j(t) = i_j, j = 0, 1, \ldots, k - 1 | I_0(0) = N_0\} = P(i_j, j = 0, 1, \ldots, k - 1; t)$ in the k-dimensional multi-event model is given by:

$$\frac{d}{dt} P(i_j, j = 0, 1, \ldots, k - 1; t) = P(i_0 - 1, i_j, j = 1, \ldots, k - 1; t)(i_0 - 1)b_0(t)$$

$$+ \sum_{j=1}^{k-1} P(i_0, i_1, \ldots, i_{j-1}, i_j - 1, i_{j+1}, \ldots, i_{k-1}; t)(i_j - 1)b_j(t)$$

$$+ \sum_{j=0}^{k-2} P(i_0, i_1, \ldots, i_j, i_{j+1} - 1, i_{j+2}, \ldots, i_{k-1}; t)i_j\alpha_j(t)$$

$$+ \sum_{j=0}^{k-1} P(i_0, i_1, \ldots, i_{j-1}, i_j + 1, i_{j+1}, \ldots, i_{k-1}; t)(i_j + 1)d_j(t)$$

$$- P(i_0, i_1, \ldots, i_{k-1}; t) \left\{ \sum_{j=0}^{k-1} i_j[b_j(t) + d_j(t)] + \sum_{j=0}^{k-2} i_j\alpha_j(t) \right\},$$

for $i_j = 0, 1, \ldots, \infty, j = 0, 1, \ldots, k - 1$.

(c) Denote by $\phi(x_0, x_1, \ldots, x_{k-1}; t) = \phi(\underset{\sim}{x}; t)$ the pgf of the probabilities $P\{I_j(t) = i_j, j = 0, 1, \ldots, k - 1 | I_0(0) = N_0\} = P(i_j, j = 0, 1, \ldots, k - 1; t)$ in the k-dimensional multi-event model as described in Example (10.2). Show that $\phi(\underset{\sim}{x}; t)$ satisfies the following partial differential equation with initial condition $\phi(\underset{\sim}{x}; t_0) = x_0^{N_0}$:

$$\frac{\partial}{\partial t} \phi(\underset{\sim}{x}; t) = \sum_{i=0}^{k-2} \left\{ (x_i - 1)[x_i b_i(t) - d_i(t)] + x_i(x_{i+1} - 1)\alpha_i \frac{\partial}{\partial x_i} \phi(\underset{\sim}{x}; t) \right\}$$

$$+ (x_{k-1} - 1)[x_{k-1} b_{k-1}(t) - d_{k-1}(t)] \frac{\partial}{\partial x_{k-1}} \phi(\underset{\sim}{x}; t).$$

Exercise 2. The Solution of Stochastic Differential Equations in Multi-event Models of Carcinogenesis

Consider the following stochastic differential equations for the state variables in the extended multi-event model of carcinogenesis in Section (10.2):

$$\frac{d}{dx}I_0(t) = I_0(t)\gamma_0(t) + \epsilon_0(t),$$

$$\frac{d}{dx}I_i(t) = I_i(t)\gamma_i(t) + I_{i-1}(t)\alpha_{i-1}(t) + \epsilon_i(t), \text{for } i = 1,\ldots,k-1.$$

(a) Let $\eta_j(t) = \int_{t_0}^{t} e^{int_x^t \gamma_j(y)dy}\epsilon_j(x)dx$ $(j = 0,1,\ldots,k-1)$. Show that the solutions of the above differential equations are given by:

$$I_0(t) = N(t_0)e^{\int_{t_0}^{t}\gamma_0(x)dx} + \eta_0(t),$$

$$I_j(t) = I_j(t_0)e^{\int_{t_0}^{t}\gamma_j(x)dx} + \int_{t_0}^{t} e^{int_x^t\gamma_j(y)dy}I_{j-1}(x)dx$$

$$+ \eta_j(x)dx \text{ for } j = 1,\ldots,k-1.$$

(b) If $\alpha_j(t) = \alpha_j, \gamma_j(t) = \gamma_j$ $(j = 0,1,\ldots,k-1$ and if $\gamma_i \neq \gamma_j$ if $i \neq j$, then $\eta_j(t)$ reduces to $\eta_j(t) = \int_{t_0}^{t} e^{(t-x)\gamma_j}\epsilon_j(x)dx$ $(j = 0,1,\ldots,k-1)$. Show that the solutions of the differential equations reduce to:

$$I_0(t) = N(t_0)e^{(t-t_0)\gamma_0} + \eta_0(t),$$

$$I_j(t) = I_j(t_0)e^{(t-t_0)\gamma_j} + \sum_{i=0}^{j-1} I_i(t_0)\left\{\prod_{u=i}^{j-1}\alpha_u\right\}\sum_{v=i}^{j} A_{ij}(v)e^{(t-t_0)\gamma_v}$$

$$+ \eta_j(t) + \sum_{i=0}^{j-1}\left\{\prod_{u=i}^{j-1}\alpha_u\right\}\left\{\sum_{v=i}^{j} A_{ij}(v)\int_{t_0}^{t} e^{(t-x)\gamma_v}\epsilon_i(x)dx\right\},$$

for $j = 1,\ldots,k-1$,
where $A_{ii}(i) = 1$ and for $0 \leq i < j \leq k-1$ and $i \leq u \leq j$,

$$A_{ij}(u) = \left\{\prod_{v=i,v\neq u}^{j}(\gamma_u - \gamma_v)\right\}^{-1}.$$

(c) Noting $E\epsilon_j(t) = 0$ for all $j = 0,1,\ldots,k-1$, obtain the expected value $E[I_j(t)]$ in cases (a) and (b) above. Assume that the covariances of $\epsilon_j(t)\Delta t$ $(j =$

$0, 1, \ldots, k-1$) are given by:

$$Cov(\epsilon_i(t)\Delta t, \epsilon_j(t)\Delta t) = \delta_{ij} V_{ii}(t)\Delta t + o(\Delta t),$$

where $\delta_{ij} = 1$ if $i = j$ and $= 0$ if $i \neq j$. Derive the covariance matrix of $I_j(t)$ ($j = 0, 1, \ldots, k-1$ for $k = 3$.

Exercise 3. The Solution of Stochastic Difference Equations in Multi-event Models of Carcinogenesis

In the extended k-stage multi-event model of carcinogenesis in Section (10.2), let $\Delta t \sim 1$ corresponding to 6 months. Then we have a discrete-time Multi-event model. For this model we have the following stochastic difference equations for the state variables :

$$I_0(t+1) = I_0(t)[1 + \gamma_0(t)] + \epsilon_0(t+1),$$
$$I_i(t+1) = I_i(t)[1 + \gamma_i(t)] + I_{i-1}(t)\alpha_{i-1}(t) + \epsilon_i(t+1), \text{ for } i = 1, \ldots, k-1.$$

(a) Show that the solutions of the above equations are given respectively by:

$$I_0(t) = N(t_0) \prod_{s=t_0}^{t-1} (1 + \gamma_0(s)) + \sum_{s=t_0+1}^{t} \epsilon_0(s) \prod_{u=s}^{t-1} (1 + \gamma_0(u)),$$

$$I_j(t) = I_j(t_0) \prod_{s=t_0}^{t-1} (1 + \gamma_j(s)) + \sum_{s=t_0}^{t-1} I_{j-1}(s)\alpha_{j-1}(s) \prod_{u=s+1}^{t-1} (1 + \gamma_j(u))$$
$$+ \sum_{s=t_0+1}^{t} \epsilon_j(s) \prod_{u=s}^{t-1}(1 + \gamma_j(u))$$

(b) Suppose that $\alpha_j(t) = \alpha_j, \gamma_j(t) = \gamma_j$ ($j = 0, 1, \ldots, k-1$ and that $\gamma_i \neq \gamma_j$ if $i \neq j$. Show that the solutions of the difference equations are given by:

$$I_0(t) = N(t_0)(1 + \gamma_0)^{t-t_0} + \sum_{s=t_0+1}^{t} \epsilon_0(s)(1 + \gamma_0)^{t-s},$$

$$I_j(t) = I_j(t_0)(1 + \gamma_j)^{t-t_0} + \sum_{i=1}^{j-1} I_i(t_0) \left\{ \prod_{u=i}^{j-1} \alpha_u \right\} \sum_{v=i}^{j} A_{ij}(v)(1 + \gamma_v)^{t-t_0}$$

$$+ N(t_0) \left\{ \prod_{i=0}^{j-1} \alpha_i \right\} \sum_{u=0}^{j} A_{0j}(u)(1 + \gamma_u)^{t-t_0}$$

$$+ \sum_{i=0}^{j-1} \left\{ \prod_{r=i}^{j-1} \alpha_r \right\} \sum_{u=t_0+1}^{t} \epsilon_i(u) \left\{ \sum_{v=i}^{j} A_{ij}(v)(1 + \gamma_v)^{t-u} \right\}$$

$$+ \sum_{u=t_0+1}^{t} \epsilon_j(u)(1 + \gamma_j)^{t-u} \text{ for } j = 1, \ldots, k-1,$$

where for $0 \leq i < j \leq k-1$ and $i \leq u \leq j$,

$$A_{ij}(u) = \left\{ \prod_{v=i, v \neq u}^{j} (\gamma_u - \gamma_v) \right\}^{-1}.$$

(b) Noting $E\epsilon_j(t) = 0$ for all $j = 0, 1, \ldots, k-1$, obtain the expected value $E[I_j(t)]$. Assume that the covariances of $\epsilon_j(t)$ $(j = 0, 1, \ldots, k-1)$ are given by:

$$Cov(\epsilon_i(t), \epsilon_j(t)) = \delta_{ij} V_{ii}(t),$$

where $\delta_{ij} = 1$ if $i = j$ and $= 0$ if $i \neq j$. Derive the covariance matrix of $I_j(t)$ $(j = 0, 1, \ldots, k-1$ for $k = 3$.

Exercise 4. Prove Lemma (10.1) by using the method of probability generating function.

Exercise 5. In human colon cancer, show that to the order of $o(Min(\alpha_{k_1-1}(t), \beta_{k_2-1}(t)))$, $Q_T(j)$ **is given by**

$$Q_T(j) = Q_1(j) + Q_2(j) + o(Min(\alpha_{k_1-1}(t), \beta_{k_2-1}(t)).$$

(see Section (10.4.3.1).)

Exercise 6. The Initiation-Promotion Bioassay For Assessing Cancer Risk of Environmental Agents.

To assess cancer risk of environmental agents, a common approach is to conduct initiation-promotion experiments using the mouse skin bioassay to test if the agent can initiate cancer and/or promote cancer. In these experiments, mouse skin is shaved and is exposed to a chemical called "Initiator" for a short time period (normally a few days or a week); immediately following initiation the exposed mice are treated by a chemical called "Promoter" weekly or twice weekly until sacrifice or termination of experiment. From studies of molecular biology of skin cancer (Weinberg [6], p 435–439; DiGiovanni [53], Hennings *et al.* [54], Yuspa [55], Missero *et al.* [56]), carcinogenesis for this bioassay experiments to generate papillomas and carcinomas can best be described by a model involving an one-stage model ($N \to I_1 \to papillomas$) and a two-stage model ($N \to I_1 \to I_2 \to Carcinomas$), see Figure (10.7); for strong initiator one has also a pathway ($N \to I_2 \to Carcinomas$).

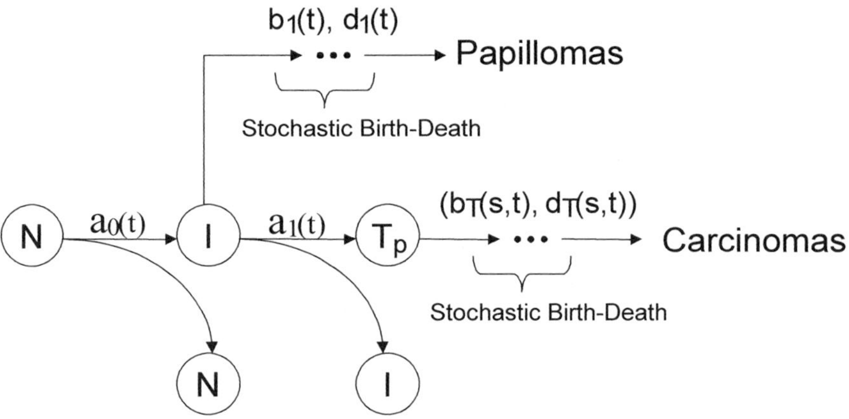

Fig. 10.6.   The multiple pathways of human colon cancer

In this model, the animal is in stage I_1 if the H-ras oncogene in a skin stem cell of the animal has been mutated (Weinberg [6], p439; Missero *et al.* [56], Brown *et al.* [57]). The I_1 stage animal is in stage I_2 if the p53 gene in an I_1 cell of this animal has been mutated or deleted or inactivated (Ruggeri *et al.* [58], Missero *et al.* [56], Weinberg [6]); the animal is in stage I_2 immediately after treatment if the agent can induce mutations of both the H-ras gene and the p53 gene simultaneously. Thus, an agent is an initiator if the agent can induce H-ras mutation in skin stem cells in animals and is a strong initiator if it can induce mutation of both the H-ras gene and the p53 gene at the same time to generate I_2 cells in some skin stem cells of mice. Promoters such as TPA (12-0-tetradecanoylphorbol-13-acetate) can not induce genetic changes but only promote cell proliferation of initiated cells (i.e. I_1 and I_2 cells; see Hennings *et al.* [54], Weinberg [6], 439; Karin *et al.* [59], Saez *et al.* [60]) have shown that promoters such as TPA act synergistically with the H-ras gene protein to activate kinase C-α (PKC-α) which in turn activates the NK-κB, the AP-1 (heterodimer of c-jun and c-fos oncogenes) and the MAPK signal pathways to drive cell proliferation; Missero *et al.* [56] and Missero *et al.* [61]) have also noticed that promoters may facilitate cell proliferation of I_1 cells by inhibiting TGF-β (Transforming Growth Factor β) activities.

In the above model, the I_r ($r = 1, 2$) cells are subjected to stochastic birth and death and each cell proceeds forward independently of other cells. Further, papillomas develop from primary I_1 cells through clonal expansion (i.e. stochastic birth-death) of these cells, where primary I_1 cells are I_1 cells generated directly by normal stem cells. Similarly, carcinomas develop from primary I_2 cells by clonal expansion (stochastic birth and death) of these cells, where primary I_2 cells are I_2 cells generated directly by mutation from I_1 cells. In initiation-promotion experiments, the initiators induce genetic changes to generate I_1 cells and/or I_2 cells. Under the action of the promoter, the I_1 cells develop into papillomas while the I_2 cells develop into carcinomas. Thus, in initiation-promotion experiments, important state variables are the number $X_P(t)$ of papillomas at time t in initiated animals and the number $X_C(t)$ of carcinomas at time t in initiated animals. In these experiments, the following two assumptions are also made: (a) Since the number of normal stem cells is very large (i.e. $10^8 \sim 10^9$) and the size of organ in adult mice is stable, it is reasonable to assume that the number $N(t)$ of normal stem cells at time t in animals is a deterministic function of t and is a constant (i.e. $N(t) = N_0$). (b) Because the spontaneous mutation rates from $N \to I_r$ ($r = 1, 2$) and from $I_1 \to I_2$ are very small, in practice one can ignore spontaneous mutations from $N \to I_r$ ($r = 1, 2$) and from $I_1 \to I_2$; thus one map practically assume that $N \to I_r$ ($r = 1, 2$) and $I_1 \to I_2$ mutations can only be generated by initiators, not by promoters; see Hennings *et al.* [54].

(a) Consider an initiation experiment in which the test agent is used as an initiator with dose level ($s_i, i = 1, \ldots, k$) and with a well-known promoter such as TPA as the promoter. Let $I_{j0}(i)$ be the number of I_j ($j = 1, 2$) cells generated by the initiator with dose level s_i during the initiation period $[t_0, t_1)$ and $\alpha_j(i)$ ($j = 1, 2$) the mutation rate from $N \to I_j$ by the initiator under dose level s_i. Show that to the order of $o(\alpha_j(i))$,

$$I_{j0}(i) \sim \text{Poisson}\{\lambda_j(i)\}, j = 1, 2, i = 1, \ldots, k,$$
$$\text{where } \lambda_j(i) = N(t_0)\alpha_j(i),$$

where $N(t_0)$ is the number of normal stem cells at time t_0 in initiated animals.

(b) Let $P_I(s,t)$ denote the probability that a primary I_1 cell generated at time s develops by stochastic birth-death process into a detectable papilloma by time t and $P_C(s,t)$ the probability that a primary I_2 cell generated at time s develops by stochastic birth-death process into a detectable carcinoma by time t. If all cells develop by the same biological mechanism under dose level s_i independently of one another, show that the conditional probability distribution of $X_P(t)$ given $I_{10}(i)$ and the conditional probability distribution of $X_C(t)$ given $I_{20}(i)$ are given respectively by:

$$X_P(t)|I_{10}(i) \sim \text{Binomial}\{I_{10}(i); P_I(t_0,t)\},$$
$$X_C(t)|I_{20}(i) \sim \text{Binomial}\{I_{20}(i); P_C(t_0,t)\}.$$

(c) Noting results (a) and (b) above and by using probability generating function method, prove the following results:

$$X_P(t) \sim \text{Poisson}\{\lambda_1(i)P_I(t_0,t)\},$$
$$X_C(t) \sim \text{Poisson}\{\lambda_2(i)P_C(t_0,t)\}.$$

(d) Assume that among all animals treated by the initiator with dose level s_i, n_i animals are sacrificed at time t $(t > t_0)$. Let N_P denote the number of animals with papillomas among the n_i sacrificed animals and N_C the number of animals with carcinomas among the n_i sacrificed animals. If all animals develop by the same biological mechanism independently of one another, prove the following distribution results:

$$N_P|n_i \sim \text{Binomial}\{n_i; 1 - e^{-\lambda_1(i)P_I(t_0,t)}\},$$
$$N_C|n_i \sim \text{Binomial}\{n_i; 1 - e^{-\lambda_2(i)P_C(t_0,t)}\}.$$

Exercise 7. Cancer Risk Assessment of Agents in Experiments in Which the Agent Is Used as a Promoter

In cancer risk assessment of environmental agents, to test if the agent is a promoter, one usually conducts an initiation-promotion experiment in which one uses a well-known initiator such as BaP (benzo[a]pyrene) as the initiator with fixed dose level but use the testing agent as the promoter with k dose levels u_i $(i = 1, \ldots, k)$. Because the I_i $(i = 1, 2)$ cells can proliferate only under promotion, if the agent is not a promoter, the number of I_i $(i = 1, 2)$ cells are very small and hence papillomas and carcinomas would not be generated. On the other hand, if the agent is a promoter, then, papillomas and possibly carcinomas will be generated at latter times.

In this experiment, if the testing agent is also a strong carcinogen besides being a promoter, the I_r $(r = 1, 2)$ cells are generated not only by the initiator during the initiation period $[s_0, s_1)$ $(s_0 = t_0)$, but also by the testing agent during promotion periods $[s_j, s_{j+1})$ $(j = 1, \ldots, m)$.

(a) Let $X_r(0)$ $(r = 1, 2)$ denote the number of I_r cells generated by the initiator during the initiation period and $X_r(i, j)$ $(r = 1, 2, i = 1, \ldots, k, j = 1, \ldots, m)$ the number of I_r cells generated by the testing agent with dose level u_i during the promotion period $[s_j, s_{j+1})$. Let ν_r $(r = 1, 2)$ be the mutation rate of $N \to I_r$ by the initiator and $\alpha_r(i)$ $(r = 1, 2; i = 1, \ldots, k)$ the mutation rate of $N \to I_r$ by the testing agent (the promoter) under dose level u_i during promotion periods. $(\nu_r \neq \alpha_r(i)$ because the initiator used is not the testing agent.) Let $\beta_1(i)$ denote the transition rate from $I_1 \to I_2$ by the promoter with dose level u_i. Assume that the testing agent is also a strong carcinogen besides being a promoter and that spontaneous mutation rates of $N \to I_r$ $(r = 1, 2)$ and $I_1 \to I_2$ are very small to be ignored. Show that for $(r = 1, 2)$:

$$X_r(0) \sim Poisson\{\omega_r\}, \text{with } \omega_r = \nu_r N(t_0),$$
$$X_r(i, j) \sim Poisson\{\lambda_r(i)\}, \text{independently for } i = 1, \ldots, k,$$
$$j = 1, \ldots, m,$$

where $\lambda_r(i) = N(s_i)\alpha_r(i)$ with $N(t)$ being the number of normal stem cells at time t.

(b) Let $X_P(t; i, 0) =$ Number of papillomas at time t derived from $X_1(0)$ cells; $X_P(t; i, j) =$ Number of papillomas at time t derived from $X_1(i, j)$ cells. Let $P_I(i; s, t)$ denote the probability that a primary I_1 cell generated at time s in animals treated by the agent with dose level u_i develops into a detectable papillomas by time t. Show that the conditional distributions of $X_P(t; i, 0)$

given $X_1(0)$ and $X_P(t; i, j)$ given $X_1(i, j)$ are given respectively by:

$$X_P(t; i, 0)|X_1(0) \sim \text{Binomial}\{X_1(0); P_I(i; s_1, t)\},$$
$$X_P(t; i, j)|X_1(i, j) \sim \text{Binomial}\{X_1(i, j); P_I(i; s_{j+1}, t)\},$$
$$\text{independently for } i = 1, \ldots, k, j = 1, \ldots, m,$$

(c) Using results from (a)-(b) above, show that

$$X_P(t; i, 0) \sim \text{Poisson}\{\omega_1 P_I(i; s_1, t)\},$$
$$X_P(t; i, j) \sim \text{Poisson}\{\lambda_1(i) P_I(i; s_{j+1}, t)\}, \text{independently for } i = 1, \ldots, k,$$
$$j = 1, \ldots, m,$$

If there are m promotion periods before time t, show that the total number $X_P(t; i) = \sum_{j=0}^{m} X_P(t; i, j)$ of papillomas at time t in animals treated by the agent with dose level u_i is distributed as a Poisson random variable with mean $\omega_1 P_I(i; s_1, t) + \lambda_1(i) \sum_{j=1}^{m} P_I(i : s_{j+1}, t)$.

(d) Let $X_C(t; i, 0)$ =Number of carcinomas at time t derived from $X_2(0)$ cells; $X_C(t; i, j)$ = Number of carcinomas at time t derived from $X_2(i, j)$ cells. Let $P_C(i; s, t)$ denote the probability that a primary I_2 cell generated at time s in animals treated by the agent with dose level u_i develops into a detectable carcinomas by time t. Show that the conditional distributions of $X_C(t; i, 0)$ given $X_2(0)$ and of $X_C(t; i, j)$ given $X_2(i, j)$ are given respectively by:

$$X_C(t; i, 0)|X_2(0) \sim \text{Binomial}\{X_2(0); P_C(i; s_1, t)\},$$
$$X_C(t; i, j)|X_2(i, j) \sim \text{Binomial}\{X_2(i, j); P_C(i; s_{j+1}, t)\},$$
$$\text{independently for } i = 1, \ldots, k, j = 1, \ldots, m.$$

Hence, show that

$$X_C(t; i, 0) \sim \text{Poisson}\{\omega_2 P_C(i; s_1, t)\},$$
$$X_C(t; i, j) \sim \text{Poisson}\{\lambda_2(i) P_C(i; s_{j+1}, t)\}, \text{independently for } i = 1, \ldots, k,$$
$$j = 1, \ldots, m.$$

(e) Let $X_C(t; i, T)$ =Number of carcinomas at time t derived from the process $N \to I_1 \to I_2 \to Carcinomas$. Let $I_1(t; i)$ denote the number of I_1 cells at time t in animals treated by the promoter with dose level u_i. Let $\beta_1(j; i) = \beta_1(i)$ be the probability of mutation from $I_1 \to I_2$ during the promotion period $[s_j, s_{j+1})$ under promotion by the agent with dose level u_i. Using methods similar to those in Section (8.3.1) in Chapter 8 and in Section (10.3), show that to the order of $o(\beta_1(i))$, $X_C(t; i, T)$ is distributed as Poisson with mean $\sum_{j=1}^{m} E[I_1(s_j; i)]\beta_1(i)P_C(i; s_{j+1}, t) + o(\beta_1(i))$, where $E[I_1(t; i)]$ is the expected number of $I_1(t; i)$].

(f) To derive $E[I_1(t; i)]$, let $\{B_1(i, j), D_1(i, j)\}$ $(j = 1, \ldots, m, i = 1, \ldots, k)$ denote the number of birth and the number of death of the I_1 cells respectively during $[s_j, s_{j+1})$ in animals under promotion by the agent with dose level u_i. Let $b_1(j; i) = b_1(i)$ and $d_1(j; i) = d_1(i)$ denote the probability of birth and the probability of death of the I_1 cells during the interval $[s_j, s_{j+1})$ in animals under promotion with dose level u_i respectively. Then, we have:

$$\{B_1(i, j), D_1(i, j)\}|I_1(s_j; i) \sim Multinomial\{I_1(s_j; i), b_1(i), d_1(i)\},$$

$$\text{independently for } j = 1, \ldots, m, i = 1, \ldots, k;$$

By the conservation law we have then the following stochastic equations for $I_1(t; i)$:

$$I_1(s_{j+1}; i) = I_1(s_j; i) + X_1(i, j) + B_1(i, j) - D_1(i, j)$$
$$= I_1(s_j; i)(1 + \gamma_1(i)) + \omega_1(i) + \epsilon_1(j + 1; i), j = 1, \ldots, m,$$
$$i = 1, \ldots, k \text{ with } I_1(s_1) = X_1(i),$$

where $\epsilon_1(j + 1; i) = [X_1(i, j) - \omega_1(i)] + [B_1(i, j) - I_1(s_j; i)b_1(i)] - [D_1(i, j) - I_1(s_j; i)d_1(i)]$.

Show that $E[\epsilon_1(j + 1; i)] = 0$ and hence show that

$$E[I_1(s_{j+1}; i)] = \omega_1(i)[1 + \gamma_1(i)]^j + \frac{\omega_1(i)}{\gamma_1(i)}\{[1 + \gamma_1(i)]^j - 1\}.$$

(g) Let $g(x, y; N, p, q)$ denote the density at (x, y) of a multinomial distribution $(X, Y) \sim Multinomial\{N; p, q\}$ and $h(x; \lambda)$ the density at x of a Poisson random variable with mean λ. Using equations in (f) above, show that the transition of $I_1(s_j; i)$ is given by:

$$P\{I_1(s_{j+1}; i) = v \mid I_1(s_j; i) = u\} = \sum_{r=0}^{u}\sum_{s=0}^{u-r} g\{r, s; u, b_1(i), d_1(i)\}$$
$$\times h\{v - u - r + s; \lambda_1(i)\}\delta(v - u - r + s),$$

where $\delta(x) = 1$ if $x \geq 0$ and $= 0$ if $x < 0$.

References

[1] F. MacDonald, C. H. J. Ford and A. G. Casson. *Molecular Biology of cancer.* Taylor and Frances Group, New York, 2004.

[2] W. Y. Tan. *Stochastic Models of Carcinogenesis.* Marcel Dekker, New York, 1991.

[3] W. Y. Tan. *Stochastic multiti-stage models of carcinogenesis as hidden Markov models: A new approach.* Int. J. Systems and Synthetic Biology 1 (2010), 313-337.

[4] W. Y. Tan, L. J. Zhang and C. W. Chen. *Stochastic modeling of carcinogenesis: State space models and estimation of parameters.* Discrete and Continuous Dynamical Systems. Series B 4(2004), 297-322.

[5] W. Y. Tan, L. J. Zhang and C. W. Chen. *Cancer Biology, Cancer models and Stochastic Mathematical Analysis of Carcinogenesis.* In: "Handbook of Cancer Models and Applications, Chapter 3." (edited by W. Y. Tan and L. Hanin) World Scientific, Sinpagore and River Edge, NJ, 2008.

[6] R. A. Weinberg. *The Biology of Human Cancer.* Garland Sciences, Taylor and Frances Group, LLc, New York, 2007.

[7] W. Y. Tan and C. W. Chen. *Stochastic Modeling of Carcinogenesis: Some New Insight.* Mathl. Computer Modeling 28(1998), 49-71.

[8] W. Y. Tan and C. W. Chen. *Cancer stochastic models.* In: Encyclopedia of Statistical Sciences, Revised edition. John Wiley and Sons, New York, 2005.

[9] K. C. Chu. *Multi-event model for carcinogenesis: A model for cancer causation and prevention.* In: Carcinogenesis: A Comprehensive Survey Volume 8: Cancer of the Respiratory Tract-Predisposing Factors. Mass, M. J., Ksufman, D. G., Siegfied, J. M., Steel, V. E. and Nesnow, S. (eds.), pp. 411-421, **Raven Press**, New York, 1985.

[10] M. P. Little. *Are two mutations sufficient to cause cancer? Some generalizations of the two-mutation model of carcinogenesis of Moolgavkar, Venson and Knudson, and of the multistage model of Armitage and Doll.* Biometrics 51(1995), 1278-1291.

[11] G. L. Yang and C. W. Chen. *A stochastic two-stage carcinogenesis model: a new approach to computing the probability of observing tumor in animal bioassays.* Math. Biosciences 104 (1991), 247-258.

[12] H. Osada and T. Takahashi. *Genetic alterations of multiple tumor suppressor genes and oncogenes in the carcinogenesis and progression of lung cancer.* Oncogene 21 (2002), 7421-7434.

[13] I. I. Wistuba, L. Mao and A. F. Gazdar. *Smoking molecular damage in brochial epithelium.* Oncogene 21 (2002), 7298-7306.

[14] W. Y. Tan, L. J. Zhang, C. W. Chen and C. M. Zhu. *A stochastic and state space model of human colon cancer involving multiple pathways.* In: "Handbook of Cancer Models and Applications, Chapter 11." (edited by W. Y. Tan and L. Hanin.) World Scientific, River Edge, NJ, 2008.

[15] W. Y. Tan and X. W. Yan. *A new stochastic and state space model of human colon cancer incorporating multiple pathways.* Biology Direct **5**(2009), 26-46.

[16] L. B. Klebanov, S. T. Rachev and A. Y. Yakovlev. *A stochastic model of radiation carcinogenesis: Latent time distributions and their properties.* Math. Biosciences **113** (1993), 51-75.

[17] A. Y. Yakovlev and A. D. Tsodikov. *Stochastic Models of Tumor Latency and Their Biostatistical Applications*, World Scientific, Singapore and River Edge, New Jersey, 1996.

[18] F. Hatim, W. Y. Tan, L. Hlatky, P. Hahnfeldt and R. K. sachs. *Stochastic population dynamic effects for lung cancer progression.* Radiation Research **172** (2009), 383-393.

[19] M. P. Little. *Cancer models, ionization, and genomic instability: A review.* In: "Handbook of Cancer Models and Applications, Chapter 5." (edited by W. Y. Tan and L. Hanin.) World Scientific, River Edge, NJ, 2008.

[20] Q. Zheng. *Stochastic multistage cancer models: A fresh look at an old approach.* In: "Handbook of Cancer Models and Applications, Chapter 2." (edited by W. Y. Tan and L. Hanin.) World Scientific, Singapore and River Edge, NJ, 2008.

[21] B. P. Carlin and T. A. Louis. *Bayes and Empirical Bayes Methods for Data Analysis (Second Edition).* Chapman and Hall/CRC, Boca Raton, FL, 2000.

[22] G. E. P. Box and G. C. Tiao. *Bayesian Inference in Statistical Analysis.* Addison-Wesley, Reading, MA, 1973.

[23] A. F. M. Smith and A. E. Gelfand. *Bayesian statistics without tears: A sampling-resampling perspective.* American Statistician **46** (1992), 84-88.

[24] B. Efron. *The Jackknife, the Bootstrap and Other Resampling Plans.* SIAM, Philadelphia, PA, 1982.

[25] K. M. Fong and Y. Sekido. *The molecular biology of lung carcinogenesis.* In: "The Molecular Basis of Human Cancer". W. B. Coleman and G. J. Tsongalis, eds. Chapter 17, 379-405. Humana Press, Totowa, NJ, 2002.

[26] R. Doll and R. Peto. *Cigarette smoking and bronchial carcinoma: Dose and time relationships among regular smokers and lifelong non-smokers.* Journal of Epidemiology and Community health **32** (1978), 303-313.

[27] P. A. Jones and S. B. Baylin. *The fundamental role of epigenetic events in cancer.* Nat. Rev. Genet. 2002, **3** (2002), 415-428.

[28] S. B. Baylin and J. E. Ohm. *Epigenetic silencing in cancer-a mechanism for early oncogenic pathway addiction.* Nature Rev. Cancer **6**(2006), 107-116.

[29] K. H. Morales, L. Ryan, T. L. Kuo, M. M. Wu and C. J. Chen. *Risk of internal cancers from arsenic in drinking water.* Env. Health Perspectives, **108**(2000), 655-661.

[30] J. Van Ryzin. *Quantitative risk assessment.* Jour. Occup. Medicine, **22**(1980), 321-326.

[31] U.S. EPA. *Proposed Guidelines for Carcinogen Risk Assessment.* Federal Register Notice **61 (79)**:17960-18011, April 23, 1996.

[32] R. Baffa, J. Letko, C. McClung *et al. Molecular genetics of bladder cancer: Target for diagnosis and therapy.* J. Exp. Clin. Cancer Res., **25**(2006), 145-160.

[33] M. Castillo-Martin, J. Domingo-Domenech, O. Karni-Schmidt *et al. Molecular pathways of urothelial development and bladder tumorigenesis.* Urol Oncol., **28**(2010), 401-408.

[34] C. Cordon-Cardo. *Molecular alterations associated with bladder cancer initiation and progression.* Scand. J. Urol. Nephrol. Suppl., **218**(2008), 154-165.

[35] A. de la Chapelle. *Genetic predisposition to colorectal cancer.* Nature Review (Cancer) **4**(2004), 769-780.

[36] R. Fodde, R. Smit and H. Clevers. *APC, signal transduction and genetic instability in colorectal cancer.* Nature Review (Cancer), **1**(2001), 55-67.

[37] R. Fodde, J. Kuipers, C. Rosenberg, *et al. Mutations in the APC tumor suppressor gene cause chromosomal instability.* Nature Cell Biology, **3**(2001), 433-438.

[38] R. A. Green and K. B. Kaplan. *Chromosomal instability in colorectal tumor cells is associated with defects in microtubule plus-end attachments caused by a dominant mutation in APC.* The Journal of Cell Biology, **163**(2003), 949-961.

[39] H. J. Hawkins and R. L. Ward. *Sporadic colorectal cancers with micro-satellite instability and their possible origin in hyperplastic polyps and serrated adenomas.* J. Natl. Cancer Institute, **93**(2001), 1307-1313.

[40] J. M. Jessup, G. G. Gallic and B. Liu. *The molecular biology of colorectal carcinoma: Importance of the Wg/Wnt signal transduction pathway.* In: "The Molecular Basis of Human Cancer". eds. W. B. Coleman and G. J. Tsongalis. Humana Press, Totowa, 2002, **Chapter 13,** pp. 251-268.

[41] P. Peltomaki. *Deficient DNA mismatch repair: A common etiologic factor for colon cancer.* Human Molecular Genetics, **10**(2001), 735-740.

[42] A. B. Sparks, P. J. Morin, B. Vogelstein and K. W. Kinzler. *Mutational analysis of the APC/beta-catenin/Tcf pathway in colorectal cancer.* Cancer Res., **58**(1998), 1130-1134.

[43] R. Ward, A. Meagher, I. Tomlinson, T. O'Connor T. *et al. Microsatellite instability and the clinicopathological features of sporadic colorectal cancer.* Gut **48**(2001), 821-829.

[44] P. Alberici, S. Jagmohan-Changur, E. De Pater, *et al. Smad4 haplo-insufficiency in mouse models for intestinal cancer.* Oncogene **25**(2006), 1841-1851.

[45] C. J. Lynch and J. Milner. *Loss of one p53v allele results in four-fold reduction in p53 mRNA and protein: A basis for p53 haplo-insufficiency.* Oncogene **25**(2006), 3463-3470.

[46] E. G. Luebeck and S. H. Moolgavkar. *Multistage carcinogenesis and colorectal cancer incidence in SEER.* Proc. Natl. Acad. Sci. USA **99**(2002), 15095-15100.

[47] S. B. Baylin and J. E. Ohm. *Epigenetic silencing in cancer-a mechanism for early oncogenic pathway addiction.* Nature Rev. Cancer **6**(2006), 107-116.

[48] K. Koinuma, Y. Yamashita, W. Liu, H. Hatanaka, K. Kurashina, T. Wada, *et al. Epigenetic silencing of AXIN2 in colorectal carcinoma with microsatellite instability.* Oncogene **25**(2006), 139-146.

[49] J. Breivik and G. Gaudernack. *Genomic instability, DNA methylation, and natural selection in colorectal carcinogenesis.* Semi. Cancer Biol. **9**(1999), 245-254.

[50] P. A. Jones and S. B. Baylin. *The fundamental role of epigenetic events in cancer.* Nat. Rev. Genet. **3**(2002), 415-428.

[51] D. Hanahan and R. A. Weinberg. *The hallmarks of cancer.* Cell **100**(2000), 57-70.

[52] D. Hanahan and R. A. Weinberg. *Hallmarks of cancer: The next generation.* Cell **144** (2011), 646-674.

[53] J. DiGiovanni. *Multistage carcinogenesis in mouse skin.* Pharmocol. Therapeut. **54** (1992), 63-128.

[54] H. Hennings, A. B. Glick, D. A. Greenhalph, *et al. Critical aspects of initiation, promotion and progression in multistage experimental carcinogenesis.* Proc. Soc. Exp. Biol. Med. **202** (1993), 1-8.

[55] S. H. Yuspa. *The pathogenesis of squamous cell cancer: Lessons learned from studies of skin carcinogenesis.* Cancer Res. **54** (1994), 1178-1189.

[56] C. Missero, M. D'Errico, G. P. Dotto and E. Dogliotti. *The molecular basis of skin carcinogenesis.* In: The Molecular Basis of Human Cancer, W.B. Coleman and G. Tsongalis, eds., pp. 407-425, Humana Press Inc., Totowa, NJ, 2002.

[57] K. Brown, A. Buchmann and A. Balmain. *Carcinogen-induced mutations in mouse c-Ha-Ras gene provide evidence of multiple pathways for tumor progression.* Proc. Natl. Acad. Sci. USA **87** (1990), 538-542.

[58] B. Ruggeri, J. Caamano, T. Goodrow, M. DiRado *et al. Alterations of the p53 tumor suppressor gene during mouse skin tumor progression.* Cancer Res. **51** (1991), 6615-6621.

[59] M. Karin, Z. G. Liu and E. Zandi. *AP-1 function and regulation.* Curr. Opin. Cell Biology **9** (1997), 240-246.

[60] E. Saez, S. E. Rutberg, E. Mueller *et al. c-fos is required for malignant progression of skin tumors.* Cell **82** (1995), 721-732.

[61] C. Missero, Y. Ramon, S. Cajal and G. P. Dotto *Escape from transforming growth factor control and oncogene cooperation in skin tumor development.* Proc. Natl. Acad. Sci. USA, **88** (1991), 9613-9617.

Subject Index